T0292708

CAMBRIDGE LIBRARY COLLECTION

Books of enduring scholarly value

Life Sciences

Until the nineteenth century, the various subjects now known as the life sciences were regarded either as arcane studies which had little impact on ordinary daily life, or as a genteel hobby for the leisured classes. The increasing academic rigour and systematisation brought to the study of botany, zoology and other disciplines, and their adoption in university curricula, are reflected in the books reissued in this series.

Niger Flora

German scientist Theodore Vogel (1812–1841) joined an 1841 expedition to the Niger as its chief botanist. He died in the course of the journey, though not before taking extensive notes about the plants that he encountered. His botanical collection and diary were passed to the botanist William Jackson Hooker (1785–1865), who had been appointed as the first full-time director of Kew Gardens in the same year. Hooker edited Vogel's diary and observations and the resulting work, *Niger Flora*, was published in 1849. Because Vogel's period in West Africa was cut short by his untimely death, much of the work looks at the flora of the places the expedition stopped at along the way – Madeira, Tenerife and the Cape Verde islands, before giving details – including numerous illustrations – about west African plants. The works also includes observations on African flora by other botanists, including Joseph Dalton Hooker, William's son.

Cambridge University Press has long been a pioneer in the reissuing of out-of-print titles from its own backlist, producing digital reprints of books that are still sought after by scholars and students but could not be reprinted economically using traditional technology. The Cambridge Library Collection extends this activity to a wider range of books which are still of importance to researchers and professionals, either for the source material they contain, or as landmarks in the history of their academic discipline.

Drawing from the world-renowned collections in the Cambridge University Library, and guided by the advice of experts in each subject area, Cambridge University Press is using state-of-the-art scanning machines in its own Printing House to capture the content of each book selected for inclusion. The files are processed to give a consistently clear, crisp image, and the books finished to the high quality standard for which the Press is recognised around the world. The latest print-on-demand technology ensures that the books will remain available indefinitely, and that orders for single or multiple copies can quickly be supplied.

The Cambridge Library Collection will bring back to life books of enduring scholarly value (including out-of-copyright works originally issued by other publishers) across a wide range of disciplines in the humanities and social sciences and in science and technology.

Niger Flora

Or, An Enumeration of the Plants
of Western Tropical Africa

EDITED BY WILLIAM JACKSON HOOKER

CAMBRIDGE
UNIVERSITY PRESS

CAMBRIDGE UNIVERSITY PRESS

Cambridge, New York, Melbourne, Madrid, Cape Town,
Singapore, São Paolo, Delhi, Tokyo, Mexico City

Published in the United States of America by Cambridge University Press, New York

www.cambridge.org
Information on this title: www.cambridge.org/9781108030380

© in this compilation Cambridge University Press 2011

This edition first published 1849
This digitally printed version 2011

ISBN 978-1-108-03038-0 Paperback

NIGER FLORA.

Sir WU. J. Hooker.

ICONES PLANTARUM.

New Series.

Vols. I--IV. Containing each 100 Plates, with Explanations.

8vo. Cloth. London. 1842—47. *Each* £1 8*s.*

THE LONDON JOURNAL OF BOTANY.

Vols. I--VI. With 24 Plates each.

8vo. Boards. London, 1842—47.

Reduced to £1. per Vol.

Professor Schleiden.

THE PLANT;

A BIOGRAPHY.

IN A SERIES OF POPULAR LECTURES ON BOTANY.

EDITED AND TRANSLATED

BY A. HENFREY, F.L.S.

With Five Coloured Plates, and Thirteen Woodcuts.

8vo. London, 1848. 15*s.*

Mexxar lith. 3. Whtington Sᵗ Strand.

CLARENCE HARBOUR, FERNANDO PO.

From a painting in the possession of Rᵗ Jamieson Esqᵗᵉ
Published by H Bailliere, Regent Street.

NIGER FLORA;

OR,

AN ENUMERATION OF THE PLANTS

OF WESTERN TROPICAL AFRICA,

COLLECTED BY THE LATE

DR. THEODORE VOGEL,

BOTANIST TO THE VOYAGE OF THE EXPEDITION SENT BY HER BRITANNIC MAJESTY TO THE

RIVER NIGER IN 1841,

UNDER THE COMMAND OF

CAPT. H. D. TROTTER, R.N., &c.

INCLUDING

SPICILEGIA GORGONEA,

By P. B. WEBB, ESQ.,

AND

FLORA NIGRITIANA,

By DR. J. D. HOOKER, R.N., F.R.S., AND GEORGE BENTHAM, ESQ.

WITH

A SKETCH OF THE LIFE OF DR. VOGEL.

EDITED BY

SIR W. J. HOOKER, K.H., D.C.L., F.R.A. & L.S.

VICE-PRESIDENT OF THE LINNÆAN SOCIETY, AND DIRECTOR OF THE ROYAL GARDENS OF KEW.

With Two Views, a Map, and Fifty Plates.

LONDON:

HIPPOLYTE BAILLIERE, PUBLISHER,

219, *REGENT STREET*,

PARIS: J. B. BAILLIERE, 13, RUE DE L'ECOLE DE MEDECINE.

MADRID: BAILLY BAILLIERE, CALLE DEL PRINCIPE.

1849.

PREFACE.

THE majority of the Plants described in the following pages were entrusted to the Editor, for the purpose of publication, by the African Civilization Society, which, as is well known, was formed in London in 1839, through the instrumentality of the late Sir Thomas Fowell Buxton. That enlightened and philanthropic statesman, deeply impressed by the aggravated horrors of the Slave Trade, was extremely anxious to try, what appeared to him to be the only remedy, to put down that iniquitous traffic by the encouragement of lawful trade and the advancement of Africa itself to a condition in which she would no longer find it her interest to furnish the slavers with supplies for their market.

Many persons of influence and sound judgment, uniting with Sir Fowell in his views, and Government having taken up the subject cordially, the " NIGER EXPEDITION" was dispatched, under the command of Capt. H. D. Trotter, in 1841.

Dr. Theodore Vogel, a German gentleman of high scientific attainments, was selected as chief Botanist to the Expedition, and with him was associated Mr. Ansell, strongly recommended by the Horticultural Society of London.

Very great and unusual pains were taken to render the service less dangerous to the health of those engaged in it, than had been the case with former attempts to explore intratropical Africa. Indeed, every precaution that could be thought of—every guard against the climate—were, as was believed, employed ;—yet, it cannot be denied, there was a failure, and it would truly appear from this, and from former voyages of a similar character, that the European constitution is incapable of withstanding the effect of that deadly atmosphere.

But while we deplore the loss of so many brave officers and men, engaged, voluntarily, in this most sacred cause, it would be unjust to shut our eyes to much good that has hereby been accomplished. It has proved to the natives the real intentions of the English, and convinced them of our sincerity in establishing mutual, and beneficial, and a wholesome commerce, and that we have no sinister ends of our own to answer. Of this, too, they were the more convinced, when they saw their friends, who had been rescued from captivity, returning with the Expedition. It further showed, that the only hope of enlightening the sons of Africa is by *native* agency : and it is with no small pride that the Editor of this Work, in the capacity of Director of the Royal Gardens of Kew, is at this moment giving in charge a considerable collection of useful Tropical plants for introduction into Africa, to two native Missionaries (recently ordained by the Bishop of London),— than whom he knows not any well educated Europeans more competent to estimate the value of such importations, or likely to feel more interest in their successful cultivation and use.

Among those who fell victims to the climate of Niger, was Dr. Vogel. Happily for science, he was not among the most early to be attacked by fever. He formed his collections with uncommon energy, while even a slight portion of health and strength remained to him ; and the number of species amassed by him, in a short space of time, and under the most disadvantageous circumstances, reflects great credit upon his memory; but the condition of the specimens shows, that the climate is as unsuited to the preservation of plants, without greater advantages than a small and crowded steamer can afford, as it is to the human constitution.—Mr. Ansell, though he fortunately survived the effects of the climate, was yet too ill, from a very early period of the voyage, to make any extensive or well-preserved collections. These facts must plead the apology for the imperfect nature of many of the descriptions. The work,

however, the Editor is sure, will be hailed by every friend of
Botany, and by every one interested in the vegetable productions of Western Tropical Africa, as a *Prodromus* of a Flora of
that region;—something upon which a more perfect super-
structure will be hereafter built : and he begs the particular
attention of Travellers in the Niger territories, and coasts
adjacent, to the subjoined "*Desiderata*" in the principal branches
of botanical science.

In the present brief, general preface it only remains for the
Editor to express his grateful acknowledgements to the distinguished Botanists who have aided in bringing out this
volume. Dr. J. D. Hooker had volunteered to describe the
whole of the plants, but unexpectedly, when considerably
advanced in the work, an opportunity offering of exploring
regions of a widely different character from those of the present
Flora, and which was embraced by him, the assistance of others
was rendered essentially necessary. Mr. P. B. Webb, already
so familiar with the vegetation of the Canary Isles, and the
opposite coasts of Africa, generously undertook to describe
all the Cape de Verd Islands Plants, Dr. Miquel the *Fici*,
Mr. Miers the *Menispermeæ*, and last, but not least in point
of extent of service, it devolved upon our inestimable friend,
Mr. Bentham, to publish the majority of the plants of the
continent of Tropical Western Africa.

EDITOR.

ROYAL GARDENS, KEW,
Nov. 1, 1849.

DESIDERATA

FROM

BOTANICAL COLLECTORS IN WESTERN TROPICAL AFRICA.

A FIRST glance at the *Flora Nigritiana* will show how very imperfect our acquaintance yet remains, not only with the luxuriant wild vegetation of West Tropical Africa, including the herbs or trees which furnish many of its most valuable products, but even with many of the plants in general cultivation there. This circumstance may in a great measure be ascribed to the want of any of those permanent botanical establishments which have afforded us so much useful information on the vegetation of the East and West Indies, and have been the means of effecting so many valuable exchanges of plants respectively cultivated in the two hemispheres. Tropical Africa has never even had a resident botanist, and all our knowledge on the subject has been derived from travellers who have either perished there before their mission has been completed, or have hastened home to avoid the effects of the deadly climate. Much is therefore now to be done by a collector who will carefully note down any authentic particulars he can learn, and any observations that occur to him, relating to the plants of which he preserves specimens.

Such information will always be the more valuable the more cautiously it is collected, avoiding as much as possible mere hearsay information, and noting down in all cases the sources whence it is derived. It is also important that the memoranda should be on labels *attached to the dried specimens,* to avoid the numerous mistakes arising from the mismatching memoranda and their specimens by the time they reach the hands

of the working botanist at home. So in regard to the specimens
themselves, an isolated leaf, a separate flower, or even a fruit
are insufficient alone to determine the species. A *perfect* dried
specimen includes *foliage, flower,* and *fruit,* with notes on size,
colour, and other points which it cannot indicate ; a flowering
branch with foliage is generally sufficient to identify a well
known species, but the description of a new one is always
imperfect, and often wholly inefficient without the fruit. Where
that is too bulky or too succulent to be laid in with the flowering and leaf specimens dried flat, and if preserved in spirits, or
dried without pressure, portions of the foliage should be invariably attached to it as the sole means of future identification
with the corresponding flowering specimen.

In a *purely botanical and systematic* point of view, any
plants not enumerated in the following Flora, and more especially those mentioned as imperfectly known would be the most
valuable. Palms, and other large *Monocotyledons,* bulbous-rooted
and smooth thick-leaved plants of the same class, aquatic plants
in general, whether floating or immersed, and cryptogamic
productions have hitherto been but seldom collected. So it is
also with *Artocarpeæ* and other large-fruited trees, *Cucurbitaceæ,*
and all plants which require a little extra care in drying, and a
little extra ingenuity in gathering, such specimens as may supply
the requisite information.

To the *geographical* Botanist, Western Tropical Africa is of
peculiar interest, as being (next to the Arctic Regions), the
point where the greatest number of species or forms belonging
to the Eastern and Western hemispheres are found to meet.
It may be consequently expected to furnish many valuable data
respecting the migration of species, either naturally or by
human aid, and the circumstances which determine the regions
of analogous forms among endemic species.

The most important information we have on this subject is
summed up in Brown's justly celebrated Appendix to Tuckey's
Congo. The facts since collected tend still farther to confirm
the supposition that the greater number of *cultivated plants*
have come to the Negroes of Western Africa from the East,

including even some of American origin, which like the *Maize*
and the *Arachis*, were so early cultivated in Asia as to give rise
to doubts whether they were not known there before the discovery of the New Continent. Among those now cultivated in
America as well as Africa, and not known, or of recent introduction into Asia, the majority (as for example, some species
of *Panicum, Amomum,* &c.) have been carried over to America
from Africa; and few only (*Manihot ?, Indigofera Anil,* some
species of *Dioscorea,* &c.) appear to have been introduced from
America into Africa, or are among the naturally indigenous
species to both countries. Additional facts tending to elucidate
these points are much wanted.

The *weeds of cultivation,* and other plants which accompany
man in his migrations, are mostly, like the cultivated plants, of
Eastern origin, although many are likewise now common in
cultivated parts of Tropical America. The principal points to
be attended to in respect to this class, are, how far they become
really naturalized by maintaining themselves and spreading
beyond the crops with which they were originally introduced.

The *cosmopolite* indigenous species are, within the tropics,
chiefly aquatic or marshy, or belonging to the glumaceous or
cryptogamic orders. The laws which are supposed to regulate
their diffusion, being deduced from the number of recorded
stations, the collector should never neglect them on the ground
of their being common elsewhere. The preserving specimens
of widely spread species is the more important, as it is often
difficult for the most experienced botanist to be certain of the
identity of plants observed at different periods, without an actual
comparison of specimens, and it is only by a careful observation
of variations of form occasioned by diversity of soil and climate
that any satisfactory judgment can be formed of the systematic
limits of species and races.

The *indigenous* species common to West Tropical Africa and
Tropical America, appear to be chiefly found near the sea, or at
any rate do not penetrate beyond the first hills; few, indeed, of
the really maritime species are Asiatic or East African, with the
exception of such as are diffused over all Tropical sea coasts.

in the nterior of the country, Eastern and Asiatic species
become much more numerous as the American ones disappear.
In regard to all these *travelled* species, we are in want of trust-
worthy data as to the stations they occupy, how far from the
sea, from the habitations of man, or from the regions of cultivation, their scarcity or abundance, the limits of the tracts they
occupy, and other circumstances tending to elucidate their mode
of transmission.

The *endemic* plants of West Tropical Africa are of the
greatest interest, as supplying data for speculations on the laws
regulating the geographical dissemination of analogous forms.
Senegal and other drier northern parts of our region, not only
have many identical species, but still more of analogous ones to
those which prevail through Nubia and Arabia, eastward to
the hot, dry plains and table-lands of India. In the moist,
close regions about the mouths and branches of the Niger and
the island of Fernando Po, some curious analogies may be
observed, with corresponding forms in Madagascar, Ceylon,
and the Malayan Archipelago. With these Asiatic forms are
mixed, in various parts of the region, African representatives of
American genera, which appear to find here their extreme
Eastern limits. European and South African forms, genera as
well as species, are more completely excluded from this than
from any other Tropical region. For all data from which any
general conclusions under this head can be formed, we must
rely entirely upon the geographical notes supplied by local
collectors.

The *practical, economical* and *commercial* botany of West
Tropical Africa is less known, perhaps, than any other branch.
Products of the greatest value have been exported during a long
course of years, without our being able to form any idea of the
plants which supply them. Every collector has sent home a
different leaf as that of the "*African Teak*," or "*Oak.*" The
learned researches of Dr. Pereira have not yet, for want of the
requisite data, solved the doubts as to what one or more species
of *Amomum* furnish the hot, acrid seeds now imported as
Guinea Grains. Similar doubts hang over the species or

varieties of *Habzelia,* whose seeds were also known as *Guinea Grains* or *Ethiopian Pepper,* and of *Cubeba,* supplying, according to Thonning, the *Ashantee Pepper.* To such points we would particularly direct the attention of the resident Naturalist, and in their case more than in any other, we look for *personal* information. The reports of the natives, as received through Europeans, are scarcely ever to be trusted, and it happens but too frequently that even the local commercial men who deal in them, either through ignorance or from mistaken views of interest, mislead scientific men in their replies to such inquiries.

Our best general information on this as on all other important points of the Botany of West Tropical Africa, is contained in Brown's already quoted "Appendix." A few additional notes on edible fruits of Sierra Leone are contained in a paper by Afzelius, in "Sierra Leone Report, 1794," and another of the late Mr. Sabine, in the fifth volume of the "Transactions of the Horticultural Society of London;" and many valuable memoranda are dispersed through Schumacher and Thonning's description of Isert and Thonning's Guinea plants, in the fourth volume of the "Transactions of the Royal Academy of Sciences of Copenhagen." This account, divided between two parts of those Transactions, has also been separately published under the title of "*Beskrivelse af Guineiske Planter,*" and has been always quoted in the following Flora, with the paging of this separate edition, the only one that we are in possession of. Dr. Vogel's collection, though full of memoranda on botanical points, which have materially assisted in the determination and description of the specimens, contains but little information on points of economical and practical botany. He was, indeed, for too short a time in the country to enable him to collect authentic data, and he well knew that mere hearsay reports from ignorant natives were of little or no value. The extent and comparative excellence of his collection show that neither zeal nor ability were wanting, so long as his health was spared, in rendering it as botanically serviceable as possible.

G. B.

Directions for Placing the Plates.

The View of Clarence Harbour, Fernando Po.—To Face the Title Page.

The Plate (Tab. I.) Representing Two Views upon the Niger.— To Face Page 54 of the "Journal of the Voyage."

The Map.—To be placed at the End of the "Journal of the Voyage." Page 72.

The Botanical Plates I—L, to be annexed at the end of the Volume.

MEMOIR OF THE LIFE

OF

DR. J. R. T. VOGEL.*

AMONGST the numerous sacrifices consequent on the unfortunate expedition to the Niger, science is not without her peculiar loss. Whatever reliance may be placed on wealth and a careful choice of means, it must be admitted that little has been accomplished by the numerous and deeply calculated plans for obtaining a more perfect knowledge of the interior of Africa. Amongst many other individuals, one of the naturalists of the expedition, to whose memory the following pages are dedicated, has succumbed to the destructive influence of the climate. If, however, according to the expression of a philosopher, it be the province of eloquence to commemorate illustrious minds, whose labours, owing to an unfortunate concurrence of circumstances, have not been productive of commensurate effects, and so, to compensate† for the want of incident, a more skilful pen than mine is requisite. I must be contented to show what the world and what science have lost, by the simple relation of a few circumstances, and by extracts from the last official records of the deceased.

Julius Rudolph Theodor Vogel, or as he frequently called himself by an abridgement of his baptismal name, Theodor Vogel, was born at Berlin, on the 30th of July, 1812. While yet a boy, he showed a decided inclination for the

---

* Translated from the German of Dr. L. C. Treviranus, in the *Linnæa*, vol. x, by the Rev. M. J. Berkeley.—ED.

† F. Hemsterhuis, Œuvres philosophiques, I. 268.

study of history, geography, and the productions of nature. No book was read by him with greater delight than Robinson Crusoe. He knew it almost by heart; and in all probability its perusal awakened in him that eagerness to visit countries yet unexplored by science, which was hereafter destined never to leave him. He received his first scientific education at the Friedrich-Wilhelms-Gymnasium at Berlin, of which Spilleké was then the director, from which establishment he was dismissed with the testimonial No. I. The Gymnasium, however, afforded no opportunity of acquiring botanical knowledge; but, under the guidance of Ruthe, upper master of the Gewerbschule, he commenced studying and collecting plants. With him he went as a pupil weekly to botanize in summer, and if he had leisure for a day or two, it was devoted to more distant excursions. At the beginning of the summer term, 1832, Vogel was a student at the University of Berlin, where he attended the philosophic and humanity lectures, but more especially those of Natural History. On the 5th of August, 1837, he was made Doctor of Philosophy : the subject of his inaugural thesis, which has also been published separately, being a Synopsis of the genus *Cassia*. The year after, he qualified himself as Private Tutor at the Berlin University in the botanical department, and in the first half of 1839, exchanged his situation for a similar one at the University of Bonn, aided by a government allowance, in consequence of which he was obliged, at the same time, after the death of Professor F. Nees von Esenbeck, to perform the duties of his office. During the latter part of his academical course, and after its completion, Vogel turned his attention principally to descriptive botany; for which the collections at Berlin, especially of Brazilian plants, furnished rich materials.

His talents were peculiarly applied to Leguminous plants. As early as the year 1837, four treatises by him appeared in the eleventh volume of the *Linnæa :* " *De Swartzeis observationes,*" " *Dalbergiearum Genera Brasiliensia,*" " *De Cæsalpineis Braziliæ,*" and " *Synopsis generis Cassiæ pars altera.*"

In these, several new genera are proposed, numerous new species are briefly but characteristically described, and many already known have received important corrections. In the year 1838, in conjunction with Dr. Schleiden, he published some greater and lesser treatises, which, for the most part, have reference to the same family, in the 19th volume of the *Acta Acad. Nat. Curiosorum*, entitled " Contributions to the history of the development of the parts of the flower in the *Leguminosæ*, and on the Albumen, especially of *Leguminosæ*," both accompanied by numerous illustrations; and in the 46th of Poggendorf's *Annalen der Chemie und Physick*, on "*Amyloid, a new vegetable substance.*" At the same time appeared, in the 12th volume of the *Linnæa*, a continuation of his labours on the leguminous plants of South America, in an Essay on Brazilian *Hedysareæ*. In the year 1839, also, he continued his task with his wonted perseverance, and in the 13th volume of *Linnæa*, we find two treatises connected with it, viz.: on Brazilian *Vicieæ*, and observations on American *Bauhiniæ*. In 1840, he received, for investigation, the *Leguminosæ* collected by the late Prof. Meyen in his journey through Brazil, Peru and China, and the results of this labour appear in an Essay in the supplement to the 19th volume of the *Acta Acad. Nat. Curios.* The 15th volume of *Linnæa* contains three of his treatises; viz. Remarks on the existence of *Amylum in Cryptogamous plants;* Additions and corrections to the Synopsis of the Genus *Cassia;* and Remarks on some species of the genera *Thymus* and *Origanum*. There is also in Buchner's " Repertorium für die Pharmacie, 1840," a Review of the species of the genus *Origanum* and a Description of *Thymus coriaceus*. Mr. Marquardt found this undescribed species of *Thymus* in many apothecaries' shops and collections of *Materia Medica*, amongst stores of *Origanum Creticum*.

As Vogel's position in Bonn rendered a perfect knowledge of the plants in its neighbourhood requisite, preparations were commenced for a Flora of Bonn; to which he devoted much time, and for which he undertook many excursions. In August, 1830, he made acquaintance with a member of the African Civilization Society, which had arisen in London under the patronage of Prince Albert, with the view of extending civilization amongst the natives of Western Africa and putting an end to the slave trade, who chanced then to be at Bonn. The British Government fitted out three steamers,* destined to run into the Niger, or Quorra, at its entrance into the Bight of Benin, on the western coast of Central Africa, to penetrate by this vast navigable river, into the interior of this little-known country, to make treaties with the inhabitants, and to establish an emporium at some suitable place. A Botanist was needed, to ascertain the vegetable productions of the country and the capabilities of the soil; and Dr. Vogel was found willing to hold this office, hoping by these means to satisfy his eager desire to explore a rich and almost unknown vegetation. He undertook, therefore, in September and October of the same year, a journey to England, to make a personal acquaintance with the committee of the society: returned for a few weeks to Germany, to arrange definitively his affairs, and finally left Bonn on the 2nd of December, 1840, to enter upon his journey, having obtained from the proper authorities a two years' leave of absence.

The departure of the expedition, which, according to the first plan, was to be in the end of January, 1841, was deferred from various circumstances and impediments to the third week in May; when, finally, the ships left Plymouth harbour and Europe, Dr. Vogel embarking in the Wilberforce. During his four months' residence in England, Vogel prepared himself in every possible way for his new destination, and in the parts for March and July of a Journal entitled "The Friend of Africa," he published an " *Essay on the Botany of Western Central Africa;*" in which the hitherto written treatises on the vegetable productions of this part of the world were reviewed. From Madeira he addressed

---

* The Albert, the Wilberforce, and Soudan.

letters to his relations and friends in Europe; but they never reached their destination. From Sierra Leone he wrote on the 30th of June, as follows:

"We sailed from Madeira by Teneriffe to St. Vincent, one of the Cape de Verd Islands, and from thence came here. At Teneriffe we remained a day; but I was able to take only a cursory glance, since I was unwell on the passage from Madeira thither, and did not venture to leave the ship. We remained a fortnight off St. Vincent: the island is small, but has an excellent harbour, and was therefore the rendezvous of the ships belonging to the expedition. Anything more comfortless than the view of this island, I never beheld: one might believe that after the formation of the world, a quantity of useless surplus stones was cast into the sea; and that thus the Island of St. Vincent arose. There is nothing but hills and mountains (some of them 2500 feet high); with small valleys, which in the broader parts are very sandy, without a plant deserving the name of tree: while the vallies themselves produce scarcely a species; for in my first excursion, I found in four hours only two species, of which one, a *lavander*, was completely dried up. What had been wanting here, namely moisture, was in a few days but too abundant. On the part of the coast where we are at present, the rainy season has begun; that is, the first portion of it, which announces itself by single thunder-storms with violent wind (tornados.) Sometimes on the passage my cabin got very wet, and what was worse, my plants. Since we have been at Sierra Leone, the weather is generally clear by day; but towards evening there comes heavy rain or a thunder-storm, and last night we had one, such as I never witnessed before.

"On entering the river at Free Town, the shore, on which the town stands, is bordered at a short distance by a range of hills, exhibiting a very pretty appearance with their gentle swelling summits and insolated lofty trees. A rich vegetation stretches from the shore upwards, which captivates the eye by its soft bright green, such as is only seen in the tropics, and gives the whole an incomparably charming character. I rushed eagerly into these woods; and much regret that the short time of our stay did not permit me to do more ; for we were obliged to proceed. The object of the colony here is to teach the Africans active habits and to christianize them : there are, I think, above 40,000 in the colony, and many of their villages are built close to the town ; so that, for miles, there is no cultivation. Since we left St. Vincent, the temperature has been nearly the same. The thermometer there was generally 81° Fahr. in my cabin: here it is about 84°, and sometimes in the middle of the day reaches 86°. This heat is not greater than with us in summer; but the slighter refrigeration of the atmosphere by night, and the power of the sun, make it seem often more intense than it is. An awning is spread over the deck, under which, when there is a breeze, it is always cool. I am very comfortable on board, except when my collections are lying about. When I return laden with plants, I have no where to prepare them ; and when they are dry, the damp insinuates itself to such a degree, that I am compelled to redry them. This is very troublesome ; and on board a ship, especially a man of war, there is no especial place for preparing or preserving plants. I am quite a nuisance to my messmates when I unpack ; and so is the servant who announces breakfast, lunch, &c. ; for the table must be cleared, and I must be off. Then I try to work on deck ; but there the wind and rain attack me ; so that I have to contend with all the elements. I am here quite amongst the negroes, for there are few white persons in the town ; and during my excursions I frequently do not see one during the whole day. I cannot, however, say that this seems altogether strange to me: on our voyage outward, we had many black sailors in our ship ; and their number has gradually increased in the course of our progress."

From Cape-Coast Castle roads, where the ships belonging to the expedition arrived on the 24th of July, Vogel writes as follows : " Our passage from Sierra Leone hither has been rather tedious. We set out from that port with but little fuel, and were therefore necessitated twice after we left Monrovia (Liberia), viz. at Grand Bassa and Cape Palmas, to cause wood to be felled, to enable us to proceed. Our voyage has been constantly along the coast; so that we have had ample opportunity for observing the remarkable nation of the Kroo : a people who dwell scattered along the coast, and often undertake long coasting voyages in small canoes. These canoes are built almost exactly in the same way as the little skiffs which at Berlin are called Seelen-verkaufer; but made of a single piece only. The natives sit in them generally naked : they use broad oars and a very small rudder ; and do not trouble themselves when the craft upsets ; for they have commonly nothing to lose, and if they carry garments with them, they are soon dried. They have mostly a piece of cloth, bound round the head, which, when they come on board, they place round the loins, and think themselves full dressed with great ivory rings round the ankles, and belts or chains round the foot or arm. We had many of their young people on board, for they are tolerably docile, and are therefore hired by the coasters, to perform such hard labours as are considered prejudicial to Europeans. When they have earned so much money by their voyage, as will enable them to buy one or more wives, they return home, establish the women, and leave them for a new expedition, until they get eight or ten more wives, who must support them ; for all field-labour, &c., is performed by females. Including these Kroos and other negroes, who are employed in various ways about the ship, we are now considerably more than one hundred men strong : frequently, therefore, when I have been for a time at that part of the vessel which they occupy and where alone smoking is allowed, and return to the quarter-deck where only the officers are, I feel quite relieved from the bustle. It is now the rainy season and we have had in Monrovia and Grand Bassa a week of continued rain ; during which the sky has been for many successive days as dark as it can be with us in autumn only. Besides, the African brooks, when they are swollen with

rain, assume the privilege of making their way down the
footpaths ; and I was therefore obliged for hours to wade up
to the knees in water. I was indeed, in general, whether at sea
or on land, as wet as it was possible to be. One advantage
accrued from the rain, it kept the decks water-tight; whereas
before, I was regularly soaked by the water when they were
washed at five o'clock in the morning; and frequently part of
my collection got damaged. At Cape Palmas we arrived at a
spot where an intermission of the rainy season takes place,
and from thence to this place we have enjoyed delight-
ful weather. The passage, however, was longer than we
expected; so that water ran very short; and one day we were
absolutely placed on half-allowance : otherwise we should
scarcely guess that we were in a foreign zone. As regards
meat and drink, we have several times a week salted beef or
pork, and in general, other kinds of meat preserved in her-
metically sealed cases. Hares, poultry, &c., prepared in
this way, often appear at table. These ship-stores are
preferred to the fresh provisions which are presented to us
on landing. My situation on board is very tolerable. The
captain gives me all the liberty possible; and I hope,
when we have once arrived at the proper field of action, to
meet with every encouragement from him. My health has
been very good; and although there cannot but be some
irksome hours to men shut up in a ship, I have yet, on the
whole, felt happy and contented, and only look forward
with impatience to the time when my own peculiar service
will begin."

The next letter from Vogel was written from Accra, on the
4th of August.

" We remain here but a few days, so that I can acquire
only a very superficial view of the vegetation of the coast.
Real forests lie at some distance in the interior, that is, about
thirty English miles :—too long an excursion, even were it
not desired that nobody should sleep on shore, for fear of
fever. Yet I have been twelve or fourteen miles into the
interior, in the district of Aquafim, to inspect a Danish

settlement. There was a geologist with me, and we were received by the Danish Governor with the greatest civility. Such a journey on foot being considered too difficult for an European, large flat baskets, used here instead of sedan-chairs, were placed at our disposal, and four negroes to carry each basket. There were, besides, a number of negroes, to take charge of our luggage; so that our caravan amounted to seventeen persons, besides ourselves. At the coffee-plantation there is a house arranged with European accommodations, where we were surrounded with all the luxury of the civilized world, and had for dinner French asparagus. The spot was lovely, pleasantly varied with hill and dale, mostly covered with savannahs; where the grass is taller and stronger than in our own meadows, and between the tufts grew little bushes, instead of flowers. I think that I saw *Blighia sapida* in cultivation, and remarked that Schumacher mentions it under a name different from that by which it is known to the natives. The negroes who accompanied us on this excursion were slaves; for the Danes still have slaves, but they seem well off, and were merry and cheerful beings. On the whole, I found in the short period of my acquaintance with them, no difference in their behaviour or dealing from the free negroes at Cape Coast Castle; except that the latter are shameless in demanding money for drink. At Cape Coast, it is absolutely necessary to keep an immoderate number of servants; and on an excursion from thence, our train of attendants consisted of thirty-six persons. There is no difficulty in this, for the blacks go as servants merely for food and clothing, which in this climate costs little: or they are sent when boys by their fathers to an European, that they may in this way learn something. The houses of Europeans here are very large, roomy, and well built, raised high above the ground to make them airy, and furnished with open verandahs for the same purpose. Europeans, however, do not in general remain long, since the climate on the coast is not suitable to their constitution. The few who are here seem to lead a miserable life: the society is very limited and monotonous, and their wishes are confined principally to making money; in which many fail. At Cape Coast, the small white shells which we use for ornamenting horses' bridles are given in exchange as coin; they are called cowries: a thousand of them are worth about a guelder, in the interior they are worth more: we have with us whole sacks of them. Gold-dust also appears at first a very curious medium of exchange; it is used especially in Cape Coast and Accra, where it is washed from the sand of the river banks which flows through the town. Every one of the market people carries a small pair of gold-scales: with which he weighs out for a silver-groschen, or perhaps for a sechser, its worth of gold-dust: they then take these very small grains with them, wrapped up in a piece of rag. All these market people are natives, and sell palm-oil, cocoa-nuts, different kinds of fruit, fish, home-woven cotton, &c. The clothing of the men consists simply of a napkin round the loins; or in addition, a long piece of cloth passed under one arm and over the other. They remove it from the shoulder when they meet a white man, and lay bare the heart by way of salutation. The women have these garments, and others in addition. The cloth round their loins is larger, and furnished behind with a monstrous bustle: the bigger this is, the more respectable is the woman, and the larger her family : in many it projects like a saddle. Little children are perfectly naked. So soon, however, as a young girl assumes a piece of cloth by way of clothing, it is furnished with a bustle, which with time is made gradually larger.

" Although I have at present had no opportunity of admiring the full splendour of tropical vegetation; yet many objects have fallen in my way which induced me to examine and to gather them. I regret very much that I have so many difficulties to overcome, in reference to my collections, from the scanty room on shipboard, and the humidity of the weather. If not attended to daily, everything is covered with mould; and even the paper in the chests becomes quite damp. Perhaps, after much pains, I am so fortunate as to get my

plants dry, with the help of the sun and steam-engine ; but I
have still to look to them again, and often find cause enough
for repeating the process. Notwithstanding all this trouble,
the specimens are bad, they fall to pieces and mould con-
tinually ; and I must sit down under the sorry consolation,
that I have effected with all my zeal as much as circumstances
would allow."

On the 9th of August, the little fleet, after it had directed
its course from Accra straight across the Bight of Benin,
reached that mouth of the Niger which is called Nun, and
Vogel writes from thence on the day of arrival, as follows :

" Last night, without any remarkable wind, there was so
strong a rocking of the ship, that I scarcely slept a wink. I
was up late for the first time, namely, after eight o'clock, and
was not present at the morning prayers; which a German
Missionary, from Sierra Leone, the Rev. Frederick Schön,
performs from half past seven to eight. Breakfast comes
between eight and nine : to-day we had ham and yams, and
as usual, coffee without milk. The atmosphere was so thick
that we could often not see half a mile, though when there
was for a moment a clear sky, we descried the mouth of a
river, which we took for the Nun; therefore we anchored
about six miles from the shore. The rain came down
in torrents, and the whole of the gun-room was flooded.
I betook myself to my cabin, from the window of which
I let down the shutter to enable me to see; but the
cabin and bedding were soon so soaked from the entrance
of the rain, that I was obliged to fly to the deck in my
mackintosh. The awning is not waterproof, and the water
stood in many places two inches deep; nevertheless, I tried
to wile away the time there till dinner. This takes place
between two and three, and, thanks to preserved meats, yams,
pastry, &c., is very comfortable. Afterwards, the carpenter
was in requisition to make my cabin water-tight. The
window-shutter was closed and the bed dried, as well as
circumstances permitted. The stove was again placed in
the gun-room; so that we had the pleasant warmth of

87° Fahr. There was enough to occupy me till tea in putting my cabin to rights. At six o'clock we have tea, without milk, and sea-biscuit. At half-past seven, evening prayers. The rain having somewhat abated, my companion and I sought for a tolerably dry place, where wrapped, in my cloak, I might smoke a cigar; and then I took a seat in the gun-room, where I am writing this letter. The violent rolling of the ship, however, still continues; and its effect is evidenced in my more than ordinarily bad writing.

" *August* 12.—We lie at anchor in the open sea, two or three miles off Cape Nun: a stately fleet, consisting of the three steamers, with a large transport and a small one, which will accompany us up the river. There is much to be done: the last stores are to be taken out of the transport, which leaves us here, and the ships supplied as much as possible with coal, that we may proceed up the Niger without delay, to its confluence with the Tschadda.

" *Sunday, August* 15, P.M.—At last we have run into the Nun, after having endured, while at anchor, the most frightful swell, and at the same time scarcity of water. The expedition, in fact, commences to-day; after long delay, we are at length arrived at the place where our observations are to begin. The vegetation, when viewed from the ship, appears extremely luxuriant, and there is something like a forest. We shall probably remain some days in this place. I only hope that the rain will permit me to make good use of the time."

At length, on the 11th of September, the expedition reached the confluence of the Niger and Tschadda, and on the 18th of September, Vogel thus writes :

" We arrived a week since at this place, which I conceived I might consider as an occasional place of rest. We reckoned on a six days' course hither, with no delay at the mouth of the Nun : instead of this, from our first arrival at the Nun (on the 10th of August), to our coming here, a month has elapsed. At Aboh and Iddáh some days were spent; till the desired contracts against the slave trade were concluded with the Kings. This stay was, however, little to my purpose, for circumstances hardly allowed me at either place to go on shore; and at the latter I was unwell, though not extremely so.

" At Iddáh, the country which was before low and flat, begins to be elevated and rises in mountains 2000 feet high, which, with occasional interruptions, extend to this place, where they are confined to the right bank of the river. Here and there, spots occur, which remind one of the Rhine: the bed of the river is, however, too broad (generally above half a mile) to be picturesque, and is often broken and enlarged by various islands. The mountains are bare, without any signs of human industry: once only I saw a village on the top of a hill, which appeared very pretty. Mount Patteh, in whose neighbourhood we lie, is a quadrangular mountain on the right bank, rising precipitously on all sides about 1200 feet high, with many patches of forest, and thickly clothed everywhere with plants. At its foot grow many slender *Oil-Palms;* so that the whole picture, painted with the fresh green which the rainy season has produced, is very lovely. As I sit under the awning on the quarter-deck, and look towards that spot, I cannot help being pleased with the view, beholding in the solitary *Baobabs,* and the *Oil-Palms,* though familiar to me now for weeks, forms which still interest me from their novelty.

" We have bought a piece of land on the right bank, extending from Mount Patteh to Beaufort Island, and at this moment are preparing a habitation for the person who is to have the charge of the station at the foot of the mountain. The land is decidedly of bad quality, and a better situation will be sought for: the other bank is far more suitable, but it has been rejected as too low; indeed, it is now under water. It is impossible for me, at present, to say any thing of the nature of the vegetation. We certainly have not here the usual exuberance of the tropics; perhaps, since I have been on the river, I have collected three hundred species. No single family gives a peculiar character to the vegetation, but this depends on a mixture of many families.

Yet it is possible I may be deceived; for scarcely any trees at present are in blossom, many have only fruit, and others are without any characteristic organ. The *Baobabs* are abundant, most of them have the habit of old thick oaks, only they are perhaps proportionally lower, but I have met with none which has answered the expectation raised by Adanson and Golberry. Among Palms, the *Oil Palm* alone is frequent along the river and in marshy places: the *Fern Palm* occurs here and there; and the *Cocoa* extends as far as Iddáh. I believe that I saw through the telescope a *Tree Fern*. *Parasitical Orchideæ* grow occasionally, though not commonly, lower down the river; here I have not met with one. A leafless *Euphorbia*, forming monstrously thick bushes, grows on Mount Patteh. *Lianes* are abundant; but their tree-like stems affect little the character of the landscape; they form, with the mother-stem, a thick vegetable mass. The most interesting are the towering and climbing herbaceous plants, which, especially along the shore, invest the shrubs and trees to a great height, often presenting real vegetable walls, adorned with the sparkling blossoms of *Convolvuli*, *Cucurbitaceæ*, and *Asclepiadeæ*. There is no fruit here adapted to European palates: the best is the *Hog-Plum*, which is worse than our *Sloes*, and its name indicates its quality. On the coasts grow good *Pine-apples, Bananas*, &c., but they are introduced : the latter alone are cultivated here, though rarely. Horses are very scarce and not larger than asses; and the oxen resemble sheep. Butter and milk are rarely or never procurable; the eggs which are brought are all set upon; we have nothing but old hens for poultry. Bearing these matters in mind, I cannot help exclaiming with Ovid :

'Heu terra felici non adeunda viro.'

The natives, who come to us from far and near, behave extremely well; they have never shewn the slightest sign of enmity, on the contrary, they are rather too confiding. They are not of that deep black hue which is observable in other Africans, and in this neighbourhood they have often very good features. They understand spinning and making cloth: they know how to work in iron, to manufacture knives, sabres, nails, &c.: they cultivate also the fields with some degree of skill. It is sad, however, to think, that they have possessed the same aptness for these arts, probably from an almost inconceivable time, without making any improvement: they lack that spiritual energy which renders every acquisition a step to further advancement. We have a daily market on the shore; whither the inhabitants of a neighbouring village resort in great numbers, to sell or barter what they possess. Small looking-glasses, framed in paper, meet with very ready purchasers; and I shall never forget the joy which beamed in the eyes of many, when they first beheld their own faces in a mirror. The women, especially, cannot be satisfied with gazing on themselves, smeared with the powder of a red wood and their short hair standing upright in little tufts, so that they appear more like horned devils than human beings. In general, however, they prefer what is useful to trifles, provided the latter be not too dazzling and enticing; as, for instance, a bright red cap edged with gold.

"We brought with us a quantity of articles of female dress, often ornamented absurdly enough, as gauze handkerchiefs, sashes, &c.; which they accept as presents with sufficient indifference; whereas they are very eager after large pocket handkerchiefs, which they wear round the loins. The men are all armed with bows and arrows. They value their arrows very highly, which are strongly poisoned: one of them, however, lately sold me all his implements of war, viz., arrows, quiver and bow, a short wooden arm-plate, a knife and an iron ring, for 2000 cowries, about a dollar and a half, which is however not above half the original price. All these things are made convenient for use, and strong, but generally without much art. The way they string their bow, which is about four feet long, is clever. In the right hand they hold a knife, with a hollow handle, through which they pass

four fingers and the middle of the hand : on the thumb they
have an iron ring, and draw between this and the handle the
bowstring; so that they cannot injure the hand. Besides,
they are ready, as soon as the arrow is discharged, to use
the knife. There is a peculiar custom in the whole of
Africa, called " dash." Before a person deals with a stranger,
a present is given, called in African-English, " dash." As
the Africans expect that strangers or Europeans give far
more than they receive, this system is a sort of indirect im-
post, and unpleasant to those who are not prepared for it;
and I have seen many a silk handkerchief given away in this
manner for nothing. The cotton ones, which I had bought,
have done me good service in this way.

" The weather has been very pleasant for travelling. The
rainy season, which we have had in perfection, bestows at present only an occasional shower : I expect therefore that the
river, now at its height, will soon begin to fall. The heat in the
afternoon is generally up to 83° Fahr., seldom so high as
87° or 90°. At night it sinks to 76° or 74°, which feels very
cold after the heat of the day. But the sun has peculiar
influence here, especially when it bursts forth gradually after
a gush of rain : it is then so burning, that I am glad to use
an umbrella and a straw hat.

" The country we have so slowly examined is pronounced
unhealthy. It is no wonder then that the African fever, or
rather fevers, kept in check during the journey itself by the
excitement, has broken out most terribly ; so that the ships
are so many lazarettos. At present we have had few deaths ;
but what may take place, it is impossible to say; for no
sickness is more deceptive, or undergoes quicker changes,
than this fever. Before the evil proclaimed itself so loudly,
the plan was as follows : One ship, the Wilberforce, was to
go up the Tschadda—this is still to be done. The two other
ships were to ascend the Niger, as far as Bussah or higher.
If they could not proceed further, two great boats were to
be manned, and, if possible, to reach Timbuctoo. Now,
however, a plan is arranged for sending the smallest vessel,

the Soudan, down the river, to convey the sick to Fernando Po. I think we shall be back here from the Tschadda in from four to six weeks; and since the rainy season will then be over, and I hope the alluvium on the shore so broad as to enable us to dwell there with comfort, I trust to be in fixed quarters and able to make wider excursions. Since being unwell a few days at Iddah, I have felt healthy and strong. The climate is, however, very injurious to an European constitution; and Sierra Leone also is considered unhealthy : I have, however, found myself quite well after strong exercise. I ascended Mount Patteh, which is about 1200 feet high, about six o'clock in the morning, without much fatigue : I was perfectly well; I botanized, returned at two, took my luncheon and rested. But the whole afternoon I found myself so extremely exhausted and incapable of doing the least thing, as I never was before : with this consolation, however, that I did not experience the slightest feeling of illness. Every one of us, who is not sick, is plagued with itching on the skin, and eruptions : this affliction, together with the mosquitos, which, however, at present have not been numerous, do not let us sleep at night. In short, it is a wretched existence for a European."

The unhappy fate of the Expedition is too well known. Vogel writes on this subject from Clarence Cove, in the island of Fernando Po, on the 22nd of October; " We were desirous of proceeding farther, to begin a real voyage of discovery; when the tropical fever, which we had long feared, but at last considered as left behind, broke out with such a degree of virulence, that in a short time almost all the Europeans were seized, and most of them suffered severely. On the same evening on which I wrote my last letter (18th of September), I fell ill of the fever, which assumed a serious aspect. The sea air being considered the best remedy against the malady, we went all together down the river to this place. First, the Soudan with the sick ; then our ship, the Wilberforce; and lastly, the Albert, after it had proceeded up the Niger for some days, was finally compelled to return, and to

c

bring all the Europeans with her. It is now the intention of the conductor of the Expedition to sail to Ascension, which is considered peculiarly healthy, there to await the perfect recovery of the sick, and in March, when a better climate for a European constitution is expected, again to ascend the Niger. I heard that I might be allowed to spend the interval at this place. They brought me on shore in a very high fever, and I have been now almost three weeks here. The fever, which on my way was almost always upon me, has left me for the last week and a half; and I am now, as I believe, out of all danger. But my strength returns very slowly, and I shall scarcely be able for these six weeks to resume my botanical investigations. At present I cannot walk, but stumble over my own feet. One of the ships, the Wilberforce, is gone to Ascension : the Albert, which arrived later, is here, and will wait for the recovery of her sick."

"Of the Island of Fernando Po itself I can say little : I have not yet been in a condition to look round me. Yet it seems rich in plants, and I hope especially that the examination of the mountains may prove productive; for they are mainly covered with thick woods; and the highest point is above 10,000 feet high. The accommodations are but limited and poor. All the houses are merely made of boards, knocked together, and are raised on strong posts, which are obliged to be frequently renewed to keep off the vermin, and to facilitate the current of air. They are constructed, principally, with a view to airiness : the windows, that is the shutters, do not close : the roof is seldom water-tight, and in the walls and floor are great holes, so that during a heavy rain, such as prevailed yesterday, our chamber is almost flooded ; and it is merely the holes in the floor which, allowing the water to escape, give some relief. The German Mineralogist belonging to the Expedition, who is somewhat more advanced than myself towards recovery, will remain here ; and we have clubbed together for our housekeeping; but even this is expensive. Anything in the shape of a kitchen is out of the question. To the open space under the house, which is beaten hard like a barn-floor, the cook brings every day his iron grate, and prepares, with a monstrous consumption of wood, in four or five iron pots, every thing that can be procured for food. There is, however, no great choice. We have fowls, and beef when ships come, but only then, and occasionally fish. Yams never fail, and they are excellent; so that I prefer them by far to our potatoes. What a pity that there is no possibility of introducing this plant at home! We can have them every day; indeed the poorer people live almost entirely upon yams. Add to this, rice, which however is not cultivated here; and it is almost all that the country can afford to set a poor invalid on his legs again; and it is little enough! If any thing else be wanted, it must be procured from Europe. For our domestic affairs, we are obliged to have two servants, of whom one is cook. Each receives daily a shilling; so that the two cost above three pounds sterling a month, and we have to keep them too. Both together do not accomplish in a day half so much as one European would. Meanwhile, my life passes in eating, drinking and sleeping; for I am fit for nothing else, and am unfit even for that. The Expedition will go up the Niger again in March, and it is hoped will be in a condition to remain there till autumn; if so, we shall return at the end of next year to Europe. Should I regain my strength by the commencement of the dry season, and be able to devote so many months to this island, I expect to reap such a harvest as will content me for some time."

Vogel's last letter is from the same place, dated the 22nd of November, and is as follows: " Since I wrote last, there has been no great alteration. My recovery is tardy, but progressive; or, rather, I have been well for some time; only my strength returns very slowly. Yet I am able to undertake moderate excursions : longer ones I must defer ; till the occasional rains cease entirely. I am most desirous of going to the mountains and to lead there for some time a really natural existence; for here there is a wretched mixture of artificial and natural. For these last five weeks, we have

c 2

had every thing in our domestic arrangements to superintend ourselves; otherwise we must have engaged more servants, and that is not only expensive, but we have quite enough to do to manage the two we have. An African servant will not listen to orders, but will do every thing out of his own head; and if his taste does not agree with his master's, the master he thinks must comply with his. If I say to the cook, "this must not be dressed so," he answers quietly, "That is how I like it;" and if my servant, contrary to my directions, goes out for the whole evening, he says coolly, "When you have got your meal, you have nothing more to do with me." It is often difficult to procure any thing for dinner: we have had no meat for two days, and there was none to be got for money. The same is often the case with bread, and if one has not a stock of ship biscuit, there is great difficulty about it. The light afforded by a palm-oil lamp is worse than that of the lamps which, in Germany, are allowed to servants, and this is very bad when we have any work to do in the evening. What I chiefly dislike is the host of ants, beetles, moths, &c., which swarm every where: they are very destructive to my collections; and I wage constant war with them. Besides the wasps, flies of all sorts, lizards, salamanders and rats pay us constant visits; so that a Zoologist ought to rejoice in having so good an opportunity to make their acquaintance."

While Vogel was busied in this manner with plans and preparations for future exertions, which bade fair to be productive, and this perhaps too early for his strength, the seeds of the last fatal malady were developed. In December, that is, at the time when the rain ceases to pour down in torrents, cold and damp weather prevailed in the island, which is highly prejudicial to Europeans.

In consequence, on the 4th of December, Vogel was seized with a dysentery which confined him to bed, and daily exhausted his strength. Dr. Thomson, surgeon of the Soudan, Dr. Mc William, of the Albert, and other physicians, paid him the greatest attention; and Herr Roscher, the companion of his journey, his fellow-lodger and friend, never left the

bed-side of the patient, who bore all the sufferings consequent on his complaint with the strength of mind peculiar to
him, and without ever losing heart. In spite of all, on the
thirteenth day of his illness he expired, and without pain,
about mid-day, on the 17th of December. His death
was calm and peaceful. He had spoken daily of the expected wanderings amongst the mountains, and even a few
minutes before his death he asked his friend if he had got
every thing ready for their excursion. His mortal remains
were committed to the earth the same evening by the side
of Captain Bird Allen, who departed before him. The ship's
company carried the coffin, which was attended by the
commander, Captain Fishbourne, Dr. Mc William, Captain
Beecroft, Dr. King, Mr. Scott and his wife, and many of his
fellow voyagers, by all of whom he was esteemed on account
of his benevolent and noble disposition, and his really Christian virtues. His little property, according to contract, came
into the possession of the African Civilization Society; and
it is to be hoped that his collections and journals, the
precious relics of an activity, which was extinguished at the
moment when a wider and more worthy field of action
presented itself, will not be lost to science.

Vogel was by nature large and well-formed: his constitution, with the exception of a slight weakness in the chest and
a tendency to rheumatism, was excellent: his countenance
serious but benevolent, and exciting confidence. Active,
without immoderate energy, he rested not till the work he
had undertaken was accomplished. As a man, he was a
fearer of God, of strict integrity, high-minded, indulgent
towards faults; warm in speech, though with a constant ob-
servance of propriety. Towards his friends, he was always
true and devoted; towards his colleagues, upright, disinte-
rested and conciliatory. As a teacher, during the short time
he was so employed, he excited approbation and love, and
much was to be expected from him had he lived longer.
What he would have done as a writer, is incontestibly shewn
by his publications.

JOURNAL

OF THE

VOYAGE TO THE NIGER.*

Wednesday, May 12.—After seven o'clock, P.M., we quitted Devonport. Knowing that I should be attacked with sea-sickness, and not be able to attend with accuracy to many matters, I determined to confine my attention during the voyage to the subject of temperature, and more especially to that of the sea, which I therefore ascertained at noon, and noted in my meteorological journal. The weather being favourable, I suffered less, although never quite free from sea-sickness, than I had expected, and became anxious to extend my observations to the temperature of the air, and to the barometer. Our instrument had not been rectified, and being without a thermometer to mark the temperature of the quicksilver, was rather useless: in our days, such imperfect observations are of little value. Nor could I find on the whole deck a place for my thermometer, without exposing it to many dangers; and the contrivance proposed by me for that purpose, and approved of by the Captain, is not yet finished.

This evening I paid particular attention to the phosphorescence of the sea. In this latitude it is not seen except in the wake of the ship. Only the waves nearest the vessel

---

* Translated from the original German Journal of Dr. Vogel, in the possession of the African Civilization Society, by the kindness of F. Scheer, Esq., of Kew Green.—ED.

were illuminated, and in fact, it appeared to me, that it was
solely the case with those actually in contact with it. If so, the
phosphorescence would seem, here at least, to be the result
of mere mechanical friction. It did not appear to be much
influenced by the moisture or dryness of the atmosphere, for
being very strong on Saturday evening, when the psychro-
meter showed a difference of 1.1° (56.1° against 55°). No
animals or plants were picked up. At a distance I descried
some dolphins, others saw *Algæ* and *Nautili*; and some tired
swallows settled on the ship.

Friday, May 21.—We reached Madeira in the morning.
The Flora of this island has become of late better known,
through Mr. Lowe, who has described many new species. I
intended to devote the few days of our stay to the study of
the indigenous plants ; but the uncertainty of our departure
did not allow of distant excursions, and obliged me to limit
myself to the vicinity of Funchal. I took immediately a
walk along the south-eastern coast, with Mr. Lowe, who
kindly pointed out the habitat of many indigenous species,
amongst which were chiefly *Mathiola Maderensis, Sideroxylon
Marmulana,* etc. On Saturday, 22nd, I was early on horseback, towards the Ribeira Frio ; where, according to Mr.
Lowe, the choicest native plants are to be found. The road
crossed Mount Church ; whose barren precipices are at first
covered with *Spartium scoparium,* higher up with shrubs of
Laurus and *Erica,* and then especially with the magnificent
Vaccinium Maderense. On the summit it was dreadfully
cold, with fog, sometimes like rain. The valley was filled
with mist, clearing occasionally, just enough to see the
Laurus trees that hung down from the surrounding steeps.
This *Laurus,* several interesting *Ferns,* and a few other plants,
were, owing to the bad weather, the unexpectedly small produce
of this trip. On the succeeding Sunday, I chiefly visited the
gardens about the town. The singularly favourable situation
of Funchal, enjoying in consequence of the protection afforded by the surrounding mountains against cold winds, an
invariably moderate temperature, has been frequently dwelt

upon. I had plenty of opportunity to perceive this; for during my stay, the weather was there constantly fine and warm, with, at the utmost, a gentle shower; but once beyond the mountains, I experienced the most furious winds, and the valleys were filled with thick mists, loaded with as much moisture as heavy rains. It were interesting to ascertain the duration and intensity of these mists, which are remarkably dense, and must be highly beneficial to a barren island and fertilizing to its valleys. Later in the season, the weather is said to be clear and settled in the interior also. The valley of Funchal receives several rivulets, and has not at this time any deficiency of water. *Chestnut-trees* abound in the valleys, and the lower declivities are frequently covered with patches of *Pines*. To the gardens at Funchal the prevailing state of the atmosphere is highly genial, and they command splendid prospects towards the town and bay. One really may fancy oneself in the East when walking, and still more when riding between these gardens, which are enclosed with stone walls, over which it is easy to behold the numerous hedges of roses full of bloom. The singular spectacle of the union of *Bananas* and *Pine-apples* with our European fruit-trees, has been frequently noticed, and is particularly attractive to any new comer. Horticulture, from what I could see, was chiefly practised for profit's sake : though in several gardens there were some choice plants, which struck, on account of their finer growth, the European traveller, who had hitherto only seen them in the greenhouse or stove. Large *Dracænas* were rare : whether this tree still occurs in an uncultivated state, I know not : no one collects the *Gum-Dragon*, except as a curiosity. Dr. Renton showed me some fine *Coffee-trees*, covered with fruit, of which the quality is said to be good. He regretted, that instead of *Festuca Donax*, the *Bamboo* was not more generally grown, as it succeeds so well; and I agree with him.

On Monday and Tuesday I made excursions in the valley called the Corral; and to the great water-fall, which yielded me, besides the common plants of Madeira, a few rare ones, viz.:

Ranunculus grandifolius, Sempervivum sp., Sinapidendron frutescens, Lowe, *Bystropogon punctatus,* Hérit., *Bupleurum salicifolium,* Sol., *Physalis pubescens,* &c. According to my limited experience, the Flora of Madeira is of a thoroughly South-European character; only a very few plants, chiefly *Dracæna,* pointing out an extra-European mixture. I do not speak of the neighbourhood of Funchal : a botanical garden there, established with proper judgment, would lead to brilliant results. A novice in travelling revels in the southern forms here first offered to his view.

Of two individual plants I will only here observe, that the indigenous *Parietaria* is that known in Germany as P. *diffusa.* Of *Cassia* I only saw *Cassia bicapsularis,* L., the true species, flowering, but not in fruit, during my stay. *Cassia ruscifolia,* which is indigenous, according to Jacquin, in Madeira, Mr. Lowe assures me, certainly does not grow in the island ; and that *Cassia occidentalis* exists only in a single garden. The history of these species remains therefore still obscure.

Tuesday, May 25.—Left Madeira in the evening. I had exposed myself too much during my last excursion to the waterfall, to the soaking rain, burning sun, and wet, in wading through brooks. The guide had committed an error ; for these people engage to conduct you any where, whether they know the place or not. I was, consequently, several days unable to move ; and when we arrived, on Friday the 28th of May, in the port of Santa Cruz, I could do no more than cast a few glimpses on the island. The next day I resumed my observations on the temperature of the sea ; but my illness, which was an entire interruption of the digestive powers, continued till we reached Cape de Verd Islands.

Thursday, June 3, we were off St. Vincent. We had mistaken the small adjoining Sta. Lucia for the former, and approached it so closely, that we could examine the nature of its shores, which gave only a prospect of wildness and sterility. Sailing along the high cliffs of the western coast of St. Vincent, I looked anxiously for some traces of vegetation, but only distinguished, far off, a few shrubs, and it was dark

ere we anchored at Porto Grande. I hastened the next
morning early on deck, impatient to survey, for the first
time, an entirely tropical vegetation. The back of the bay is
flat and sandy, with a few cottages on the north-east side :
beyond the shore rise hills overtopped in the distance by
mountains. I could clearly descry two main valleys, reaching
far inland, and exhibiting the same white sand as the beach.
Every place was burnt up and bare of vegetation, except a
few shrubs in one of the valleys, whither I directed my first
walk, and found these were *Tamarix Senegalensis*, a shrub
mostly 6 to 7 feet high, but sometimes a small tree, being
the only plant, I might almost say the only object, which
in these valleys affords any shadow. After a search of four
hours, climbing several hills and crossing as many valleys,
I only met with two plants, the same *Tamarix*, and a low
shrub-like *Labiata*, (*Lavandula formosa ?*) almost dried up,
with few leaves and some blossoms just opening. I found
subsequently, that this plant spreads over the whole island.
The Great Desert, whose horrors are so eloquently described
by travellers, cannot exhibit a more desolate aspect than
this part of St. Vincent. Yet the soil ought to be fertile, for
it is a conglomerate of large and small bits of basalt, in
a loamy and chalky soil, closely covered in many places with
dried grass, forming natural hay and furnishing scanty
fodder to cattle and goats, when they have not the *Tamarix*
to nibble at. This soil only wants water, and we may guess,
from these remnants of its vegetation, how fertile it must
be, when supplied during the brief rainy season with some
moisture.* To the above-mentioned plants of the plain, (if I
may so express myself, where there is only hill and dale), I
could add subsequently very few more. A small *Euphorbia*,
perhaps *prostrata* or *serpyllifolia*, but appearing new to me,
a few littoral plants, especially *Zygophyllum album* or *simplex* ; and on the shore, *Cassia obovata*, just then in blossom

---

* According to the natives, the wet season lasts from the beginning
of August to the middle of October, pretty regularly ; but sometimes very
little rain falls.

and fruit, and extending about 600 feet (German?) up the mountains. This scanty harvest induced me to explore the higher regions for more botanical treasures; but even there I found frequently the same barrenness. The mountain chain, which borders the western side of the chief valley, rising frequently to 1500 feet, only afforded me a dozen species on its northern declivity. Two spots however were more productive, viz : the highest ridge and the next highest, situated rather more towards the middle of the island. The former is undoubtedly the richest, and hence goes by the name of "Monte Verte." It is a basaltic rock, topping a gradually ascending table-land, according to my barometrical admeasurement, as high as 2500 feet. It is the only mountain in the island, having its summit always enveloped in clouds ; consequently there are, on its upper half, many well watered spots, whilst every thing else is burnt by drought.

Of the difference between the lower and mountain vegetation I can hardly speak ; but it seemed clear that many plants, flourishing on the mountain, did also grow in the lower country, though now dried up. With the *Tamarix* of the plain, grows a shrubby *Euphorbia* (I believe the only frutescent *Euphorbia* of the island) commonly 2 or 3 feet high ; but sometimes a small tree, with twenty or thirty leaves amongst the blossoms at the ends of its branches, it is characteristic of the mountains and gives an agreeable verdure to the clefts, abounding in the upper valleys and reaching to the very top of Monte Verte. It appears to be the same found by Brunner at St. Jago, and mentioned as *Euph. genistoides?* I think it is an undescribed species. A spreading, creeping, branching, completely leafless *Asclepiadea*, occurred frequently, at 500 feet, on small flats, or pendent from rocks, sometimes with white flowers at the tips of the branches. A handsome *Statice*, a *Campanula* (related to *dulcis*) a *Labiata* with red flowers and coriaceous leaves, (*Lavandula?*), a *Sida*, which I am inclined to think new, with a *Linaria, Borago Africana, Echium, Tribulus terrestris, Achyranthes aspera, Lotus* sp., half a dozen *Com-*

positæ, a shrubby *Urtica*, a flowerless *Sempervivum*, and a few *Gramineæ* and *Cyperaceæ*, formed in this region a pleasant spectacle ; such as one would hardly have expected on an apparently desert island. The general aspect of vegetation was very European, enhanced by *Samolus Valerandi, Nasturtium officinale*, and *Plantago minor ?* To these situations were some cultivated plants ; but they looked, at least just now, very poorly : Beans (especially *Lablab*) *Maize, Cucumbers*, a few *Bananas, Cotton, Ricinus*, and *Batatas*, seemed to be the chief, but hardly in sufficient quantity for the six or seven hundred inhabitants. The *Bananas*, furnished to us, were said to come from St. Antonio. There were also a few *Sycamore figs*, and *Jatropha Curcas*; there are said to be some *Guavas* and *Papayas*. A creeping *Convolvulus* is much grown, and in reply to my repeated inquiries, I was always assured that it was used as thatch.

In *Cryptogamia* this island is proportionally still poorer. Four *Ferns*, all at above 400 feet, a few *Confervæ*, perhaps three or four *Mosses*, on the top of Monte Verte, all without fructification, and *Algæ* on the sea coast very sparingly. Of insects, I found chiefly flies and grasshoppers; few beetles. On the whole I have collected here about eighty or ninety *Phanerogamia* in flower.

Wednesday, June 6.—Towards the evening we quitted this, certainly most sterile island, after a stay of thirteen days. I had been most anxious to visit the adjacent island, which from all accounts appeared more interesting; but this wish could not be indulged, the uncertainty of our departure rendering such an excursion hazardous. However, the Wilberforce had now to go there, (to Terrafal Bay) for water. We anchored off St. Antonio on the same evening, without exactly recognising the spot where we were. In the morning we descried the green shore, proving to be the valley which was to furnish the water. There is a plantation intersected by a clear brook, full in the upper part, but caught in ponds near the coast, for the purpose of irrigating the grounds ; and as the distance is considerable, the ground very dry, and the conduits ill contrived, much valuable water is lost. It would be better to conduct the water in the exact direction of the chief valley, which would shorten its way to the coast considerably. The map of Vidal, however excellent, is not quite correct as regards this valley. Our short stay did not allow me to visit the whole valley. Close to the shore were many plants of *Asclepias gigantea,* whose shining coriaceous leaves attracted notice, even from on board ship. The plantation consists chiefly of some *Sugar-cane, Cotton, Papaya, Citrons, Limes, Guavas, Ricinus, Curcas,* and *Figs.* Higher up the valley *Bananas* are chiefly grown, with *Cassia occidentalis, Cocoa* and *Capsicum.* Amongst the plants on the sandy shore, there were frequently *Argemone Mexicana, Heliotropium,* a *Sonchus,* several *Grasses,* &c. The other indigenous plants correspond mostly with those of St. Vincent, but grow more luxuriantly. The same *Sida* was common : the usual *Euphorbia (prostrata?), Cassia obovata, Tribulus terrestris,* the leafless *Asclepiadea* of St. Vincent, *Borago Africana* and *Tamarix Senegalensis* were also found here : in the part of the plantation nearest the shore grows an *Indigofera* (near *Ind. Anil*) a new species of *Phaca, (Phaca micrantha)* ; and a *Plumbago,* which if it be P. *scandens,* mentioned as belonging to St. Jago, must be indigenous on these islands. The brook in the main valley was full of *Bamboo,* which looked very pretty, especially where intertwined with *Convolvolus* near a small cascade. Along the stream there I also noticed an *Epilobium, Plantago, Cyperus* and *Samolus Valerandi. Orchil* is chiefly exported from this island.

Friday, June 18.—Left St. Antonio at noon. Unfavourable winds and the rolling of the sea made me sick for several days; and I found it not a little disagreeable to be every morning soaked with water, dripping into my cabin; when the deck was washed overhead. The first days, especially, it poured through in absolute streams, and swamped every thing. Of course, my plants suffered not a little, and many things were so spoiled, that I was absolutely forced to throw them overboard. If I were a surgeon in the Royal Navy, I would make most humble supplication that more care should be devoted to the construction of ship's decks, and recommend their being water-tight, which surely cannot be difficult; and if I were not attended to, I would add, like a second Cato Censorius, to every report a "*ceterum censeo*," that the decks be rendered water-tight. It must be surely extremely injurious to health to lie in wet beds.* On my recovery, (Tuesday 22) I recommenced my observations on the temperature of the sea, and was surprised to find it in this latitude still so high. It, however, soon decreased, and towards the coast became very irregular. With regard to the observation of Tuesday, June 22, at half past three, P.M., of 86⁰ 1′, I will here especially observe, that every care had been taken to avoid any chance of error.

Saturday, June 26.—We anchored towards evening at Free Town, Sierra Leone, which presents a very charming appearance. From the Cape of Sierra Leone to the town, gentle undulations, bordered by a mountain chain, on which one may distinguish isolated trees, run close to the shore of the river; while the intermediate space, and even far up the ascent, is covered with the most luxuriant vegetation, brilliantly shining in the full tropical freshness of the rainy season, which has just set in. Between the shrubs, many negro villages, full of closely set cottages with pointed roofs, are sprinkled up to the town, and beyond it along the river. The town itself has a very pleasing appearance : though laid out in regular streets, the houses stand as yet singly amongst trees and shrubs. Probably the aspect of the country may not always be so agreeable : we are now at the end of the tornadoes, when the land has been considerably invigorated by rains : a few months earlier it probably looked very different. Some turns of the mountain-road afford indeed most splendid prospects. The vegetation of Sierra Leone has been so often described, that my observations, limited

---

* This defect, it is well known, does not occur generally in men-of-war, and seldom except in man-of-war steamers.—(H. D. Trotter).

as they were by our short stay, can hardly be worth notice. What may perhaps not be generally known, is the fact, that *Orchidaceæ* occur here frequently : at Mr. Whitfield's I saw a collection of more than thirty species, which he means to take to Europe in a living state. The edible fruits, so interesting through Sabine's publication of Brown's Remarks in Don's Collection, were not just now to be met with, and it requires, in fact, more local knowledge than can be acquired in a few days, to get them together. I inquired a great deal after the somewhat mystical *Cream-fruit* of Afzelius. The name was unknown ; and several persons, even Mr. Whitfield, guessed from my description, that it must be a fruit they called *Bird-lime;* of which the said gentleman gave me a dried, nearly ripe specimen. It is not eaten readily by any body. Although there are here discrepancies, I must after all believe, that we have yet to learn whether *Cream-fruit, Bird-lime* and Don's sweet *Pishanin* are, or are not, identical. The *Oil Palm (Elais Guineensis)* is the only one occurring often near Free-Town. It is monoecious ; the male flower growing above the female. It produces fruit (perhaps not always) when only 7 feet high ; and before the lowest ribs have decayed. I also saw a *Leguminosa,* belonging, as far as I could judge from the fruit, to the genus *Afzelia,* but if so, it would form a separate division. Though a rich flora, it was not, either near the town or in the mountains, by any means so luxuriant as the descriptions had led me to expect. The soil is a close clay, impregnated with iron, and cannot therefore be fertile. It having been soon ascertained that the land near the town could not yield so much as had been expected, the attention of the earlier emigrants was already directed towards other parts of the vicinity; I know not with what success. It is singular, that this thickly peopled colony should not produce any thing fit for exportation : the trade in teak or camwood seems only a waste of the rich endowments of nature. This surely is a matter worth consideration. The Africans, collected here in such multitudes, furnish abundant and cheap labourers, and yet there is no cultivation on an enlarged scale. Much diligence is used to convert and educate the "liberated Africans;" but without any beneficial influence on the neighbouring tribes. This is not very satisfactory, and shows that if it was intended to extend civilization to these parts, great faults must have been committed; and also proves that the Africans are not inclined to follow a good example. The liberated Africans, on their arrival at Sierra Leone, are apprenticed with a planter till their twentieth year; after that, a piece of land is apportioned to them, from which they raise a scanty maintenance. On the whole, their villages appeared to me, as far as I saw them, clean and cheerful (of course *cum grano salis*). But the total want of hospitality, for we often found it impossible to get anything to eat, was painful.

During the few days that we spent here, the weather was mostly fine : the sky generally bright, with a hot sun, though sometimes clouded : towards evening tornadoes occurred, bringing frequently several hours' rain. After having abundantly enjoyed the noise of African tongues and the offensive exhalations of their persons, especially on Thursday, when the Kroomen and negroes were engaged, we left Free Town on Friday, July 2, about noon. Taking the " Soudan" in tow, we made but slow progress, and only got to Monrovia,* on Monday, July 5, and cast anchor in the bay. The few hours which I devoted to a walk towards the head of Cape Mesurado, taught me, that the vegetation is very similar to that of Sierra Leone. *Sarcocephalus esculentus* grew abundantly ; and the fruit called pomegranate by Don, occurred sparingly. A *Poivrea*, with beautiful red flowers, seems new. *Cassia occidentalis, Borreria Kohautiana,* and an herbaceous *Phyllanthus* grew in abundance. Around the dwellings *Coffee-trees* had been planted, but left to grow too freely; *Limes, Figs, Curcas, Guavas, Ananas, Anona muricata* and also *Cytisus Cajan* and *Arrow-root* were

* Monrovia is the capital of the American colony of Liberia.—(H. D. Trotter).

cultivated; *Bananas* and *Oil Palms* occurred of course. The plantations were no doubt extensive, but during my short stay, I could not see more than those of *Cotton* and *Sugar*. *Anona muricata* is much eaten, both here and at Sierra Leone, under the name of *Soursop ;* and I was assured that it is considered the finest fruit of all; but I could not taste it without disgust: altogether I cannot join in the praise of African fruit. The land was not very rich. On the shore there is the same iron clay as at Sierra Leone, and somewhat higher up to the Cape it also prevails (according to Rosher), only finer-grained and firm. In several places water (rain?) has percolated, and caused it to assume singular shapes, almost models of mountain ridges.

Monrovia Town has a pleasing appearance, many of the houses are large. Few white people are seen. The coloured population, with hardly any of whom I had intercourse, appeared inquisitive, obtrusive, and fond of idleness : no traces of hospitality, but an eagerness to make money, and a desire to affect importance. The connection* between Liberia and the United States I could not make out very clearly. There is but one flag flying in Monrovia, that of the United States, viz: on the house of the Governor; professedly because he is the American Consul. The school-house is a large hall, hung with maps of Africa and America; there were also near the raised desk some philosophical instruments, used by the missionary, who had some prepared heads of animals, as he told me, to exhibit during his lectures. The boys and girls are taught in the same room ; but as I was there only during the free hours, I could not witness the method of instruction practised. It is singular, that instead, as we hear, of Liberia being on good terms with the natives, it is always at war with them. The last war ended

* The settlement of Liberia is under the control of a Society, in the United States: the Superintendent being appointed by the Society and not by the United States' government. By the laws of the United States, the Federal Union cannot possess colonies beyond the seas.—(H. D. Trotter.)

about six months ago. The inhabitants allege the destruction of the slave factories as the cause.

The rainy season had now fairly set in, and my cabin being so damp that I could not dry either plants or paper, to form a collection became impossible, and I carried away but a few single specimens.

Near Monrovia, is a Kroo town; whence fishermen, in their small canoes and with angling lines, came paddling about our ship. Except a slight covering on the head, they were quite naked: in warm weather, this was probably the fittest attire for them.

Towards the evening of Tuesday, July 6th, we left Monrovia, and until Thursday evening, were in tow of the Albert. We then proceeded, by ourselves, to Grand Bassa, where we anchored on Friday morning, for the purpose of taking in fuel. We stayed several days; not one of which passed without rain, sometimes most violent throughout the entire day. This, and other circumstances, limited my researches to the immediate vicinity of the shore; where, however, I found more plants than I was able to preserve. I made a collection of about a hundred specimens, at the risk of losing everything by the wet. Many plants, especially the *Monocotyledone*, were not yet in flower; and I regretted this most especially in the case of the numerous parasitical *Orchideæ*. The shore is flat and sandy; and the sand has drifted so far inland, that I never got beyond it. There were no forests, only bushes, intermingled with isolated high trees; which I could not determine, for they were all without blossom or fruit. The African *Bombax* appeared amongst them, and the same *Spondias* as at Sierra Leone, forming a considerable tree; respecting which I feel doubtful whether it be identical with *S. Myrobalanus*. The pride of this coast is the *Elais*, often growing in clumps of twelve or more, exhibiting under different circumstances a different habit, and giving a considerable variety of aspect to the country. This *Palm* is of generally moderate height, and constitutes with various *Fici*, the chief masses of wood. The underwood consists of

close-growing shrubby *Rubiaceæ*, with shining leaves, intermingled with *Gloriosa superba, Cissi, Leguminosæ, Banisteriæ*,
as creepers, leaving hardly room for *Melastoma* and other
low plants that peep through with their fine blossoms. It is
a very interesting sight, that of a few *Oil Palms* growing in
a clump ; the ribs of the lower leaves still adhering to the
stems, which are clothed with a fresh verdure of parasitical
Ferns and *Orchidaceæ ;* whilst other parasites, such as *Ferns,*
Pothos, Anonæ, Commelinæ, small *Rubiaceæ* and *Leguminosæ*,
choose the airy shelter of the foliage for their habitation.
Of single plants one might specify *Sarcocephalus*, which
occurs frequently, the same *Phyllanthus* as in Liberia,
Schmidelia Africana, a genus of *Apocyneæ*, apparently new
and near *Tabernæmontana*, remarkable for its double fruit as
large as a child's head, the seeds nestling in the almost woody
pulp, wild *Sugar-cane*, not in blossom, *Conocarpus erectus*,
var. *β.* a small shrub, a probably new *Cassytha, Scævola*
(really different from *S. Lobelia ?*), *Indigoferæ* sp. *Cannæ* sp.
Cassia occidentalis (cult[d].), *Borreria Kohautiana*, &c. The
Stylosanthes forms a close jungle, with its erect and much
branched stem, about 1½ foot high, along the sandy shore.
A few open spaces amongst the shrubby woods were covered,
as if cultivated, with *Cyperaceæ ;* amongst which a species of
Eriocaulon is frequent. A few more watered spots showed
Grasses, with a beautiful *Orchidea* 2 or 3 feet high. Near the
village, I found *Euphorbia drupifera*, Schum. An excursion to
the river enabled me to examine the *Mangrove woods*, where
a *Rhizophora* (different from *R. Mangle ?*), but not yet in
ripe fruit, formed the bulk of the woods : amongst it an
Avicennia, judging by the leaves, different from that at Sierra
Leone (*nitida ?*), was frequent ; and the shrubby *Conocarpus*
racemosus (is it not identical with an American species?), which
so far as I know, has not yet been enumerated amongst African
plants, but inhabits similar situations at Sierra Leone. Intermixed with these, *Drepanocarpus lunatus* rendered my progress very difficult. *Pandanus Candelabrum*, without leaves,
occurs here for the first time, in swamps. An *Anona* (a tree

10 to 12 feet high), in fruit, and apparently very similar to *chrysocarpa*, Lepr., if not the same, was not uncommon in these swamps. Leguminous trees seem rare, and do not attain a large size: there are no *Mimosæ* or *Cæsalpiniæ*. Of cultivated plants, the *Sweet Cassada* is most valued and grown; also *Rice*, various sorts of *Capsicum, Papaw* and *plantains*, and *Holcus* here and there, with *Ananas* in large quantities amongst the shrubs.

Our anchorage was between a town belonging to Liberia, called Idine (according to the pronunciation), and the River Keún, but nearer the latter. The jungle begins with the flat shore; and the native villages, consisting of a few huts, are situated amongst it. The Kroomen live near the shore : the natives are of another race. The cottages of the former which I visited, were neat and clean, built of mats, square, with pointed roofs; and generally a raised floor, 1½ feet above the ground, composed of plaited palm-ribs. The Kroomen themselves appeared rather intelligent; and they pleased me by their straightforward and modest behaviour, touching none of my things without permission, which might have served as a good example to the people of Liberia.

Wednesday, July 14.—We left in the afternoon, and anchored on Friday, July 16th, about ten o'clock, A.M., off Cape Palmas, to take in a fresh supply of fuel. The Cape is formed by a narrow projection into the sea; on the foremost part of which, the houses of the American colony have been built. The dwellings of the fishermen are situated on the part nearest the main land. Their huts are very different from those of the Kroomen of Grand Bassa, being without raised floors, and having much more pointed roofs. The buildings of the American colony are straggling, and they extend, I was told, about four miles into the interior. There are none but people of colour at the Cape; the only whites, if I understood rightly, being a few missionaries, who devote all their attention to the natives. At this colony, the soil is very bad: the rock, frequently protruding through it, consists of hornblende (micaceous slate). The soil is a very hard iron-clay, in small clumps, originating, according to Rosher's statement, in the *débris* of decomposed granite veins traversing the rock; but to me it appears that the rock itself has much to do with the formation. Further up the stream, the land is said to be good. North of Cape Palmas, the river, according to the statement of the Governor, is navigable for seven miles with canoes, and empties itself into the sea, through several mouths. From a distance, the Cape has an agreeable aspect : the isthmus is well clothed with vegetation, and beyond it the beautiful forms of the *Oil* and *Fan-palms* are seen.

My excursions were limited to the isthmus and nearest parts. On the isthmus grows *Phœnix spinosa*, Th., a low shrub : beyond the river it is said to produce flowers and fruit. A few *Cocoas*** had been planted, some years back, and were still small, as were the trees of *Anona muricata*. The plants chiefly cultivated seemed to be *Cassava, Sweet-potato, Bananas, Plantains, Indian corn,* and *Rice ;* while *Cassia occidentalis* was seen in every cultivated spot : the same *Spondias* as before grows also here : *Coffee* had been introduced from Monrovia : here and there the indigenous species of *Cotton* had been raised : *Arachis hypogœa (Africana ?)* I found planted in one place. In the native Flora, which, however, I have hardly seen, *Rubiaceæ, Convolvulaceæ, Leguminosæ* were chiefly conspicuous. The same *Anona* (near *chrysocarpa*) as in Grand Bassa grew here : *Pandanus Candelabrum* on dry ground; several sorts of *Figs*, amongst which is the small fruited kind of Grand Bassa : *Jatropha Curcas* was frequently employed for fences. Amongst the underwood I found a small shrubby tree, related to *Belvisia (Napoleona)*, and probably a distinct genus nearly approaching it, it bore blos-

---

* The inhabitants believe, that whoever plants a Cocoa-palm will die, before it produces fruit (*i. e.* in about seven years). The Chief of the fishermen yielded at last to the entreaties of the American Governor, and put some Cocoa-nuts on the ground : he then drove cattle over the spot, that he might not incur the consequences of planting and covering them with earth !

som and fruit; the latter convinced me that I had seen the
same, and a species but little differing from it, at Grand
Bassa.

Sunday, July 18.—We left Cape Palmas about 2 P.M., and
were off Cape Coast Castle on the evening of Saturday, the
24th. On Sunday, Captain Trotter issued a circular, prohibit-
ing any one belonging to the expedition from remaining all
night on shore : the unhealthy season here having begun. The
Gold Coast was of the greatest importance to me, the plants
described by Schumacher forming a sort of standard for the
African Flora; but I deemed it best to be careful, and to
decline all friendly invitations to stay on land; although this
would have been of infinite advantage in collecting, and in
fact was almost indispensable.

The vicinity of the town exhibits no great fertility : granite
and gneiss, often naked, extending to the coast.   A few miles
inland, a fine black loam prevails, apparently very favourable
for cultivation; and further inland still, the soil is said to be
extremely fertile, consisting probably of vegetable mould.
On account of the heavy surf, it is impossible to land, otherwise than in canoes; and in this and every case where you
are obliged to depend on negroes, punctuality is out of the
question; and much time was always lost. A trip to the Model
Farm, five miles inland, now under the superintendence of
Mr. Wilson, promised to make this place very interesting.
The major part of this plantation lies on the declivity of a
hill, consisting of indifferent soil, (decomposed granite) ; whilst
before and beyond it, the land is excellent.   They call this
plantation " Napoleon."   The dwelling-house is on the top
of the hill, and commands a very interesting prospect.   The
plantations consist chiefly of *Coffee-trees*, only a few years
old : some, covered with fruit, were, according to Mr. Wilson,
of only seven months growth, which seems truly wonderful, for
in the West Indies, *Coffee* bears no fruit even in the best soils
under eighteen months.   Besides *Coffee, Bananas, Plantains,
Arrow-root, Yams, Limes, Lemons, Oranges* and *Indian corn*
were much cultivated.   In the grounds of the natives, *Indian*

corn, Bananas, Plantains and *Yams,* were conspicuous, but
no *Holcus* (!) From the *Indian corn* they prepare a very
sour bread, which with *Bananas,* constitutes their chief food.
Palm-soup, a native dish, when made of boiled *Palm-nuts*
only, is very well flavoured. They pick the nuts off those
young stems of the *Elais Guineensis* which have not yet lost
any of the leaves, and consider these as superior to the fruit
of older plants, and cut them also down, to collect palm-wine.
Besides this *Palm,* there is the *Cocoa;* which frequently
assumes a singular aspect from the multitude of birds'
nests appended to the mid-rib of the leaves, and which
might be taken at a distance for fruit, and had formerly
puzzled me in drawings. The birds hang their nests in this
position to protect them against the cats! The *Fan-palm*
grows too at Cape Coast Castle, but apparently is less
frequent. To judge by parts of the stems which I met
with, *Calamus* must occur further in the interior.

Another excursion was about six or seven miles inland, to
Orange Town and Quowprath. Here the soil was fertile,
with good vegetable mould and extensive plantations of *Indian
corn; Bromelias* skirting the former plantations. The best
habitations of the natives resemble those of the Ashantees,
and have a square court in the middle, its four sides
surrounded by buildings.

It is almost impossible to travel in European clothes; especially during this season, when the water collected in the
roads reaches often up to the middle. Besides, great exertion or exposure to the sun is dangerous, and occasionally
fatal to new comers. The residents go out in small carriages,
drawn by four negroes, or travelling-chairs carried by two. The
former can only be used on tolerably good roads, and the latter
have also their inconveniences. For instance, I was myself
upset in the middle of a puddle; because my bearers slipped;
but I happily fell on an adjacent dry grass-plot. It is a great
inconvenience for persons who, like me, travel *ex professo,*
that at such places as Cape Coast Castle, it is impossible to
hire the necessary vehicles, but you must be dependant on

the kindness of others. I had the good fortune to find, in Mr. Henry Smith, a man who anticipated all my wants with the utmost affability, assisting me, in fact, in every possible manner.

There is much less of botanical interest near the town than I had expected, the number of plants increasing materially with the distance from it. The present season, immediately subsequent to the rains, is not very favourable: the rain had nearly ceased on the coast, and only a few showers fell now and then; but a few miles inland, much rain prevails about this time; and on my trip to Quowprath, about six miles, I got thoroughly soaked. I saw many plants without flowers or fruit; but not one that was *Monocotyledonous*, though many are said to occur with splendid flowers. The difference of the vegetation from what we had last visited, was very striking. Here *Leguminosæ* were predominant, and *Rubiaceæ* less so ; *Mimosæ*, with their characteristic foliage, which I had hitherto seen but rarely, became conspicuous. The country is varied with hill and dale, and covered with shrubs 6 or 7 feet high, intermingled with single lofty trees, particularly *Bombax*, in leaf, but without blossom or fruit, which the inhabitants call *Iron-wood*.

I found another single tree of considerable height, with flowers and fruit: it seems to be a new genus related to *Crescentia*. The fruit is filled with solid firm pulp, 2 feet long, 1½ foot broad hanging downwards, as also does the flower, by a long pedicel. About the town, and in its vicinity, grows a half-shrubby *Cassia*, similar to *occidentalis*, but with a round divided fruit which might be taken for that of *C. Sophora*. The true *Cassia occidentalis* occurs likewise. *Poinciana pulcherrima*, just coming in flower, prettily lined the roadsides ; and in the jungle grows a yellow *Composita* (I only saw two *Compositæ* in flower) which often adorned great parts of the way, and seems diffused over the whole coast. *Sarcocephalus* was seen in blossom and fruit. The new genus of *Apocyneæ*, with large fruit, did not occur. A beautiful avenue of *Hibiscus populneus*(?) planted at the west

end of the town, forms one of the marked features of Cape
Coast Castle. As we were about to proceed to Accra, I
thought it important to avail myself of the opportunity and
to visit, if possible, the Danish settlements, founded in the
interior by Isert, and to obtain information respecting them,
which had not been received at all of late. The Wilberforce
was not ready for sea; but the Albert left on Friday, the
30th July; and Captain Trotter allowed me to make the
passage in this vessel, thus saving much of my time. We
anchored on Saturday afternoon, at British Accra; but it was
late before I got on shore; for the surf would not let us land
without canoes, which, as at Cape Coast Castle, are made
pointed at one end, and provided with a high bulwark.

As my excursions led into the mountains, Dr. Stanger
offered to accompany us ; and Mr. McLean, who went with
us on shore, kindly provided us with quarters for the night,
it being too late to proceed to Danish Accra. Sunday
morning, the 1st of August, we set out in two little carriages,
each drawn by four negroes, (here also the common way of
travelling for Europeans) for Danish Accra ; where we called
on Mr. Richter, a Danish merchant, and accompanied by
him visited the Danish Governor, Mr. Dall; to whom Mr.
Richter and Mr. M'Lean introduced us.

The fortifications here are not important: they consist of
a few large houses, with lofty, airy rooms surrounded by a
wall and breastwork, and are inhabited by the Europeans.
They are white-washed and conspicuous at a great distance.
The Danish fort is classic ground for a botanist, for here
Isert and Thonning made the collection, through which we
became acquainted with this Flora. The humane spirit of
Isert, so warmly expressed in his writings on behalf of the
negroes, rendered this place highly interesting to me; and
the more so, as we were engaged in an enterprize, aiming at
the objects which he had endeavoured to attain during the
latter years of his life. I inquired anxiously after his establishments in the interior, but could obtain no official information about them. After Isert's decease, they had gone

to decay. Mr. de Khon, who is said to have assumed the
management, and introduced the plough, and is represented
in various works which I have read, to have effected so
much, never came here, as Mr. Richter and the Danish
Governor positively assured me! Since his time, indeed,
no one took any trouble about these plantations; and about
1808, they were altogether given up. Every thing is now a
wilderness, and the place not to be recognised. Flindt established about this time, another plantation on the River
Volta near the Fort, the main object being distillation: but
this was soon discontinued. About ten years ago, I believe
another plantation was formed at the foot of the mountain
in Aquafim, named " Frederic's Gau;" and as we wished to
visit it, Mr. Dall had the kindness to indulge us; but he
told us it was not extensive, and the superintendent
being ill, it could not be in a very satisfactory state. The
distance is fourteen or fifteen miles: the only mode of getting
there is in a sort of palanquin or basket, carried by two
poles, on the head of two or four negroes. Mr. Dall, by
providing abundantly for all our wants, caused our *cortège*
to amount to about sixteen persons. The direction, according to compass, was almost exactly N. by E. We started
at half past eleven o'clock. The first and greatest part of
the way leads through Savannahs, covered with *Grasses* and
Cyperaceæ, intermixed with many species of shrubby and
half-shrubby *Leguminosæ*, besides a few *Malvaceæ*, and
some tall, but more generally only moderately high trees,
viz: *Bombax*, the genus which I mentioned at Cape Coast
as perhaps related to *Crescentia*,* *Ficus*, *Fan-palms*, *Euphorbia
drupifera*, very conspicuous from its naked spur-like branches,
bearing only a few stiff inversely spathulate leaves at the
extreme points, and near the villages and huts *Tamarinds*
and *Hibiscus populneus*. Towards the coast, the soil is sandy
like decomposed sandstone; but soon improves, from the
culture of *Indian Corn*, *Cassava*, *Yams*, *Arachis*, various
sorts of *Cucumbers*, and *Bananas*. *Cocoas* are little culti-

* May it not be the *Bignonia tulipifera*, Schumacher?

vated here ; or in any part of Africa, which I have seen.
We crossed several ridges of hills affording pleasant views
over the surrounding country, covered with fresh green, and
struck then into the jungle ; where the shrubs, common
on this coast, grew abundantly, about a man's height,
and closely interwoven with creepers. *Leguminosæ* dimi-
nished, and *Rubiaceæ* increased. *Sarcocephalus*, described by
Schumacher as *Cephalina esculenta*, Th., is not uncommon.
We arrived at the settlement towards six o'clock, P.M., too late
to see much. The house of the superintendent lies half-way
up the mountain ridge, and is roomy and comfortable, and
being white-washed is conspicuous far off. At the foot of
the mountain is a negro village and the plantation. Monday,
August 2nd, having passed the night, in consequence of the
friendly care of Mr. Dall, most comfortably, and supplied
with every convenience, we were off at dawn of day ;
thermom. $73\frac{1}{2}°$ Fahr. The mountain is a quartz rock, covered
in many places, and often to the depth of several feet, with
vegetable mould, overgrown, where not cultivated, with
Brushwood. The site of the house was at an elevation of
about 1000 feet, and 100 above it grew a high *Oil Palm*.
The brushwood consisted chiefly of *Rubiaceæ*, interwoven
with *Convolvulus :* few in flower and none remarkable. In
the plantation were the usual edible plants of this country :
the settlement consists of a coffee ground, of no great
extent. Governor Dall told us that about three years back,
the trees had been destroyed by an insect, and they were
now very small, 3 to 4 feet high, but thriving and bearing
abundantly. The soil is excellent and rich ; but the establishment looked neglected, which must be ascribed to the
absence of the superintendent. Close by is another coffee
ground belonging to Mr. Richter ; but none of our companions speaking English, I only heard of it after our return.
Near these grounds is an avenue of *Soursops* (*Anona muricata*) and *Oranges*, and close by several trees just now bearing
ripe fruit, clearly the *Akee*, or *Blighia sapida*. They seem
to have been planted ; but on looking into Schumacher's

description of Guinea plants, I found a *Cupania edulis*, mentioned as an indigenous tree, which I dare say, is identical with the above.

As we had only leave of absence until sunset of this day, we were obliged to content ourselves with the slight survey of a few hours, and after enduring an hour's heavy rain, we started at eleven o'clock and came back by the same road, though being down-hill, we got on faster ; and having returned sincere thanks to Governor Dall and Mr. Richter for their obliging and liberal assistance, we arrived in good time at British Accra ; where we found that the hour of departure had not yet been fixed.

As soon as I got on board the Wilberforce, my first care was to shift my entire collection, especially the plants gathered since we arrived at Cape Coast Castle; but though I had taken all possible care, much was spoilt and almost everything in a bad state. It has been my lot with almost all my collections on this coast, that after endless labour, I could only get together ill-conditioned plants ; for dampness and want of room are obstacles impossible to be overcome, and which forced me at last to satisfy myself with the miserable consolation, that I have done all the circumstances would admit. I mention this, on purpose, that in case my collection comes into other hands, I may not be accused of negligence. I have sacrificed every convenience to gain room, and spared no trouble to overcome the dampness of the ship and of the atmosphere, but without success. The general arrangements of a man-of-war do not give much opportunity for such experiments. When will the time arrive, that expeditions, whose result must depend on the observations of naturalists, will afford them, from the outset, the appropriate and necessary accommodation ? At present, the vessels are fitted up for other purposes, and it is left to chance, to discover a little nook for the philosopher. I was now obliged to devote the two days remaining which we spent at Accra, to the drying of my collection ; that all might not be lost.

Thursday, August 5.—We left Accra after midnight, and cast anchor on Sunday, the 9th, at the mouth of a river, supposed to be the Nun. The weather was gloomy, and a dense rain falling all day, caused the wet to make its way through the shutters, so that it was difficult to find a dry place, even for standing room. We stayed there the whole day, and sailed next morning for the mouth of the Nun, anchoring about nine miles off it, alongside the Albert.

Friday, August 13.—The want of water, already felt the day before, was now more severely experienced; although we had collected some rain on Monday. How such an Expedition came to be unprovided with water, especially when we consider that, on no account, ought we to have made use at first of the Niger water, is incomprehensible to me! It had been easy to obtain abundance of good water at Danish Accra.

Sunday, August 15.—We quitted our anchorage at half past eleven, A.M., and crossed without difficulty the bar; beyond which we cast anchor beside the Albert, at about a quarter to two, P.M. Here we stopped four days; during which I could only examine the right bank of the river, because I had no boat to get to the opposite side; where the greater extent of land and a village seemed to offer more interest. The river is here perhaps 10,000 yards wide; and the stream carries down a great deal of sand. The tide showed itself very distinctly, running perhaps three or four knots an hour, and the current seeming to set more on the left shore, which appears to be a mere sandbank, or sandy foreland, than on the right, which is covered with jungle, immediately beyond the sandy strand. The mouth of the Nun looks like a Delta, on a small scale; at least now, during the rainy season, being intersected by many shallow watercourses, forming, further on, low lands covered with *Mangroves*, similar to what I observed at Bassa Cove (Grand Bassa). The *Avicennia* appeared to prove, that the one hitherto seen, with quite naked leaves (A. *nitida ?*) at Grand Bassa, is but a variety of that at Sierra Leone. In these

Mangrove swamps, the *Oil palm* often grew, covered with
parasitical Ferns, (I found only two species of Ferns besides
those, which are terrestrial), and on somewhat higher ground,
Drepanocarpus lunatus, Ormocarpus verrucosus, a few shrubby
Rubiaceæ, and a few *Mimoseæ.* Of the trees, intermixed with
the Mangroves, little can be said : they were not many, and
all covered, to the very top, with parasites. Some belonged to
the genus *Bombax.* This land, if it can be so called, was but a
few feet above high-water mark, and consisted of sea-sand and
vegetable remains. The beach was quite flat, hardly higher
than the sea, covered in many places with water, and formed
of sand, mixed with mica, probably carried down by the
Niger, and giving its shores a shining and peculiar appearance.
In some places, the strand is clothed with jungle close to the
sea, consisting of *Chrysobalanus Icaco* and *Ecastophyllum
Brownei* ; the fruits of the former, of a beautiful red, were very
conspicuous. Intermingled with these grew *Melastomaceæ,
Diodia maritima,* Th., some other small *Rubiaceæ,* and *Scoparia dulcis* ; while the border, towards the higher woods,
was frequently ornamented with the beautiful yellow flowers of
Hibiscus tiliaceus. Amongst these shrubs, spots might be seen,
here and there, covered with tall rough Grass and *Cyperaceæ,*
to the height of a man, and higher, bound together by *Convolvuli, Cassytha,* and other *Lianes,* rendering them perfectly
impenetrable. I found several places closely matted with
Stylosanthes Guineensis, forming carpets ; upon which one
might cross pools without observing them. The most
barren and sandy places were much overgrown with a *Te-
leianthera,* R. Br., (*Illecebrum obliquum,* Schum. ?) an *Euphorbia* (*trinervia,* Schum.?) but especially with a yellow-flowered
creeping *Dolichos* and *Convolvulus Pes Capræ,* (*rotundifol.*
Schum.), which latter is diffused over the whole coast from
Monrovia. An *Umbellifera* (*Hydrocotyle interrupta,* β.
platyph. DC.), grew every where on the beach amongst
the Mangroves, and seems to overspread the whole coast.
A species of *Malaghetty Pepper,* differing from that in
Grand Bassa by the long beak of the fruit, was frequent.

On one spot, amongst the Mangroves, I noticed, on the de-
caying roots, a delicate white plant, having white scales instead
of leaves, and three flowers : it was a parasite on the roots,
but sent forth roots of its own. I have preserved a few
specimens in spirits. Upon the whole, I have seen too little
of the vegetation here, to compare it with that of any place
hitherto visited on the coast. On the opposite shore, they
cultivate *Cocoa Palms*, of which the natives brought us
the nuts : on the right bank, where we did not now see any
inhabitants, the *Cassada* showed traces of abandoned plantations. The scenery is not remarkable. At the entrance,
the left side presented a pleasant prospect, from the familiar
forms of the forest and *Cocoa Palm* : on the opposite shore,
beyond the forest and brushwood, there appeared a sort of
lagoon ; while behind that, the Mangroves rose into an erect
and lofty-stemmed wood.

Of the natives, I saw only few, and none very near. They
seemed to be well-formed, robust men, with their hair
frequently shorn in a crest shape, but having nothing particular in their dress. I was told that they have a language of
their own (Bassa language). The weather was changeable, alternate rain and sunshine, the former moderate and the
heat never oppressive. By day and night, but especially
during the day, a fresh sea-breeze prevailed.

Friday, August 20.—At break of day, we proceeded up
the river, and although it rained violently, every one was in
high spirits at our at last moving onwards, and beginning,
after so much detention, the Expedition itself. A little
above the bar, the river, dividing into creeks and branches,
is very wide ; resembling a lake ; but the only branch deep
enough for the steamers, at present known to unite with the
upper part, called " Louis Creek," is narrow in proportion,
at one part only sixty to eighty English yards wide. So
far, the shore is covered with Mangrove (*Rhizophora*),
which, with its roots descending from the branches, has
a singular appearance; but this is only the case with old
trees ; for the young Mangroves often form woods of dense

foliage, now in the full splendour of green leaves—a glorious sight! Only in a few places, I saw *Ferns* spring out of the water amongst the Mangroves. A little beyond Louis Creek, the character of the vegetation underwent a marked change : although the country was still much covered with Mangroves, they receded to the back-ground, and the stream itself was lined with young, still bushy, *Oil-Palms : Pandanus Candelabrum* showed, not seldom, its grass-like leaves; while, here and there, other trees mixed with them; until, near Sunday Island,* (about thirteen English miles from the sea), the *Mangroves* and *Pandanus* disappear. Then the shore was lined with small trees and shrubs, with fresh glossy foliage, backed by the tall and elegant forms of fully grown *Oil palms*, a view which can never tire our sight. These Palms are 60-80 feet high. The stems are thickest in the middle; but the contraction towards the bottom is hardly perceptible. The top is rounded. The leaves are long, their tips somewhat pendent ; the lower leaves more so, which causes the cylindrical shape.

Hitherto we had met few natives ; but they now began to show themselves, more and more numerous, in their small canoes. Their thatched huts, close to the river, were surrounded by plantations of *Pisang,* descending apparently into the water. I saw occasionally *Bombax* trees, or *Leguminosæ* and *Mimoseæ,* easily distinguishable by their peculiar foliage ; and some other trees, which might have been taken for species of *Ficus.* The trees increased in number : towards evening, we passed shores covered with tall *Reeds,* beyond which thick forests extended; but under no circumstances was there a deficiency of *Oil Palms.* Alternating with reeds, we observed plantations of *Pisang* and *Sugar-cane,* completely in water ; close to small villages which became very numerous. After sunset, we anchored in the midst of the stream. From Alburka Island we reckon to have made thirty-five English miles, (or forty from the sea.)

* The influence of the tide extends only as far as " Sunday Island."—(H. D. Trotter.)

Saturday, August 21.—We proceeded, the three ships in company, at day-break. The vegetation resembled, on the whole, what we had seen yesterday: the trees often descending close to the water, and exhibiting a mass of parasites of most singular forms. Sometimes I saw flowers, and fruit, which only made me regret, that I could not examine them closer. In Madeira I botanized on horseback, at Cape Coast Castle out of a carriage, and at Accra in a basket; but from a steam-ship it was impracticable. The villages became very frequent: in the plantations we saw (through the telescope) besides *Pisang* and *Sugar*, occasionally *Cassada, Maize* and *Yam*; to which may be added the *Oil palm* and the *Cocoa*, similar to the latter, but (here at least) not so slender, being rather short and of vigorous growth. But whilst the *Oil palm* grew every where, the *Cocoa* showed itself only near villages: a sure proof of its not being indigenous. Soon after noon, an attempt was made to proceed by another than the usual branch of the river, round an island; but we found that it did not speedily join the main stream; and we were separated from the other vessels, which had taken the eastern branch. After sunset we anchored, having come about thirty-six miles. Soon after entering the western branch, we perceived on the right shore a village of clay cottages, from whence a chief came off to us: the village was called Otuo. The men in the canoes were a robust race, and, like others who visited us in the course of the day, had a line or mark drawn over the forehead down to the nose. Their clothes showed nothing remarkable; but the hair of some was divided into squares, like a chess-board; while others wore it plaited, in numerous little tails, which stood erect on the head like so many horns. They spoke the Bassa language. The shore was generally very low, rising but little above the river, at the most elevated part perhaps 4 feet; while the bared roots of plants made me think that the water is sometimes higher than at this time.

Sunday, August 22.—Proceeding at break of day, we soon

E

perceived on the left side a town : the first we had yet seen, situated on an elevation of 6 to 10 feet above the river, and containing clay cottages, each with a covered court-yard; while higher up were some magazines or warehouses. I saw here no *Cocoa palms ;* but in the course of the day a few single ones occurred. The natives, who assembled on the shore, to the number of several hundreds, it was fancied, mentioned the name of the town as " Amasuma" and that of the river as " Oguberri." Further on we came to two equally wide branches of the river, with equally strong currents, joining together : after some consideration, the easternmost was chosen, and at two o'clock we arrived at a similar place, but where the western channel was very narrow. We proceeded a short way upwards ; and Captain Allen caused two plants to be fetched by the boat, which was towing. One is probably a new *Dalbergia,* and one a Creeper, which I had watched eagerly ever since " Sunday Island." It climbs up the trees along the shore, to their very summits, and then drops many thread-like stalks, 6 feet long, covered at the top with bundles of yellow flowers, which often reach the ground. It appears a new genus, closely allied to *Mucuna ;* and I call it provisionally *Mucuna flagellipes.* Both plants were unfortunately without fruit. Returning down this branch, we saw, close to the fork on the left side, a village, the name of which we understood to be Haddi, i. e. *small box.* Towards sunset, we arrived again at the eastern or main branch, left on Saturday, which is, at the place of separation, a river of 3 to 4000 feet wide : its shores are elevated some feet and covered with reeds and shrubs : on the left bank, immediately opposite to the fork, stands a village, or rather three small ones, somewhat apart and consisting of clay huts, and magazines, raised on posts. The name of the last of the three sounded like " Obokriga." Not far beyond this we anchored, when it got dark. The general character of the country was the same as yesterday; but the shores being somewhat higher, I was able generally to see the soil,

though frequently the shrubs and plants were immersed up
to their lower leaves. The vegetation appeared the same as
before.

Monday, August 23.—Again in movement at break of day.
On the shore, which was lower than on the previous day,
we noticed a few villages ; and some negroes came alongside
in canoes and on board. They wore not only the streak
down the forehead, but mostly three parallel lines on each
cheek-bone. Towards ten o'clock we arrived at a village on
the right shore, named in Laird's expedition " Ibu," and
" Little Ibu" in Allen's chart :* the inhabitants called it Ocro-
tombi or Korotumbi; but it was some time, before we could
clearly hear the name. The chief, who came on board, wore
an old blue European jacket, and a perfectly new green cap,
with tassel strings. It had rained in the morning : towards
noon the weather cleared, and a boat going on shore to take
the sun's meridian, I joined it, and we landed at a plantation,
where the ground, about 4 or 5 feet above the level of the
water, consisted of good vegetable soil, mixed with clay and
sand, and cultivated with *Cocoa trees, Yams,* and *Capsicum. Sorghum (rubrum ?)* grew apparently indigenous, and
formed grassy forests, 10 or 11 feet high. The geographical
latitude was found to be 5° 14′ N. The spot was a little
lower down than that called Ofitulo on Allen's map.
Towards ten o'clock we approached Stirling's Island, and on
account of the violent rain, we cast anchor there for a short
time : the rain felt very cold (refer to my Meteorol. Journal).
We proceeded about three o'clock ; the rain continuing
till night, with variable violence. Shortly before dark we
passed a place on the right shore, called, according to

* Lieut. Allen's chart of the River Niger or Quorra, published by
Bate, in the Poultry, London.—Lieut. William Allen, who surveyed the
river in 1832—3, in the Alburka steamer, under Messrs. Lander and
Laird, was second in command on the Niger Expedition, and Commander of H.M. Ship Wilberforce, the steamer in which Dr. Vogel ascended
the river.—(H. D. Trotter).

Mr. Brown, "Ingliana." Near it I noticed an extensive plantation of *Bananas;* and soon after this, we cast anchor. The borders of the rivers were every where covered with forests, reaching to the water's edge, or with intervening high grass, (*Sorghum saccharinum?*) Amongst these, there were frequently places cleared for plantations, or they might be natural open spots in the forests, where high trees would stand singly. A great inconvenience and misfortune it is that we are obliged to drink such bad water: it has not only a dirty colour, but owing to its being saturated with decomposed vegetable and animal matter, a sickening taste, which, though somewhat lessened, is not removed by filtration.

Tuesday, August 24.—At eight o'clock we passed so near the shore, that I could botanize; and I observed the blossoms of a high tree (*Mimosa*) and of a climber, a *Tetracera,* perhaps not different from *T. Senegalensis (obovata ?)* Towards ten o'clock we came to the Benin (Warree) branch. On the point of land, between the two arms of the river, a signal-post was erected, and this gave me the opportunity of visiting the shore for a few minutes, and I found it covered with the *Sorghum,* previously noticed. An *Æschynomene, Cassia mimosoides* and a *Malvacea,* were all I could pick up in the hurry. Though, from on board ship, the shore had appeared swampy, it proved firm to the water's edge; and I am inclined to believe that spots, looking marshy at a distance, are not really so. Perhaps some swamps may be formed in dry weather by the receding of the waters; but since our quitting the Mangrove country, I have not observed any absolute morasses: on the contrary, the land appears every where to rise 2 or 3 feet above the water, though what are now creeks may become swamps in the dry season. We descended the Benin (Warree)* branch for a few

---

* The branch is erroneously called the Benin branch in Allen's Chart. It leads to Warree or Warri, and ought therefore to be called the Warree branch.—(H. D. Trotter).

miles: it nowise differs from the main river, except that the stream is somewhat narrower. By four o'clock we returned to the point of junction; and during our short stay, a great many canoes assembled about us. Some were large and carried twelve or sixteen persons, others fewer; and some had only one in them. The canoes are the same as before, with a high and broad stern. One man stood steering with a paddle. There were perhaps sixteen canoes, containing about one hundred and ten people, who had come mostly from Obiah, on the right shore of the river. Their dress had nothing very peculiar. The main difference consisted in the various coral and pearl strings, or ivory and brass rings, which they wore on arms and legs, and in the manner of dressing the hair. The latter struck us particularly, now that so many individuals had collected, and we could look down on their heads, from the deck of our ship. Some had cut their hair so round and formally, that it bore the most deceptive semblance to a wig: some shaved their heads quite bald; while others only kept a portion of hair behind, or a large portion forming a narrow ridge across, or it was allowed to grow high in the middle of the head, like a small steeple. Some whimsical fellows exhibited merely a narrow strip of hair from behind to the front, looking like the crest of a helmet, or perhaps an oblong square; or it was cut in chequers, and the remaining portion twisted into numbers of little tails; while others wore their hair like our European dandies, arranged in various ways on the sides of the head.

The river,* at the separation of the Benin (Warree) branch,

* The branch which here separates from the Nun or main branch of the Delta of the Niger, runs to the sea by the town of Warree or Warri, falling into the Bight of Benin to the north-west of the mouth of the Nun river. Captain Becroft of the Ethiope, Mr. Jamieson's steamer, was the first to ascend the Niger by this branch, in 1840. Lieutenant Allen had previously conjectured it to be the Benin river, with which, however, there is only a communication by creeks. This accounts for Dr. Vogel calling it the Benin branch in his Journal.

Above the separation of these two branches, the river may be properly

is about a mile wide: the commencement of this branch measured 696 yards. At five o'clock we quitted the Benin (Warree) branch, returning into the main stream, which has here a lake-like appearance, surrounded with high trees: many of the canoes followed, spreading over the water, and greatly enlivening the scene by zealously rowing to keep up with us. Towards sunset we cast anchor. The weather was very cheerless, being generally rainy, except at noon.

Wednesday, August 25.—Proceeded at the usual time. Much rain and therefore several stoppages. At noon we reached a place, marked on Allen's map, Egaboh, but now called " Ulok." The sun showing itself, and an attempt to make observations following, I was enabled to land for a short time. The grass along the shore was not a *Sorghum*, but some other genus. Close to the water-side grew a fig-tree, with very small fruit. The neighbouring chief, an old leprous man, came on board: he wore a drummer's jacket given him at the time of Laird's expedition (he seemed to have taken great care of it) and carried an iron staff divided at the top and ornamented with brass rings. After some detention, occasioned by heavy rains, we pursued our course, the stream being generally about half a mile wide, and the vegetation the same as heretofore. Approaching the creek that leads to Ibu (Abòh)* the current proved so strong, that we

called the Niger, the name by which it has been so long known in the civilized world. The natives have no name for the river, excepting the general appellation of " Water," which varies with the different languages spoken on the banks. Mungo Park found it called " Joliba" in the higher parts of the river. In the Houssa country it is called " Quorra."—(H. D. Trotter).

* Schön says the proper name of this town is not Ibu, but " Abòh." The town had hitherto been called by Europeans " Ibo" or " Eboe," and was generally supposed to be the capital of the whole of the Ibo country; but we ascertained that its proper name is " Abòh," and that it is the principal town of the territory of the same name, which forms a part only, and that probably the most western, of the Ibo country. (H. D. Trotter).—See Captain Trotter's Report to Lord Stanley; Parliamentary Papers relating to the Niger Expedition, p. 91.

Tab. 1.

DELTA OF THE NIGER.— LEFT BANK.— BELOW ABOH.

VILLAGE ON LEFT BANK OF THE DELTA.

London. Published by H Bailliere, Regent Street.

could hardly make way against it: on the preceding day it had only been one and a half or two knots an hour. Towards half past seven we cast anchor at the Ibu (Abòh) creek, abreast of the creek leading to the town of Abòh.

Thursday, August 26.—Early in the morning, the Captain and myself rowed about in the Ibu (Abòh) creek, and collected a few plants. This creek, at present very wide, is without a current: the main channel measures perhaps 100 yards. The right shore is now inundated; the shrubs being altogether covered with water, and the grasses immersed to their ears, on which snails, ants and small beetles had settled, by way of refuge, in great numbers. We had taken on board, on the previous day, a man who wanted to go as pilot to Abòh : he seemed to be a careful and clever person. Granby, our interpreter " for Brass and Ibo," recognised him as an old acquaintance, he (Granby) having lived here a long while before being sold to the Europeans. The Ibo man was rejoiced to see him again, and expressed his astonishment, that a man sold to the Europeans should return; it being the general opinion that such slaves were used for food!

Large canoes were fastened in the jungle: they had come from the Brass country, chiefly to purchase palm oil, for which purpose, large casks lay on board, under roofs of matting. Abòh is on the opposite side of the shore, here intersected by several small creeks : otherwise it is covered to the water's edge with brushwood, behind which are the huts. I gathered on this occasion a few *Mimoseæ, Sapindaceæ*, and *Rubiaceæ;* but the most interesting was a shrub (*Polyand. Pentag.,* fruct. placentis 5 parietalibus) apparently a new genus of *Bixaceæ.* In the main stream, and even in the smaller creeks was a *Pistia*, perhaps *Pistia Stratiotes:* it does not, however, seem to grow here, but to float down the Niger, where it may be seen drifting in large masses. Some specimens were in flower : fruit I could not discover. In the morning we had a visit from King Obi's son : towards noon he came himself, with a lot of noisy followers, and henceforth we were constantly surrounded by many canoes. These people wear either a piece of cloth round the loins, or portions of European dresses; only King Obi had both coat and trowsers. Obi is between fifty and sixty, with a true Negro face, but cunning. The son is a finely formed, strong, powerful young man. King Obi brought with him one of his wives, a very young person, and a daughter, dressed in African style, i. e. *sans géne*. When this was observed, Commissioner Cook gave to the wife a red, and Captain Allen to the daughter, a coloured gown; but the latter was not pleased with hers. One might mention several peculiarities about their attire; but such things, and their smoking pipes, &c., did not particularly interest me. Several women wore enormous ivory rings round the legs. The account I have before given of the various ways of dressing their hair might be extended. The desire to possess whatever they saw, was unequivocal; but I heard of no thefts. There were a good many tools scattered about on deck, which in the confusion might easily have been taken. The weather was rainy and very uncomfortable.

Friday, August 27.—Through incessant rains the ground got swampy, in fact so muddy, that it became impossible to make any extensive excursions. Besides some plants previously mentioned, I collected *Cucurbitaceæ, Apocyneæ*, a *Ficus* and a species of *Malaghetty pepper*, which, judging by the leaves and fruit, is identical with that at the mouth of the Nun River: a fine *Costus* was very common : a *Salvinia*, not rare in the creeks, and a *Ceratophyllum*, which I had seen before in Abòh creek. On the stems of trees grew three species of *Mosses :* on the ground none. Whoever may have the good fortune to investigate these creeks in a boat, would probably find many *Cryptogamia*, new to the African Flora.

Saturday, August 28.—I had yesterday seen a tree, about thirty feet distant from the water's edge, of moderate height, with three long straight branches, closely appressed at the top.

and bearing a corymb of rose-coloured blossoms, rising from
the terminal cluster of leaves.   Having noticed this object
through the telescope from the deck, I of course wished to
obtain the flower, and landing, I asked two negroes (from
Sierra Leone) who accompanied me, whether they would
procure it; but they both declared it impracticable, because
of the high grass. I therefore cut a way with my knife; but on
reaching the tree, I found it too lofty for me to get to the top
without loss of time; the period for which the boat was lent
me having expired. To-day, I succeeded again in obtaining
the boat for a short while; and I found fortunately one
amongst the negroes who climbed the tree, about 16 feet high,
and gathered a few branches with an iron hook. I record
this circumstance here intentionally, as an instance of my
nearly daily difficulties. Amongst the few plants which
I collected, there were many that occur along the whole
coast; as, for instance, *Sarcocephalus*. According to what
Mr. Schön told me, the name of this place, which I had
considered to be Ibu, is Abòh. In the afternoon we left Ibu
(Abòh,) and steamed it by moonlight till eleven o'clock, when
we cast anchor. Sunday, 29th, we did not move. Weather
very bad.

Monday, August 30.—Started by day-break. Neither the
country nor the river offered any thing new.

Tuesday, August 31. — I had twice an opportunity of
visiting the shore for a short while. The first time, I found a
terrestrial *Orchidea*, 4 feet high : a great part of the jungle
on the right shore consisted of a *Fig-tree*, with long branchlets, covered with fructification shooting out from the old
wood : its white bark was visible at a great distance. The
ants were here dreadfully troublesome. At two o'clock, when
passing an island, we perceived a strong very sweet smell,
(almost like the *Tetracera* which I had collected on the 24th),
but I could not descry any flowers through the telescope.
In the afternoon we saw, at a distance, on the left shore, the
first low hills, and soon afterwards a water-course on the

same side; apparently quite still, for the current of the Niger ran in a sharply distinct line athwart it. This part, including the hills and river, is said to be called " Oredtha;" it is opposite Kirro market, (so named in Allen's chart.) In this branch of the river grew many *Pistiæ*; but higher up the Niger, we also met them floating in large quantities. This plant appears to have been displaced, by rising waters, from its tranquil domicile, as is frequently the case with others: for we pass many small floating islands of grass and other plants, clumps of rolled-up grass, and stems of huge trees, appearing in the distance, with their roots and branches partly emerging from the water, exactly like canoes. The river, since we left Ibu, (Abòh), continues about half-a-mile in width, sometimes more: the water very muddy, and of a clay colour: the shores low, covered with brushwood, inter-twined with so many creepers as to form, sometimes for great distances, a vegetable wall. This wall was particularly remarkable on the left side of the said still water; behind it rose a few hillocks, with much cultivation, (*Sorghum vul-gare?*) amongst which single trees were interspersed. A peculiar feature of this part consists in the small huts raised on poles along the shore; from which the natives, according to Brown, drop their fishing-lines into the river.

Wednesday, September 1.—This morning the river was very wide, in one part above a mile, and covered with *Pistiæ*. There were hills, especially on the left side, but they ceased before we reached Damugu.* Of this place we only discerned a few huts, the first round ones, with a pointed overhanging grass roof. On the whole we saw to-day but few villages: if there are more, they must lie beyond the jungle. Nor did we observe any Cocoa palms, which had occurred in several places on the previous day. About Damugu, the country seems covered with high forests: hitherto, there had been only low woods. Towards evening, we saw isolated high trees, apparently covered with blossoms; but through the

* Or Addá-Mugu.—(H. D. Trotter.)

telescope we descried these fancied flowers to be white birds, (*Egrets ?*) of which several stalked, here and there, along the shore.

Thursday, September 2. — Beyond Damugu, the land appears again lower and covered with jungle. I think that the shores of the main river are mostly lined with forests, and the islands covered with grass and underwood. Towards noon we came to finely wooded hills; and in the evening, King William's Mountain appeared, (see Allen's chart.) I had twice the opportunity of going for a short time on shore. First to an open place, covered with grass; where I found *Cassia Absus, mimosoides ?* a *Psoralea,* some *Gramineæ, Malvaceæ* and *Schmideleæ :* a *Sarcocephalus* grew likewise here. The second time was near a village; where the cottages are round, and plaited of palm-leaves and grass. Storehouses, raised on poles, are filled with *Indian corn.* A *Tephrosia* (*toxicaria*), almost arborescent, was planted about the huts, which a Krooman told me, was used to benumb the fish. A fine red flower, on a high tree, could not be procured : it appeared to be Beauvois' *Spathodea ;* and I fancied I had seen it several times in the Delta.

Friday, September 3.—We can quite overlook the country from on board our vessel. On both sides, the river is margined at some distance with hills : further off, towards the north, rise mountains, enveloped with blue mist. Only on the left side, the hills approach the shore, and are, for the space of about a quarter of a mile, quite abrupt to about 100 feet high, of red sandstone, visible, because of its bright colour, at a great distance. The top is often covered with overhanging vegetation. On this hill stands the town of Attah*, (Iddah), surrounded by cultivated grounds. In the distance grow *Cocoa palms* and *Baobab trees,* the latter bearing long pendent fruit. This morning I had another

---

* Attah is the name of the chief, and not of the town ; or rather, Attah is the *title* of the chief, who is styled the Attah or King of Egarra, or more generally "the Attah." The town is called *Iddah*. — (H. D. Trotter.)

opportunity of going on shore. The ground in front of the hill, and down the river, is now quite covered with water. Some way up, I found a *Baobab tree*, apparently consisting of several stems joined : it was by no means low, perhaps 30 feet high to the branches, and altogether 70 to 80 feet high. The fruit is remarkable, suspended from stalks 1½ foot long; but I could only collect a few specimens, being obliged to return. We moved to the right shore, where the " Soudan" already was, to cut fire-wood, the "Albert" remaining behind, and lay close to the shore; of which a considerable breadth was inundated. In the afternoon, a number of natives arrived to see what we were doing; especially, (as they said), because the people of the Attah sometimes come here to make slaves. They appeared never yet to have been in contact with Europeans : they wore the country cloth round their loins, and were armed with bows and arrows, the latter with only wooden points. The quivers seemed to be formed of goatskins. Their town is said to be five miles inland, and is called " Waápa." The country is called *Angori*, and is under the chief of this town.

According to one of our free negroes, a native of these parts, this district belongs to " Benin Country," which extends to the sea. The "Great King" of it sacrifices daily three human beings. (!) It was singular that none of the Angori people had canoes, although their plantations came down to the edge of the river. One, of Yams (*Dioscorea sativa*) and *Maize*, was situated close to our vessel : amongst these plants grew a few *Tephrosiæ*, which, a "Nufi man" told me, were used in his country for catching fish, and are seen both wild and cultivated. The brushwood near the river consisted chiefly of *Quisqualis obovata*, (Schum.), which, whether bearing white or red flowers, had a beautiful appearance;—and a *Porina, Spondias, Sarcocephalus,* a few *Oil palms, Lonchocarpus formosa,* &c.

Saturday, September 4.—A trip into the interior showed me that the soil on the hills is much mixed with sand, owing to the decomposed sandstone. I could not get far; the land being chiefly savannahs, the remnants of decayed forests : *Tamarinds*, and other *Leguminosæ*, a *Banisteria*, (?) and *Bombax* were conspicuous, besides other trees, already mentioned. Of herbaceous and shrubby plants I found, amongst the *Cyperaceæ* and *Grasses*, chiefly *Leguminosæ*, *Desmodium*, *Cassia*, *Malvaceæ*, *Euphorbiaceæ*, (*Phyllanthus*, *Tragia*). Near the shore, in water-holes, grew frequently a *Lemna*,* now in flower. A flowering *Loranthus*, with verdigris-coloured fruit, was parasitical on a *Leguminosa*, now almost under water.

The burning sun, which came out after rain, gave me a violent head-ache. Towards evening, we proceeded a few miles up the river, and staid there during Sunday the 5th of September, in company with the other vessels, keeping the Sabbath as a day of rest. The current ran here extremely strong, about three knots and a half per hour.

Monday, September 6.—I felt very unwell; and towards noon slight fever came on, which exhausted me much. In the evening we followed the "Albert" to Iddah, and grounded near the eastern inundated part of English Island. Here we remained till Wednesday, September 8, in the evening, when we succeeded in getting afloat again, and proceeded a few miles upwards.

Thursday, September 9.—Till mid-day I felt unwell and weak, but then got better. We approached the mountains, which proved to consist of small ridges, 1,000 to 2,000 feet high; and the scenery was sometimes very pretty, the mountains being overgrown with trees to the top. The hills, which we passed first, and then the mountains, seemed to form several (more than two?) basins; through which the river had forced its way, as is frequently the case with mountain streams. We proceeded along the eastern branch, to the Bokweh Island. The foremost mountains of King's Peak (so called in Allen's chart) came down to the river, and we could clearly distinguish large strata in the declivity

---

* Is it different from L. *minor,* of Europe? The leaves are distinctly striated ; which, so far as I recollect, is not the case in our plant.

and down to the bottom. At the northern end of the island,
a beautiful prospect was suddenly disclosed, upon the
mountains on the right shore, from Mount Jervis to Mount
Saddleback, (see Allen's chart), contrasting, at the moment
we came out of the channel, most distinctly with the horizon,
then strongly illuminated by the setting sun. I observed no
great change in the vegetation; unless perhaps less grass
prevailed on the right shore. We never before saw so many
canoes descending the river as to-day: some very large: all
had a small scaffolding in the middle; and in some of them
were horses, no bigger than donkeys. The current, where we
anchored a little above Bokweh Island, was three knots and
a half.

Friday, September 10.—To-day we passed the mountains,
most of which rise in elongated ridges; but others are isolated,
their slopes covered with large boulders, between which is a
thick brushwood. The scenery is very pretty: mountains
often like those of the Rhine; but castles and vineyards
are wanting, and the rivers too wide and full of island and
swamps. About noon, we stopped near a small island,
beyond Mount Soracle (in Allen's chart); the name of
which, according to some natives who came on board, was
Dagore. I was again unwell and could not go on shore; but
Roscher, who did, found the island of granite formation, and
he brought me a few plants. Between Mount St. Michael and
Mount Franklin in Allen's chart, stood a village, situated
on a partly isolated hill; the first, which I had observed here,
built on a considerable elevation; most of the villages being
close to the river, so that, because of the unusual rise of
water, a portion of the huts are under water. A *Leguminosa*
with the habit of *Robinia*, and violet blossoms now in full
splendour, struck me: I also saw here and there a *Baobab*
with fruit: yesterday I noticed many *Cocoas*, to-day none.
Near a village, on the right shore, a little above Maconochie
island, grew some *Fan palms*; and we subsequently met with
more: before this, I had only seen one in the Delta. We
anchored about half-way between Mount Franklin and the

confluence of the Niger and Chadda. The current runs two and a half knots.

Saturday, September 11.—Before eight o'clock we cast anchor off Adda-Kuddu, the place which had been preliminarily fixed upon for the model-farm. The river expands here to a lake; while, to the extreme left, the confluence with the Chadda is seen. Mountains above 2000 feet high are visible in every direction at a distance. The landing-place was remarkable for the many boulders, lying one over the other, surrounded and partly overgrown with shrubs and trees. In one conspicuous place I found a *Baobab*, looking much like an old *Oak*. Close by, were several others, one quite denuded, the rest with a little foliage, but all showing their characteristic pendent fruit. Being still poorly, I took Captain Trotter's advice and went on shore. The ruins of Adda-Kuddu surrounded the place, and were already covered with vegetation.

Cylindrical holes, several feet deep, and 2 feet in diameter, and bricked for making dyes, were still visible. The ruins of African towns offer nothing picturesque. We hurried to some spot; from whence we might survey the country. About the town, the habitations of which had been round clay huts, lies a level valley bounded by low hillocks, which promised the territory best fit for cultivation. To get at it, we had to pass a place, where seemed to have been something like a ditch and wall. The valley itself had evidently been cultivated at one time, but is now covered with *Gramineæ*, *Cyperaceæ*, a few small *Euphorbiæ*, *Malvaceæ*, and particularly *Leguminosæ;* amongst which two *Tephrosiæ*, one 5 or 6 feet high, were the most remarkable plants, rendering our progress very difficult by their woody stalks. The valley was nearly dry, with only a few puddles of rain water; and the ground is pretty well cleared, with here and there a few large pieces of broken rock. The soil consisted of decomposed granite, and if it ever had been mixed with vegetable earth, it is exhausted by former cultivation. Quartz remained abundantly in it, in the shape of coarse sand, and I could not help condemning the soil as extremely indifferent. The inhabitants of Adda-Kuddu, upon their town being destroyed by the Felahtis, removed to the opposite side of the river, and built there the town called " Schimri," (afterwards I heard other names for the new Adda-Kuddu) close to the shore. It is now, by reason of this year's unusually high water, quite inundated ; and therefore the people have erected another new city. The chief or governor (or Aneidjo) appointed by the King of Iddah, paid us a visit. His companions wore the Nufi Toba, an under-dress with wide sleeves, reaching to the knuckles. He was decorated with large bells on the wrists ; and a slave fanned him with a leathern fan. In the afternoon we proceeded up the Niger, to Stirling Hill, to examine the country : it was difficult to learn at whose disposal it was ; but at last we were assured, that an independent tribe, said to be very savage, dwelt on the mountain. I was requested, towards sunset, to examine the soil in the valley, and found it no better, than at Adda-Kuddu. There were plantations of *Maize* and *Yams*. Mr. Carr had, in the meantime, been on the hill, and detected a rich vegetable soil. We returned immediately to Adda-Kuddu, which we reached at dark. The current here is two knots. The natives had brought cocoa-nuts on board, and on my inquiry, they said, the tree grew on the other shore ; but afterwards they asserted, that it was not found here at all. Mr. Brown had brought me from thence a *Unona* (!) and an apparently entirely new genus of the family of *Leguminosæ*, with a fruit similar to *Swartzia*, and I subsequently found this little tree every where on the shore about Stirling.

Sunday, September 12.—We remained quietly at anchor.

Monday, September 13.—I went on shore to botanize amongst the ruins of Adda-Kuddu; but the hot sun quickly forced me back. *Papaws* are here still frequent; also some sorts of *Cucurbitaceæ*, which, with *Asclepiadeæ* and *Creepers*, have overgrown the ruined huts. A *Lemna* growing in a puddle was the same as I had seen at Iddah. I observed here but a single *Pistia* float by; whilst the day before, we

met with them in abundance, floating on the Quorra (Niger).
In the afternoon I went again to Stirling hill, and explored
it for a short time; but found the soil to consist of sandstone, impregnated with iron, and therefore bad. A few
spots only exhibited vegetable soil, formed of decomposed
plants.

Tuesday, September 14. — At six o'clock we climbed
Mount Pattéh. It is rather steep, difficult of ascent, and
covered with many boulders of red iron sandstone. The
pea-like formation is remarkable. There were single strata
of quartz. The cultivation of *Yams, Capsicum, Guinea-grain,*
(now without blossom or fruit) a bean or *Dolichos,* and a few
Bananas, continued to the summit. A streamlet, running
down from somewhere about midway of the mount, had a
bed of clay, which is also more or less mixed with the soil
generally; and along this channel the chief brushwood grew.
Largish isolated trees are met over the whole declivity, probably remnants of former forests. It looks as if the useful
trees had been preserved. Four species occurred particularly
often; *Baobab; Parkia,* now without fruit or blossom, but
with foliage ; *Sarcocephalus,* sometimes a stately tree, but
with long branches showing a disposition to climb ; and the
Hog-plum (*Spondias*), but this chiefly at the summit. The
barometer gave 1200 feet, according to a hasty calculation,
(subsequently 1150), above the level of the Niger. On the
top is table-land (level plateau) much cultivated, and covered
often with brushwood and a tree with yellow flowers, I think
Beauvois' *Spathodea ;** another tree, of which blossom† and
fruit are preserved in acid, a shrubby *Mimosa* and species of
Ficus, without fructification. A species of *Tephrosia* was frequently cultivated. I saw no *Palm.* The natives appeared, as
yet, to have had no communication with Europeans : they were
armed with bows and arrows, much like those of the country

* A handsome tree, with dark scarlet flowers, of the same genus, was
frequent on the declivity.

† A high, much branched, leafless *Euphorbia,* the juice of which is
said to cause blindness.

F

near Angori: their arrows are said to be poisoned; and
their clothes consisted of stuffs, manufactured by themselves.
They were of a gentle nature; and the mere word " scanu"
was sufficient to conquer their diffidence. For some presents
which we gave them, they expressed their thanks by bowing
to the ground, and strewing repeatedly dust on the forehead,
perhaps twelve times : the women uncovered the bosom and
put dust on it. Decency amongst the women seemed to
require, that the upper garment should be tightly fastened
above the bosom, so as to cover it completely. The boys
we saw were circumcised.

Towards two o'clock I returned, not feeling well, for I
had exerted myself too much. The sun had been clouded,
and I had latterly protected myself with an umbrella; nevertheless in the afternoon and evening I felt so tired, and yet
so heated and restless, that I cannot recollect ever having
been so uncomfortable and disabled, without absolute illness. Every exertion seems now to produce more or less
this effect. Restlessness and exhaustion, burning of the skin
and eruptions, become quite insufferable.

Tuesday, September 14.—To-day I had to take care of the
plants, which I gathered yesterday, and wished to arrange
my collection, for which purpose I had been unable to
obtain either room or a case, and was therefore obliged to
preserve them, as best I could, in bundles in my cabin :
a plan which was good neither for them, nor for myself.
My assistant, now somewhat trained, was unfortunately the
best linguist, and our intercourse with the natives being very
great, I could hardly ever avail myself of his aid. At a
distance this all appears trivial; but to a traveller in my
situation the frequent repetition of such trials is extremely
disheartening. The natives, perceiving our wishes, brought
chiefly arms on board, some apparently made in a hurry for the
occasion; also calabashes, mats and sacks of plaited grass,
honey, palm-wine, stuffs of their own manufacture, reels of
cotton, earth-nuts, yams, goats, sheep, poultry and fat. In
return they took cowries, cloth, wearing-apparel and particu-

larly looking-glasses: the latter being chiefly bought by the
women. The women are often beautifully painted with red
Camwood (?) pulverized and made into balls as large as a fist,
and thus sold: the eyelids they paint with antimony, which
they brought with them on board in very neat cylindrical
cases made of skins.

Wednesday, September 15.—The intercourse with the
natives continued. They bring, besides the things mentioned,
tobacco, which they call taba, in flat rolled disks; also a chalk-
like substance, prepared from burnt bones, with which they
rub the fingers when spinning, it is called Effu in the Aku
language, Alli in Houssa; they kept this in small calabashes,
or in masses like elongated dice: whips of hippopotamus
skins, called Uoji: some rice, grown on the left shore,
and a few Limes. The process for discharging their arrows
seemed to me ingenious. They have a knife with a somewhat broad handle into which they insert the hand,* and pull
up the string of the bow with the back of the handle, being
thus sure not to hurt the hand, and are thus ready to
kill with the knife whatever the arrow may have hit. On
the left upper arm they carry arrows for their immediate use
in a wooden quiver.

Thursday, September 16.—Captain Trotter wished me to
visit the left shore. The current on the right side,
where we were at anchor, was 1 and 1½ knots; but towards
the middle it ran much stronger; and in some places the boat
could hardly make way against it. We kept therefore, after
reaching the left bank of the Niger, close to the jungle, (I
must not say *shore;* for every thing was under water).
Amongst different things, I noticed a rather thick tree,
30 feet high, which attracted my attention by its large fruit:

---

* In Treviranus' Memoir occurs the following quotation from a letter of
Dr. Vogel's, more clearly showing their manner of using the bow. " In
the right hand they hold a knife with a hollow handle, through which
they place four fingers in the middle of the handle. On the thumb they
have an iron ring, and draw between this and the handle the bowstring,
so that they cannot injure the hand."—(*See* Memoir, p. 15.)

it is apparently an *Artocarpus*. The Kroomen call it *Oqua*, and told me that they eat the boiled seed. I saw only fruit and female blossoms : no male flowers. The tree contained much milky juice. Besides this I found here a seemingly new species of *Anona*, and the above-mentioned genus of *Leguminosæ*, occurring often as a small branchy tree, with white flowers, remarkable for its bright red terminal leaves. In those nooks, where the current was weak, the *Pistia* grew in large quantities, mixed with *Ceratophyllum*, without fructification, and the *Salvinia*, and *Jungermunnia* (?) of Ibu. At last we reached a bit of dry land, deep in the bush; where some negroes had pitched their tent-like straw huts for temporary dwellings. They told me that they had come from the opposite side (from Dgaggu ?) to plant this place, against the rainy season; but they had not yet begun. The ground, now inundated, would be cultivated in the dry season, for it all consisted of rich vegetable soil.

On my return, I could find no place but the deck for my plants. I then went on board the " Albert," to make my report to Captain Trotter, but was obliged to stop there a long time, for want of a boat to return. In the mean time, we had a heavy shower of rain, and on my subsequent arrival in the " Wilberforce" I found not a few of my plants spoiled, or quite lost, amongst them the *Anona*; and I was unable to care for the rest, every nook that I could use having been filled long since, and my cabin was crammed nearly full. During the last four weeks, for want of suitable boxes in which to preserve my collections, I was unable to do almost anything in Botany.

Friday, September 17.—I bought to-day a complete set of arms of Adghó for 2000 cowries. Captain Allen purchased an ox for 30,000 cowries, from the son of a former chief of Adda-Kuddu, whom he called Mallen Katab, and who had poisoned old Pascoe and the Kroomen. This son, Machmakal, was one of the handsomest negroes I ever saw; but he wanted to give his father's name differently. He made me a present of a pair of shoes of antelope hide, very well made. He understood a little Arabic, though he could not pronounce it according to Müller's notions, but he wrote it; and singularly enough, he put the paper not in the customary oriental manner before him, nor writing the letters from the top downwards, but so, that they must be read in the usual manner. I have his name and mine written by him. I had understood his name as Makola. According to Müller, what he wrote, is in the Algerine dialect, meaning: Machmakal.*

Saturday, September 18.—The number of sick increases considerably; and the " Soudan" is to take them to-morrow down to the sea. I, *therefore, wrote letters to-day.* I continue unwell; *head-ache and fever.*

WRITTEN LATER, AT FERNANDO PO.

Sunday, September 19.—Decided, but slight fever. The " Soudan" leaves for the sea.

Monday, September 20.—It is settled that the " Wilberforce" shall also proceed to sea with the sick, which have much increased in number; and my first resolve was to remain here; but our circumstances on shore were such, that as an invalid, I could hardly hope to be comfortable, and I therefore take Captain Allen's advice, which is to go down to sea in the " Wilberforce," and stop at Fernando Po.

Tuesday, September 21.—At six o'clock in the morning we proceeded down the river, I becoming daily worse. We arrived at Fernando Po on the 1st of October, and I earnestly entreated to be put on shore; for the vessel was to proceed to Ascension Island, and stop there several months; which would have been for me worse than a prison. On leaving the ship I had still violent fever, which only quitted me after a week and a half. In the landing of my collection I was kindly assisted by Mr. Forster. Of several of the most interesting fruits, however, which, until disabled, I had kept on deck to dry, nothing was to be seen. I regret especially the fruit of *Adansonia,* ripe fruit of *Artocarpus,* a fruit,

* Vogel's Private Journal.

the blossom of which I have never seen, from Mount Pattéh, being amongst the most interesting, with many more. Captain Allen had the goodness to order us a lodging at Mr. White's, the agent of the West African Company; and Mr. Roscher having also determined to remain here, he and I agreed to live together. The house intended for us not being quite ready, Mr. White was so kind as to give us, in the mean time, quarters in his own dwelling. We found soon how difficult it was to procure on this island the necessary provision; and as we had to be our own housekeepers, we asked for some articles from on board ship, that we might not at the outset be quite bare.

On the 5th of October we landed. They sent us from our mess a few necessary utensils, cups, plates, &c., which were not to be obtained any how at Fernando Po, and for which we felt very grateful; but time forbade their furnishing us with the least provisions, the "Wilberforce" sailing on Saturday. On Monday, October 18, we quitted Mr. White's house; to make room for the sick which had arrived on the previous day, by the "Albert." I had to be carried to our new residence, for we were in miserable plight; and to get a piece of bread for money on the island, was actually impossible. If acquaintances had not obligingly supplied us in some degree, we should have had to fast this and the next day, in the strictest sense of the word. We, therefore, addressed Captain Trotter, who made arrangements, by which we were at least spared the necessity of running about in the heat of the day for provisions; as all those, who have no stores of their own, are obliged to do.

Here I stop. My recovery proceeds but slowly; to-day (October 25), I am not yet able to walk for half an hour. What concerns our stay at Fernando Po must be written hereafter.

These are the concluding words of the *Botanical Journal.* In Dr. Vogel's *private Journal* there are some few entries after this date, referring mostly to personal affairs, despatches, provisions, and the like.

It would appear that, towards the end of November, he felt strong enough to begin his botanical excursions, and says : " The heat is too great to allow convalescents who are still very weak, to work much. Besides plants, I have now taken to collecting insects. Roscher has quite a mania for sporting ;"—and again :—

December 2.—" We had intended to proceed this week into the mountains, to the tent which had been erected for Captain Trotter; but ever since Sunday, Roscher has been ill, probably in consequence of his sporting, often in the heat of the sun; and Thomson, who during the absence of the " Albert," remains here as doctor, attends him. There are several cases of fever : amongst them White, the storekeeper, and the doctor : all people who have been here for some time ! The weather is certainly not genial to European constitutions. Mornings and evenings are dull and foggy; though not so thick but that one can see the country: noon and afternoon changeable, a few hot hours, with west and south wind. Because of Roscher's illness I must attend to our housekeeping, which comes rather awkward to me. In the meantime, I continue my previous way of living, *i. e.* I make excursions from three o'clock till dusk (6 o'clock), but am very anxious to get into the mountains. Yesterday I went towards the farm; to seek for the *Calamus* which Roscher had seen, but could not find it."

With these words Dr. Vogel's *private Journal* ends; and we may here introduce an extract from the Report of Captain Trotter, addressed to the Right Honourable Lord Stanley, Principal Secretary of State for the Colonies, dated March 15, 1843.

" We found at Clarence Cove, Fernando Po, on our return in the Albert from the Niger, Dr. Vogel and Mr. Roscher. These indefatigable gentlemen, of whose zeal on

all occasions it would be impossible to speak too highly,
had fallen sick at the confluence, and were obliged to
descend the river in the " Wilberforce;" but they declined
going to Ascension for the re-establishment of their health,
hoping to be able to pursue their scientific researches in
Africa. Dr. Vogel lived only to the 17th December following; but his memory will be cherished, as long as Botany
remains a science."

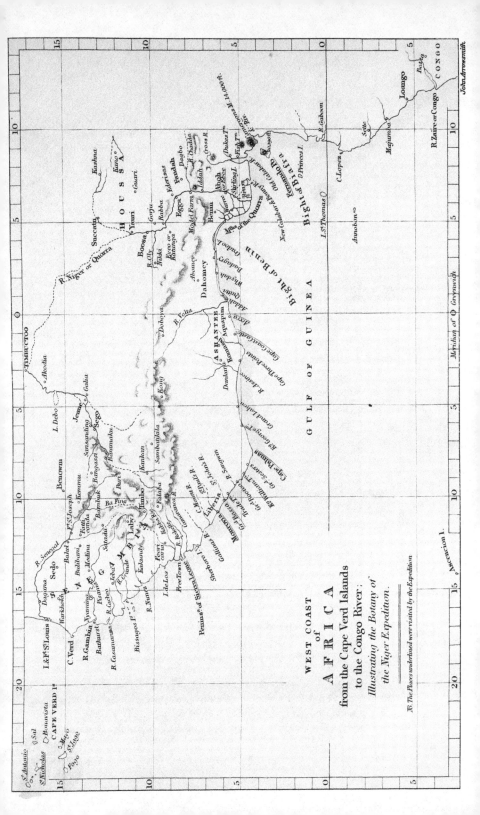

WEST COAST
of
AFRICA
from the Cape Verd Islands
to the Congo River;
Illustrating the Botany of
the Niger Expedition.

N: The Places underlined were visited by the Expedition

John Arrowsmith.

BOTANY

OF THE

NIGER EXPEDITION.*

NOTES ON "MADEIRA PLANTS."

So great was Dr. Vogel's zeal in the cause of Botany, that his collections were commenced before leaving England, during the few days spent by the Niger Expedition in Plymouth Sound. The plants in question consist principally of *Algæ*, and being only the common South of England species, and foreign to the object of this Memoir, need no further notice.

During his four days' stay at Madeira, although unable to make any distant excursions, Dr. Vogel formed a very excellent Herbarium, having been assisted in his investigations by the Rev. Mr. Lowe. These plants we deem worthy of enumeration ; as shewing what future voyagers may expect to obtain during an equally short visit; and facilitating the troublesome task of determining their names by those general works on Botany in which alone the Madeira plants are described. The names of those collected by the Antarctic Expedition† on its outward voyage are added to this list: the majority of which, having been gathered (in October) at a very different season, were not met with by Dr. Vogel.

All the species have been determined by Dr. Lemann ;

* By Sir W. J. Hooker and Dr. J. D. Hooker.

† The smallness of this collection is to be attributed to the temporary ill health of the Botanist of the Antarctic Expedition during the ships' ten days' sojourn at Madeira.

whose botanical accuracy and acquaintance with the Floras of S. Europe, Madeira, and the Canaries, entitle us to place great reliance on the authenticity of the nomenclature. That gentleman has also favoured us with some notes on the Botany of Madeira, as compared with other neighbouring islands, which we beg to acknowledge most heartily, and which are embodied in the following remarks.

The Island of Madeira contains 672 species of flowering plants and Ferns, of which 85 are absolutely peculiar, and 480 common to Europe ; 280 are common to Madeira and the Azores (whose Flora is estimated at 425 sp.) ; 312 (or probably more) to Madeira and the Canaries ; and 170 to the neighbourhood of Gibraltar (where 456 have been collected.)

It is remarkable that out of 400 European, and these Mediterranean species, indigenous to Madeira, not more than 170 occur in Gibraltar : for it were natural to suppose that the majority of 480 species are very widely dispersed throughout the S. Europe, and must have migrated by way, as it were, of Gibraltar, if transported across the ocean to Madeira. It is further worthy of observation, that the Azores, though very far to the westward, and the Canaries to the south, both contain many more of the Mediterranean plants seen in Madeira, than does Gibraltar.

A considerable number of the Madeira plants belong to genera not found in the adjacent continent,* but in the Canaries, Azores, or Cape de Verd Islands; thus indicating a botanical affinity between these groups and confined to them.†

* Except, possibly, on the hitherto unexplored Atlas Mountains on the Morocco coast.

† The following are some of the leading features of the N. Atlantic Island Flora, as distinguishing it from the continental.

1. Genera confined to the four groups, and represented in two or more of the islands, are :—

    Melanoselinum, (Madeira and Azores.)
    Æonium,
    Aichryson, } (Madeira and Canaries.)
    Sinapidendron, (Madeira and Cape de Verd Islands.)

The evidence of this relationship is very decided, from the peculiarity of the genera or species giving rise to it. Though comparatively few in number, their characters are so prominent and so widely different from the Mediterranean plants which accompany them, that the latter, though numerically much the greatest, seem superadded, and, as it were, intruders on the former.

The Canaries and Madeira, from their central position and various other causes, are the centre of this Botanical region, called by Mr. Webb the " Macaronesian," and exhibit more peculiarity than the Cape de Verds, (as far as they are at present known), or the Azores. There can be little doubt Madeira was even more peculiar in its vegetation than now, previous to the destruction by fire of the luxuriant forests, of which, almost clothing the lower parts of the island, we have historic evidence. Not only does such a catastrophe destroy species, but their place is afterwards occupied by strong-growing imported weeds, which prevent the re-appearance of the native plants by monopolizing the soil.

With very few exceptions, the Mediterranean are the only plants found in Madeira and the Canaries besides what are confined to those islands : in the Azores, on the other hand, more Northern European species are associated with

Heberdenia, } (Madeira and Canaries.)
Phyllis, }

Campylanthus, (Canaries and Cape de Verd Islands.)

2. Orders represented by closely allied, but peculiar genera :—

SCROPHULARINEÆ.

Isoplexis, (Madeira,) and *Callianassa,* (Canaries.)

CAMPANULACEÆ.

Musschia, (Madeira,) and *Canarina,* (Canaries.)

which are further represented by the singular *Campanula Vidalii* in the Azores, and the equally distinct *C. Jacobæa* in the Cape de Verd Islands.

Other instances of representation by peculiar species are found in the *Seneciones* and *Sonchi,* and in the curious *Euphorbiæ* of the Canaries and the Cape de Verds, and several other genera.

them. In the Cape de Verds, far to the south, W. African and W. Indian plants replace those of the Mediterranean. The Island of Madeira participates in the Flora of the W. Indies to a much greater degree than does any part of the adjacent continent :—that this is in a great measure due to the dampness of its insular climate, is clear, from the plants in question being almost entirely Ferns, viz. :—

> Acrostichum squamosum, *Sw.*
> Aspidium molle, *Sw.*
> Asplenium monanthemum, *Sw.*
>   „   furcatum, *Sw.*
> Trichomanes radicans, *Sw.*

species found nowhere on the continent of Europe, or in N. Africa. The presence of a plant belonging to the otherwise exclusively American genus, *Clethra*, is striking, because indicating a further relationship with the Flora of the New World, but of a very different character from the above.

The *Helichrysa* of Madeira are allied in rather a remarkable degree to the S. African species of that genus : a fact which reminds us that the *Myrsine Africana*, a Cape of Good Hope plant, is a native of the Azores, but of no intervening latitude on the West coast of Africa or the Atlantic Islands, nor indeed anywhere else but Abyssinia. Though not a subject falling immediately within the province of the pure Botanist, it may not be amiss here to state, that the four Island-groups in question have been conceived by my friend, Professor Forbes, to be the exposed remains of one continuous and extended tract of land, which formed the western prolongation of the European and African shores. He points to the specific identity of these islands and Europe, as affording Botanical evidence of this ingenious theory, which, however, he chiefly rests on geological grounds. Regarded in this light, the question will resolve itself, in the opinion of most Botanists, into one concerning the power of migration, and the probability of transport having taken place, to a very great extent, over the Atlantic Ocean, and against the prevailing direction of the winds. It may be contended that such a migration would have peopled these islands solely, or mainly, with certain of the more transportable classes of plants; and that the result must be, that the number of species belonging to each natural order would be great in proportion to the facility with which they bear transportation : while only those orders could be numerous, which possess that faculty in an eminent degree. But such are not the characteristics of the Mediterranean plants found in Madeira.

On the other hand, the existence of such a continent, during the period when these islands bore the plants which they now produce, would argue the former presence of a very large Flora belonging to the type which now distinguishes the islands in question from the Mediterranean ; and of whose previous existence the remaining species, peculiar to them, are the indication. Against this theory it might be urged, that more specific identity between the plants of the several insular groups, would then be the natural consequence, than now is seen : for the affinity of vegetation between the different islands consists, not in identical species, but in representatives. The same agent, in short, which effected the peopling of the several groups with the plants of continental Europe, would also have distributed more equally the non-European species over the same area.

It is, however, to the lofty peaks of Atlas that we must look, if any where, for the continental representatives of those peculiar plants which mark the North Atlantic Insular Floras. Thus, we expect to find the productions of the Galapagos Archipelago on the higher levels of the Cordillera; and the mountains of St. Thomas, Fernando Po and the Cameroons, on the west coast of Tropical Africa, may yet exhibit to us the Botanical features of St. Helena. Outlying and high islands commonly partake in the peculiar vegetation of a climate cooler than belongs to the low lands of the adjacent continent ; though, in the case of Juan Fernandez, they sometimes exhibit genera equally isolated in botanical affinities as their habitats are in geographical position.

CATALOGUE OF "MADEIRA PLANTS."*

1. Ranunculus grandifolius, *Lowe.*—Ribiera Frio, *Vogel.*
2. R. repens, *L.*—Ribiera Frio, *Vogel.*
3. Papaver dubium, *L.*—Curral, *Vogel.*
4. Fumaria media, *Loisel.*—Curral, *Vogel.*
5. Matthiola Maderensis, *Lowe.*—Funchal, *Vogel* & *J. D. H.*
6. Cheiranthus mutabilis, *L'Hér.*—Curral, *Vogel.*
7. Nasturtium officinale, *R. Br.*—Funchal, *J. D. H.*
8. Arabis albida, *Stev.*—Ribiera Frio, and Grand Waterfall, *Vogel.*
9. Cardamine hirsuta, *L.*—Grand Waterfall, *Vogel.*
10. Teesdalia Iberis, *DC.*—Grand Waterfall, *Vogel.*
11. Sinapidendron frutescens, *Lowe.*—Curral, *Vogel.*
12. Raphanus Raphanistrum, *L.*—Funchal, *Vogel.*
13. Viola Maderensis, *Lowe.*—Road to the Curral, *J. D. H.*
14. V. sylvestris, *Lam.*—Ribiera Frio and Grand Waterfall, *Vogel.*
15. Linum angustifolium, *Huds.*—Funchal, *Vogel* & *J. D. H.*
16. Malva parviflora, *L.*—Ribiera Frio, *Vogel.*
17. Sida carpinoides, *DC.*—Funchal, *J. D. H.*
18. S. rhombifolia, *L.*—Funchal, *J. D. H.*
19. Hypericum humifusum, *L.*—Funchal, *J. D. H.*
20. H. perforatum, *L.*—Funchal, *Vogel* & *J. D. H.*
21. H. glandulosum, *Ait.*—Curral, *Vogel.*
22. H. grandifolium, *Chois.*—Curral, *J. D. H.*
23. Erodium Botrys, *Bertol.*—Grand Waterfall, *Vogel.*
24. Geranium rotundifolium, *L.*—Curral and Grand Waterfall, *Vogel.*
25. Oxalis corniculata, *L.*—Funchal, *J. D. H.*
26. Mesembryanthemum nodiflorum, *L.*—Funchal, *J. D. H.*
27. Polycarpon tetraphyllum, *L. fil.*—Funchal, *Vogel, J. D. H.*
28. Cerastium glomeratum, *Thuill.*— Curral and Funchal, *Vogel, J. D. H.*

---

* Collected by the Botanist of the Niger Expedition; to which are added those of the Antarctic Expedition, drawn up by C. Lemann, Esq., Cantab. F.L.S. &c. &c.

29. Cerastium triviale, *Link.*—Curral, *Vogel, J. D. H.*
30. Stellaria uliginosa, *Murr.*—Curral, *Vogel, J. D. H.*
31. S. media, *Sm.*—Ribiera Frio, *Vogel.*
32. Silene Gallica, *L.*—Grand Waterfall, *Vogel.*
33. Ulex Europæus, *L.*—Ribiera Frio, *Vogel.*
34. Genista virgata, *DC.*—Curral, *Vogel, J. D. H.*
35. G. Maderensis, *Webb.*—Ribiera Frio, *Vogel.*
36. Lathyrus sphæricus, *Retz.*—Curral, *Vogel.*
37. Lotus glaucus, *Ait.*—Funchal, *J. D. H.*
38. Medicago tribuloides, *Lam.*—Funchal, *Vogel, J. D. H.*
39. Psoralea bituminosa, *L.*—Funchal, *Vogel, J. D. H.*
40. Vicia sativa, *L.*—Curral, *Vogel.*
41. Scorpiurus subvillosus, *L.*—Funchal, *Vogel.*
42. Ornithopus perpusillus, *L.*—Grand Waterfall, *Vogel.*
43. Cassia bicapsularis, *L.*—Funchal, *Vogel, J. D. H.* (introduced?)
44. Acacia Farnesiana, *Willd.* — Funchal, *Vogel, J. D. H.* (introduced?)
45. Chamæmeles coccinea, *Lindl.*—East Coast, *Vogel.*
46. Alchemilla arvensis, *Scop.*—Ribiera Frio, *Vogel, J. D. H.*
47. Poterium verrucosum, *Ehr.*—Funchal, *Vogel, J. D. H.*
48. Fragaria vesca, *L.*—Ribiera Frio, *Vogel*; Curral, *J. D. H.*
49. Lythrum Græfferi, *Tenore.*—Curral, *J. D. H.*
50. Sempervivum glutinosum, *Ait.*—Funchal, *Vogel.*
51. S. villosum, *Ait.*—Curral, *Vogel, J. D. H.*
52. S. aizoides, *Lam.*—Funchal? *Vogel.*
53. Umbilicus pendulinus.—Hab.? *Vogel.*
54. Saxifraga Maderensis, *Don.*—Curral, *Vogel.*
55. Bupleurum salicifolium, *Solander.*—Curral, *Vogel.*
56. Crithmum maritimum, *L.* β. latifolium.—East Coast, *J. D. H.*
57. Sambucus nigra, *L.*—Ribiera Frio, *Vogel.*
58. Galium Aparine, *L.*—Ribiera Frio, *Vogel.*
59. Sherardia arvensis, *L.*—Ribiera Frio, *Vogel, J. D. H.*
60. Phyllis Nobla, *L.*—Curral, *Vogel.*
61. Ageratum conyzoides, *L.*—Funchal, *J. D. H.*
62. Phagnalon saxatile, *DC.*—Hab.? *Vogel.*

63. Eclipta prostrata, *L.?*—Funchal, *J. D. H.*
64. Bidens leucantha, *Willd.*—Funchal, *Vogel, J. D. H.*
65. Chrysanthemum pinnatifidum, *L. fil.*—Ribiera Frio, *Vogel.*
66. Artemisia argentea, *L'Hér.*—Hab. ? *Vogel.*
67. Helichrysum obconicum, *DC.*—Sea-coast, *J. D. H.*
68. H. melanophthalmum, *Lowe.*—Grand Waterfall, *Vogel.*
69. Gnaphalium luteo-album, *L.*—Funchal, *Vogel, J. D. H.*
70. Calendula arvensis, *L.*—Curral, *Vogel, J. D. H.*
71. Galactites tomentosa, *Mœnch.*—Hab.? *Vogel.*
72. Tolpis pectinata, *DC.*—Funchal, *J. D. H.*
73. T. crinita, *Lowe.*—Hab.? *Vogel.*
74. T. umbellata, *Bertol.*—Curral, *Vogel, J. D. H.*
75. Thrincia nudicaulis, *Lowe.*— Curral, *Vogel*; Funchal, *J. D. H.*
76. Sonchus ustulatus, *Lowe*, (leaves.)—South-east coast, *J. D. H.*
77. Campanula Erinus, *L.*—Curral, *Vogel, J. D. H.*
78. Centranthus Calcitrapa, *Dufr.*—Curral, *Vogel.*
79. Vaccinium Maderense, *Link.*—Ribiera Frio, *Vogel;* Pico Ruivo, *J. D. H.*
80. Erica arborea, *L.*—Curral, *J. D. H.*
81. E. scoparia, *L.* — Ribiera Frio, *Vogel*; Pico Ruivo, *J. D. H.*
82. Clethra arborea, *Ait.*—Ribiera Frio, *Vogel.*
83. Heberdenia excelsa, *DC. fil.* (leaves.)—Curral, *J. D. H.*
84. Sideroxylon Marmulana, *C. Sm.*—Funchal, *Vogel.*
85. Convolvulus althæoides, *L.*— Hab.? *Vogel.*
86. C. solanifolius, *Lowe.*—Ribiera Frio, *Vogel.*
87. Plantago Lagopus, *Hall.* α. β. Lusitanica. — α. Grand Waterfall, β. Ribiera Frio, *Vogel.*
88. P. Coronopus, *L.*—Funchal, *J. D. H.*
89. P. arborescens, *Poir.*—South-east coast, *J. D. H.*
90. Globularia longifolia, *Ait.* — South-east coast, *Vogel;* Funchal, *J. D. H.*
91. Echium plantagineum, *L.*—Grand Waterfall, *Vogel.*
92. E. fastuosum, *Jacq.*—Hab? *Vogel.*
93. Myosotis repens, *Don.*—Ribiera Frio, *Vogel.*

94. Lavandula viridis, *Ait.*—Funchal, *Vogel, J. D. H.*
95. L. pinnata, *L. fil.*—Hab. ? *Vogel.*
96. Bystropogon punctatus, *L'Hér.*— Hab. ? *Vogel.*
97. Origanum virens, *Link.*—Curral, *J. D. H.*
98. Micromeria varia, *Benth.*—Curral, *J. D. H.*
99. Melissa Calamintha, *L.* β. villosissima, *Benth.*—Curral, *J. D. H.*
100. Prunella vulgaris, *Mœnch.*—Grand Waterfall and Ribiera Frio, *Vogel.*
101. Cedronella triphylla, *Mœnch.*—Grand Waterfall, *Vogel.*
102. Stachys hirta, *L.*—Curral, *Vogel.*
102. S. arvensis, *L.*—Curral, *Vogel.*
102. S. Betonica, *Benth.*—Hab. ? *Vogel.*
103. Clinopodium vulgare, *L.*—Curral, *Vogel, J. D. H.*
104. Sideritis Massoniana, *Benth.*—Curral, *Vogel.*
105. Teucrium abutiloides, *L'Hér.*—Curral, *J. D. H.*
106. Lantana aculeata, *Ait.*—*J. D. H.*
107. Antirrhinum Orontium, *L.*—Grand Waterfall, *Vogel.*
108. Sibthorpia peregrina,—Hab. ? *Vogel.*
109. Veronica acinacifolia, *L.*—Ribiera Frio, *Vogel.*
110. V. Anagallis, *L.*—Curral, *Vogel.*
111. V. arvensis, *L.*—Hab. ? *Vogel.*
112. Odontites Holliana, *Benth.* (fruit.)—Ribiera Frio, *Vogel.*
113. Physalis pubescens, *L.*—Funchal, *Vogel, J. D. H.*
114. Hyoscyamus Canariensis, *Ker.*—Funchal, *J. D. H.*
115. Vinca major ? not wild.—Funchal, *Vogel.*
116. Olea (Phillyrea, *D.C.*) Lowei, *DC.*—Maritime spots, *J. D. H.*
117. Jasminum odoratissimum, *L.*—Funchal, *Vogel.*
118. Chenopodium ambrosioides, *L.*—Funchal, *J. D. H.*
119. Suæda laxifolia, *Lowe.*—East coast, *J. D. H.*
120. Rumex Maderensis, *Lowe.*—Curral, *Vogel.*
121. R. Acetosella, *L.*—Hab. ? *Vogel.*
122. R. aculeatus, *L.*—Curral and Ribiera Frio, *Vogel.*
123. Polygonum maritimum, *L.*—East coast, *J. D. H.*
124. Mercurialis annua, *L.* var. β. (M. ambigua, *L. fil.*)—Funchal, *Vogel, J. D. H.*

G

82   BOTANY OF THE

125. Euphorbia Peplus, *L.*—Funchal, *Vogel, J. D. H.*
126. E. hypericifolia, *L.*—Funchal, *Vogel, J. D. H.*
127. Persea Indica, *Spr.*—Curral, *Vogel, J. D. H.*
128. Oreodaphne fœtens, *Nees.*—Ribiera Frio, *Vogel.*
129. Apollonia Canariensis, *Nees.*—Ribiera Frio, *Vogel.*
130. Myrica Faya, *Mx.*—Mr. Veitch's garden, *J. D. H.*
131. Parietaria Lusitanica, *L.*? (P. Maderensis, *Rchb.*)—Funchal, *J. D. H.*
132. Ephedra alata, *Dcne.*—Funchal, *J. D. H.*
133. Peristylus cordatus, *Lindl.*—Hab.? *Vogel.*
134. Himantoglossum secundiflorum, *Lindl.*—Ribiera Frio, *Vogel.*
135. Amaryllis Belladonna, *L.*—Road to Curral, *J. D. H.*
136. Ruscus Hypoglossum, *L.*—Hab.? *Vogel.*
137. Commelina communis, *L.*—Funchal, *J. D. H.*
138. Juncus glaucus, *Sm.*—Hab.? *Vogel.*
139. J. filiformis, *L.*—Grand Waterfall, *Vogel.*
140. Isolepis Saviana, *Schult.*—Grand Waterfall, *Vogel.*
141. Carex divulsa, *Gooden.*—Curral, Grand Waterfall, Ribiera Frio, *Vogel.*
142. Panicum vaginatum, *Swtz.*—Funchal, *J. D. H.*
143. P. repens, *L.*—Funchal, *J. D. H.*
144. Pennisetum cenchroides, *Rich.*—Funchal, *J. D. H.*
145. Lagurus ovatus, *L.*—Curral, *Vogel.*
146. Cynosurus echinatus, *L.*—Ribiera Frio, *Vogel.*
147. C. elegans, *Desf.*—Ribiera Frio, *Vogel.*
148. Dactylis glomerata, *L.*?—Hab.? *Vogel.*
149. Melica ciliata, *L.*—Curral and Grand Waterfall, *Vogel.*
150. Poa megastachya, *Koel.*—Funchal, *J. D. H.*
151. Briza minor, *L.*—Grand Waterfall, *Vogel.*
152. B. major, *L.*—Curral and Ribiera Frio, *Vogel, J. D. H.*
153. Aira præcox, *L.*—Grand Waterfall, *Vogel.*
154. A. caryophyllea, *L.*—Ribiera Frio, *Vogel.*
155. Avena hirtula, *Lag.*—Curral, *Vogel.*
156. Bromus maximus, *L.*—Curral, *Vogel.*
157. Festuca bromoides, *L.*—Curral, Grand Waterfall and Ribiera Frio, *Vogel, J. D. H.*

158. Festuca jubata, *Lowe.*—Curral, *Vogel.*
159. Andropogon Halepensis, *Sibth.*—Funchal, *J. D. H.*
160. A. hirtus, *L.*—Funchal, *Vogel, J. D. H.*
161. Polypodium vulgare, *L.*—Curral, *Vogel.*
162. Gymnogramma Lovei, *Hook.* and *Grev.*—Ribiera Frio, *Vogel.*
163. Notholæna lanuginosa, *Desv.*—Funchal, *J. D. H.*
164. Grammitis Ceterach, *L.*—Funchal, *Vogel, J. D. H.*
165. Adiantum reniforme, *L.*—Hab.? *Vogel.*
166. A. Capillus Veneris.—Funchal, *Vogel, J. D. H.*
167. Pteris aquilina, *L.*— Curral and Ribiera Frio, *Vogel, J. D. H.*
168. P. arguta, *Vahl.*—Ribiera Frio, *Vogel.*
169. Lomaria Spicant, *Desv.*—Grand Waterfall, *Vogel.*
170. Athyrium Filix-fœmina, *Roth.*—Ribiera Frio, *Vogel.*
171. Asplenium Adiantum-nigrum, *L.* (A. *productum*, Lowe). Curral, *Vogel;* Funchal, *J. D. H.*
172. A. monanthemum, *Sm.*—Ribiera Frio. *Vogel.*
173. A. anceps, *Soland.*—Curral, *Vogel.*
174. A. palmatum, *Swtz.*—Ribiera Frio, *Vogel.*
175. Cystopteris fragilis, *Bernh.*—Funchal, *Vogel, J. D. H.*
176. Nephrodium molle, *Br.*—Funchal, *J. D. H.*
177. Aspidium angulare, *Sm.*—Curral, *Vogel.*
178. A. elongatum, *Swtz.*—Ribiera Frio, *Vogel.*
179. A. falcinellum, *Swtz.*—Ribiera Frio and Curral, *Vogel, J. D. H.*
180. Davallia Canariensis.—Ribiera Frio and Curral, *J. D. H.*
181. Lycopodium denticulatum, *Willd.* — Curral, *Vogel, J. D. H.*

TENERIFFE.

The next point visited by the Niger Expedition, after leaving Madeira, was the island of Teneriffe : where the vessel in which Vogel had embarked remained but a few hours. The same island, and the same port, Santa Cruz, had been touched at by the Antarctic Expedition during the previous winter. Teneriffe is always held to be classic ground by the Naturalist, as the opening scene of the labours of Humboldt, who there first appreciated, in their full extent, the laws governing the geographical distribution of plants. His life-like pictures of the natural phenomena, observed during an ascent of the famous Peak, have given to many succeeding scientific travellers that impulse which has turned their thoughts and steps from closet studies and the pursuit of Natural History at home, and induced them to seek far distant scenes, in the West, the East and the South.

The Peak itself is seldom descried: one hurried glimpse of its very apex, from upwards of sixty miles' distance, was all we obtained: it then appeared like a little short and broad cone high in the clouds, or rather as an opaque triangular spot on the firmament. It is difficult to imagine this to be the " culminating point;" that mighty mass, at whose base the toil-worn traveller pauses ; who, having surmounted four-fifths of the mountain, finds his heart quail at beholding a " Pelion upon Ossa piled," so stern, so stony and so steep.

Much and deeply did the officers of Captain Ross' and Trotter's Expeditions deplore the necessity of hurrying from this spot, most interesting to the sailor; being the point for which every circumnavigator first steers, and from whence, with chronometers carefully corrected at its well-determined position, he takes his departure. For years, too, this was the prime Meridian, distance in longitude at sea having been at one period reckoned from Teneriffe, as zero, by all the seafaring nations of Europe; and by some it is so still. From the days of the earliest circumnavigators, to the present, the words " we sighted the Peak of Teneriffe," indicate that page in the narrative, from which all that is interesting in the voyage commences.

In the History of Geology, the Canary Islands hold a conspicuous position. Von Buch developed his theory of craters of elevation from what he there observed: his name too recalls, and most appropriately, that of his fellow-la-bourer on the same shores, Christian Smith, the amiable and gifted Swede, who first, after Humboldt, explored their

Botany. Christian Smith returned to Europe to embark
in the ill-fated Congo Expedition : when he again saw the
Peak of Teneriffe, he welcomed it as a familiar object, and
bade it adieu, rejoicing that a still more novel field of inquiry
was opened to him, beyond this scene of his early exertions.
A few short months terminated his life and hopes : like Vogel,
he fell a victim to the dread fever of the pestilential coast of
Africa : like him, too, he was a martyr in the cause of Botanical Science.

Fraught with so many and such touching associations, no
naturalist-voyager can see the Fortunate Isles rising, one
by one, on the horizon of the mighty Atlantic, without a
feeling of melancholy, while he reflects on the fate of these
his two predecessors—both accomplished Naturalists of their
age and day—whose prospects and hopes were in every
respect as bright, perhaps brighter, than his own.

The excellent and beautiful work of Mr. Webb, on the
Natural History of the Canaries, leaves little to be said,
especially of their Botany, and renders even an enumeration
of the few species gathered by Vogel and the Botanist of the
Antarctic Expedition unnecessary ; for they were all collected
within a very few miles of Santa Cruz, during a very hurried
walk, and scarcely include a dozen kinds. This locality is
one of the most barren of the whole group, especially in the
immediate neighbourhood of the sea. The broad frontage of
cliff and mountain, reaching upwards for several thousand
feet above the town, and fore-shortened to the view from
seaboard, presents a progressive increase of verdure from
the water's edge to the mountains. At this season, when
the vines are out of leaf, nothing green meets the eye.
The trees, either standing singly or in very small clumps,
dot the alternate ridges and steep gullies with which the
slopes are everywhere cut like the edge of a saw, producing
that spotty effect in the landscape so admirably rendered
in the phytographical illustrations of Mr. Webb's work,
and which is eminently characteristic both of the Canaries
and Madeira.

The *Kleinia, Euphorbia* and *Plocama* are three plants
which the voyager recognizes long before reaching the shore;
and they are so singular, whether as regards habit, locality,
or botanical characters, that the opportunity of seeing them
in a wild state, even from the sea, must be deemed a privilege by the Botanist.

CAPE DE VERD ISLANDS.

The voyage, from the Canaries to the Cape de Verd
Islands, generally presents a hiatus in the journals of those
sea-faring Naturalists who have followed this route. Before
arriving at the Canaries, landsmen have scarcely recovered from
the novelty of ship-board and its effects; nor has there been
time, since leaving those islands, to become thoroughly inured
to the monotony of a sailing life. At first sight, the Cape
de Verd Islands are very disappointing. It is true that we
had passed from an extra-tropical latitude to far within the
tropics; but the change in position was not accompanied
with a corresponding difference, still less with luxuriance,
in the vegetation and scenery. Yet these apparently barren
islands have associations of great interest; and their examination yields both pleasure and profit. They afforded us
the first glimpses of the fever-smitten coast of Africa, and of
slavery. Even here the black man, deprived of freedom,
and an alien to the land in which, though guiltless, he is
a prisoner for life, is apt to be regarded as a mere object
of Natural History by his Caucasian fellow-creature; who,
before he has time for reflection, may perhaps be excused for
pausing to consider, whether a being so different in features
and social position, be really of the same origin as himself;
whether, in short, the poor African is a race of the same
stock, or a species apart.

There are many other circumstances, connected with
these islands, which keep the mind busy while in their
neighbourhood. They form the western extreme of the Old
World, of what was the whole world to civilized man, till

within the last very few hundred years; and hence, with the North Cape and Cape of Good Hope, they constitute the three salient points in the geography of the eastern Atlantic. In many of their physical features, they form a continuation of the great Sahara desert, that mysterious blank on our maps, upon whose sea of sand so many of our venturous countrymen have embarked, to be heard of no more. The hitherto unexplored mountains of the Cape de Verds rise 8000 feet and upwards above the sea, in serried ridges and isolated peaks; promising a rich harvest to some Botanist, who may in those higher and cooler parts of the islands rely on immunity from disease and a temperate climate. There he may expect to find new types of plants; for the Mountain Flora of Western Tropical Africa is wholly unknown; and of its probable nature even we can form no guess. To conclude, the Linnæan axiom of " semper aliquid novi ex Africa" has never yet proved false. A Naturalist cannot see the shores of that continent without feeling that no other spur is required to exertion, in a field to which such a motto still applies with so much force.

SPICILEGIA GORGONEA;

OR A

CATALOGUE OF ALL THE PLANTS

AS YET DISCOVERED IN THE

CAPE DE VERD ISLANDS.

FROM THE COLLECTIONS OF

J. D. HOOKER, ESQ. M.D. R.N., DR. T. VOGEL,

AND OTHER TRAVELLERS.

BY

P. BARKER WEBB.

ναίυσι πέρην κλυτῦ Ὠκεανῦιο
᾽Εσχατιῇ πρὸς νυκτὸς.

HESIOD. THEOG. 274.

PREFACE*

THE collections from which the following catalogue was composed were formed in a hurried manner by different Botanists while on their way to more fertile regions, during the short stay made by the vessels in which they sailed, at one or the other of the Cape de Verd Islands. They were confided to me for publication by Dr. J. Dalton Hooker, on account of the supposed affinity of the Flora of these islands to that of the Canarian Archipelago. They do not probably contain more than a small portion of the coast vegetation, with a sprinkling of that of the neighbouring hills, to the height, in general, of less than 3000 feet. The mountains of the interior of the larger islands and the lofty Island of Fogo, supposed to attain nearly 8000 feet, still remain unvisited, and are, botanically speaking, unknown. It may be added, as some palliation for possible errors in this list of names, that a portion of the materials were in a state which almost defied the powers of the most intrepid nomenclator : those, for example, of the *Sapoteæ*, from which, through his intimate knowledge of the Order, M. Decaisne was enabled to elicit a new species, contained

* The Introduction to the following Florula of the Cape de Verd Islands, drawn up by the author of the Florula itself, explains the nature of the collections and the motives that induced us to request his aid in this portion of the "Flora of the Niger Expedition." We here tender our sincere thanks for the generous manner in which he undertook the task, and for the extreme care he has devoted to the accurate determination of the plants. The ability Mr. Webb has displayed, and the classical polish with which the whole is executed, speak for themselves.—W. J. H.

but a single flower. Such a result, derived from so much previous experience, cannot have existed in every case ; and some perhaps of our new species, founded upon imperfect materials, or even those assimilated to species already known, may not carry with them equal authenticity. But so much care has been taken in ascertaining their identity, essential for botanico-geographical data, that it is hoped that no grave errors have occurred. Collected, too, in different seasons, deficiencies of one set were occasionally made up by the better state of those supplied by other travellers.

The most complete collections were those of Dr. J. D. Hooker and the much lamented Theodore Vogel. They may be said to form the groundwork of the *Spicilegia ;* though both were formed in the dry season, that of Dr. Hooker in November 1839, and that of Vogel in June 1841. Another very interesting set, gathered by Forbes, in a much better season, March and April 1822, and most liberally communicated by the Horticultural Society, has afforded many species in the best order. To these must be added a small collection made by that accomplished naturalist, Mr. Darwin, (the property of the Cambridge University Museum) and generously confided for publication to Dr. Hooker by Professor Henslow : it has added some interesting species to the catalogue. The rules of the British Museum forbid the loan of the treasures contained within its walls ; and it is through the collation by Dr. J. D. Hooker of the specimens of the other herbaria with that of Christian Smith, preserved in that establishment, that I have been enabled to cite his plants and to quote the catalogue, contained in his journal, published in Captain Tuckey's Voyage to the Congo. Those of Forster I have not been able to see.

The late Dr. Brunner, of Berne, on his return from Senegal, visited several of the Cape de Verd Islands. The species he there gathered appeared in his *Ergebniss,* originally published in the " Flora od. Bot. Zeitung." I was obligingly furnished by him with a nearly complete set of his plants, and by this means have been enabled to cite his synonomy with certainty.

Lastly, the Professor Administrators of the Museum of Natural History of Paris, confided to me, with their well-known liberality, a collection, formed probably for the Portuguese government, and brought from Lisbon in 1808 by M. Geoffrey St. Hilaire. This was accompanied by no written document by which the native country of the plants could be ascertained; but the identity of the very great majority of the species with those of the British collections from the Cape de Verd Islands, leaves no reasonable doubt of its origin. The specimens are generally satisfactory, and were evidently picked up at a moment when vegetation was starting into life and in its most florid state. It has added some highly interesting and characteristic species to our list. With this collection, there was brought at the same time from Lisbon, another, supposed to have come from Brazil. The following circumstance, which might lead us to imagine that some confusion may have taken place between these two sets of plants, has been pointed out to me by M. Adrien de Jussieu. The genus *Asteranthos*, taken from the latter set by Desfontaines, belonging as it does to the strange African Order of the *Napoleoneæ* beautifully illustrated by the descriptions and drawings of this distinguished naturalist, never having since been met with in America may very possibly have wandered to the Brazilian set from that formed in the Cape de Verd Islands, and thus be in reality, what from analogy might be supposed a denizen, not of America, but of Africa. At any rate it is useful to call the attention of travellers to the existence of this geographical doubt, in order that it may be investigated and cleared up.

The present catalogue, compiled from these several sources, owing to the causes referred to above, contains only 250 species of ferns and flowering plants. Of these 204 belong to the Dicotyledonous orders, only 31 to Monocotyledones, and 13 to *Equisetaceæ* and Ferns. It is probable, however, that the proportion of Dicotyledons to Monocotyledons cannot be entirely depended upon, as it stands in this list; but that the latter should be rated somewhat higher; because Professor Parlatore, who kindly undertook to describe the grasses, left several in the English collections untouched, and did not inspect those contained in the Cape de Verd collection of the Museum of Paris. Add to this, the fugacious nature of many Monocotyledonous genera renders their collection by a casual visitor doubtful; whereas of many Dicotyledons and Ferns some remains may be found at all seasons.

Of our 250 species, upwards of 48, or nearly a fifth, are either found in the Canaries or belong to decidedly Canarian genera and forms: about 25, or a tenth, belong to the Arabico-Nubian region: the Mediterranean series is represented by about a twelfth. The remainder are either common to most tropical regions, or Senegambian, or belonging to the islands. It is singular that in a country contiguous to the Old World, and amongst so restricted a number of plants, nearly a third should turn out to be species previously undescribed; although Mr. Bentham had already published two *Labiatæ* and three *Scrophularineæ* from the collections of Forbes and Brunner. These prefatory remarks may be terminated with one word in justification of our title of *Spicilegia Gorgonea*. It must certainly be conceded, of all that lay that beyond the " Fortunate Isles" the geographical knowledge of the ancients was exceedingly vague. Nevertheless the text of Pliny shows that they had a competent notion even of the Niger, its divergence into many streams or ἐκτροπὰς, as mentioned by Ptolemy, and its gradual rise, like that of the Nile, after the tropical rains, which is recorded by Pliny. After the Canaries, Ptolemy speaks of the promontory called Gannaria or Cape Blanco: near it is the island of Arguin, supposed to be the Cerne of Pliny. The next promontory mentioned is the Hesperian Ceras, which can hardly be any other than that of Cape Verd, where the continent is most protruded towards the west; and the isles opposite to it will be the Gorgades, or Isles of the Gorgons. " Contra hoc promontorium Gorgades insulæ narrantur, Gorgonum quondam domus, bidui navigatione distantes a continente."* It is right however to mention, that D'Anville places the Gorgades at the Bissagos, probably too far to the south, opposite to no promontory, and close to the continent.

P. B. W.

* Plin. lib. vi. c. 36.

SPICILEGIA GORGONEA.

I. Anonaceæ, Juss.

1. Anona *squamosa,* Linn. *Sp. Pl.* p. 757. Dun. *Monogr.* p. 69. DC. *Syst.* 1. p. 472. *Prodr.* 1. p. 85. Brunn. *Ergebn.* p. 15. —Ic. Rumph. *Amb.* 1. t. 46. Sloane, *Hist. Jam.* t. 221. Rheed. *Mal.* 3. t. 29. Jacq. *Obs.* t. 6. f. 1. Hab. Arbor 20-pedalis, quæ in ins. *S. Jacobi,* sylvis ut plurimum destituta, nemora ad summitatem collium vallis *S. Dominici* efficit. (*J. Dalton Hooker,* n. 131. November, 1839. *sp. fructifera.*)

2. Anona *Senegalensis,* Pers. *Syn.* 2. p. 95. Guill. *et* Perr. *Fl. Sen. Tent.* p. 5. Brunn. *Ergebn.* p. 14.—Ic. Deless. *Ic. Sel.* 1. t. 86. Hab. In vallibus *S. Dominici* et *Organorum ins. S. Jacobi* (*Brunn.* l. c.) *Anonæ tripetalæ,* Linn. in catalogo Smithiano (Tuck. voy. p. 250) enumeratæ nulla extant in herb. Mus. Brit. specimina (*J. Dalton Hooker,* in litt.)

II. Menispermeæ, Juss.

3. Cocculus *Leæba,* DC. *Syst.* 1. p. 250. Richard, *in Tent. Fl. Sen.* 1. p. 13. Leæba, *Forsk. Fl. Æg.-Arab.* p. 108 (ex specimine Forsteriano Mus. Brit. et scheda sua inscripta *St. Jago, Cape de Verd* (nec maris Australis ut Candolleus credebat.) (*J. Dalton Hooker,* in litt.) Menispermum Leæba, *Del. Fl. d'Eg. descr. des pl. p.* 140. Menispermum ellipticum, *Poir. Suppl. p.* 657. Cocculus Epibaterium, *DC. Syst. p.* 530, *et* Cocculus ellipticus, *ejusd. ibid. p.* 526. Smilacina anomala, *genus forte novum,* Chr. Smith, *l. c. p.*

H

249! (*J. Dalton Hook.* in litt.) — Ic. Forst. *l. c. t.* 54. Delile, *Fl. d'Eg. t.* 51. *f.* 2 *et* 3. The discovery, made by Dr. J. D. Hooker, of the identity of the *Leæba* of Forskål with the *Epibaterium* of Forster, has cleared up an error which might have long remained a blot on science. Though evidently the same species as the Egyptian, our plant is smoother, the young shoots alone being pubescent. The name *Leæba* was published a year previous to that of Forster: we may therefore retain it in preference to *pendulus,* by which no species of the genus can be distinguished from another, they being all climbers and pendulous.

III. PAPAVERACEÆ, *Juss.*

4. Argemone *Mexicana*; Linn. *Sp. Pl. p.* 727. Chr. Smith, *in Tuck. voy. p.* 250! (*J. Dalton Hook.* in litt.)—a. floribus luteis, stylo subnullo.—Ic. Bauh. *Prodr. t.* 92. Curt. *Bot. Mag. t.* 243. Wight, *Ill. Ind. bot.* 1. *t.* 11.—β. floribus pallide luteolis, stylo brevi. A. *ochroleuca,* Sweet, *Brit. flow. Gard.*—Ic. Sweet, *l. c. t.* 242. Lindl. *Bot. Reg. t.* 1343. HAB. In arvis *Gossypio* satis et ad apicem usque Montis *Verdo* ins. *S. Nicolai,* necnon in sinu *Tarrafal* sive *Tamaricum* ins. *S. Antonii* (*Forbes,* n. 27, 11, et 7, April, 1822). In arena maris ins. *S. Jacobi* (*J. Dalton Hooker,* n. 135. November 1839). In maritimis ins. *S. Antonii* copiosa, (*Th. Vogel,* n. 43. Junio 1841.)

5. Papaver *Rhœas,* Linn. *Sp. Pl. p.* 726.—Ic. *Engl. bot. t.* 645.
HAB. In herb. ins. Cap. Vir. (*Mus. reg. Paris.*)

6. Papaver, *sp. nov.* ? P. orientali *affin.*
HAB. In herb. ins. Cap. Vir. (*Mus. reg. Par.*)

This specimen of Poppy is in a very imperfect state, so that we are unable to describe or refer it to any known species. The flower-stalks, somewhat hispid with erect appressed hairs, arise at once from the root, and are naked and monanthous. What remains of the leaves shows them to have been very hispid and like those of P. *Rhœas* on a large scale. The summit of the peduncle has the annular swelling large and papillated with the scars of the stamens. The capsule measures about six lines in length, and three in breadth : it is cylindraceo-turbinated, nearly glabrous, transversely torulose, and divided longitudinally into eight or nine ribs, which are themselves striated lengthways on the back.

IV. CRUCIFERÆ, *Juss.*

7. Nasturtium *officinale,* R. Br. *Hort. Kew. ed.* 2, 4. *p.* 110. Sisymbrium Nasturtium, *Linn. Sp. Pl. p.* 916.—Ic. *Engl. Bot. t.* 855. HAB. In ins. *S. Jacobi* (*Chr. Smith herb.!* fid. *J. D. Hooker* in litt.) In montibus ins. *S. Vincentii,* ad alt. 1500 ped. (*Th. Vogel,* n. 33. Junio 1841 : specimina macilenta.)

8. Sinapis *nigra,* Linn. *Sp. Pl. p.* 933. *Engl. Bot. t.* 969. HAB. In ins. *S. Jacobi* (*Chr. Smith!* in herb. Mus. Brit. fid. *J. D. Hooker* in litt.) In eadem ins. ad apicem montis cujusdam vallis *S. Dominici* (*J. Dalton Hooker,* n. 139. November 1839. spec. florida.) In eadem ins. (*C. Darwin,* spec. sine flore et fructu.)

9. Sinapidendron (Podocarpica, *Webb,*) *gracile,* Webb ; ramis elongatis lignosis gracilibus albidis, foliis ovato- vel spathulato-lanceolatis tenuibus glabris margine sparse vel obsolete grosse dentatis denticulis muticis vel spinuloso-apiculatis, spicis ad apicem ramorum elongatis gracillimis, pedicellis filiformibus, calycis foliolis apice pilosis, petalis longe unguiculatis, ovario gynophoro setaceo-filiformi duplo longiore, stylo brevi, stigmate capitato subdiscoideo, siliqua lineari, valvis tenuibus glabris subtrinerviis podocarpio gracili insidentibus, seminibus 1-seriatis pendulis testa (immersa) mucilaginosa, cotyledonibus incumbentibus conduplicatis.—Ic. (TAB. I.) Hook. *Ic. Pl. t.* 751. HAB. In vallibus ins. *S. Nicolai* (*Forbes,* n. 30, die 29 Martii, 1822, spec. florida et fructifera) et in herb. ins. Cap. Vir. (*Mus. reg. Par.!*) TAB. I. *Fig.* 1. flower ; *f.* 2. petal ; *f.* 3. ovarium ; *f.* 4. siliqua;

H 2

f. 5. seed; *f*. 6. embryo; *f*. 7. transverse section of the same ;—all *magnified*.

10. Sinapidendron *Vogelii*, Webb; ramis crassis nodosis, foliis rotundatis ovatisque crassis subtus nervosis pilis brevibus strigoso-hirtis margine dentatis vel demum crenato-dentatis, basi integris cuneatis petiolatis, spicis basi hirsutis, calyce subsaccato strigoso, petalis amplis flavis aurantiaco-maculatis ? in unguem attenuatis, ovario lato gynophoro tenui triplo longiore, stigmate capitato subsessili, siliqua . . . .—Ic. (Tab. II.) Hook. *Ic. Plant. t.* 752.

Hab. In ins. *S. Vincentii*, (*Vogel*, n. 32. Junio 1841, spec. florida.) Scientiæ martyris manibus hanc plantam dicatam voluimus.

The region to which the genus *Sinapidendron* belongs we have elsewhere called *Macaronesian*. The two *Sinapidendrons* of the Cape de Verd islands differ from the Madeira and Canarian species in having a long slender support to the ovary, and fruit analogous to that of the *Capparideæ*. No other characters, however, of any value, present themselves, so as to authorize their separation as a genus.

Tab. II. *Fig.* 1. flower; *f*. 2. petal; *f*. 3. stamen; *f*. 4. siliqua : *magnified*.

11. Koniga *intermedia*, Webb; Lobularia intermedia, *ejusd. Phyt. Can.* 1. *p.* 92.—Ic. Hook. *Lond. Journ. of Bot. v.* 5. *t.* 6. ubi in errorem inscripti *Koniga Brunonis*.

Hab. In declivibus aridis ins. *S. Nicolai* (*Forbes*, die 27 Martii, spec. florida et fruct.!) In montosis ins. *S. Vincentii* (*Th. Vogel*, n. 70. Juni 1841, spec. fructifera.) Hujus plantæ in herb. ins. Cap. Vir. (*Mus. reg. Par.*) specimina extant procera : an sp. diversa ?

The specimens from the Island of St. Nicholas do not appear essentially distinct from what we formerly gathered in Teneriffe. Those from St. Vincent are very dwarfish, and have small regularly spathulate and very hairy leaves ; differences which may perhaps be caused by the season in

which they were collected. It occurs in the Cape de Verd
herbarium of the Paris Museum, with much broader leaves
and larger flowering stems; but I cannot venture to separate
these specimens from the plant described in the " Phyto-
graphia Canariensis;" for, in a genus where the species vary
so exceedingly, and at the same time are so alike, it is only
by the observation of the living plant under cultivation that
such doubts can be effectually removed.

V. RESEDACEÆ, DC.

12. Caylusea *canescens*, St. Hil. 2ème mém. *sur les Resedacées*
(Montpell. 1837) *p.* 29. Reseda canescens, *Linn. Syst.
Veg. ed.* 12. (1767) *p.* 33, *non ejusd. Sp. Pl. ed.* 1. (1753)
p. 448. *nec ed.* 2. (1764) *p.* 644. *Vahl, Symb.* 2. *p.* 52. *Willd.
Sp. Pl.* 2. *p.* 817. *excl. patria et præter Forsk. syn. omnibus.
Lamrk. encycl.* 6. *p.* 158. *excl. patria* Salmantica *et syn.
omnibus præter* Vahl *et* Forsk.! Reseda Mediterranea, *Linn.
Mant. p.* 564 *ex herb.!* Reseda hexagyna, *Forsk. Pl. Æg.
p.* 92. Reseda podocarpos, *Viv. Pl. Æg. dec. p.* 7.—Ic.
Viv. l. c. t. 2. *f.* 3.

HAB. In aridis ins. *S. Jacobi*, (*J. Dalton Hooker*, n. 165.
November 1839, sp. florida et fruct.)

The above composes the very intricate synonymy of this
curious plant. The name *canescens* was originally given by
Linnæus to a species of *Reseda*, mentioned by Clusius,
belonging to the actual genus *Astrocarpus*. This species has
been admirably elucidated by M. Gay in the "Archives de
la Flore de la France et d'Allemagne" of Schultz, 1842,
p. 35, and named A. *Clusii*, which includes both the original
Reseda canescens, L., and the R. *purpurascens*, ejusd. A few
years afterwards, in the "Systema Naturæ," Linnæus, probably from forgetfulness, without advertising his readers or
informing them from whom he had received it, substituted
for his original *R. canescens*, and under the same name, a
totally different plant, sent to him probably from Palestine
or Egypt. This plant was our present species. Hence has
arisen the greatest confusion, most of the authors who

followed him confounding the two indiscriminately. Vahl
was the first to doubt whether the *R. canescens* of the Syst.
Pl., which he recognized as the *R. hexagyna* of Forskåhl, was
not distinct from the plant called by the same name in the
two editions of the Species Plantarum. M. de Tristan
(Mém. du Mus. 18. p. 395) pointed out the peculiar nature
of the fruit of this species; and M. de St. Hilaire, having still
further investigated its structure, raised it to the rank of a
genus, without however unravelling its synonyms.

The *R. Mediterranea* of Linnæus, Mant., another very
doubtful species, must, I consider, likewise merge in this.
The specimen preserved under that name in the Linnean
herbarium, I found, on examination, to be certainly a *Caylusea*, and I believe a cultivated specimen of *C. canescens*,
St. Hilaire. M. Gay, however, for whom I inspected it,
and who has since seen it himself, thinks that it may possibly be a fragment of *Caylusea Abyssinica* (*R. Abyssinica*,
Fresen. Museum Senck. 2. p. 106) distributed by the
Esslingen un. itin. in the Abyssinian collection of Schimper
under the No. 103.

VI. CAPPARIDEÆ.

13. Gynandropsis *triphylla*, DC. *Prodr.* 1. 238. Cleome
triphylla, *Linn. Sp. Pl. p.* 498.—Ic. *Herm. Lugd. p.* 565.
Descourtilz, *Fl. des Ant. t.* 44.

HAB. Circa *Porto Praya* ins. *S. Jacobi*, (*Hooker fil.* n. 196.
November, 1839). Planta est annua circa tropicos sparsa
quæ Fortunatas non attingit.

VII. CISTINEÆ.

14. Helianthemum *Gorgoneum*, Webb; caule fruticoso, ramis
pilis floccosis asperatis albidis, foliis latis ovatis lanceolato-ovatisque acutis pannosis subenerviis cinereopubescentibus petiolatis, stipulis lanceolato-linearibus inconspicuis caducissimis petiolo duplo triplove brevioribus,
calycibus floccoso-tomentosis, foliolis latis rhomboideoovatis subacutis demum obtusissimis crassis coriaceis,

nervis obsoletis 3-5 costato-nervosis, petalis aurantiacis
basi purpurascentibus, staminibus ovarium globosum villosum vix excedentibus, filamentis crassis purpureis, stylo
elongato, capsula rotunda pubescente.
HAB. In herb. ins. Cap. Vir. (*Mus. reg. Par.*)
Affine est *H. Canariensi* sed omnibus partibus major fortior-
que, stipulæ crassiores, breviores, foliola calycina latiora
apice haud contorsa, crassa, aurantiaca, oculo purpureo
nec luteo, stamina crassiora ovario vix longiora nec pistillum integrum subæquantia, capsulæ valvis late ovatis
nec ellipticis. Cum tot sint et tales differentiæ stirpes,
ut probabile est, diversas, confundere voluimus. Generis
zonam temperatam incolentis species est ultima versus
æquatorem protensa.

VIII. POLYGALEÆ.

15. Polygala *erioptera*, DC. *Prodr*. 1. *p*. 326. Guill. et Perott.
Flor. Sen. Tent. p. 38.—Io. Deless. *Ic. select. t.* 15.
HAB. In insula *S. Antonii* (*Vogel,* n. 50. Junio, 1841) in
ins. *S. Vincentii* locis saxosis et arenosis (*Vogel,* n. 58,
Junio, 1841) in planitie circa *Porto Praya* ins. *S. Jacobi*
(*J. D. Hooker,* n. 134. November, 1841).
An huc referenda *P. obtusata,* Brunner, circa Porto Praya
observata ex speciminibus quæ sub oculis habemus in
eodem loco a cl. Hookero fil. lectis? Plantæ sunt ambæ
Senegalenses atque Ægyptiaco-arabicæ nondum in Fortu-
natis lectæ.
16. Polygala *micrantha,* Guill. et Perr.! *Fl. Sen. Tent. p.* 39.
HAB. In herb. ins. Cap. Vir. (*Mus. reg. Par.*) specimen
unicum cujus flores aliquantulum majores quam in planta
Senegalensi, capsula ovata, semina tota villosa elongato-
olivæformia.

IX. FRANKENIACEÆ.

17. Frankenia *ericifolia,* Chr. Smith in *Buch, Beschr. der
Can. Ins. p.* 154. DC. *Prodr.* 1. *p.* 350. *Phyt. Can.* 1.

p. 132. Brunn. *Ergebn. p.* 73. *excl. sy. Desf.* —Ic. *Phyt. Can. t.* 45 *et t.* 17.

HAB. Ad salinas ins. *Sal (Forbes,* n. 1. die 26 Martii, 1822) in clivis maritimis arenosis ins. *S. Vincentii (Vogel,* n. 8). In herb. ins. Cap. Vir. *(Mus. reg. Par.)*

X. CARYOPHYLLEÆ.

18. Mollugo *bellidifolia,* Ser. in DC. *Prodr.* 1. *p.* 391. Pharnaceum spathulatum, *Swartz, Fl. Ind. occ.* 1. *p.* 568. Pharnaceum bellidifolium, *Poir. Encycl.* 5. *p.* 262.—Ic. Plum. *Pl. Amer. t.* 21. *f.* 1.

HAB. Occurrit hæc planta in apricis aridis circa *Porto Praya* ins. *S. Jacobi (J. D. Hooker,* n. 163. November, 1839). Planta est circa tropicos sparsa quæ et in Nubia obvia, *(Kotschy,* n. 119. die 3 Octobr. 1839).

19. Polycarpia *nivea,* Nob. Achyranthes nivea, *Hort. Kew. ed.* 1. 1. *p.* 286. *ed.* 2. 2. *p.* 57. Polycarpia candida, *Webb, Phyt. Can.* 1. *p.* 138. Helichrysum? an Phagnalon? *Brunn. Ergebn. p.* 77. Polycarpia glauca, *Chr. Smith, Tuck. p.* 250. *(Hook. fil.* in litt.) Polycarpia candidissima, *Bert. Miscell.* 111.—Ic. *Phyt. Can. t.* 21. Bert. *l. c. t.* 1. *f.* 1.

HAB. Valde vegeta crescit in arenis ins. *S. Antonii (Th. Vogel,* n. 2) occurrit quoque ad dimidium Monte *Verede,* ins. *S. Vincentii* sed minor *(Th. Vogel,* n. 54, Junio, 1841.) In ins. *Sal (Brunner* in herb. nostro).

The Cape de Verd specimens of this plant are much stronger than those of the Canaries, and the flowers are somewhat larger; but during a careful examination as well of the flowers as of the fruit, no really distinctive characters whatever presented themselves. The specimen of *Illecebrum gnaphalodes,* sent to Desfontaines by Schousboe himself, is in a different state of vegetation, and offering besides some distinction in the forms of the floral teguments, it requires to be seen in other stages before its real position can be decided.

20. Polycarpia *Gayi,* Webb.

a. helichrysoides; fruticosa, ramis fuscis nodosissimis, inter-

mediis brevibus, ramulis tomentosis albis, foliis oppositis
vel verticillato-aggregatis rotundato- vel ovato-spathu-
latis subobtusis crassis tomento brevi molli albo undique tomentosis subsericeis basi petiolatis, stipulis liberis
minutissimis lanceolatis acutis, floribus ovato-cylindra-
ceis, laciniis calycinis subtomentosis late scariosis margine
ciliato-laceris, petalis lanceolatis acutis erosis, staminibus
cum petalis et urceoli parapetaloidei lobis crenulatis alternantibus, ovario longius stipitato, stylo brevi, stigmate
2-lobo.

HAB. In ins. *Sal* petrosis *Brunner* (n. 172). In herb. ins.
Cap. Vir. (*Mus. reg. Par.*)

β. *halimoides;* suffrutex, ramis elongatis cortice papyraceo
fusco lacero tectis, junioribus sordide cinereo-pubescentibus vix nodosis, internodiis distantibus, foliis spathulato-
lanceolatis oppositis vel verticillato-aggregatis obsolete
undulatis tomento brevissimo albido argentatis basi in
petiolum brevem crassum attenuatis liberis, stipulis liberis
minimis late scariosis caducis, cymis confertis, floribus
cylindraceis, laciniis calycinis late scariosis, petalis oblongo-
lanceolatis obtusis, staminibus cum petalis et urceoli
parapetaloidei lobis emarginatis alternantibus, ovario depresso ovato-3-angulari breviter stipulato, stylo brevi,
stigmate orbiculari capitato.

HAB. In herb. ins. Cap. Vir. (*Mus. reg. Par.*)

γ. *lycioides;* suffrutex, ramis rectis duris divergentibus,
ramulis pubescentibus, internodiis brevibus, foliis oppositis vel verticillato-aggregatis lanceolatis in petiolum
brevem attenuatis pube stellata puberulis mox glabrescentibus ciliatis, stipulis minimis scariosis, cymulis crassis
depressis pubescentibus, laciniis calycinis lanceolatis exterioribus papposo-scariosis, petalis ovato-lanceolatis acutiusculis, staminibus cum petalis et urceoli parapetaloidei
lobis rotundatis alternantibus, stylo brevi, stigmate amplo
capitato.

HAB. In herb. ins. Cap. Vir. (*Mus. reg. Par.*)

We had originally considered as species these three very

distinct forms; but our indefatigable friend, M. J. Gay, who
is preparing with his well-known accuracy a most interesting
monograph of the group, informs us, that after a laborious
and minute examination of the flowers, which occupied eight
days, he has been led to reduce them to simple variaties
of a common species, which, though intimately allied to
Polycarpia nivea, nob., differs essentially from that plant.

21. Paronychia *illecebroides*, Webb; caule prostrato ramosissimo, ramis filiformibus pubescentibus, stipulis foliis
lineari-lanceolatis vel linearibus acutis puberulis dimidio
brevioribus, bracteis flore brevioribus, calycibus brevissime mucronulatis, mucrone recto vel inflexo.—Herniaria
illecebroides, *Chr. Smith, in Tuck. voy. p. 250 ! ex herb.
Mus. Brit. (J. D. Hooker,* in litt.) — Ic. (Tab. VII.)
Hook. *Ic. Plant. t.* 756.

Hab. Communis est in insulis *Gorgoneis.* In sinu *Tarrafal*
sive *Tamaricum* ins. *S. Antonii (Forbes,* n. 24. die 2 Aprilis,
1822, spec. florida). In ins. *S. Jacobi,* vulgaris (*J. Dalton
Hooker,* n. 112. November, 1839, spec. flor. et fruct.) In
ins. *S. Vincentii* ab alt. 500 ped. usque ad cacumen *Montis
Verede (Th. Vogel,* n. 25. Junio, 1841, spec. deusta.)

Radix lignosa ; ramis filiformibus prostratis ; stipulis hyalinis,
oblongis, ciliatis, apice setaceis, folio duplo triplove brevioribus. *Folia* oblongo-linearia vel lineari-lanceolata,
angusta, brevissime petiolata, acuta, pubescentia. *Flores*
omnes axillares ; bracteis hyalinis, ovatis, calyce brevioribus. *Calyx* cylindraceus, hirsutus, foliolis oblongis angustis
costato-trinerviis brevissime mucronulatis, mucrone crassiusculo inflexo, margine vix scariosis. *Ovarium* globu-
losum, hirtum. *Capsula* exigua, ovato-rotundata. *Semen*
glabrum. *Embryo* vermicularis, hemicyclica.

Affinis est hæc species P. *polygonifolia,* DC., a qua plurimis
notis differt, stipularum scilicet et bractearum quoad folia
et flores longitudine, calycis forma et mucrone. A
P. *argentea,* Lamck. cui flores ut plurimum capitati longius
quoque recedit.

Tab. VII. *Fig.* 1. flower, included within the bracteæ ;

f. 2. bractea; *f.* 3. flower removed; *f.* 4. same cut open ; *f.* 5 . ovarium ; *f.* 6. seed ; *f.* 7. embryo ;—all *magnified.*

XI. SILENEÆ.

22. Silene *Gallica,* Linn. *Sp. Pl. p.* 595. Silene Anglica *et* Lusitanica, *ejusd. ibid.*—Ic. *Engl. bot. t.* 1178. HAB. Specimina valde vegeta hujus plantæ extant in herb. ins. Cap. Vir. (*Mus. reg. Par.*)

XII. MALVACEÆ, *Juss.*

23. Gossypium *nigrum,* Ham. *var. punctatum.* Gossypium punctatum, *Guill. et Perr.! Fl. Sen. Tent. p.* 62. *Brunner Ergebn. p.* 75.—Ic. nulla ex toto varietatem nostram exprimit: folia habet fere *G. micranthi,* Cavan. *Diss.* 6. *t.* 193. sed valde tomentosa, involucrum autem et florem *G. Peruviani, ej. ib. t.* 168. HAB. In ins. *S. Jacobi* (*Chr. Smith!* in herb. Mus. Brit., *J. D. Hooker,* in litt.) Ad sinum *Tarrafal* sive *Tamaricum* ins. *S. Antonii* (*Forbes,* n. 12. die 2 April, 1822, spec. flor. et fruct.) In eadem ins. *S. Antonii* et in *Monte Verede* ins. *S. Vincentii* (*Th. Vogel,* n. 55 et 5. Junio, 1841, spec. flor. et fruct.)

24. Malva *spicata,* Linn. *Sp. Pl. p.* 967. Cav. *Diss.* 2. *p.* 80. Malva ovata, *ejusd. ibid. p.* 81.—Ic. Cav. *l. c. t.* 20. *f.* 2 et 4. In ins. *S. Jacobi* (*Chr. Smith!* in herb. Mus. Brit. *J. Dalton Hooker* in litt.) In valle *S. Dominici* et in planitie oppidi ejusd. ins. (*J. D. Hooker,* n. 185 et 187. November, 1839, spec. flor. et fruct.)

25. Sida *spinosa,* Linn. *Sp. Pl. p.* 960. DC. *Prodr.* 1. *p.* 460. —Ic. Pluk. *Phytogr. t.* 9. *f.* 6.

β. Foliis ovato-subrotundis.—Sida alba, *Linn. l. c. DC. l. c. Guill. et Perr. Fl. Sen. Tent. p.* 74. *Wight et Arn. Fl. Pen. Ind. or. p.* 58. Sida repens, *Chr. Smith, in Tuck. voy. p.* 250! (Herb. Mus. Brit. fid. *J. D. Hooker,* in litt.) HAB. In ins. *S. Jacobi* (*Chr. Smith*) ibid. (*J. D. Hooker,* n. 189. et β. n. 194. November, 1839, spec. flor. et fruct.)

26. Sida *stipulata*, Cav. *Diss.* 1. *p.* 22. DC. *Prodr.* 1. *p.* 460.
—Ic. Cav. *l. c. t.* 3. *f.* 10.

HAB. In arvis *Gossypio* consitis ad sinum *Tarrafal* ins.
S. *Antonii* (*Forbes*, n. 11. die 2 Aprilis, 1822, spec. fl. et
fruct.) In valle S. *Dominici*, ins. S. *Jacobi*, (*J. D. Hooker*,
n. 190. November, 1839, spec. fl. et fruct.)

27. Sida *rhombifolia*, Linn. *Sp. Pl. p.* 961. Cav. *Diss.* 1. *p.* 23.
DC. *Prodr.* 1. *p.* 462. Webb, *Phyt. Can.* 1. *p.* 36. Sida
canescens, *Cav. l. c. p.* 23. Sida Canariensis, *Willd. Sp.
Pl.* 3. *p.* 755. *DC. Prodr.* 1. *p.* 462.—Ic. Cav. *l. c. t.* 3.
f. 12 *et t.* 8. *f.* 3.

HAB. In valle S. *Dominici* ins. S. *Jacobi* (*J. D. Hooker*,
n. 186. November, 1839, spec. flor. et fruct.) In ins.
S. *Antonii* (*Th. Vogel*, n. 37. die 17 Junii, 1841, spec. flor.
et fruct.)

28. Sida *cordifolia*, Linn. *Sp. Pl. p.* 961. DC. *Prodr.* 1.
p. 464. Wight et Arn. *Fl. Pen. Ind. or.* 1. *p.* 58. Sida
herbacea, *Cav. Diss.* 1. *p.* 19. *DC. Prodr.* 1. *p.* 463. Sida
rotundifolia, *Cav. ibid. p.* 20. *DC. ibid. p.* 464. Sida althæifolia, *Swartz, Prodr. p.* 101, *atque auct. omnium.* Sida
Africana, *Pal. de Beauv. Fl. d'Ow.* 2. *p.* 87.—Ic. Cav. *l. c.
t.* 3. *f.* 2. *t.* 13. *f.* 1. *t.* 194. *f.* 2. *Pal. de Beauv. l. c. t.* 116.

HAB. In ins. S. *Jacobi* vulgaris, (*J. Dalton Hooker*, n. 184,
197 et 198, November 1839, sp. flor. et fruct.) In eadem
ins. (*Darwin*) et cop. in herb. ins. Cap. Virid. (*Mus. reg.
Par.*)

29. Sida *urens*, Linn. *Sp. Pl. p.* 963. DC. *Prodr.* 1. *p.* 465.
Guill. et Perr. *Fl. Sen. Tent. p.* 73. Sida micans, *Chr. Smith,
in Tuck. Journ.* 250! (*J. D. Hook.* in litt.)—Ic. Cav. *Diss.*
1. *t.* 2. *f.* 7.

HAB. In ins. S. *Jacobi*, (*J. Dalton Hooker*, n. 188. November
1839, sp. fl. et fruct.) et in herb. ins. Cap. Virid. (*Mus.
reg. Par.*)

30. Abutilon *periplocæfolium*, G. Don, Wight et Arn. *Pr. Fl.
Pen. Ind. Or.* Sida periplocæfolia, *Sp. Pl. p.* 962. *DC.
Prodr.* 1. *p.* 467.—Ic. Pluk. *Phyt. t.* 74. *f.* 7. Dill. *Hort.
Elth. t.* 3. *f.* 2. Cav. *Diss.* 1. *t.* 5. *f.* 2. mala.

HAB. In ins. *S. Jacobi (J. Dalton Hooker,)* n. 192. November
1839.) In eadem ins. *(Darwin)* et in herb. ins. Cap. Virid.
(Mus. reg. Par.)
31. Abutilon *glaucum,* Webb ; Sida Asiatica, *Cav. Diss.* 1.
p. 31. *quoad plantam* Senegalensem *et t.* 7. *f.* 2. *non ejusd.*
Diss. 5. *t.* 128. *f.* 1. Guill. et Perr. ! *Fl Sen. Tent. p.* 67. *non*
Linn. Sida glauca, *Cav. Ic.* 1. p. 8. t. 11. Sida mutica,
Del. ! Ill. Fl. Eg. Voy. de Caill. p. 60. n. 45. *Brunn.*
Ergebn. p. 113. Sida polycarpa, *Chr. Smith, l. c. p.* 250.
(J. D. Hooker, in litt.)
HAB. In ins. *Sal. (Brunner !)* In ins. *S. Jacobi* rarior, *(J.*
Dalton Hooker, n. 196. November 1839, sp. fl. et fruct.)
Ad dimidium montis *Verede* ins. *S. Vincentii,* et in ins.
S. Antonii, (Th. Vogel, n. 85 et 30, spec. fl. et fr.)

Brunner was perfectly right in considering this to be the
true *Sida mutica* of Delile. A specimen from Senegal, identical with the *S. Asiatica* of the Flor. Sen. Tent. and with
the Cape de Verd plant, is so named in the herbarium of
Desfontaines by that distinguished botanist himself. It
is likewise the plant sent by Despréaux from the Canaries ;
and which we named *Abutilon Indicum* in the Ann. des Sc.
nat. (2ème sér. 13. p. 132), the specimens being without
flower or fruit; because the authors of the Pr. Flor. Pen. Ind.
Or. seem to consider the *A. Indicum* hardly distinct from the
A. Asiaticum, of which Dr. Wight has given a good figure in
his Icones. The carpels in the Indian plants are however acute,
whereas in ours they are rounded at the apex. Hence the
excellent name of Delile, which, however, must give place
to the earlier appellation of his predecessor, Cavanilles. The
confusion which has arisen in the species is owing entirely
to the latter author, who originally confounded Adanson's
specimen from Senegal, in the Jussiæan herbarium, with the
Indian plant of Plukenet and Linnæus ; and thus De Candolle
and the authors of the Fl. Sen. Tent. were led astray.
Besides the form of the carpels in the Indian plant, which is
distinctly marked in the figure of Plukenet and by Cavanilles
himself, Diss. 5. t. 128. f. 1. *e* et *f,* the seeds of this species

are smooth, with a few hairs about the hilum; whereas in the
Cape de Verd and Egyptian plant they are entirely covered
with hair. After having merged the present plant in his *Sida
Asiatica*, Cavanilles cultivated it from Senegalese seeds, and
reproduced it under the name of *Sida glauca*, thereby adding
to the confusion; though this name must necessarily be
adopted by us. It is not only a question of nomenclature,
but important geographically; as we thus obtain a purely
African species in the place of an Asiatic plant, reappearing
somewhat unaccountably in Egypt, the Canaries, the Cape
de Verds, and Senegal.

32. Adansonia *digitata*, Linn. *Sp. Pl. ed.* 1. *app. p.* 1190.
Adans. *Mém. de l'Acad. Roy. des Sc.* 1761. *p.* 218. Chr.
Smith *in Tuck. voy. p.* 249! (fid. *J. D. Hooker*, in litt.)
Guill. et Perr. *Fl. Sen. Tent. p.* 76.—Ic. Prosp. *Alp. Pl.
Æg. t.* 67. Adans. *l. c. t.* 6 *et* 7. Cav. *Diss.* 5. *t.* 157.
Gaertn. 2. *t.* 135. Lamck. *Ill. t.* 588. Juss. *Fl. des Ant.* 3.
t. 33 *et* 34.

HAB. Prope *Portum Praya*, ins. *S. Jacobi*, arbor unica. (*J.
Dalton Hooker*, n. 141. November 1839, spec. floridum.)

XIII. BYTTNERIACEÆ, *R. Br.*

33. Waltheria *Indica*, Linn. *Sp. Pl. p.* 941. Waltheria Americana, *ejusd. ibid.* Waltheria microphylla, *Cav. Diss.* 6.
p. 317. Waltheria elliptica, *Cav. l. c. Phyt. Can.* 1. *p.* 41.
—Ic. Cav. *l. c. t.* 170. *f.* 2.

HAB. In ins. *S. Jacobi* (*C. Darwin.*)

The figure of Cavanilles, cited above, comes nearest the
specimen collected by Mr. Darwin. For the numerous
forms and synonyms of this polymorphous plant see Wight
and Arn. Prodr. Fl. Pen. Ind. Or. p. 67; to which may be
added *Waltheria arborescens*, Cav. *l. c. t.* 170.*f.* 1.

XIV. TILIACEÆ, *Juss.*

34. Melhania *Leprieurii*, Webb; Brotera Leprieurii, *Guillem.
et Perr. Fl. Sen. Tent. p.* 85.—Ic. (TABS. IV. V.) Hook.
Ic. Plant. *t.* 753 *et* 755.

HAB. In planitie aprica circa *Portum Praya*, ins. *S. Jacobi*, (*J. Dalton Hooker*, n. 195. Novembre 1835, sp. florida et fructifera.) In eadem ins. (*Darwin*, n. 301); vidi quoque specimina plura in herb. ins. Cap. Vir. (*Mus. reg. Par.*) OBS. The authors of the Fl. Pen. Ind. Or., in their remarks on *Melhania incana*, Heyne, observe that it exceedingly resembles the figure of *Brotera ovata*, Cav., said to have distinct styles. This, indeed, is the only character which separates *Melhania* and *Brotera*. A careful examination, however, of the species which constitute this latter genus, has proved to me that no such character in reality exists, and that it must therefore merge in *Melhania*. In the *Brotera ovata*, formerly cultivated by Desfontaines in the Paris Garden, probably from seeds sent by Cavanilles himself, and preserved in his herbarium, as well as in the *B. Leprieurii* and *B. bracteosa* of the Fl. Sen. Tent. in the Delesserian herbarium, I have found a very distinct style, not so conspicuous as in the species formerly placed under *Dombeya*, equally so, however, with those of the group of true *Melhanias*, to which *M. ovata* and *M. Leprieurii* belong, *M. bracteosa* will form another group with *M. Kotschyi*, Koch. The error of Cavanilles and of the authors of the Fl. Sen. Tent. arose from the shortness of the style, and from its being frequently masked at a certain age by the ascendant hairs of the ovary; so that its divisions seem to be separate to their base, and to be seated directly on that body, though they are really connected below, and form a single column, as visible as in many of the neighbouring genera. The species of this genus are still exceedingly puzzling; and it is not impossible but that the *M. Leprieurii* may hardly be distinct from the *M. velutina*, Forsk., or the *M. incana*, Heyne: it seems, too, to bear a remarkable resemblance to *Melhania ovata*. But I have not sufficient data by which to determine their identity or their difference. From a herbaceous plant, as it appears at first, *M. Leprieurii* becomes in time a low woody shrub.

112   SPICILEGIA GORGONEA.

Tab. IV. *Fig.* 1. unexpanded flower; *f.* 2. ditto with the
sepals expanded.
Tab. V. *Fig.* 1. petal; *f.* 2. portion of staminal column;
f. 3. ovarium; *f.* 4. transverse section of ditto; *f.* 5. ovule;
f. 6. ripe fruit; *f.* 7. transverse, and *f.* 8. longitudinal
section of ditto; *f.* 9. fruit burst open; *f.* 10. seed; *f.* 11.
vertical section of ditto; *f.* 12 and 13. embryo: *all more
or less magnified.*
35. Corchorus *trilocularis,* Linn. *Mant. p.* 77. Chr. Smith,
Journ. in Tuck. voy. p. 251. DC. *Prodr.* 1. *p.* 504. Guill.
et Perr. *Fl. Sen. Tent. p.* 88. Wight et Arn. *Prodr. Fl. Pen.
Ind. p.* 72.—Ic. Jacq. *Hort. Vind. t.* 173.
Hab. In valle umbrosa ad orientem *Portus Prayæ,* ins. *S.
Jacobi,* (*J. Dalton Hooker,* n. 168. November 1839, sp.
fl. et fruct.) Chr. Smith places this species in his list
amongst the " Plantæ boreali-Africanæ quæ simul Cana-
rienses;" but it has never, that I know of, been found in
those islands.
36. Corchorus *olitorius,* Linn. *Sp. Pl. p.* 746. DC. *Prodr.* 1.
p. 504. Guill. et Perr. *Fl. Sen. Tent. p.* 87. Wight et Arn.
Fl. Pen. Ind. Prodr. p. 73.—Ic. Gaertn. *t.* 64. Lamck. *Ill.
t.* 478.
Hab. In valle *S. Dominici,* ins. *S. Jacobi,* rarius, (*J. Dalton
Hooker,* n. 156. Nov. 1839, sp. flor. et fruct.)
37. Corchorus *tridens,* Linn. *Mant. p.* 566. DC. *Prodr.* 1.
p. 505. C. Burmanni, *ejusd. ibid.* C. trilocularis, *Burm. Fl.
Ind. p.* 123. Guill. et Perr. *Fl. Sen. Tent. p.* 89.—Ic. Pluk.
Phyt. t. 127. *f.* 4. Burm. *l. c. t.* 37. *f.* 2.
Hab. In humidiusculis ins. *S. Antonii,* (*Th. Vogel,* n. 42.
Junio 1841.
38. Corchorus *Antichorus,* Rœuschel, *Nomencl. Bot. ed.* 3.
(1797) *p.* 158, nomen, Antichorus depressus, *Linn. Mant.
p.* 64. *et Regn. Veg. ed.* 13. (cur. Murray) *p.* 297. DC.
Prodr. 1. *p.* 504. Brunn. *Ergebn. p.* 161. Jussiæa edulis,
Forsk. Fl. Æg. Ar. p. 210. Caricteria, *Scop. Introd. p.* 255.
Corchorus sect. Antichorus, *Endlich. Gen. p.* 1008. Cor-

chorus fruticulosus, *Vis. Pl. Æg. et Nub. p.* 21.—Ic. Linn.
f. Dec. fasc. 3. t. 2. Vis. *l. c. t.* 3. *f.* 2.
HAB. In planitie *Porto Prayensi,* ins. *S. Jacobi,* vulgaris, (*J. Dalton Hooker,* n. 166. Nov. 1839. In ipsis oppidi plateis (*Brunner* l. c.) In ins. *S. Vincentii* et *S. Nicolai,* (*Th. Vogel.*)

This species, first found by Forskåhl in Arabia, and described by Linnæus in the Mantissa, though it reappears in these islands, belongs essentially to Nubia and the Arabian peninsula. Aucher found it at Mascato (exsicc. n. 4286), Schimper near Djedda (exsicc. n. 813), Brocchi and Kotschy (it. Nub. n. 342) near Chartum, at the confluence of the White and Blue Rivers, which unite there in the latitude of the Cape de Verd Islands to form the Nile, and the latter likewise at Tekele, on the borders of Cordofan. It cannot certainly be separated generically from *Corchorus.* Professor Visiani justly describes it as frequently pentamerous; and probably on that account he did not recognize it as the *Antichorus* of Linnæus, of which it scarcely forms a separate division, but should be placed in the section *Coretoides* of De Candolle. Both Forskåhl and Brocchi say it is edible, like its congener, *C. olitorius,* L., a circumstance not mentioned by our voyagers.

39. Triumfetta *Lappula,* Linn. *Sp. pl. p.* 637. T. Plumieri, *Gaertn. p.* 137.—Ic. Plum. *ed. Burm. t.* 255. Descourt. *Fl. des Antilles,* 2. *t.* 101 *et* 102. Da Arr. *Fl. flum. t.* 5.
HAB. In valle *S. Dominici,* ins. *S. Jacobi,* (*J. Dalton Hooker,* n. 191, Nov. 1839.)

40. Triumfetta *pentandra,* Rich. *in Guill. et Perr. Fl. Sen. Tent. p.* 93.—Ic. Rich. *l. c. t.* 19.
HAB. In locis umbrosis vallis *S. Dominici,* ins. *S. Jacobi,* (*J. Dalton Hooker,* n. 177. Nov. 1839.)

The Cape de Verd specimens differ from the plant figured in the Flora of Senegambia in having the leaves entire and not 3-lobed. The reduction of the stamens to five, alternating with the petals, and that of the glands to the smallest proportion, with the almost entire suppression of

I

114   SPICILEGIA GORGONEA.

the urceolate appendix to the torus, occur equally in our plant.

41. Grewia *echinulata*, Del. *Cent. des plantes de Caill. p.* 82. *n.* 70. Grewia corylifolia, *Guill. et Perr. Fl. Sen. Tent. p.* 95.—Ic. Guill. et Perr. *l. c. t.* 20.

HAB. Arbuscula 15-pedalis ad apicem collium alt. 1000 ad 2000 ped. in ins. *S. Jacobi, Avellanæ* sylvulas mentiens (*J. Dalton Hooker*, n. 175, Nov. 1839), specimina flor. et fruct. In ead. ins. (*T. R. H. Thomson*) in herb. Chr. Smith, Mus. Brit. ex *cl. J. D. Hooker*, in litt.

Mr. Arnott (in Ann. des Sc. Nat. 2ème sér. Bot. 2. p. 236) considers this plant not specifically distinct from the *G. pilosa* of the Fl. Pen. Ind. Or. p. 79. et Hook. Comp. bot. Mag. t. 10.) It is safer, however, to keep the two apart, till their positive identity has been fully confirmed. The African plant is everywhere smaller, more particularly its flowers; its leaves are much more villous, the leaflets of the calyx are thinner and transparent, with nerves more conspicuous and divaricated.

XV. OLACINEÆ, *Mirb.*

42. Ximenia *Americana*, Linn. *Sp. Pl. p.* 497. DC. *Prodr.* 1. *p.* 533. X. multiflora, *Jacq. Amœn. p.* 106. Heymassoli spinosa, *Aubl. Guyan. p.* 324.—Ic. Lmck. *Ill. t.* 297. *f.* 1 et 2. Jacq. *l. c. t.* 297. *f.* 31. Aubl. *l. c. t.* 125.

HAB. In ins. *S. Jacobi,* (*Chr. Smith,* in herb. Mus. Brit. fid. *J. D. Hooker* in litt.)

XVI. SAPINDACEÆ, *Juss.*

43. Cardiospermum *Halicacabum*, Linn. *Sp. Pl. p.* 925. Cardiospermum hirsutum, *Chr. Smith in Tuck. Journ. p.* 249. (fid. *J. D. Hook.* in litt.—Ic. Rheed. *Mal. t.* 28. Rumph. *Amb. t.* 24. *f* 2. Lamck. *Ill. t.* 317. Camb. *Mém. du Mus.* 18. *t.* 1. *f.* A.

HAB. In ins. *S. Jacobi,* (*J. Dalton Hooker,* n. 160. spec. fl. et fruct.)

Melia Azederach is likewise in the collection (herb. *C.*

Darwin), but can scarcely perhaps be considered indigenous.

XVII. Oxalideæ, *DC.*

44. Oxalis *corniculata*, L. *Sp. Pl. p.* 624.—Ic. Jacq. *Ox. t.* 5. Hab. In rupibus Montis *Verede*, ins. *S. Vincentii,* (*Th. Vogel,* n. 35. Junio 1841. spec. pusillum.) In ins. *S. Jacobi,* (*Chr. Smith* in herb. Mus. Brit. ex Cl. *J. D. Hooker.*) In herb. ins. Cap. Vir. (*Mus. reg. Par.*)

XVIII. Zygophylleæ, *R. Br.*

45. Tribulus *terrestris*, Linn. *Sp. Pl. p.* 554. Guill. et Perr. *Fl. Sen. Tent. p.* 139.—Ic. Reich. *Ic. Fl. Germ. et Helv. t.* 161. *t.* 4821. Hab. In ins. *S. Jacobi,* (*J. Dalton Hooker,* n. 157. Nov. 1839, spec. florida et fructifera.)

46. Tribulus *cistoides*, Linn. *Sp. Pl. p.* 554. — Ic. Pluk. *Phyt. t.* 67. *f.* 4. Jacq. *Hort. Schœnb. t.* 103. Hab. In rupestribus sinu *Tarrafal* (*Forbes,* n. 20, die 2 Aprilis, 1822, sp. flor. cum fructu maturo.) In ins. *S. Vincentii* (*Forbes,* n. 4. sp. flor.) In planitie lapidosa ins. *S. Jacobi* (*J. D. Hooker,* n. 159. November, 1839, sp. fruct. cum floribus, et *Chr. Smith,* ex Cl. *J. D. Hooker.)* In ins. *S. Vincentii* ad alt. 500 ped. (*Th. Vogel,* n. 34. Junio, 1841.) In ins. *S. Antonii* (Id. n. 31. sp. flor.) This is a true *Tribulus,* and not a *Kallstrœmia,* and therefore perfectly distinct from the *K. cistoides,* Endl. (*Syn. Fl. Ind. Occ. in Ann. des Wien. Mus.* 1. *p.* 184.)

47. Fagonia *Cretica*, Linn. *Sp. Pl. p.* 553. Brunn. *Ergebn. p.* 69. In sinu *Tarrafal* ins. *S. Antonii* (*Forbes,* n. 23. die 2 Aprilis, 1822, sp. fl. et fruct.); in rupestribus ins. *S. Antonii* (*Th. Vogel,* n. 41. Junio, 1841, sp. fructifera); in ins. *Salis* (*Brunner,* l. c.)

48. Zygophyllum *Fontanesii,* Webb, *Phyt. Can.* 1. *p.* 17. Zygophyllum album, *Desf. Fl. Atl.*—Ic. *Phyt. Can. t.* 1. Hab. In arena maris ins. *S. Vincentii,* frutex ramosus diffusus (*Vogel,* n. 121. Junio, 1841, spec. fruct.)

116   SPICILEGIA GORGONEA.

49. Zygophyllum *simplex*, L. *Mant. p.* 68. Z. portulacoides, *Forsk. Fl. Æg. Ar. p.* 88. Z. stellulatum, *Chr. Smith! in Tuck. Journ. p.* 250 qui zonæ temperatæ per errorem civem dixit cum sit ex toto littoralis (*J. D. Hooker* in litt.) Fagonia prostrata, *Brunn.! Ergebn. p.* 69. — Ic. Forsk. *ic.* 12. In ins. *Salis* (*Forbes* sine n. *Brunner!) Circa Portum Praya* ins. *S. Jacobi* planta maritima. (*J. D. Hooker,* n. 179. November, 1839, sp. fl. et fruct.) In ins. *S. Antonii* (*Th. Vogel,* Junio, 1841, sp. unicum sole ustum.) In ins. *S. Jacobi* (*Darwin,* sp. fl. et fruct.)

XIX. RHAMNEÆ, *Juss.*

50. Zizyphus *Jujuba*, Lamck. *Encycl.* 3. *p.* 318. Rhamnus Jujuba, *L. Sp. Pl. p.* 282. Z. sororia, *Schult. Syst.* 5. *p.* 338. Z. insularis, *Chr. Smith in Tuck. Journ. p.* 250, spec. pessimum (fid. *J. D. Hooker* in litt.)—Ic. Rheed. *Mal.* 4. *t.* 41. Rumph. *Arab.* 3. *t.* 36. Pluk. *Phyt. t.* 312. *f.* 4.

The true Z. *Jujuba*, remarkable for its tawney down, and not the Z. *orthacantha,* DC. of Senegal, which Brunner says he saw in San Tiago (*Ergebn. p.* 127), of which the down is white in the specimens given me by Mr. Perrottet. A specimen, apparently of the latter, occurs without flower or fruit, in the herb of the Paris Museum.

XX. LEGUMINOSÆ, *Juss.*

51. Crotalaria *Senegalensis*, Bacl. in DC. *Prodr.* 2, *p.* 133. Guill. et Perr. *Fl. Sen. Tent. p.* 165, excl. syn. Del. In rupestribus ins. *S. Jacobi* (*J. D. Hooker,* n. 146 et 147, November, 1839, sp. flor. et fruct.) ibid. (*Chr. Smith,* in herb. Mus. Brit. excl. *J. D. Hooker.)* Our plant appears perfectly distinct from the C. *macilenta,* Del. *Cent. Pl. Afr. p.* 35. *t.* 3. *f.* 2, with which it has been associated by the authors of the *Fl. Sen. Tent.* The leaves, as well as the whole plant, are more pubescent and the leaflets more oblong and obovate and not elliptic. The flowers

are half and sometimes less than half the size of those of
the Nubian species, the calyx and its teeth shorter, the
standard much more ample, very hairy and not pubescent
only outside, nor entirely yellow, but marked with longitudinal streaks of red; it is about the same length as the wings
and not longer; the wings are narrow and oblong, not wide
and spathulate, the plicatures all placed much nearer the
apex; the keel is longer than the wings and standard,
instead of being about the length of the wings and shorter
than the standard; the fruit is oblong-oval and when ripe
the base of the style which forms the apex is placed nearly
in a straight line with the upper suture; whereas in the
Nubian plant the ripe fruit is nearly round, and the apex
forms an angle with the suture. Notwithstanding the great
similitude of these plants, presenting as they do so many
differences, it is more prudent to keep them apart. Our
comparison was made with the plant of Kotschy (Iter Nub.
n. 24.)

52. Crotalaria *microphylla,* Vahl., *Symb.* 1. *p.* 52. Benth.
Enum. Legum. in Hook. Lond. Journ. of Bot. 2. *p.* 573.
C. pumila, *Hochst. et Steud. Exsicc. Arab. Schimp.* 1837.
n. 778.

HAB. In rupibus maritimis ins. *S. Antonii* (*Forbes,* die
2 Aprilis, 1822, sp. unicum fructiferum.)

I had at first distinguished this plant specifically under
the name of *C. trigonelloides;* but after a careful examination
it seems scarcely possible to separate it from *C. microphylla,*
Vahl; though the branches which are stouter appear less
procumbent and the leaves are more approximated. The
upper leaflets in both, as in *C. humilis,* Eckl., frequently
become linear and elongated and sometimes simple. The
pubescence and the stipules are identical; the segments of the
calyx in our plant are however somewhat narrower and more
sharply pointed. Unfortunately, we have no perfect flowers;
but the keel is apparently of the same form: the fruit, ripe
in our specimen, though otherwise similar to the unripe fruit
of that of Schimper, is a little larger.

118   SPICILEGIA GORGONEA.

53. Lotus *Jacobæus,* Linn. *Sp. Pl. p.* 1091. DC. *Prodr.* 2,
p. 210.—Ic. Commel. *Hort. Amst.* 2. *t.* 83. Curt. *Bot.
Mag. t.* 79.

HAB. In zona temperata ins. *S. Jacobi (Chr. Smith,* l. c.
p. 250. In rupestribus maritimis ad sinum *Tarrafal* ins.
S. Antonii (Forbes, n. 18. die 2 April, 1822, spec. florida
valde hirta, foliis latioribus, carina et vexillo pallidis, alis
atropurpureis, specimina his similia sed graciliora in
eadem ins. legit *Th. Vogel).* In vallibus ins. *S. Nicolai,*
(Forbes, n. 18. die 29 Maii, 1822, sp. florida, alis et vexillo
atropurpureis, carina pallida.) In palmetis ins. *S. Jacobi*
(Forbes, n. 5. die 5 April, 1822, spec. florida et fructifera,
alis atropurpureis, carina et vexillo pallidis. Ad apicem
montis cujusdam acuti vallis *S. Dominici* ins. *S. Jacobi* ad
alt. 2000 ped. (*J. D. Hooker,* n. 153. November 1839, spec.
fructifera et florida, alabastris novellis undique prodituris,
gracilia foliosissima, foliis angustis, alis et vexillo atropurpureis, carina pallida.)

It appears that this plant grows from the sea-coast to an
elevation of 2000 feet, varying according to its station in
the breadth and hairiness of its leaves and in the colour of its
petals ; though these are never entirely yellow, like those of
Lotus Brunneri.

54. Lotus *melilotoides,* Webb ; caule frutescente, ramis elongatis diffusissimis molliter pilosis, foliis sessilibus foliolis
stipulisque elongatis lineari-lanceolatis vel linearibus basi
attenuatis pilosis viridibus, junioribus rufo-hirtis, bracteis
linearibus flores vix excedentibus, calycis turbinati hirti
dentibus lanceolatis apice setaceis, alis lanceolatis carinæ
longitudine vexillo acutiusculo vix brevioribus. Flores
videntur rosei ; affinis est *L. anthylloidi,* Vent., sed foliorum
forma et pubescentia differt.

HAB. In herb. ins. Cap. Vir. (*Mus. reg. Par.*)

55. Lotus *purpureus,* Webb ; caule frutescente, ramis diffusis,
foliis appresse pilosis virentibus, stipulis ovatis cum acumine, foliolis latis ovatis vel obovatis obcordatisque, calyce
urceolato distincte 2-labiato, dentibus lineari-lanceolatis

apice setaceis labii superioris longioribus, inferioris dente intermedio lateralibus duplo fere longiore, carina vexillo obtusiusculo subbreviore alis oblongis breviore.—Icon. Tab. VI. Hook. *Ic. Plant. t.* 757. Hab. In arvis et in *Euphorbiæ Tuckeranæ* sylvis ins. *S. Nicholai* (*Forbes*, die 30 Martii 1822, spec florida.) This species, though very distinct, is allied to *L. macranthus*, Lowe, (*L. Portosanctanus*, Nob. in Steud. Nomenclat.) of which the flowers are pale purple. Those of our plant have the keel and standard rose-coloured and the wings tipped with deep purple. Some of the leaflets are as much as 4 lines long by 2½ wide.

Tab. VI. *Fig.* 1. flower ; *f.* 2. vexillum ; *f.* 3. ala ; *f.* 4. carina ; *f.* 5. ovarium : —*magnified.*

56. Lotus *coronillæfolius*, Webb ; caule frutescente, ramis elongatis gracilibus diffusissimis glabellis, stipulis lanceolatis, foliis petiolatis foliolisque petiolulatis latis obovatis obcordatisque parce et appresse pubescentibus viridibus, junioribus fulvis, calycis turbinati pubescentis dentibus subæqualibus basi ovatis apice lineari-setaceis, vexillo glabro, carina acuta alis obtusis longiore, stylo breviter 1-dentato.

Hab. In herb. ins. Cap. Vir. (*Mus. reg. Par.*)

This species in its general appearance resembles the *Lotus spartioides*, Nob. (Phyt. Can.), but is easily distinguished by the form of its leaves and of the teeth of its calyx.

57. Lotus *Brunneri*, Webb ; caule fruticoso procumbente, foliis pube cinerea adpressa sericeis, stipulis lanceolatis sessilibus petiolo brevioribus caducis foliolis obovato-lanceolatis, pedunculis paucifloris, calycibus urceolato-campanulatis sericeo-pubescentibus, dentibus lanceolatis acutis, corolla lutea, vexillo elliptico subacuto carina sublongiore. Lotus anthylloides? *Brunn. Ergebn. p.* 86! non Vent.! ex sp. Malmaisonensi herb. Desfont.—Icon. Tab. III. Hook. *Ic. Plant. t.* 754.

Hab. " In insulæ *Salis* lapidosis magna copia sed constanter

flore luteo nunquam atropurpureo." (*Brunner*, in scheda
speciminis nobiscum communicati.)
A plant, perfectly distinct from the *L. anthylloides*,
Vent., of which Brunner had seen no authentic specimen
when he originally associated them, nor of *L. Jacobæus*
of which he afterwards considered it a mere variety. It
differs from the latter by the form of its leaves, by its
appressed and not villous and patent pubescence, by its
short sessile stipules, by the teeth of the calyx being merely
acute and not more or less filiform at the extremity and by
the form and colour of the corolla.

TAB. III. *Fig.* 1. flower; *f.* 2. vexillum; *f.* 3. ala; *f.* 4. carina; *f.* 5. stamina and ovaria; *f.* 6. ovarium; *f.* 7. pod;
f. 8. seed :—*magnified*.

58. Lotus *glaucus*, Hort. Kew, *ed.* 1, 3, p. 92. Chr. Smith *in*
Tuck. voy. p. 250. quanquam nulla extant in herb. Mus.
Brit. specimina (ex cl. *J. Dalton Hooker*, in litt.) *Phyt.*
Can. 2. *p.* 84.—Ic. *Phyt. Can. t.* 61.

HAB. In regione temperata, ins. *S. Jacobi*, (*Chr. Smith, l. c.*)
In ins. *S. Vincentii*, (*Th. Vogel*, n. 75. Junio 1841, sp.
fructifera, et n. 74. parce florifera.)

The leaves of the Cape de Verd plant are much broader than
those of the Canarian and Madeiran specimens, and resemble
those of *L. Lancerottensis*, Nob., from which, however, as
well as the typical form it is distinguished by its one or two
flowered capitules and by the narrow teeth of its calyx.
Christian Smith mentions an *L. lanatus*, sp. n. Tuck. voy.
p. 251.; but Dr. J. D. Hooker informs us that he has
not identified such a plant in his herbarium at the British
Museum.

59. Indigofera *hirsuta*, L. *Sp. Pl. p.* 1062. Lamck, *Encycl.*
3. *p.* 246. Guill. et Perr. *Fl. Sen. Tent. p.* 174. Walp.
Repert. 1. *p.* 660.—Indigofera Guineensis, *Thonn. et Schum.*
K. Darsk. Vid. Selskap. Afhandl. 4. *p.* 140.—Ic. Astra-
galus spicatus, siliquis pendulis hirsutis, foliis sericeis,
Burm. Thes. Zeyl. t. 14. Hemispadon pilosus, *Endl. Atakt.*

t. 3. ex Walp. l. c. sed planta macrior quam in speciminibus nostris.

HAB. In herb. ins. Cap. Vir. (*Mus. reg. Par.*)

60. Indigofera *tinctoria*, L. *Sp. Pl. p.* 1061. *a.* macrocarpa, *DC. Prodr. p.* 244. *Guill. et Perr. Fl. Sen. Tent.* 1. *p.* 178. —Ic. Pluk. *Phytogr. t.* 165. *f.* 5. HAB. In ins. *S. Antonii,* frutex tripedalis, (*Th. Vogel,* Jun. 1841, sp. fruct.)

61. Indigofera *viscosa,* Lamck. *Encycl.* 3. *p.* 247. DC. *Prodr.* 2. *p.* 227. Guill. et Perr. *Fl. Sen. Tent.* 1. *p.* 180. Indigofera glutinosa, *Perr. in DC. l. c.* non Vahl!—Ic. *Wendl. Sert. Han. t.* 12. HAB. In locis graminosis planitiei ins. *S. Jacobi,* (*J. Dalton Hooker,* Nov. 1839, sp. macilenta fruct. et florida, flores coccinei.)

62. Indigofera *Senegalensis,* Lamck. *Encycl.* 3. *p.* 248. Guill. et Perr. *Fl. Sen. Tent. p.* 183. Indigofera tetrasperma, *Vahl in herb. Desf.! Pers. Syn.* 2. *p.* 325, non Thonn. et Schum. Pl. Guin. nec DC. Prodr. HAB. In herb ins. Cap. Vir. (*Mus. reg. Par.*)

63. Indigofera *linearis,* Guill. et Perr. *Fl. Sen. Tent. p.* 184. HAB. Vulgatissima circa *Portum Praya,* ins. *S. Jacobi,* (*J. D. Hooker,* n. 151. Nov. 1839, sp. fl. et fruct.)

64. Tephrosia *bracteolata,* Guill. et Per. *Fl. Sen. Tent. p.* 194. HAB. In vallis Dominici, ins. *S. Jacobi,* (*J. Dalton Hooker,* Nov. 1839, sp. fl. et fruct.)

65. Tephrosia *anthylloides,* Hochst. *in Kotsch. Sched. It. Nub.* 1841, n. 1841. et *Schimp. It. Abyss.* 1842, *n.* 721. HAB. Specimina 2 fructifera in ins. parva *Coturnicum,* prope ins. *S. Jacobi,* a cl. J. Dalton Hooker lecta, aliaque flor. et fruct. in herb. ins. Cap. Vir. (*Mus. reg. Par.*) huc referenda opinor.

66. Sesbania *punctata,* DC. *Prodr.* 2. *p.* 265. Guill. et Perr. *Fl. Sen. Tent. p.* 198. HAB. In palmetis ins. *S. Jacobi,* (*Forbes,* n. 8. die 5, April, 1822, sp. florida.)

67. Zornia *angustifolia,* Smith *in Rees Cycl. n.* 1. DC. *Prodr.*

2. *p.* 316. Guill. et Perr. *Fl. Sen. Tent. p.* 203. Hedysarum
diphyllum, *a. Linn. Sp. Pl. p.* 747.—Ic. Rheed. *Malab.* 9.
t. 82.

HAB. In ins. *S. Jacobi,* (*J. Dalton Hooker,* n. 143. Nov.
1839, sp. pusillum.)

68. Desmodium *tortuosum,* DC. *Prodr.* 2. *p.* 332. Humb. et
Kunth! *Nov. Gen. et Sp.* 6. *p.* 521. Hedysarum tortuosum,
Sw. Prodr. Veg. Ind. Occ. p. 107. Desmodium terminale,
Fl. Sen. Tent.! p. 307. *an DC.* Desmodium ospriostreblum,
Steud. in sched. Schimp. It. Abyss. sect. 2. *n.* 1039!—
Ic. Sloane, *Hist. of Jam.* 1. *t.* 116. *f.* 9.
HAB. In ins. *S. Antonii,* (*Th. Vogel,* June 1841, sp. fl. et
fr.) In vallibus et in pascuis siccis ins. *S. Jacobi,* (*J. Dalton
Hooker,* n. 148 et n. 149, Nov. 1839, sp. fl. et fr.)
Apparently a common plant in tropical regions, particularly those of the New World. It begins to flower at
a very early stage, and varies exceedingly in the strength of
its stem and the size of its leaves and pods, on which
account it will probably be found to have a synonymy as
wide as its geographical range. The specimens of *Hedysarum
tenellum,* Kunth! (in the herb. of the Museum of Paris)
appear to be a delicate form, and *Hed. Cumanense* ejusd.!
a robust individual, with larger and more elliptic leaves, of
this species, very similar to the specimens of Vogel, gathered
in St. Antonio. I have not, however, ventured to unite
these species, the authentic specimens not admitting of
sufficient examination.

69. Mysicarpus *vaginalis,* DC. *Prodr.* 2. *p.* 353. Guill. et
Perr. *Fl. Sen. Tent. p.* 210. Wight et Arn. *Pr. Fl. Pen. Ind.
Or.* 1. *p.* 233. Hedysarum vaginale, *Linn. Sp. Pl. p.* 1052.
—Ic. Pluk. *Phyt. t.* 59. *f.* 3.
HAB. In herb. ins. Cap. Vir. (*Mus. reg. Par.*)
70. Arachis *hypogæa,* Linn. *Sp. Pl. p.* 1040. DC. *Prodr.*
2. *p.* 474. Arachidnoides Americana, *Nissole Mém. de l'Ac.
des Sc.* 1723. p. 387. Arachis Africana et A. Asiatica, *Lour.
Fl. Coch.*—Ic. Pluk. *Phyt. t.* 60. *f.* 2. Rumph. *Amb.* 5.
t. 156. *t.* 2. Trew, Ehret, *t.* 3. *f.* 3. Nissol. *l. c. t.* 19.

HAB. Extat ex ins. Cap. Vir. in herb. *Mus. reg. Par.*

71. Sœmmeringia *psittacorhyncha*, Webb; ramis robustis
striatis glaberrimis fulvis, stipulis amplis coriaceis oblongis
latioribus rotundatis sessilibus amplexicaulibus integerrimis
striato-venosis glabris, foliis abrupte pinnatis 5-jugis,
foliolis brevissime petiolulatis, petiolulis crassis stipellatis,
stipellis caducis ovato-flabellatis obliquis obtusis e basi
8-nerviis reticulato-nervosis pellucido-puncticulatis coriaceis margine membranaceis integerrimis, spicis dichotomis basi stipulatis imbricatim bracteatis, bracteis am-
plissimis membranaceis sessilibus connatis nervoso-reti-
culatis glaberrimis integerrimis imbricatis obcordatis florem
foventibus, floribus in axillis bractearum solitariis pedunculatis, pedunculo parce piloso bistipulato, stipulis scariosis
ovatis sessilibus apice ciliolatis persistentibus, calyce
bilabiatim bipartito corolla vix breviore chartaceo striato,
labio inferiore compresso carinato 3-dentato dentibus
acutis pilosulis, superne 2-dentato basi ad unguem vexilli
gibbo, corollæ papilionaceæ vexillo oblongo apice assur-
gente medio incumbente basi utrinque auriculato ungui-
culato, ungue reflexo plicato, alis vexillo incumbente involutis rectis oblongis ab apice ad basin plicaturis deorsum
spectantibus corrugatis unguiculatis per auriculas connatis carinæ petalis liberis deflexis recurvis psittaci rostrum
aduncum referentibus duplo longioribus, staminibus 10
in phalanges 2-5 andras intra carinæ petala recurvas
connatis, ovario breviter stipitato elliptico recurvo 2-
ovulato hirto, stylo ascendente glabro, stigmate apicillari,
legumine * * * *, seminibus * * * *.

HAB. In herb. ins. Cap. Vir. (*Mus. reg. Par.*)

Though this plant differs in many particulars from the
original *Sœmmeringia*, in the absence of fruit it is not
possible to separate them. The corolla appears to be tawney
or yellow, with the tip of the keel purple.

72. Phaca *Vogelii*, Webb; cinereo-villosa, ramis gracilibus
elongatis foliosis decumbentibus, foliis subsexjugis, foliolis

parvis ovatis, spicis axillaribus densifloris folio demum
subduplo longioribus, floribus exiguis sessilibus, calyce
campanulato pilosissimo, dentibus linearibus subæqualibus,
corolla calycem vix excedente, legumine minimo inflato
elliptico-ovato acutiusculo pilosissimo.—Ic. (Tab. VIII.)
Hook. *Ic. Plant. t.* 758.
Hab. In rupibus maritimis ins. *S. Antonii* (*Forbes*, n. 2. die
2 April. 1822, sp. florifera cum fructu immaturo.) Ibid.
(*Th. Vogel*, n.46. Junio 1841, sp. florifera et fructifera.)
Herba diffusa, lignescens; ramis 1-2-pedalibus, decumbentibus, crassitie pennæ columbinæ, pube ascendente villosulis,
foliosissimis. *Folia* impari-pinnata, pedicello brevi nudo,
subsexjuga, jugis approximatis, foliolis exiguis sessilibus,
ovato-lanceolatis, obtusiusculis vel mucronulatis, integerrimis, utrinque pilis subappressis cinereo-villosis. *Spicæ*
e foliorum fere omnium axillis prodeuntes, 12-16-floræ,
folio demum subduplo longiores, filiformes, duriusculi,
incurvuli, post foliorum casum persistentes. *Flores* minimi
inconspicui, glomerati, subsessiles, bracteis filiformibus,
hirtis, calyce duplo brevioribus subtensi. *Calyx* campanulatus, hirtus, persistens, dentibus brevibus linearibus,
acutis, subæqualibus, superioril us latioribus. *Corolla* flava
calycem vix excedens, post fecundationem ab ovario inflato
protrusa. *Vexillum* oblongo-ellipticum, apice rotundatum
emarginatum, concavum, alæ oblongæ vexillo breviores,
carina lata obtusa incurva valde concava longiores et cum
ea per seculos connexæ. *Stamina* 10, filamento vexillari
libero. *Ovarium* 1-loculare, 1-ovulatum, ellipticum, pi-
losissimum. *Stylus* brevis, arcuatus, mox antrorsum recurvus. *Stigma* capitatum, demum antrorsum decline.
Legumen parvum, inflatum, ovato-ellipticum, apice acutiusculum, 1-loculare, 2-spermum, demum ad suturam
superiorem dehiscens. *Semina* elongato-reniformia, scrobiculata, nigerrima. *Embryo* perispermio mucilaginoso
immersus, cotyledonibus ovato-ellipticis, petiolulatis, supra
radiculam claviformem incumbentibus.

TAB. VIII. *Fig.* 1. flower ; *f.* 2. vexillum ; *f.* 3. ala ; *f.* 4. carina ; *f.* 5. ovarium ; *f.* 6. lateral ; *f.* 7. front view of legume; *f.* 8. transverse section of ditto :—all *magnified.*

73. Dolichos *Daltoni,* Webb ; caule annuo gracillimo volubili pilis patentibus mollibus hirtulo, stipulis oblongis acutis nervosis, pedicellis elongatis gracilibus pilosis, foliolis elliptico-obliquis acutis appresse pubescentibus demum glabris ciliatis stipellis 2 minimis filiformibus munitis, bracteis filiformibus basi dilatatis, floribus 1-2 in axillis foliorum pedunculo brevissimo insidentibus, calycis tubo brevi, laciniis elongatis filiformibus ciliatis, leguminibus compressis falcatis (immaturis) pilosulis 4-5-spermis, seminibus planis nigris.

HAB. In pascuis ins. *S. Jacobi,* (*J. Dalton Hooker,* Nov. 1839, sp. unicum macrum. In herb. ins. Cap. Vir. (*Mus. reg. Par.*) sp. adulta florida et fructifera.

74. Lablab *vulgaris,* Sav. *Diss. p.* 15. DC. *Prodr.* 2. *p.* 401. Wight et Arn. *Fl. Pen. Ind. Or.* 1. *p.* 250. Dolichos Lablab, *L. S. Pl.* Lablab niger, *Mœnch, Meth. p.* 153. Dolichos Benghalensis, *Jacq. H. Vind.* et D. purpureus *ejusd. fragm. p.* 45. Pro syn. reliquis cf. W. et Arn. l. c.—Ic. Rumph. *Amb. t.* 136, 137, *et* 1841. *f.* 1. Jacq. *Vind.* 2. *t.* 124. *fragm. t.* 55. Smith, *Ex. Bot. t.* 74. Wight *in Hook. Bot. Misc.* 2. *Suppl. t.* 15. Ker, *Bot. Reg. t.* 830. Sav. *l. c. f.* 8-9.

HAB. In ins. *S. Jacobi* (*J. D. Hooker,* sp. fruct.) In herb. ins. Cap. Vir. (*Mus. reg. Par.*)

75. Voandezeia *subterranea,* Du Pet. Thou. *Gen. Mad.* n. 77. DC. *Prodr.* 2. *p.* 474. et *Mém. Lég. p.* 464. Guill. et Perr. *Fl. Sen. Tent.* 1. *p.* 254. Arachis Africana, *Burm. Fl. Cap. Prodr. p.* 22. non Lour. Glycine subterranea, *Linn. Dec. p.* 37. *Mant. p.* 442. *Syst. ed. Murr. p.* 548. *Willd. Sp. Pl. p.* 1053.—Ic. Linn. *Dec. t.* 17. DC. *Mém. t.* 20. *f.* 106.

HAB. Extat hæc planta in herb ins. Cap. Vir. (*Mus. reg. Par.*) ubi vulgo " *Obizendanbanbi*" ex scheda.

76. Cajanus *Indicus,* Spreng. *Syst.* 3. *p.* 248. Wight et Arn.

126   SPICILEGIA GORGONEA.

Prodr. Fl. Pen. Ind. 1. *p.* 256. Cytisus Cajan, *L. Sp. Pl.*
p. 1041. *Syst. ed. Murr. p.* 555. Variat vexillo nunc toto
luteo nunc extus striis purpureis picto. Cajanus flavus et
Cajanus bicolor *DC. Cat. Hort. Monsp. p.* 85. *Prodr.* 2.
p. 406.—Ic. Pluk. *Phyt. t.* 231. *f.* 3. Rheed. *Malab.* 6.
t. 13. Jacq. *Obs. t.* 1. et Cytisus pseudo-Cajan, *hort.*
Vind. 2. *t.* 119.
HAB. In herb. ins. Cap. Vir. (*Mus. reg. Par.*)
77. Rhynchosia *minima*, DC. *Prodr.* 2. *p.* 385. Guill. et Perr.
Fl. Sen. Tent. 1. *p.* 214. Dolichos minimus, *L. Sp. Pl.*
p. 1020. Glycine, *Chr. Smith,* in herb. ins. Cap. Vir.
Mus. Brit. (*J. D. Hooker* in litt.)—Ic. Jacq. *Obs. t.* 22.
HAB. In ins. *S. Jacobi (Chr. Smith).* In ins. *Salis* humi-
diusculis (*Brunner* in herb. nostro, *Ergebn. p.* 109). In
herb. ins. Cap. Vir. (*Mus. reg. Par.*)
78. Rhynchosia *Memnonia,* DC. *Prodr.* 2. *p.* 386. Glycine
Memnonia, *Del. Fl. d'Eg. p.* 100.—Ic. Del. *l. c. t.* 38. *f.* 3.
HAB. In ins. *S. Jacobi* (*J. D. Hooker,* November, 1839, sp.
fruct.) In herb. ins. Cap. Vir. (*Mus. reg. Par.*)
79. Abrus *precatorius,* Linn. *Syst. p.* 533. Glycine Abrus,
Sp. Pl. p. 1025.—Ic. Pluk. *Phyt. t.* 214. *f.* 5. Rheed.
Malab. 8. *t.* 39. Rumph. *Amb.* 5. *t.* 32.
HAB. In herb. ins. Cap. Vir. (*Mus. reg. Par.*) In ins.
S. Jacobi, valle *Organorum* (*Brunner*).
80. Cassia *obtusifolia,* Linn. *Sp. Pl. p.* 539. (excl. syn.
Rumph.) DC. *Prodr.* 2. *p.* 493. Guill. et Perr. *Fl. Sen.*
Tent. p. 260. Vog. *Syn. p.* 24. Torr. et Gray, *Fl. N. Am.* 1.
p. 394.—Ic. Dill. *Hort. Elth. t.* 62. *f.* 72. Sloane, *Hist. Jam.*
2. *t.* 180. *f.* 5. Plum. *ed. Burm. t.* 76. *f.* 2.
HAB. In ins. *S. Jacobi* (*J. D. Hooker,* n. 146. November,
1839, sp. flor. et fruct.)
81. Cassia *occidentalis,* L. *Sp. Pl. p.* 540. Collad. *Mon. p.* 107.
DC. *Prodr.* 2. *p.* 497. Wight et Arn. *Prodr. Fl. Pen. Ind.*
Or. p. 290. Vog. *Syn. p.* 21. Torr. et Gray, *Fl. N. Am.* 1.
p. 294. Pro syn. reliquis cf. Walp. *Repert. bot.* 1. *p.* 816.
—Ic. Comm. *Hort. Amst.* 1. *t.* 26. Ker, *Bot. reg. t.* 83.

HAB. In ins. *S. Jacobi (J. D. Hooker*, n. 150. November, 1839, sp. flor. et fr.) Ibid. (*Chr. Smith* et *Thomson* in herb. ins. Cap. Vir. *Mus. Brit. J. D. Hooker*, in litt.)
82. Cassia *bicapsularis*, Linn. *Sp. Pl. p.* 538. DC. *Prodr.* 2. *p.* 494. Vog. *Syn. p.* 18. Cassia sennoides, *Jacq. Collect.* 1. *p.* 74. *DC. l. c. Brunner, Ergebn. p.* 37 !—Ic. Plum. *ed. Burm. t.* 76. *f.* 1. Jacq. *Fragm. t.* 58. et *Ic. rar.* 1. *t.* 170.
HAB. In ins. *Brava (Brunner* in herb. nostro !)
83. Cassia *obovata*, Collad. *Monogr. p.* 92. DC. *Prodr.* 2. *p.* 492. Guill. et Perr. *Fl. Sen. Tent. p.* 260. Vog. *Syn. p.* 36.—Ic. Burm. *Fl. Ind. t.* 33. *f.* 2. C. Senna, Nect. *Voy. Eg. t.* 1. Hayne, *Arzneigew. t.* 42.
HAB. In alveis siccis rivorum ins. *Salis (Forbes*, n. 4. 26 Maii, 1822, spec. florida.) Ad oram maritimam et in montibus ins. *S. Vincentii* usque ad altitudinem 500 ped. (*Vogel*, n. 6. Junio, 1841, sp. florida, fructu immaturo.)
84. Cassia *micrantha*, Guill. et Perr. *Fl. Sen. Tent.* 1. *p.* 262. Walp. *Repert.* 1. *p.* 834.
HAB. In herb. ins. Cap. Vir. (*Mus. reg. Par.*)
85. Cassia *microphylla*, Willd. *Sp. Pl.* 2. *p.* 529. DC. *Prodr.* 2. *p.* 505. Cassia geminata, *Vahl*, herb. *Schum. Beskr. Guin. Plant.* 2. *p.* 228. *Guill. et Perr. Fl. Sen. Tent.* 1. *p.* 263.
HAB. In herb. ins. Cap. Vir. *(Mus. reg. Par.)*
86. Dicrostachys *nutans*, Benth. *in Hook. Journ of Bot.* 4. *p.* 352. Caillea Dichrostachys, *Guill. et Per. Fl. Sen. Tent.* 1. *p.* 240. Desmanthus trichostachys *et* nutans, *DC. Prodr.* 2. *p.* 445—6. *Brunn.! Ergebn. p.* 54.—Ic. DC. *Legum. t.* 67.
HAB. In ins. *S. Jacobi* montosis *(Brunner* in herb. nostro).
87. Acacia *albida*, Guill. et Perr. *Fl. Sen. Tent.* 1. 245. Brunn. *Ergebn. p.* 4. A. albida β, Senegalensis, *Benth. in Hook. Lond. Journ of Bot.* 1. *p.* 505.
HAB. In ins. *S. Jacobi (Brunner*, l. c.)
88. Acacia *Arabica*, Guill. et Perr. *Fl. Sen. Tent.* 1. *p.* 250. A. Arabica, *a* tomentosa, *Benth. l. c. p.* 500.

HAB. In ins. *S. Jacobi*, arbor 30-pedalis (*J. D. Hooker*, n. 145, November, 1839, sp. sine flore et fructu.)

89. Acacia *Farnesiana*, Willd. *Sp. Pl.* 4. *p.* 1083. Benth. l. c. *p.* 494. Mimosa Farnesiana, *Linn. Sp. pl. p.* 1506. Vachellia Farnesiana, *Wight et Arn. Prodr. Fl. Pen. Ind. or.* 1. *p.* 272. Farnesia odora, *Gasp. Descr. di uno nuov. gen. p.* 5.—Ic. Ald. *Hort. Farn. p.* 2 *et* 7. Pluk. *Phyt. t.* 73. *f.* 3. Descourt. *Fl. des Antil. t.* 1. Gasp. *l. c.*

HAB. In ins. *S. Jacobi* (*J. D. Hooker*, November, 1839) circa *Portum Praya* et in valle *S. Dominici* (*Brunner, Ergebn. p.* 5.)

XXI. TAMARISCINEÆ, *A. de St. Hil.*

90. Tamarix *Gallica*, Linn. *Sp. Pl.* Webb, *Obs. in Hook. Lond. Journ. of Bot.* 3. (1840) *p.* 429, *et in Ann. Sc. Nat.* 2ème sér. 16. (1841) *p.* 264. T. Canariensis, *Willd. Act. Ber.* (1812—13) *ed.* 1816. *p.* 77. *ex DC. Prodr.* 3. *p.* 96. *Webb, Phyt. Can.* 1. *p.* 171. T. Senegalensis, *DC. l. c. Guill. et Perr.! Fl. Sen. Tent.* 1. *p.* 309.—Ic. *Phyt. Can. t.* 25. Webb *in Hook. Lond. Journ. l. c. t.* 15.

HAB. In ins. *S. Vincentii* ubi sæpe arbor fit mediocris (*Th. Vogel*, n. 7. Junio, 1841, sp. fl. et fr.) In ins. *S. Jacobi* (*C. Darwin, Thomson*, in herb. ins. Cap. Vir. Mus. Brit. *J. D. Hooker*, in litt.) In herb. ins. Cap. Vir. (*Mus. reg. Par.*)

XXII. ONAGRARIEÆ, *Juss.*

91. Epilobium *parviflorum*, Schreb. *Spic. p.* 146. Mert. et Koch, *Deutschl. Fl.* 3. *p.* 14. *Phyt. Can.* 2. *p.* 7.—Ic. *Fl. Dan. t.* 347. *Engl. Bot. t.* 795.

HAB. Ad rivulos in ins. *S. Antonii* (*Th. Vogel*, n. 24. Junio, 1841, sp. floriferum.) In herb. ins. Cap. Vir. (*Mus. reg. Par.*) (sp. procera valde hirsuta.)

XXIII. CUCURBITACEÆ, *Juss.*

92. Citrullus *Colocynthis*, Schrad. *in Eckl. et Zeyh. enum.*

SPICILEGIA GORGONEA. 129

p. 279. Linnœa, 12. p. 414. Phyt. Can. 2. p. 3. Cucumis Colocynthis, *L. Sp. Pl. p.* 1435. *DC. Prodr.* 3. *p.* 302. *Chr. Smith in Tuck. Voy. p.* 251. *Wight et Arn. Prodr. Fl. Pen. Ind. or.* 1. *p.* 342. *Brunn. Ergebn. p.* 50.—Ic. Nees, *Pl. off.* 12. *t.* 11. Turp. *Fl. méd. t.* 128. Wight, *Ic.* 2. *t.* 498.

HAB. In arvis *Gossypio* satis ins. *S. Jacobi, (Forbes,* n. 12. die 5 April, 1822.) Ibid. (*J. D. Hooker,* n. 133 et *Darwin*) in ins. *S. Antonii, (Th. Vogel,* n. 23.)

93. Momordica *Charantia,* L. *Sp. Pl. p.* 1433. DC. *Prodr.* 3. *p.* 311. Wight et Arn. *Prodr. Fl. Pen. Ind. or. p.* 348. Brunn. *Ergebn. p.* 90. Momordica Senegalensis, *Lamck. Encycl.* 4. *p.* 239. *Chr. Smith in Tuck. Journ. p.* 249. Momordica muricata, *Willd. Sp. Pl.* 4. *p.* 602.—Ic. Rumph. *Amb.* 5. *t.* 151. Rheed. *Mal.* 8. *t.* 10. Hill. *Sex. syst. class* 21. *ord.* 10.—Ic. Wight, *Ic.* 2. *t.* 504.

HAB. In sylvis *Phœnicis dactyliferœ,* ins. *S. Jacobi (Forbes,* n. 9. die 5 April, 1822, sp. fl. et fruct.) Ibid. (*Chr. Smith,* in herb. ins. Cap. Vir. *Mus. reg. Par.*) in valle *S. Dominici* ins. *S. Jacobi (J. D. Hooker,* n. 162. November, 1839, sp. fl. et fructifera.)

XXIV. PORTULACACEÆ, *Juss.*

94. Portulaca *oleracea,* Linn. *Sp. Pl. p.* 638. *Phyt. Can.* 1. *p.* 169.—Ic. Lob. *Ic. p.* 388. Turp. *Fl. méd.* 5. *t.* 283.

HAB. In ins. *S. Jacobi (J. D. Hooker,* n. 109. November, 1839, sp. jam diu fructifera.) Ibid. (*Chr. Smith,* in herb. ins. Cap. Vir. *Mus. Brit.* et in herb. *Mus. reg. Par.*)

95. Aizoon *Canariense,* Linn. *Sp. Pl. p.* 700. DC. *Prodr.* 3. *p.* 453. *Phyt. Can.* 1. *p.* 207.—Ic. Pluk. *Phyt. t.* 503. *f.* 4. Niss. *Acta Act. Par.* 1711. *t.* 13. *f.* 1. DC. *Pl. grasses t.* 136.

HAB. In campestribus ins. *Salis (Forbes,* n. 5. die 26 Martii, 1822, sp. juniora). In ins. *S. Jacobi (C. Darwin* sp. fructifera).

96. Umbilicus *horizontalis,* DC. *Prodr.* 3. *p.* 400. Cotyledon horizontalis, *Guss! Ind. sem. in Bocc.* 1826. *Prodr. Fl.*

K

130   SPICILEGIA GORGONEA.

sicc. 1. *p.* 517. *Presl, Fl. sicc.* 1. *p.* 517.—Ic. Ten. *Fl.*
Neap. t. 234. *f.* 1.

HAB. In herb. ins. Cap. Vir. *(Mus. reg. Par.)*

XXV. MELASTOMACEÆ.

OSBECKIA, L. Char. gen. reformatus.*—*Calyx* campanulatus,
4 sæpius 5-fidus; divisuris simplicibus (id est interiore
membrana non duplicatis ut in plurimis *Melastomacearum*
generibus) ovatis vel triangularibus, acutis; tubo hirsuto
vel piloso. *Corollæ petala* 4-5, obovata, mucrone piliformi
vel setis aliquot fasciculatis haud raro terminata inter-
dumque ciliolata. *Stamina* 8-10, alternative majora et
minora; filamentis glabris; antheris lineari-subulatis, apice
uniporosis; connectivo infra loculos longe producto et in
insertione filamenti varie conflato, sæpius biauriculato.
Ovarium 4-5-loculare, ovatum, ad medium usque tubo
calycino vittis 8-10 antheras in præfloratione intus reflexas
separantibus adhærens, apice setis styli basin cingentibus
coronatum. *Stylus* filiformis vel utrinque subulatus.
Stigma subcapitellatum aut punctiforme. *Ovula* numerosa, placentis 4-5 subtriquetris centralibus affixa. *Capsula* calyce persistente vestita, loculicide 4·5-valvis. *Semina* minuta cochleata.—Osbeckiæ *omnes frutices suffruticesve 1-3-pedales, calidiorum partium* Americæ australis,
Africæ, Indiæ *nec non insularum quarumdam* Oceani Atlan-
tici Indicique *incolæ, nec, quod mirum est, e* Nepaliæ *montibus temperatis omnino exules.*

97. Osbeckia *Princeps,* Dec.; fruticosa, ramis dense hirto-
tomentosis rufescentibus, foliis 1-3-pollicaribus petiolo
4-8 lineas longo instructis. DC. *Prod.* 3. *p.* 140. *Rhexia
Princeps,* Bonpl. *Rhex. tab.* 45.

The genus *Osbeckia* is probably the only one in the
family of *Melastomaceæ* which is found both in the New
and the Old World; unless botanists prefer, which is fre-

---

* We are indebted to M. Ch. Naudin, who has paid much attention to
the difficult family of the *Melastomaceæ,* for the description of this species
of *Osbeckia.*

quently the case, to divide it into as many genera as there
are parts of the globe or even islands wherein it occurs;
though I doubt whether there are sufficient characters for
doing so. But if the genus itself is cosmopolite, such is not
the case with its species, which seem generally to be included
in restrained limits; yet, the group being widely disse-
minated it is not surprising that in the present instance,
contrary to the usual rule, our plant differs from the few
species hitherto found on the neighbouring continent
of Africa. However this may be, it certainly does not
differ from the *O. Princeps* of Bonpland and Decandolle found
in Brazil. This we have ascertained by attentively examining
the specimen from the same country in the herbarium of the
Museum of Paris. We have grounds to suppose its true
native country to be Africa, whence it may have been
brought into the New World by the negroes.

XXVI. UMBELLIFERÆ, *Juss.*

Trib. nov. *Tetrapleureæ*, Parl.

Fructus a dorso lenticulari-compressus. Mericarpiorum *juga
4 prominentia, æqualia.*—Tribus ad *Umbelliferas* orthos-
permas pertinens, fructus forma ad *Angeliceas* et *Peuce-
dineas* accedit, sed ab utrisque omnino differt jugis 4 prominentibus, quorum 2 ex secundariis formata, nec marginalia in alam expansa.—*Parl.*

TETRAPLEURA, Parl.; *Calyx ... Petala ... Mericarpia* jugis
primariis 5, lateralibus marginantibus prominentibus, duobus aliis dorsalique obsoletis fere nullis, secundariis 4,
duobus dorsali proximis elevatis marginantium magnitudine, duobus aliis subnullis, unde mericarpia quadrijugata.
Vittæ solitariæ sub jugis quatuor prominentibus, commis-
sura 2 vittata, vittæ omnes filiformes. *Carpophorum* bi-
partitum. *Semen* complanatum. *Albumen* carnosum, carnosum, planiusculum.—*Parl.*

98. Tetrapleura *insularis*, Parl.

HAB. In insula *S. Vincentii* (*Th. Vogel.*)

132   SPICILEGIA GORGONEA.

Hujus species ramum pessimum tantum possideo, sic ejus
descriptionem mihi ullo modo non licet adumbrare. *Parl.*
99. Tetrapleura . . . . sp. ?
HAB. In montosis ins. *S. Jacobi (J. D. Hooker.)*

XXVII. RUBIACEÆ, *Juss.*

100. Hedyotis (Oldenlandia) *Burmanniana,* R. Br. *in Wall.
Cat. n.* 868. Wight et Arn. *Prodr. Fl. Pen. Ind. Or.* 1. *p.*
415.—Ic. Rheed. *Mal.* 10. *t.* 35. Burm. *Thes. Zeyl. t.* 11.
HAB. In umbrosis ins. *S. Jacobi,* (*J. Dalton Hooker,* spec.
duo florida et fructifera cum sequente sub eodem numera
commixta.)

It is not without much hesitation that I introduce the
present species here; but I cannot find any perceptible
difference between the two specimens named above, and
others from India, under the same name, sent either by
Roxburgh or Wallich to Lambert. Our plant may perhaps
prove not distinct from the *H. longifolia,* Schum., a little
known species. From the following it differs, not only
in having its peduncles two-flowered, but likewise by its
much larger fruit and the wider lanceolate teeth of its calyx,
which are more subulate at their apex.

101. Hedyotis (Oldenlandia) *corymbosa,* Linn. *Sp. Pl. p.* 274.
DC. Prodr. 4 *p.* 426.—Ic. Plum. *ed. Burm. t.* 212. *f.* 1.
ex DC. l. c.
HAB. In locis umbrosis circa Portum Praya, ins. *S. Jacobi,*
(*J. Dalton Hooker,* n. 172. Nov. 1839, sp. fructifera.)
Christian Smith cites in his catalogue (Tuck. voy. p. 252)
the *Hed. Capensis:* no specimen, however, of this plant is
found in his herbarium at the British Museum, (*J. Dalton
Hooker* in litt.)

102. Hedyotis (Oldenlandia) *virgata,* Willd. *Sp. Pl.* 1. *p.* 567.
DC. Prodr. 4. *p.* 425.
HAB. In herb. ins. Cap. Vir. (*Mus. reg. Par.*)

103. Hedyotis (Kohautia) *stricta,* Smith *in Rees Cycl.* 17.
n. 21. *DC. Prodr.* 4. *p.* 430.
HAB. In herb. ins. Cap. Vir. (*Mus. reg. Par.*)

104. Borreria *Kohautiana,* Cham. et Schlect. *in Linnæa,* 1828. *p.* 311. DC. *Prodr.* 4. *p.* 541. Spermacoce verticillata, *Linn. Sp. Pl. p.* 148. quoad plantam Africanam *Chr. Smith l. c. p.* 249. (ex cl. *J. D. Hooker* in litt.)—Ic. Dill. *Hort. Elth. t.* 277. *f.* 358. HAB. In arvis *Gossypii,* (*Forbes,* n. 1. die 1 April, 1822, spec. fructifera.) In campis apertis ins. eadem, fruticulus parvus, floribus albis capitatis. (*J. Dalton Hooker,* n. 174. Nov. 1839, sp. florida et fruct.) Spermacoce " diversi generis videtur," *Chr. Smith,* l. c. p. 249. (in herbario suo deest *J. D. Hooker,* in litt.)

105. Mitracarpum* *Senegalense,* DC. *Prodr.* 4. *p.* 572. Staurospermum verticillatum, *Thonn. ex Schum. Act. Hafn.* 2. *p.* 93. HAB. In herb. ins. Cap. Vir. (*Mus. reg. Par.*)

106. Galium *Aparine,* Linn. γ. *scaberrimum,* Webb, *Phyt. Can.* 2. *p.* 183. G. hispidum, *Willd. Enum.* 1. *p.* 154. G. scaberrimum, *Hornem. Hort. Hafn.* 1. *p.* 135. HAB. In herb. ins. Cap. Vir. (*Mus. reg. Par.*)

107. Galium *rotundifolium,* Linn. *var. villosum,* Webb, *Phyt. Can.* 2. *p.* 185. G. rotundifolium, *a. Linn. Sp. Pl.* 1. *p.* 156. G. hirsutum, *Nees, et Bach. Hor. phys. Ber. p.* 113. G. Neesianum, *Req. in DC. Prodr.* 4. *p.* 600.—Ic. Bocc. *Sicc. t.* 9. *f.* 1. Moris. *Hist.* 2. *s.* 9. *t.* 21. *f.* 3. Barr. *Ic.* 304. Nees, *l. c. t.* 22. HAB. Prope apicem Montis *Gurdo,* ins. *S. Nicolai,* (*Forbes,* d. 30 Martii, 1822, spec. florida cum fructu juniore.)

XXVIII. GLOBULARIEÆ, DC.

108. Globularia *amygdalifolia,* Webb; caule fruticoso, foliis lanceolatis ovato-lanceolatisque in petiolum brevem attenuatis 1-nerviis divaricato-nervulosis, capitulis ad apicem ramulorum axillaribus approximatis subumbellatis, pedunculis folio vix brevioribus pilosis, bracteis paucis oblongis distantibus, involucri squamis oblongis acutis ciliatis, calycis tubulosi dentibus basi lanceolatis apice

* *Mitracarpum,* sed non bene, scripsit generis cl. auctor.

subulatis ciliatis bracteola sublongioribus, corollæ calycem paullo excedentis labio superiore subnullo inferiore trifido laciniis linearibus, genitalibus corollæ longitudine vel breviter exsertis.

HAB. In herb. ins. Cap. Vir. (*Mus. reg. Par.*) This plant approaches very nearly the *G. salicifolia*, Lamck. but differs in the shape of its leaves, in the length of its peduncles, in the form of the teeth of the calyx, and its much shorter stamens and style. In none of the specimens from Madeira, or from the Canary Islands, is there any approximation towards the Cape de Verd plant: this identity of form is so remarkable that I am induced to consider it specifically different.

XXIX. COMPOSITÆ, *Juss.*

109. Vernonia *cinerea*, Less. *in Linnæa*, 1829, *p.* 291. *et* 1831, *p.* 673. DC. *Prodr.* 5. *p.* 24. Conyza cinerea, *Linn. Sp. Pl. p.* 1208. Chrysocoma violacea, *Schum. Pl. Guin.* 158, *ex DC.*—Ic. Pluk. *Phyt. t.* 243. *f.* 3. Rumph. *Amb.* 6. *t.* 14. *f.* 1. Burm. *Thes. Zeyl. t.* 96. *f.* 1.

HAB. In campis *Gossypio* consitis ad sinum *Tarrafal*, ins. *S. Antonii,* (*Forbes,* n. 10. die 2 April. 1822.) In vallibus umbrosis ins. *S. Jacobi,* (*J. Dalton Hooker,* n. 200. Nov. 1839.) In cultis ins. *S. Antonii,* (*Th. Vogel,* n. 35. Jun. 1841.) Spec. omnia florida et fructifera.

110. Erigeron *varium,* Webb; suffrutex, ramis divaricatis vel rectis pubescentibus, foliis ovatis lanceolatis oblongisve dentato-serratis serraturis apiculatis utrinque pubescentibus in petiolum brevem basi attenuatis, panicula laxa vel conferta, pedicellis filiformibus hispidis, capitulis parvis, involucri squamis linearibus pubescentibus, pappo rufo denticulato, ligulis brevissimis discum haud superantibus 2-3-dentatis, florum hermaphroditorum styli ramis lineari-lanceolatis acutis.

HAB. Passim in ins. *S. Nicolai,* forma major, foliis ovatis, (*Forbes,* n. 36. die 27 Martii, 1822.) In ins. *S. Antonii,* forma eadem sed panicula valde conferta (*Forbes,* sine n.) In ins.

S. Vincentii, a medio ad apicem *Montis Verede, (Th. Vogel,*
n. 48 et n. 49, forma parva, foliis oblongis.) In ins. *S.
Antonii, (Th. Vogel,* n. 9. foliis fere linearibus apice tantum
dentatis, capitulis minimis.)

111. Conyza *pannosa,* Webb; caule lignescente erecto piloso,
ramis junioribus setosis, foliis ovatis obtusis grosse crenato-
dentatis inferioribus petiolatis superioribus basi attenuatis
sessilibus dentibus usque ad caulem protractis, panicula
corymboso-cymosa, pedunculis pedicellisque hirsuto-pubescentibus, involucri squamis linearibus vel lineari-lanceolatis apice apiculatis margine scariosis glabrescentibus
pappo rufulo subscabro brevioribus, floribus radii fœmineis
filiformibus, corolla apice minutissima denticulata denticulis
inæqualibus, stigmate corollam duplo excedente, antheris
acutis ecaudatis, styli ramis lanceolatis dorso papillosis,
achænio sub-complanato ad peripherium papilloso parce
piloso albido.
Hab. In ins. *S. Vincentii* ad partem tertium superiorum
Montis Verede, ubi copiosissima, (*Vogel,* n. 52. Jun. 1841.
spec. fructifera et florifera.)

112. Conyza *odontoptera,* Webb; caule elato tenuiter pubescente cum ramis per totam longitudinem alis runcinato-
dentatis, foliis oblongis dentatis acutis glanduloso-puberulis,
panicula racemosa patula, capitulis amplis, involucro pauci-
seriali squamis linearibus acutis punctulato-glandulosis,
receptaculo plano punctulato nudo, pappo albido subscabro, floribus radii plurimis filiformibus denticulato-
truncatis, disci 5-dentatis, stylo papilloso, achenio 4-angulo
subcompresso.
Hab. In herb. Mus. reg. Par. sp., floridum et fruct.
Though this species has all the appearance of the section
Pterocaulæ of the genus *Blumea,* and comes near the *B. odontoptera* (male *Pterodonta*) of De Candolle, its anthers, entirely
without *caudæ* or *appendices,* necessitate its being placed in
the genus *Conyza.* These artificial sections, however, of
very similar plants, require revision.

113. Phagnalon *melanoleucum,* Webb; fruticulus, ramis te-

nuibus incurvis tomento pannoso albo vestitis, foliis
alternis lanceolatis majoribus sæpe pollicaribus 4 fere lin.
latis margine revoluta undulata in petiolum brevem attenuatis, junioribus sæpe in axillis fasciculatis supra tomento
albo araneoso deciduo tectis mox atrovirescentibus subtus
albo-tomentosis, pedunculis ad apicem ramorum vel in
axillis supremis vel oppositifoliis solitariis vel geminatis
2-3-chotomis filiformibus tomentosis apice nudis, involucri
campanulati glaberrimi nigrescentis squamis 5-serialibus
exterioribus ovatis mediis oblongis interioribus linearibus
margine scariosis denticulatis apiculatis, capitulis paucifloris heterogamis, floribus omnibus tubulosis, fœmineis pluriserialibus filiformibus 5-dentatis, stylo exserto
ramis elongatis obtusis, hermaphroditis sub 8 campanu-
latis glabriusculis, antheris basi attenuatis ecaudatis, styli
ramis exsertis superne claviformibus apice subtruncatis
stigmatoso-papillosis, receptaculo angusto, acheniis oblongis
subcompressis erostris, pappo albo pilosiusculo florum
fœmineorum 3-4-setoso hermaphroditum 5-setoso.—Ic.
(TAB. IX.) *Hook. Ic. Plant. t.* 759.

HAB. In *Monte Verede,* ins. *S. Vincentii,* ultra alt. 1000 ped.
usque ad apicem, (*Th. Vogel,* n. 37. Junio 1841, spec.
florida et fructifera.)

TAB. IX. *Fig.* 1. hermaphrodite; *f.* 2. female flower, both
magnified.

114. Phagnalon *luridum,* Webb; fruticulus lignosus durus
multirameus foliosus, ramis fuscis nigrisque striatis superioribus gracilibus tomento parco fusco indutis, foliis alternis
lineari-lanceolatis basi attenuatis sæpe 2-pollicaribus 2-3
lin. latis inferioribus petiolatis petiolo tenui margine revolutis eroso-dentatis, junioribus pube fusco-cinerea parca
obsitis demum glabratis lucidis nigris, pedunculis terminalibus 2-3-chotomis filiformibus nigris junioribus fusco-
puberulis apice parce squamigeris, involucro turbinato-
campanulati glaberrimi nigrescentis squamis scariosis
inferioribus ovato-oblongis mediis lineari-oblongis interioribus linearibus angustissimis acutis apice subfimbriato-

ciliatis, capitulis paucifloris heterogamis, floribus fœmineis
pluriserialibus filiformibus apice setoso-denticulatis styli
ramis elongatis setaceis obtusis, hermaphroditis paucis
cylindraceo-campanulatis laciniis glabris acutis stylo sub-
exserto ramis cylindraceis apice subclavatis truncatis stigmatoso-papillosis, receptaculo plano foveolato, acheniis ovato-oblongis compressiusculis erostris pilosis, pappo albo
superne scaberulo fl. fœm. 2-3-setoso herm. 5-setoso.
HAB. In *Monte Verede*, ins. *S. Vincentii*, ultra alt. 1000 ped.
(*Vogel*, n. 51. Junio 1841. spec. fructifera et quodam
florida.)

115. Pluchea *ovalis*, DC. *Prodr.* 5. *p.* 450. Pers. *Syn.* 2.
p. 424.
HAB. In ins. *S. Vincentii*, ubi austrum et favonium spectat
ad alt. 500 circiter pedum, frutex ramosus 2-3-pedalis,
(*Vogel*, n. 45. Junio 1841, spec. florida et fructifera.)

116. Inula (Limbarda) *leptoclada*, Webb; caule erecto,
ramis gracilibus, pilis crispulis superne hirtulis fuscis, foliis
distantibus oblongis lingulatis acutis basi auritis semiamplexicaulibus margine dentatis pubescentibus dentibus
quandoque subobsoletis, capitulis ad apicem ramulorum
confertis subcymosis, pedunculis filiformibus, foliis minimis
stipatis, involucri squamis anguste linearibus acuminatis
glanduloso-puberulis, receptaculo subplano, floribus omnibus hermaphroditis, radii 3-dentatis aliquando subligu-
latis, disci 5-dentatis, dentibus brevibus ovatis subacutis,
genitalibus inclusis, styli ramis brevibus, antheris 2-setis,
acheniis (immaturis) cylindraceis leviter apice constrictis
hirsutis, pappo denticulato albo basi libero.
HAB. In herb. ins. Cap. Vir. (*Mus. reg. Par.*)
Valde affinis est sp. nostra plantæ Æthiopicæ a cl. Kotschyo
olim lectæ (1839—38, n. 26), sed folia habet magis denticulata, ramos decumbentes, capitula in paniculo laxiore
disposita, involucri squamis lævioribus, acheniis compressiusculis et sub apice vix ac ne vix constrictis. *Pulicariam*
quoque *Arabicam* refert sed pappus duplex et aliæ a *Pulicaria* notæ differentiales.

117. Pegolettia *Senegalensis,* Cass. *Dict.* 38. *p.* 230. DC. *Prodr.* 5. *p.* 481. Brunn. *Ergebn. p.* 97.
HAB. In ins. *Salis, (Brunner,* l. c.)

118. Francœuria *crispa,* Cass. *Dict. des Sc. Nat. p.* 44, 38 et 374. DC. *Prodr.* 5. *p.* 475. Schultz, *Bip. in Phyt. Can.* 2. *p.* 222. Aster crispus, *Forsk. Fl. Æg. Arab. p.* 150. Inula crispa, *Pers. v.* 2. *p.* 450 (excl. syn. Vent.) Francœuria diffusa, *Shuttlew. in Brunner Ergebn. p.* 72.—Ic. Inula crispa, *Del. ! Fl. d'Eg. t.* 45. *f.* 2.

HAB. In ins. *Salis* planitiebus siccis (*Forbes,* n. 3, die 26 Maii, 1822). Provenit magna copia in lapidosis ins. *Sal.* (*Brunner,* ms. in herb. nostro.)

Our plant, described by Shuttleworth as a distinct species, is identical with the Senegambian specimens in the herb. of Desfontaines, and of M. Gay, described by Cassini in the Dict. des Sc. Nat. vol. 38, as *F. crispa.* It is undoubtedly of stronger growth than the Egyptian plant: the capitules are much larger; and, instead of being merely ciliated, the scales of the involucrum are covered with down. This is likewise the case with the Canarian specimens; but their capitules are not larger than those of the Egyptian plant; and as after a minute inspection of the inflorescence and fruit no other tangible difference is discernible, this plant can scarcely be considered specifically distinct.

119. Odontospermum, Neck. C. H. Schultz, *Bip. in Phyt. Can.* 2. *p.* 231.

We formerly remarked to our friend and collaborator, Dr. C. H. Schultz, of Deux Ponts, that Lessing, and after him De Candolle, by adopting the errors of Mœnch, who misunderstood the genus *Asteriscus* of Tournefort, and by giving that name to the well assorted group called *Nauplius* by Cassini, had been the unintentional cause of considerable confusion. The *Asteriscus* of Tournefort belongs in reality to the genus *Pallenis,* Cass.; since from *Buphthalmum spinosum,* L., its sole occupant, both the character and the figure of that genus were taken by him; to which he appended two other species, *Buph. maritimum* and *aquaticum,* which do not accord either with the character or figure of his genus. One of these was unadvisedly considered by Mœnch to constitute the Tournefortian genus, and to it he attaches the name and cites the figure of that author. Dr. Schultz adopted our views and has shewn besides that the *Odontospermum* of Necker, composed of the two species wrongly appended by Tournefort to his genus *Asteriscus*, is identical with *Nauplius*, and not with *Borrichia* as De Candolle imagined.

It might have been preferable to retain the names of Cassini for genera he has so well defined; but the question having been once mooted and more ancient names erroneously applied, the law of priority must now be fully carried out. *Pallenis*, Cass. must reassume its name of *Asteriscus*, Tourn., and *Nauplius* that of *Odontospermum* and the two genera will stand thus :

Asteriscus, Tourn. non *Mœnch*, nec *Less*., nec *DC.* Pallenis, *Cass.*

Odontospermum, Neck. Asteriscus, *Mœnch*, *Less. DC.*, non *Tourn.* Nauplius, *Cass.*

We here give three new species of the latter genus. The first, *O. Smithii*, resembles so closely in appearance the *O. sericeum*, C. H. Sch. Bip. as to be easily taken for it; but the pappus and the teeth of the corolla are perfectly distinct. The second, *O. Daltoni*, has very much the aspect of *O. intermedium*, ejusd.; the third, *O. Vogelii*, that of some forms of *O. stenophyllum*, ejusd.; but they are in reality quite different plants.

120. Odontospermum *Smithii*, Webb; fruticulus robustus, ramis crassis fuscis, foliorum cicatriculis rugulosis junioribus sericeo-albidis, foliis latis ad apicem ramorum congestis spathulato-lanceolatis in petiolum dilatatum attenuatis sericeis albis, involucri foliolis spathulatis, interioribus oblongis basi concretis, capitulis amplis, floribus exterioribus ligulatis, ligulis elongatis apice 3-dentatis, dentibus ovatis acutis, floribus disci tubo cylindraceo medio constricto basi coriaceo albido laciniis lanceolatis acutiusculis subtus papillosis membrana marginali destitutis,

antheris basi caudatis caudis laceris, styli ramis elongato-
lanceolatis, fl. radii acheniis triquetris angulis pubescentibus, pappo illic sublongiore, disci 4-angulatis, pappo
æquali paleis achenia amplectentibus concavis dorso carinatis carina denticulata superne hirsutis apice in setam
fuscam productis.

Hab. In rupibus ins. *S. Nicholai* (*Forbes*).

121. Odontospermum *Daltoni,* Webb ; fruticulus erectus,
ramis virgatis dichotomis inferioribus rufis superioribus
pubescenti-hirtis subalbidis, foliis sparsis distantibus lineari-spathulatis in petiolum longiusculum sensim attenuatis, involucri oblongi squamis exterioribus basi atque
interioribus totis inter se concretis, capitulis mediocribus,
floribus exterioribus ligulatis, ligulis linearibus apice
breviter ovato 3-dentatis, disci tubulosis, tubo cylindraceo
superne constricto laciniis lanceolatis acutis margine membrana denticulato-fimbriata auctis, antheris basi caudatis,
caudis breviusculis basi sublaceris, styli ramis spathulatis
obtusiusculis, acheniis setosis, ligularum triquetris, pappo
setaceo vix denticulato subæquali ad angulos vix longiore,
paleis achenia amplectentibus concavis dorso acutis integris,
apice glabellis denticulatis obtusiusculis cum acumine.

Hab. In rupestribus sinus *Tarrafal* ins. *S. Antonii* (*Forbes,*
die 2 April. 1822, sp. juniora florida). In collibus alt.
1000 ad 2000 ped. que vallem *S. Dominici* obvallant in ins.
S. Jacobi (*J. D. Hooker,* n. 204, November, 1839, sp. fl.
et fructifera.)

122. Odontospermum *Vogelii,* Webb ; fruticulus diffusus,
ramis dichotomis albis, foliis sparsis lineari-spathulatis in
petiolum attenuatis utrinque sericeis, capitulis parvis ovatis,
involucri squamis lanceolatis imbricato-appressis inter se
concretis.

Var. *β, Darwini,* foliis elongato-linearibus confertis, capitulis
majoribus rotundatis, corollæ dentibus hirsutioribus ligulis
longiusculis valde papillosis breviter et acute dentatis.

Flores disci tubulosi, tubo cylindraceo medio vix constricto
basi crasso colorato, laciniis lanceolatis obtusiusculis mem-

brana margine subintegerrima auctis. *Antheræ* breviter caudatæ, caudis basi laceris. *Styli* rami ovato-claviformes obtusi. *Achenia* exteriora 3-quetra pappo ad angulos multo longiore, disci 4-gona, striata, subglabra, ad angulos setosa. *Pappus* setaceo-paleaceus, basi dilatatus, distans, brevis, seta unica ad angulum internum longe producta. *Paleæ* achenia amplectentes concavæ, apice acutæ setoso-hirtæ dorso carinatæ, carina fimbriata.

HAB. In ins. *S. Vincentii (Forbes,* n. 2. die 1 Aprilis, 1822, sp. unicum floridum.) Inter rupes Montis *Verede,* ins. *S. Vincentii* ab alt. 800 circiter ped. usque ad cacumen. ' Fruticulus pulcher, ramis decumbentibus, ramulis arrectis.' (*Th. Vogel,* n. 46 et 80. Junio, 1841, sp. fl. et fruct.) β. *Darwini,* in ins. *S. Jacobi (Darwin,* sp. florida.)

123. Blainvillea *Gayana,* Cass! *Dict.* 47. *p.* 90. DC. *Prodr.* 5. *p.* 492.

HAB. In rupestribus ins. *S. Jacobi* vulgaris (*J. D. Hooker,* n. 182. November, 1839, sp. flor. et fruct.)

Cassini, in his description of this plant (l. c. p. 91), says that the pappus is formed of three squamules, between which there exist some rudiments of smaller imperfect squamellules. The fact is that the pappus is biserial, the exterior composed of two or three aristæ continuous with the nerves of the angles of the achenium, the interior shorter, of many aristæ fringing the margin of the disk (*pulvillus,* Cass.), which surrounds the base of the corolla and nectarium. This latter organ is slightly elongated in the form of a beak, more so indeed in this species, than in any other of the genus.

124. Zinnia *pauciflora,* Linn. *Sp. Pl. p.* 1269. Rudbeckia foliis oppositis hirsutis ovato-acutis, *Zinn. hort. Gætt. p.* 409. Zinnia lutea, *Gærtn.* 2. *p.* 459.—Ic. Zinn. *l. c. t.* 1. Gærtn. *l. c. t.* 172.

Var. β. *multiflora.* Zinnia multiflora, *Linn. Sp. Pl. p.* 1269.— Ic. Linn. fil. *Dec. t.* 12.

HAB. Var. α. in ins. *S. Jacobi (Darwin).* Var. β. ad apicem collis alt. 2000 ped. in valle *S. Dominici* ins. *S. Jacobi* (*J. D. Hooker,* n. 206. November, 1839. sp. flor. et fruct.)

Ambæ ut credibile est formæ cum *Tagete patula*, Linn.
ex hortis urbanis olim erant transfugæ et nunc e civibus
Americanis Africæ metœcæ factæ sunt.
125. Sclerocarpus *Africanus*, Jacq. *Act. helv.* 9. (1786) *p.* 34.
DC. *Prodr.* 5. *p.* 566.—Ic. Jacq. *l. c. t.* 2. *f.* 1. et Ic. rar.
1. *t.* 176.
Hab. In petrosis ins. *S. Jacobi* (*J. D. Hooker,* n. 183. Nov.
1839. sp. fruct.) In herb. ins. Cap. Vir. (*Mus. reg. Par.*)
sp. florifera.
126. Bidens *bipinnata*, Linn. *Sp. Pl. p.* 1166. DC. *Prodr.* 5.
p. 603.—Ic. Moris. *Hist. s.* 6. *t.* 7. *f.* 23. Herm. *Parad.*
t. 123.
Hab. In herb. ins. Cap. Vir. (*Mus. reg. Par.*)
127. Bidens *pilosa*, Linn. C. H. Schultz. Bip. *in Phyt. Can.*
2. *p.* 242.
Var. *a. radiata,* C. H. Sch. Bip. *l. c.* Coreopsis leucantha,
Linn. Sp. Pl. p. 1282. Bidens leucantha, *Willd. Sp. Pl.* 3.
p. 1719. *DC. Prodr.* 5. *p.* 598. *Brunn. Ergebn. p.* 26.
Var. *β. discoidea,* C. H. Sch. Bip. *l. c.* Bidens pilosa, *Linn.*
Sp. Pl. p. 1166. *Willd. Sp. Pl. v.* 3. *p.* 1719. *DC. Prodr.* 5.
p. 1597.
Hab. In ins. *S. Jacobi* vulgatissima, ubi achenia uncis armata
ambulantibus mense Novembre valde molesta (*J. D. Hooker,*
n. 201. Nov. 1839, sp. florida et fruct. discoidea, unicum
radiatum). In ins. *S. Vincentii* et *S. Antonii* (*Th. Vogel,*)
spec. discoidea et radiata in speciminibus autem *Viriden-*
tibus radius et quasi exstincturus nec pulchre et conspicue
explicatus ut in *Canariensibus* et *Maderensibus.*
128. Tagetes *patula*, Linn. *Sp. Pl. p.* 1249. DC. *Prodr.* 5.
p. 643.—Ic. Dod. *Pemph. p.* 255. Dill. *Hort. Elth. t.* 279.
Hab. In ins. *S. Jacobi* (*Darwin*). In planitie *Porto Prayensi*
(*J. D. Hooker,* n. 205. Junio, 1839, specimina florifera).
129. Artemisia (Absinthium) *Gorgonum*, Webb; caule frutescente, ramis robustis fulvo-tomentosis, foliis flavide
tomentosis, supra sulcatis subtus 1-nerviis 2-3-pinnati-
partitis, pinnis latis oblongis versus apicem 3-5-dentato-
lobatis, paniculis thyrsoideis, capitulis mediocribus nutan-

tibus, involucri squamis late ovatis apice rotundatis
margine scariosis laceris dorso leviter tomentosis, receptaculi convexi setis brevibus latis acutis, floribus glabris
tubulosis laciniis lanceolatis acutis, radii paucis fœmineis,
disci hermaphroditis, genitalibus inclusis, antheris oblongis
apice longe acuminatis acutissimis loculis basi subacutis,
styli ramis truncatis apice breviter et parce papilloso-peni-
cellatis.

HAB. In herb. ins. Cap. Vir. (*Mus. reg. Par.*)
Ramis robustis, involucri squamis, corollæ antherarum et
stigmatum forma, ab affini *A. Canariensi* differt. Generis
borealis conturmalis extremus æquinoctii æstus subit.

130. Gnaphalium *luteo-fuscum*, Webb; herbaceum, totum
albido-tomentosum, radice lignescente nigro, foliis inferioribus spathulatis petiolatis superioribus oblongis apice
spathulatis sessilibus subamplexicaulibus obtusiusculis,
capitulis corymboso-cymosis heterogamis, floribus fœmineis
multiserialibus, hermaphroditis paucis, involucri squamis
scariosis glabris acutissimis fusco-luteis demum sordide
fuscis, acheniis nigris ovatis costatis glabris.

HAB. In petrosis supra medium Montis *Verede* ins. *S. Vincentii* (*Vogel*, n. 38. 55. 56. Junio, 1841, sp. florida et
fructifera.)

131. Gnaphalium *luteo-album*, L. *Sp. Pl.* 1196. DC. *Prodr.* 6.
p. 230.—Ic. *Engl. Bot. t.* 1003. *Fl. Dan. t.* 1763.

HAB. In declivibus umbrosis ins. *S. Nicolai* (*Forbes*, n. 26.
die 27 Martii, 1822) et in arvis *Gossypinis* sinus *Tarrafal*
ins. *S. Antonii* (Id. n. 5. die 2 Aprilis, 1822, spec. florida.)

132. Centaurea *Melitensis*, Linn. *Sp. Pl. p.* 1297. Var. *a.*
conferta, *C. H. Schultz, Bip. Phyt. Can.* 2. *p.* 360.—Ic.
Bocc. *Pl. sic. et Mel. t.* 35.

HAB. In herb. ins. Cap. Vir. (*Mus. reg. Par.*) Spec. floridum.

133. Schmidtia *farinulosa*, Webb; caule fruticoso brevi crasso,
foliis lanceolatis et lineari-lanceolatis sessilibus margine
sparse dentatis apice attenuatis integerrimis cum panicula
juniore tomento albo deciduo farinoso coronata, panicula

brevi subumbelliformi nuda squamigera, involucri squamis
exterioribus brevissimis, interioribus filiformibus glabrescentibus margine scariosis, acheniis turbinatis costulatis
subpapillosis, pappo 4-5-setoso, squamulis intermediis
plurimis interjectis.

HAB. In summo cacumine Montis *Verede* ins. *S. Vincentii*
(*Th. Vogel*, n. 53. Junio, 1841, spec. 2. florifera et fructifera).

134. Urospermum *picroides*, Desf. *Cat. hort. Par. ed.* 1.
p. 90. DC. *Prodr.* 7. *p.* 116. Tragopogon picroides, *L.*
Sp. Pl. 1111.—Ic. Lamck. *Ill. t.* 646. *f.* 3.

HAB. In herb. ins. Cap. Vir. (*Mus. reg. Par.*)

135. Lactuca *nudicaulis*, Murr. *N. comm. Gœtt.* 3. *p.* 74.
C. H. Schultz, Bip. *in Linnæa*, 15. *p.* 725. Chondrilla nudicaulis, *Linn. Mant. p.* 278. Microrhynchus nudicaulis, *Less.*
Syn. p. 139. DC. *Prodr.* 7. *p.* 180. excl. var. *β.*—Ic. Murr.
l. c. t. 4.

HAB. In saxosis ins. *S. Jacobi (J. D. Hooker*, n. 202. Nov.
1839. spec. fructifera.)

Dr. Schultz rightly observes that the original *M. nudicaulis*, Less. is a mere section of *Lactuca*, distinguished
solely by its achenia having a beak shorter than in the
other species. We shall show hereafter that the remaining
species, placed by DC. in this genus, together with his variety
β. of the original species, belong to the genus *Rhabdotheca*,
Cass. *Microrhynchus* therefore must be entirely abandoned.

136. Sonchus *oleraceus*, *a.* et *β.* Linn. *Sp. Pl. p.* 1116.
Sonchus oleraceus, *Koch, Syn.* (*ed.* 2.) 2. *p.* 497.—Ic.
Hayne, *Arzneigen, t.* 48.

HAB. Ad apicem montis cujusdam in valle *S. Dominici* ins.
S. Jacobi (J. D. Hooker, n. 203. Nov. 1839, spec. fructi-
ferum) in cultis ins. *S. Vincentii* ad alt. 500 ped. (*Th.
Vogel*, n. 68. sp. flor.) et in cultis ins. *S. Antonii* (*Th.
Vogel*, n. 36. sp. fructiferum, Junio, 1841.)

137. Sonchus *Daltoni*, Webb; caule brevi crasso lignoso
apice foliosissimo, foliis anguste lanceolatis apice attenuatis runcinato-lobatis lobis latis rotundatis denticulatis

glabris utrinque viridibus, petiolo basi incrassato coriaceo
amplexicauli, caulinis oblongis acutis basi in appendicem
rotundato-cordiformem amplexicaulem dilatatis, inflorescentia umbelliformi, capitulis rotundatis, involucri ovato
rhomboidei squamis acutis glabris margine minutissime
denticulatis.

Sonchum congestum, Link, refert sed involucri squamæ valde
diversæ.—Ic. Tab. nost. X.

Docto atque indefesso Josepho Dalton Hooker orbem An-
tarcticam jam visuro ejusque floram illustraturo stirpem
inter primas quas Britannia relicta compulit sacram dica-
tamque voluimus.

HAB. In cacumine collis abrupti alt. 1500 ped. in valle
S. Dominici ins. *S. Jacobi (J. D. Hooker,* n. 199. Nov.
1839) et in ins. *S. Vincentii* copiosus ad apicem Montis
Verede (Th. Vogel, Junio, 1841. sp. flor.)

TAB. X. *Fig.* 1. floret; *f.* 2. achænium :—both *magnified.*

RHABDOTHECA, *Cass.*

Great confusion has all along existed in the classification
of the plants attributed by DC. to the genus *Microrhynchus,*
founded originally by Lessing on the second species of
Lomatolepis, Cass. viz. *L. (Chondrilla,* L.) *nudicaulis,* Cass. To
the genus thus constituted by Lessing from this single plant,
De Candolle appended in his first section three others,
Sonchus divaricatus, Desf., as a mere var. of *M. nudicaulis,*
Less., (we made the same mistake in the *It. Hisp.*), *M. patens,*
DC. and *M. asplenifolius,* ejusd. The original *M. nudicaulis,* and perhaps *M. patens,* alone truly belong to the
genus. *M. nudicaulis,* β *divaricatus (S. divar.* Desf.) is in
reality a very distinct species, in which, as in the cognate
species, *M. asplenifolius,* the pappus is entirely sessile, there
being at no time any appearance of beak either in the ovary
or fruit. These plants, therefore, require to be removed from
their present position; and we must see to what genus they
can be conveniently attached.

The first was erroneously supposed by De Candolle

L

to be the plant described by Cassini as his *Rhabdotheca sonchoides.* He was led astray probably by the article itself of the Dict., in which the genus *Rhabdotheca* is described; where the author says that the plant on which it is founded was ticketed *S. divaricatus* in the herbarium of M. Gay. By an inspection, however, of the plant itself, I find it to be the *Sonchus chondrilloides,* Desf. (*Zollikoferia chondrilloides* of DC.), and it is the type of his genus *Zollikoferia,* as well as of the more ancient genus *Rhabdotheca,* Cass. This latter name must therefore prevail. Under this genus I consider that our present plants ought to be placed. It is distinguished from *Sonchus,* by its aspect approaching more to that of the *Lactuceæ* than the *Soncheæ,* by its capitulum not swelled at the base, by the scales of the involucrum being usually bordered by a wide scarious margin, the outer ones remarkably shorter, and by its tetragonal or rarely sub-5-gonal, sharp or rounded, not compressed achenia, rarely though sometimes attenuated towards the summit or base, and more or less papillated or scabrous. The following are the species which compose the genus *Rhabdotheca* thus considered.

R. chondrilloides,—R. sonchoides, Cass., *Zollikoferia chondrilloides,* DC., and probably likewise his second species *Rhabdotheca (Zollikoferia) pumila.*

R. divaricata,—Sonchus divaricatus, Desf. *Ann. Mus.* 2. *p.* 212. *t.* 46. *M. nudicaulis,* Webb, *It. Hisp. excl. syn.* non Less. *Microrhynchus nudicaulis, β divaricatus,* DC.

R. asplenifolia!—Prenanthes asplenifolia, Willd. *Microrhynchus asplenifolius,* DC.

To these I add two new species and the *Sonchus spinosus,* DC., which cannot remain with the section *Atalanthus* of *Sonchus* where De Candolle has placed it, with which it has so little affinity. Its capitulum, achenia, and general aspect, approach much more closely those of our present group; though its admission renders the genus less uniform.

138. Rhabdotheca *picridioides,* Webb ; caule basi suffrutescente foliorum cicatricibus superne annulato apice subrosulato-folioso, foliis oblongis in petiolum basi dilatatum

amplexicaulem angustatis apice rotundato-obtusis margine
spinulosis glaberrimis, scapo longissimo terminali tereti
striato glabro fusco apice vel rarius per totam longitudinem
ex axillis squamarum floriferis, inflorescentia subramosa,
pedicellis squamis sessilibus ovatis acutis margine undu-
lato-scariosis cum involucro continuis dense obsitis, involucri squamis inferioribus brevissimis conformibus ovato-
lanceolatisque interioribus elongatis subscariosis capitulum
cylindraceum efformantibus, flosculis inferne pilosis, antheris
basi caudatis, pappo pluriseriali exteriore simplici interiore
denticulato, acheniis elongatis 4-gonis angulis obtusis apice
vix attenuatis interioribus sublævibus exterioribus squa-
moso-scabridis.—Ic. Tab. nostr. XI.
The achenia of this species are somewhat longer than those
of the original *R. chondrilloides*, with the angles blunt, or as
Cassini expresses it subcylindraceous; but the granulation is
nearly the same.

HAB. In fissuris rupium ins. *S. Nicolai*, (*Forbes*, n. 34. die
27 Martii, 1822, spec. florida et fruct.) In Monte *Verede*
ins. *S. Vincentii* ab alt. 1000 ped. usque ad apicem (*Th.
Vogel*, n. 43. 44. Junio, 1841 spec. florifera et fruct.)
To this species the description of another, though not
within the province of our Flora, may be appended.*

TAB. XI. *Fig.* 1. floret; *f.* 2. achænium:—both *magnified*.
139. Rhabdotheca *spinosa*, Webb; "Thorny shrub sow-
thistle of Africa." Parkins, *Theatr. p.* 804. Prenanthes
spinosa, *Forsk. Fl. Æg. Arab. p.* 144. *Brunner, Erg. p.* 104.
Lactuca spinosa, *Lamk. Encycl.* 3. *p.* 408. Sonchus spinosus, *DC. Prodr.* 7. *p.* 189.—Ic. Parkinson, *l. c. Phytogr.
Can.* 2. *t.* 125.

* Rhabdotheca *Brunneri*, Webb; fruticulosa, ramis glabris horizontalibus,
foliis ramorum oblongo-linearibus linearibusque elongatis integerrimis,
inflorescentia corymbosa, pedicelli squamis ovatis, capitulis subcylindraceis squamatis, involucri glabri foliolis exterioribus lanceolatis
anguste scariosis, acheniis 4-gonis elongatis apice haud attenuatis
angulis subacutis striatis interioribus dense squamulosis.—Sonchus
"ex ins. *Sor* et palude *Limnutt*," Brunn. *Ergebn. p.* 116. Sonchus an
Prenanthes? ex siccis juxta paludem *Limnutt; ejusdem in herb. nostro!*

L 2

HAB. In siccis ins. *Boa Vista* (*Brunner* in schedis herb. nostri.)

Achenia quædam 4-gona, plura nunc obsolete nunc omnino 5-gona, angulis obtusiusculis, transverse rugulosis, spatio inter angulos sulcato.—Planta est quoad ordinationem difficilis neque *Soncho* nec *Prenanthi* uti nunc constituentur conjungenda, atque huc potius inter concives *Africanos* facie et patria similes melius collocanda est, quanquam achenia potius *Sonchorum* habeat virorum, sed involucrum *Rhabdothecæ.*

XXX. CAMPANULACEÆ, *Juss.*

140. Campanula (Medium) *Jacobæa,* Chr. Smith, *in Tuck. Voy. p.* 251. (Herb. Mus. Brit. ex *J. D. Hooker*); fruticulosa, caule noduloso lignescente cavo, ramis diffusis albidis junioribus fuscis strigoso-hirtis foliosis, foliis spathulatoovalibus lanceolato-ovatis obtusiusculis strigoso-hirtis subtus pallidis nervosis basi attenuatis caulinis breviter petiolatis superioribus semiamplexicaulibus, calycis tubo brevi cyathiformi laciniis anguste lanceolatis strigoso-ciliatis, corolla campanulata æquali calycis laciniis 3-plo longiore, filamentis plano-filiformibus basi dilatatis glaberrimis, capsula depressa, seminibus ovatis.   Variat floribus cœruleis albisve.—Ic. nost. tab. XII.

HAB. In ins. *S. Nicolai* fissuris rupium (*Forbes,* n. 35. die 27 Martii, 1822, spec. florida.)   In ins. *S. Antonii* (*Forbes,* n. 4. April, β alba spec. flor.)   In rupibus collis acuti *Campanula* sp. n. perpulchra in vallis *S. Dom.,* ins. *S. Jacobi* ad alt. 2000 ped. (*J. D. Hooker,* November, 1839, spec. flor. et fruct.)   In monte *Verede,* ins. *S. Vincentii* ab alt. 1500 ped. usque ad summitatem (*Th. Vogel,* n. 73. Junio 1841, specimina adusta.)   In ins. *S. Jacobi* (*Darwin.*)

TAB. XII. *Fig.* 1. stamen; *f.* 2. pistillum:—both *magnified.*

XXXI. CYPHIACEÆ, *D.C.*

141. Cyphia *Stheno,* Webb; caule elongato sarmentoso tenui flexuoso herbaceo glaberrimo vel superne parcissime

piloso basi squamulato, foliis distantibus glabris sparsim
glanduloso-dentatis acutis inferioribus 3-angularibus, superioribus linearibus, floribus axillis foliorum summorum
solitariis pedicellatis, pedicello filiformi puberulo supra
medium bibracteolato, bracteolis lineari-spathulatis integerrimis vel denticulatis, calyce turbinato 5-fido, tubo
brevi inter costulas pubescente, laciniis tubo duplo longioribus lineari oblongis obtusis infra medium laciniato-
dentatis glabratis, corolla calyce 3-plo longiore apice
purpurea subtus lutescente ultra medium tubulosa, tubo
leviter incurvo 2-labiato, laciniis æqualibus, 3 superioribus
arrectis lanceolatis acutis extus glabris pilis albidis pubescentibus venosis, petalis mox basi solutis, staminibus 5,
tubo corollæ inserta subdimidio breviora, filamentis liberis
basi latioribus, superioribus apice pilosulis, stylo staminibus breviore crasso glabro complanato stigmate simplici
rotundato sublaterali, ovulis a placentis ad apicem loculi
cujusque sitis pendulis.

HAB. In herb. ins. Cap. Vir. (*Mus. reg. Par.*)

XXXII. ASCLEPIADEÆ, *Juss.*

142. Sarcostemma *Daltoni*, Dcne.; ramis teretibus aphyl-
lis, umbellis terminalibus, pedicellis glabris, corollæ laciniis
ovatis ex oblique acuminatis glaberrimis, coron. staminea
exter. plicata sinubus subæqualibus obtusis, folior. coronæ
inter. basin æquantibus, fol. coronæ inter. rotundato-ovatis
gynostegio incumbentibus, stigmate pentagono medio ma-
milloso, folliculis lineari-lanceolatis glabris. *Decaisne*, MSS.
—Ic. Tab. nostra XIV.

Sarcostemma *nudum*, Chr. Smith, in Herb. Mus. Brit. (ex cl.
J. D. Hooker.)

HAB. Ad apicem collium et in rupestribus maritimis ins. *S. Jacobi*, " caulis haud volubilis" (*J. D. Hooker*, Nov. 1839, sp. fl.
et fruct.) Ibid. (*Forbes*, n. 11. die 5 Aprilis, 1822.) In
ins. *S. Antonii* (*Th. Vogel*, n. 22. Junio, 1841.) In ins.
S. Vincentii, " Asclepias, caule basi lignoso, ramis diffusis
teretibus viridibus procumbentibus vel pendulis. Latex

albus, flores rari, folia nulla vel pauca marcida." (*Th. Vogel.*)

Tab. XIV. *Fig.* 1. flower; *f.* 2. corona:—both *magnified.*
143. Calotropis *procera*, R. Br. *in Hort. Kew. ed.* 2. *p.* 78.
Decaisne, *in DC. Prodr.* 8. *p.* 535.—Ic. Apocynum Syriacum, *Clus. Hist.* 2. *p.* 87. Lindl. *Bot. Reg. t.* 1792.]
Hab. In insula parva *Coturnicum Portus Praya,* ins. *S. Jacobi* (*J. D. Hooker,* n. 207. Nov. 1839.) Ibid. (*Chr. Smith,*) in Herb. Mus. Brit. (ex cl. *J. D. Hooker.*) In ins. *S. Antonii,* (*Th. Vogel,* sine num.)
144. Periploca *lævigata,* Hort. *Kew.* 1. *p.* 301.—Ic. Periploca punicæfolia, Cav. *Ic.* 3. *p.* 91. *t.* 217. Periploca angustifolia, La Billard. *Dec.* 2. *t.* 7. Periploca rigida, Viv. *Pl. lib. spec. t.* 6. *f. f.* 3. 4. *h.*
Hab. In herb. ins. Cap. Vir. (*Mus. reg. Par.*)

XXXIII. Gentianeæ, Juss.

145. Erythræa *ramosissima,* Pers. *Syn.* 1. *p.* 283. "Centaurium minus palustre ramosissimum, flore purpureo." Vaill. *Bot. Par. p.* 32. Gentiana pulchella, *Swartz, Act. Holm.* 1783.—Ic. Vaill. *l. c. t.* 6. *f.* 1, Swartz, *l. c. t.* 3. *f.* 8 *et* 9. *Fl. Dan. t.* 1637. *Engl. Bot. t.* 458.
Hab. In ins. *S. Jacobi* (*Chr. Smith* ex sp. Mus. Brit. a cl. *J. D. Hooker,* viso.)

XXXIV. Bignoniaceæ, Juss.

146. Sesamum *radiatum,* Schum. *Guin.* 282, *ex DC.* Sesamopteris radiata, *DC. Prodr.* 10. *p.* 251. excl. syn. Endl. *Heudel. exsicc. Seneg.* n. 547!
Nolo huc *S. gracilis,* Endl. iconem cum celeb. DC. adducere, nam plantæ nostræ caulis nec gracilis, dense villosus nec glaberrimus, folia nunquam trisecta aut glabra, calyx corolla capsula hirsutissima. *Caulis* superne 4-gonus, pilis crispis villosissimus, foliosus. *Folia* villoso-pubescentia, subtus glaucescentia, inferiora rotundato- vel ovato-rhomboidea irregulariter et grosse dentata, superiora lanceolata subintegerrima acuta. *Calyx* persistens laciniis linearilanceolatis extus villosis. *Corolla* conspicua, purpurascens,

villosa. *Capsula* crassa 4-gona, pollicaris vel pollice brevior, 2 vel 2½ lin. lata brevissime acuminata. Semina nigra, testudinea, hinc plana illinc convexa, pyriformia, sulcis e medio ad marginem radiantibus ornata. *Testa* crustacea per totum marginis peripherium in laminas binas quarum altera plana, altera testudinis modo gibba, solubilis ; tegmen chartaceum albescens, apice chalaza basi pilo notatum. *Cotyledones* ovato-rotundatæ, compressæ, carnosæ ; radicula cotyledonibus triplo brevior.

XXXV. Convolvulaceæ, *Juss.*

147. Rivea *tiliæfolia,* Chois. *Conv. Or. p.* 25, *et in DC. Prodr.* 9. *p.* 325. Convolvulus tiliæfolius, *Desrouss. Encycl.* 3. *p.* 544.

Hab. In ins. *S. Vincentii.* "Caulis sarmentosus et volubilis, sæpe quoque colitur tantum modo sicut visum est ut sarmenta ad tecta straminis modo tegenda inserviant." (*Th. Vogel,* n. 27. Junio, 1841, sp. florida).

148. Batatas *paniculata,* Chois. *in DC. Prodr.* 9. *p.* 339. *a* lobata. Convolvulus paniculatus, *Linn. Sp. Pl. p.* 223. Ipomœa Mauritiana, *Jacq. Hort. Schœnb.* 11. *p.* 73. Ipomœa gossypiifolia, *Willd. enum. p.* 208. Ipomœa paniculata, *R. Br. Prodr. p.* 486. Convolvulus roseus, *Kunth, Syn. p.* 222.—Ic. Jacq. *l. c. t.* 200. Ker, *Bot. Mag. t.* 62. Hab. Ad dimidium et inde ad apicem usque Montis *Verede,* ins. *S. Vincentii,* caulis prostratus. (*Th. Vogel,* n. 64. Junio, 1841, spec florida cum ramis aliis jam diu fructiferis.)

149. Batatas *pentaphylla,* Chois. *Conv. Or. p.* 54. *et in DC. Prodr. p.* 339. Wight, *Ic. Pl. Ind. Or. p.* 3. Convolvulus pentaphyllus, *Linn. Sp. Pl. p.* 223. Ipomœa pilosa, *Cav. Ic.* 4. *p.* 11.—Ic. Cav. *l. c. t.* 353. Jacq. *l. c. t.* 319. Wight, *l. c. t.* 834. Hab. In rupestribus ins. *S. Jacobi* (*J. D. Hooker,* n. 138. Nov. 1839, flores albi, spec. fructifera apice florida.) Ibid. (*Chr. Smith,* in herb. Mus. Brit. ex cl. *J. D. Hooker.*)

150. Ipomœa *Pes-capræ,* Sweet, *Hort. sub Lond. ed.* 2. *p.* 289.

152   SPICILEGIA GORGONEA.

Chois. *in DC. Prodr.* 9. *p.* 349. Convolvulus Pes-capræ
et Convolvulus brasilianus, *Linn. Sp. Pl. p.* 226. Ipomœa
maritima et Ipomœa carnosa, *R. Br. Prodr. p.* 486. Convolvulus retusus, *Coll. hort. rup. app.* 3. *p.* 31.—Ic. Ipomœa
maritima, Curt. *Bot. Mag. t.* 319. Coll. *l. c. t.* 8.
HAB. In ins. *S. Jacobi* ad pagum Villa do Rio (*Forbes*, n. 10.
d. 5 April. 1822, spec. florida.) Ad littora ins. *S. Jacobi*
frequens (*Brunn.* sp. flor. in herb. nostro.) Ibid. (*Darwin*,
sp. sine fl. et fruct.) In ins. *S. Jacobi* (*Chr. Smith*, in herb.
Mus. Brit. ex cl. *J. D. Hooker*.)
151. Ipomœa *leucantha*, Jacq. *Coll.* 2. *p.* 280. Chois. *in DC.
Prodr.* 9. *p.* 382.   Convolvulus leucanthus, *Desrouss.
Encycl.* 3. *p.* 541.—Ic. Jacq. *Ic. rar.* 2. *t.* 318.
HAB. In rupestribus ins. *S. Jacobi*, flores dilute carnei
(*J. D. Hooker*, n. 137. November, 1839, spec. fructifera et
juniora florida.)
152. Ipomœa *Coptica*, Roth, *n. sp. p.* 110. Chois. *in DC.
Prodr.* 9. *p.* 384.   Convolvulus copticus, *Linn. Mant.
p.* 559.
HAB. In valle *S. Dominici*, ins. *S. Jacobi*, (*J. D. Hooker*,
n. 161. Nov. 1839. spec. fl. et fruct.)
153. Ipomœa *Cairica*, Linn. *Sp. Pl. p.* 222. (sub Convolvulo).
Ipomœa palmata, *Forsk. Fl. Æg. Arab. p.* 43.—Ic. C. Bauh.
Prodr. p. 134. Moris, *Hist. s.* 1. *t.* 4. *f.* 5. (cum ic a C. B.
mutuata.) Barrel. *Ic. t.* 319 (cum ic. eadem Bauhiniana)
et t. 320 (cum ic. propria). Convolvulus Ægyptius, Vesling.
Obs. t. 74. *Bot. Mag.* 699.
HAB. In ins. *S. Jacobi* (*J. D. Hooker*.) Ibid. (*Chr. Smith*,
in herb. Mus. Brit. ex cl. *J. D. Hooker*.)
154. Ipomœa *sagittata*, β diversifolia, Chois. *in DC. Prodr.* 9.
p. 372.   Convolvulus diversifolius! *Vahl, MSS. in herb.
Desf.*
HAB. In herb. ins. Cap. Vir. (*Mus. reg. Par.*) sp. unicum
sine flore et fructu.

The leaves of our plant are softly downy and the auricles
somewhat rounded at the point.   It is perhaps distinct; but
in the absence of flower and fruit it is not possible to decide

this. The plant of Vahl has no hairs, like those of Desfontaines and Michaux; but more than two species are probably confounded under this name.

155. Ipomœa *muricata*, Ker, *Bot. Reg. in notulis ad calc. v.* 4. Convolvulus muricatus, *Linn. Mant. p.* 44. I. bona-nox, var. β purpurascens, *Ker, Bot. reg. t.* 290. Calonyction speciosum, β muricatum, *Chois. in DC. Prodr.* 9. *p.* 345. Notwithstanding the admonitory note of Linnæus, repeated by Ker, this plant is still referred by M. Choisy to his *Calonyction bona-nox*, of which the *I. muricata*, Jacq. is really a variety. Not so our plant; if we rightly refer to this species, on account of its annual stem and small capsules, a fragment collected by Vogel, and of which there exists a specimen in the herb. ins. Cap. Vir. (*Mus. reg. Par.*)

156. Evolvulus *linifolius*, Linn. *Sp. Pl. p.* 392. Chois. *in DC. Prodr.* 9. *p.* 449. Convolvulus herbaceus erectus &c. *P. Br. Jam. p.* 152.—Ic. P. Browne, *l. c.*

Hab. In herb. ins. Cap. Vir. (*Mus. reg. Par.*) et ex ins. *S. Jacobi,* (*Chr. Smith,* in herb. Mus. Brit. ex cl. *J. D. Hooker.*) In ins. *S. Jacobi,* (*Darwin.*)

XXXVI. Boragineæ, *Juss.*

157. Pollichia *africana*, Med. *Bot. Beobacht.* (1784) *p.* 248. *num.* 223. *Philosophisch. Bot. pars* 1. *p.* 32. "Cynoglossoides Africana verrucosa et hispida." *Isn. act. ac. Par.* (1718) *p.* 256. Borraginoides angustifolia, flore pallescente, cæruleo. *Boerhaav. ind. alt. p.* 188. Borrago africana, *Linn. Sp. Pl. p.* 197. Trichodesma africana, *R. Br. Prodr. p.* 496. Borago, *Chr. Smith,* in herb. Mus. Brit. ex cl. *J. D. Hooker,* Borago gruina, ejusd. *Tuck. Voy. p.* 250.— Ic. Isnard. *l. c. t.* 11. Boerhaav. *l. c. t.* 9. sed sine numero.

Hab. In rupestribus sinus *Tarrafal* sive *Tamaricum* ins. *S. Antonii,* (*Forbes,* n. 17. d. 2 April. 1822, spec. florida et fructifera.) In eadem ins. (*Th. Vogel,* n. 20. sp. floridum.) In ins. *S. Vincentii* ad alt. 500 ped. (*Th. Vogel,* n. 67. Junio, 1841, sp. florida et fructifera.)

Ut Medorum et Persarum ita Botanicorum leges stabiles

firmæque servandæ; hoc tantum modo inextricabilis fugienda confusio. Nulla autem jure antecessionis sive ut dicitur *prioritatis* sacriorem fuisse legem necesse est. Hoc lege *Pollichiæ Medici,* jam diu a celeberrimo Roberto Brown deletæ, pristini honores reparandi, Solandriana delenda. Hoc ill. Brownium effugisse videtur cum *Trichodesmam* suam, nunc ex albo eradendam, designaverit. Oculatus enim Medicus non tantum in opere, cui titulus, non plane immeritus, *Philosophische Botanik,* magno scrutatore citatum, anno 1789 divulgatum (eodem scilicet quo in Hort. Kew. suam Solander) sed jam ab anno 1783 in observationibus suis (*Botanische Beobachtungen,*) uti ipse advertit pro *Boraginibus Indica* et *Africana* constitutam, *Pollichiam* omnium primus juris publici anno 1784 fecerat et species fusius descripserat, quod ex libro facile videndum. Pro *Pollichia* Sol. *Meerburgia campestris* scribenda quod nomen ætate provectius.

158. Heliotropium *hispidum,* Forsk. (sub Lithospermo.) *Fl. Æg. Arab. p.* 38. Heliotropium undulatum, *Vahl, Symb. p.* 13. Heliotropium crispum, *Desf.! Fl. atl.*

159. Heliotropium *undulatum,* β ramosissimum, *Lehm.! Asperifol.* 1. *p.* 57. *Ic. et Descr. p.* 24. forma elongata macra. *Schult. Syst.* 4. *p.* 30. *et* 728. Heliotropium plebeium, *Chr. Smith,* in herb. Mus. Brit. (ex cl. *J. D. Hooker.*)— Ic. Desf. *l. c.* Lehm. *l. c. t.* 40.

HAB. In rupibus sinus *Tarrafal* sive *Tamaricum* ins. *S. Antonii,* (*Forbes,* n. 21. d. 2. April. 1822. spec. flor. et fruct.) In ins. *S. Vincentii* (Idem, n. 3.) In rupibus prope *Portum Praya* in ins. *S. Jacobi,* flores pallide purpurei (*J. D. Hooker,* n. 124. Nov. 1839, spec. flor. et fruct.) In cultis *S. Vincentii* et *S. Antonii* (*Th. Vogel,* n. 23. 33. et 69. Junio, 1841, sp. fl. et fruct.) In ins. *S. Jacobi,* (*Darwin.*)

Flores quanquam hujus speciei pallide cærulescentes scripserit cl. J. D. Hooker, albos Forskåhl et Desfontaines, non aliam credemus nostram, nulla enim apparet differentia nisi cocca forsan magis angusta ac rugosiora, sed et

hoc variabile, extant enim specimina quæ Ægyptiaca ex
toto referant.

160. Echium *stenosiphon*, Webb ; caule fruticoso strigoso,
ramis fuscis pilis aculeatis strigosis foliis strigosissimis
ovato- vel rhomboideo-rotundatis margine sinuatis acutis
vel obtusis inferioribus sinuato-lobatis breviter petiolatis
superioribus sessilibus, spicis ramosis foliosis basi nudis
apice floridis, floribus densis secundis, bractea oblonga
calycis laciniis oblongis vel linearibus subæqualibus longiore, corolla hirsuta cœrulescente calyce 4-plo vel 5-tuplo
longiore tubo cylindraceo subarcuato ima basi squamarum
annulo instructo, fauce vix ampliato, lobis brevibus ; staminibus exsertis infra medium tubi insertis, stylo staminum longitudine apice glabro sub apice usque ad basin
pilosissimo, ovarii lobis angustis acutis glabris.—Ic. Tab.
nostra XV.

HAB. Undique in insula *S. Nicolai* (*Forbes*, n. 32. die 29
Martii, 1822, spec. floridum minus scabrum.)   In monte
Verede, ins. *S. Vincentii* ultra 1000 ped. alt. frutex bipe-
dalis ramosus (*Th. Vogel*, n. 81. Junio, 1841, spec. florida.)

TAB. XV. *Fig.* 1. flower :—*magnified.*

161. Echium *hypertropicum*, Webb ; caule fruticoso, ramis
robustis fuscis cicatricibus foliorum notatis, ramulis pilis
strigosis appressis cinereis, foliis ovato-lanceolatis basi
attenuatis sessilibus utrinque pilis crebris brevibus tenuibus appressis e bullio lato plano prodeuntibus strigoso-
incanis nervis prominentibus margine planis ciliatis, panicula thyrsoidea ramosa, bracteolis oblongis subfalcatis
calycis longitudine cinereo-hirsutis apice strigosis, floribus
carneis vel albidis, calycis tubo brevissimo laciniis 4 lineari-
lanceolatis quinta lanceolata, corolla calyce subduplo longiore campanulata vix incurva extus pilosiuscula, tubo
crasso, lobis ovato-lanceolatis acutis, lacinia infima minore,
staminibus exsertis subincurvis glabris purpureis(?) stylo
leviter piloso.

HAB. In herb. ins. Cap. Vir. (*Mus. reg. Par.*)

Affinis est *E. giganteo* differt tamen foliorum forma et pubescentia et floris characteribus.   Non minus ab *E. Descaisnei*

diversum est cui flores lactei nec cœrulescentes ut in Phyt.
Can. (v. 2. p. 49.) falso diximus quod nuper ex speciminibus, insulæ Lancerottæ pulchre florescentibus cl. Bour-
geau cognovimus.

XXXVII. Labiatæ, *Juss.*

162. Ocymum *Basilicum,* Linn, (1764, ed. 3.) 2. *p.* 833·
Benth. *Lab. Gen. et Spec.* (1832-36.) *p.* 4. Basilicum indicum, *Burm. Herb. Amboin.* (1747) *pars* 5. *p.* 266. Soladi
Tirtava. *Rheed Hort. Malab.* (1690) *pars* 10. *p.* 171. Ocymum Americanum, *Jacq. Hort. Vind.* (1766) 3.—Ic. Burm.
l. c. t. 93. mediocris. Rheed, *l. c. t.* 87. mediocris. Jacq. *l.
c. t.* 86. bona.
Hab. In ins. *S. Jacobi* in valle *S. Dominici,* (*J. D. Hooker,*
n. 121. November, 1839, spec. fructifera.)
Hab. hæc species in regionibus calidioribus orbis veteris et
novi sed illic si indigena incertum. Occurrit in oris utrisque Lybiæ calidioris (ex Benth.) in insula Mauritii (ex
Benth.) in regionibus Indiæ intra Gangem Mace! Wight!
in insula Taprobana, (ex Benth.), in insula Java Commerson! in insulis Philippianis Commerson! ad insulas Antillicas in ins. Jamaica, Murray! in Brasilia, Martius!
Calycis tubus intus pilosus. *Mericarpia* oblonga, lævia, ni-
grescentia, humectata mucilaginosissima. *Cotyledones* oblongæ, obtusæ, cordatæ, radicula parva, crassiuscula.
163. Ocymum *suave,* Willd. *Enum.* (1809) 2. *p.* 629. Benth.
Lab. Gen. et Sp. (1832-36) *p.* 7.
Hab. In insula *S. Jacobi,* (*Darwin,* n. 276. specimen floridum.)
Legitur hæc planta in Senegambia prope urbem Kandoniæ,
Heudelot! (exsicc. n. 769 in herb. Delessertiano) atque
inde in insulam Madagascariæ, (ex Benth.) in insulam
Sanctæ Mariæ, Richard! (exsicc. n. 23. in herb. Delessertiano), et in insulam Anjouan, Richard! (exsicc. n. 239.
in herb. Delessertiano) procurrit.
Calycis tubus intus nudus. *Mericarpia* rotundato-ovata,
excavato-punctata, fusca, humectata non mucilaginosa.

Cotyledones mericarpio conformes cordatæ, radicula parva crassiuscula.

164. Hyptis *spicigera,* Lamk. *Dict.* 3. *p.* 185. Benth. *Lab. Gen. et Spec.* (1832-36) *p.* 78. Nepeta maxima, *Sloane Hist. of Jam.* (1707) *t.* 8. bona.

HAB. In herb. ins. Cap. Vir. (*Mus. reg. Par.*) Occurrit hæc species sed infrequens in vetere orbe et in novo in Senegambia per ripas Casamancæ, Le Prieur et Perrottet ! in herb. Delessertiano in Madagascaria, (ex Benth.), in ins. Luçon, prope Manillam, (ex Benth.) ; in ins. Antillicis in herb. Desf. ! in ins. Jamaica in uliginosis circa urbem St. Jago de la Vega, (ex Sloane) ; in Peruvia et Brasilia, (ex Benth.)

In hac specie *folia floralia* parva primum ovata, integra, mox 3-4 lineari-partita et ab hanc causam bracteolas esse plures celeb. Bentham apparuit. *Calyx* (8 m. metr. longus) tubulosus, 10-striatus, 5-dentatus ; dentibus tubo dimidio brevioribus. *Corolla* apice villosa. *Filamenta* parum exserta, exteriora longiora. *Stylus* inclusus. *Stamina* introrsa adæquans. *Mericarpia* oblongo-ovata, obtusa, ad basin exteriorem cicatricula minuta fungosa notata. *Cotyledones* ovatæ, obtusæ, cordatæ, radicula parva crassiuscula.

165. Lavandula *rotundifolia,* Benth. *Lab. Gen. et Sp.*

HAB. Ubique in insula *S. Nicolai,* (*Forbes,* n. 33. die 27 Martii, 1822, sp. florida et fructifera) ; ad medium montem *Verede,* ins. *S. Vincentii,* (*Vogel,* n. 77. Junio, 1841, spec. fructifera et florida.) In ins. *S. Antonii,* (*Vogel,* n. 5.) et in *herb. Mus. reg. Par.*

Frutex ramis lignosis, glabris, elongatis, basi foliosis. *Folia* petiolata, late oblongo-ovata, laciniato-dentata, basi cuneata, glabra, coriacea, rugoso-nervosa ; floralia scariosa, ovata, acuta, 5-striata, cinereo-puberula, calyce breviora, in spicam adpressam crassiusculam ramosam disposita. *Cymulæ* uniflores. *Calyx* sub 2-labiatus, tubuloso-ovatus, 15-striatus, 5-dentatus, dentibus ovatis acutis subæqualibus, fructiferis recurvis. *Corolla* calycem sub 2-plum

158   SPICILEGIA GORGONEA.

superante; tubo angusto, pilis reflexis intus piloso, labio
superiore bifido, inferioris 3-lobi lobo medio multo majore.
Antheræ ciliatæ.   *Stylus* glabrescens, lobis stigmatosis
ovatis latiusculis.   *Mericarpia* atro-purpurascenta, dorso
basi cicatricula magna notatis, humectata dense mucilaginosa.
166. Lavandula *coronopifolia*, Poir.! *Dict. Suppl.* (1813) 111.
p. 308. De Gingins, *Hist. Nat. des Lav.* (1827) *p.* 160.
t. 9. bona. Benth. *Lab. Gen. et Spec.* (1832-36) *p.* 151.
Lavandula multifida, *Burm. Fl. Indic.* (1768) *ic.* 38. *(mala.)*
Lavandula stricta, *Delile! Fl. Ægypt.* (1813) 1. *p.* 94.
ic. 32. *f.* 1. (*optima.*)
HAB. In ins. *S. Antonii,* (*Vogel,* n. 48.); et in collibus et
declivitatibus in *S. Vincentii,* (*Vogel,* n. 24. Junio, 1841.)
In Ægypto ad sinum arabicum, Delile! in herb. Mus. Paris.
In Arabia Petræa, prope montem Sinai, N. Bove! exsicc.
n. 59. prope Wadi Hebran, W. Schimper! exsicc. n. 141.
prope Djeddam Botta! in herb. Mus. Paris.
This plant certainly is the *L. coronopifolia* of Poir. and the
L. stricta of Delile. Although the spikes are more ramified,
its general appearance agrees with what Delile mentions in his
description, " cette espèce est principalement caractérisée par
ses longs épis linéaires."
    The Baron Gingins de Lassaraz, in his clever "Histoire
Naturelle des Lavandes," has attached great importance to
the forms of the lobes of the style; and I firmly believe
with this botanist that they form good specific characters.
167. Lavandula *dentata*, Linn. *Spec.* (1764) 11. *p.* 800.
Desf. *Flor. Atl.* (1796) 11. *p.* 14. Ging. de Laz. *Hist. Nat.
Lav.* (1826) *p.* 138. Benth. *Lab. Gen. et Spec.* (1832-36)
p. 148. Webb. *Phyt. Can.* (1845) *p.* 57.—Ic. Ging. de Laz.
l. c. t. v. f. i. (ic. pulchr.)
HAB. In herb. ins. Cap. Vir. (*Mus. reg. Par.*)
Occurrit hæc species in Europa per regiones maris interni
et in plagis rupestribus Africæ occidentalis inde per insulas
Maderensem Fortunatosque usque ad insulas Cap. Viridis
devenitur.

Observandum corollam calyce vix longiorem, filamentis pilosulis, stylo apice piloso, lobis oblongis, obtusis, meri-
carpiis oblongis, obtusis, cotyledonibus cordatis, radicula
parva crassiuscula.

168. Salvia *Ægyptiaca*, Linn. *Spec.* (1762) *p.* 33. Desf. *Flor.
Atl.* (1796) 1. *p.* 19. (ubi perperam annua dicitur.) Delile,
Flor. Ægypt. illustr. (1812) *p.* 49. Benth. *Lab. Gen. et Sp.*
(1832-36) *p.* 309. De Noel, *in Phytog. Canar.* (1845) 2.
p. 6. Salvia pumila, *Benth. l. c. p.* 726. *Cambess. Voy. de
Jacquemont, Bot. p.* 128.—Ic. Cambess. *l. c. t.* 133. De
Noe, *l. c.* bona.

HAB. In insula *S. Vincentii*, in arenosis (*Forbes*, specimina
n. 1. fructifera, 1⁰ die Aprilis, 1822.) In campis siccis
insulæ *S. Jacobi* frequens, (*J. D. Hooker*, specimina fruct.
n. 123. November, 1839.) In insula *S. Antonii*, (*Vogel*,
specimina n. 29.) nec non in vallibus arenosis *S. Vincentii*,
(n. 9. Junio, 1841.)
Ab Oceano Atlantico stirps usque ad Pentapotamidem Indiæ
Borealis tractum agri Cachemyriani confinem. Per trans-
versam Africam Asiamque excurrit inter lat. bor. grad. 22
et 33 ubique inclusa occurrit. In agro Tunetano interiore
circa Cafsam (Desf.!) In collibus magnæ Syrteos (ex
Viviano, sub nomine *Thymi hirti.*) In insula Teneriffæ,
(Webb!) In desertis Ægypti Inferioris circa Cahiram,
(Forsk. et Delile) et Suez (Delile!) In Arabia circa
Djeddam (Schimper! exsicc. n. 820.) Ad sinum Persicum,
(Aucher! exsicc. n. 5216.) In collibus gypsaceis et
salinis Indiæ Borealis ad Hydaspem Flumen, circa Pen-
dadenkhan, ad viam inter Lahore et Cachemyr, (Jacquemont! herb. quæ *S. pumila*, Benth.)

169. Micromeria *Forbesii*, Benth.! *Lab. Gen. et Spec.* (1832-
36) *p.* 376.
HAB. In insula *S. Nicolai*, in saxosis montis *Gourdo*, (die
Martii 30, 1822, *Forbes*, n. 5. spec. florida et fruct.) In
insula *S. Antonii*, in rupestribus ubi ex alto desilit aqua
(Junio, 1841, *Th. Vogel*, n. 93. spec. florida et fruct.)
Radix robusta. *Caules* diffusi, hirtuli. *Folia* breviter petio-

lata, rotundato-ovata, acuta, integra, utrinque hirtula, superiora elliptica. *Cymulæ* pedunculatæ, folio breviores,
bracteolatæ; bracteis linearibus, 3-6 floribus. *Calyces* tubulosi, hirtuli, 13-striati, intus ad basin dentium pilosi.
Corolla extus pilosa, calyce subduplo longior, labium
superius bifidum, inferius longius, lobis rotundatis, medio
aliquando emarginato. *Mericarpia* oblonga, obtusa. fusca,
humectata parum mucilaginosa.

170. Stachys *arvensis,* Linn. (1764) 1. *p.* 814. Benth. *Lab.*
Gen. et Spec. (1832-36) *p.* 550.—Ic. Curt. *Flor. Lond.*
1817, 1. bona. Reich. *Icon. Bot.* (1832) *tab.* 967. bona.
HAB. In herb. ins. Cap. Viridis, (*Mus. reg. Par.*)

171. Leucas *Martinicensis,* Br. *Prod. Flor. Nov. Holl. et ins.*
Van Diem. (1810) 1. *p.* 504. Benth. *Lab. Gen. et Spec.*
(1832-36) *p.* 617. Leucas Schimperi, *Hochst.! exsicc. Ab.*
n. 15.—Ic. Clinopodium Martinicense, *Jacq. Stirp. Americ.*
Hist. (1763) *p.* 173. *t.* 177. *f.* 75. (calyx solus).) Phlomis
Caribæa, *Jacq. Ic. Plant. Rar.* (1781-86) 1. *p.* 11. *t.* 110.
pulch.

HAB. In herb. ins. Cap. Vir. (*Mus. reg. Par.*)
Occurrit hæc species ad oras utrisque Africæ calidioris in
Senegambia prope Bakel (Heudelot! exsicc. n. 121.) In
insula Madagascariæ, per provinciam Emirnensem, (Bojer!)
Ad Adoam, in Abyssinia, (Schimper! (exsicc. n. 15.) In
regionibus Birmanicis, prope montem Toang Dong, (ex
Benth.) Ad insulas Antillicas, in insula Martinense,
(Goudot!) In insula Hispaniola, (Poiteau!) Ad plagos
Boreales Australesque Americæ. Rio Janeiro, (Commerson!) etiam in Canada, (herb. Vaillant!) (*Herb. Mus. reg.*
Par.)
Corolla apice incurva, dentes calycis subæquans, labium superius emarginatum, extus et intus hirsutissimum, inferius
vix longius, 3-lobum, lobis lateralibus oblongis truncatis
emarginativse, medio majore latioreque bifido. *Filamenta*
longe viscoso-pilosa, inclusa. *Stylus* longitudine staminum
introrsorum. *Mericarpia* oblongo-ovata, obtusa, dorso
superne glandulosa, ceterum lævia, atro-fusca. *Cotyledones*

oblongæ obtusæ cordatæ, radicula longiuscula crassiuscula.

172. Ajuga *Iva*, Schreb. *Plant. Vert. unilab. Gen. et Spec.*
(1774) *p.* xxv. Benth. *Gen. et Sp.* (1832-36) *p.* 698. Teucrium Iva, *Linn. Spec.* (1764) *p.* 787. *Desf. Fl. Atl.* (1796)
11. *p.* 3.—Ic. Lobel. *Plant. Hist.* (1576) *p.* 208. bona.
Sibth. *Flor. Græc. t.* 525. (optima.)

HAB. In insula *S. Vincentii*, ad montium basin, (*Vogel*,
n. 20. Junio, 1841, spec. macilenta.)
Hæc planta in Europa a gradu boreali 45, usque in insulas
Cap. Viridis. In Gallia, (Maille!) In Dalmatia, (Petter!)
In Lusitania, (Welwitsch!) In Hispania, (Chaubord!)
In Græcia, (Despreaux!) In Algeria, (Durieu!) In insulis
Canariensibus, (Webb!)

XXXVIII. VERBENACEÆ, *Juss.*

173. Verbena *officinalis*, Linn. *Sp. Pl. p.* 29.—Ic. *Fl. Dan.*
t. 628. *Engl. Bot. t.* 767. *Turp. Fl. Méd.* Savi, *Mat. Med.*
Tosc. t. 52, *fig. dextra.*

HAB. In valle *S. Dominici*, ins. *S. Jacobi*, (*J. Dalton Hooker*,
n. 120. Nov. 1839, spec. floriferum.) Ad rivulos vallis
Pico (*Brunner*, Ergebn. p. 123.)

XXXIX. SOLANACEÆ, *Juss.*

174. Physalis *Alkekengi*, Linn. *Sp. Pl. p.* 262.—Ic. Matth.
(*ed. Valgris*, 1665) *p.* 1070. Blackw. *Herb. t.* 161. Savi,
Mat. Méd. Tosc. t. 59.

HAB. In ins. *S. Jacobi*, (*Darwin.*)

175. Physalis *somnifera*, Linn. *Sp. Pl. p.* 161. Physalis
flexuosa, *ejusd. ibid.*—Ic. Matth. (*ed. Valgris*, 1685) *p.*
1071. Clus. *Hist.* 2. *p.* 85. Barr. *Ic.* 149. Cav. *Ic.* 2.
t. 103. Sibth. *Fl. Græc. t.* 233.

HAB. In arvis *Gossypii*, ins. *S. Jacobi*, et circa sinum *Tarrafal*, ins. *S. Antonii*, (*Forbes*, n. 2 et 27. d. 2 et 5 Aprilis,
1822, spec. flor. et fruct.) Ad radices montium *S. Vincentii*, et in vallibus, arbustum pedale, (*Th. Vogel*, n. 29 et

M

162   SPICILEGIA GORGONEA.

59. Junio, 1841, spec. flor. et fruct.) Circa *Porto Praya,*
ins. *S. Jacobi,* (*J. Dalton Hooker,* n. 119. Nov. 1839.) In
ins. *S. Jago* et *Brava,* (*Brunner,* in herb. nostr.)
176. Capsicum *frutescens,* var. *a.* Linn. *Sp. Pl. p.* 271. Capsicum frutescens, *Willd.* 1. *p.* 1052. *Fingerh. Monog. p.* 17.
Brunn. Ergebn. p. 35.—Ic. Rumph. *Amb.* 5. *t.* 88. *f.* 3.
Fingerh. *l. c. t.* 4. *f. d.*
HAB. In ins. *S. Antonii,* " suffrutex, ramis ascendentibus vel
procumbentibus," (*Th. Vogel,* n. 16. Junio, 1841, spec.
flor. et fruct.)
177. Capsicum *microcarpum,* DC. *Cat. Hort. Monsp. p.* 86!
(ex specim. Candolleano, herb. Mercier.) Fingerh. *Monogr.*
p. 19—Ic. Fingerh. *l. c. t.* 4. *f. b.*
HAB. In valle *S. Dominici,* ins. *S. Jacobi.* " Flores albi,
rami sarmentosi arbusculis dependentes." (*J. D. Hooker,*
n. 116. Nov. 1839, sp. fl. et fr.)
178. Datura *Stramonium,* Linn. *Sp. Pl. p.* 225.—Ic. Turp.
Fl. Méd. 6. *t.* 332.
HAB. In valle *S. Dominici,* ins. *S. Jacobi, J. Dalton Hooker,*
n. 118. Junio, 1839, spec. florida.
179. Datura *Metel,* Linn. *Sp. Pl.* 256.—Ic. Fuchs. *Hist.*
p. 690.
HAB. In ins. *S. Jacobi,* (*Darwin.*)
180. Lycopersicum *cerasiforme,* Dunal, *Hist. des Solan.*
p. 113.   Solanum pomiferum, fructu rotundo parvo molli,
nunc luteo, nunc rubro, *Moris. Hort. H. bles. p.* 195.—Ic.
Dun. *l. c. t.* 3. B.
HAB. In valle *S. Dominici,* ins. *S. Jacobi,* (*J. Dalton Hooker,*
n. 164. Nov. 1839, spec. flor. et fruct.)   In monte *Verede,*
ins. *S. Vincentii,* ab alt. 1000 pedum et superne (*Th. Vogel,*
n. 36. Junio, spec. flor. et fruct.)
181. Solanum *nigrum,* Linn. *Sp. Pl. p.* 266. Koch, *Syn. Fl.*
Germ. ed. 2. *p.* 584. Solanum Guianense, *Brunn.! Ergebn.*
p. 115. *non Lamck.* — Ic. Reichb. *Pl. Crit. t.* 954. *et*
Solanum pterocaulon, *t.* 955.
HAB. In arvis *Gossypii,* sinus *Tarrafal,* ins. *S. Antonii,*
(*Forbes,* n. 16. d. 2 Aprilis, 1822, spec. florida.)   In ins.

S. Antonii, (*Th. Vogel*, n. 8. Jun. 1841, sp. fl. et fr.) In
ins. *S. Jacobi*, (*J. Dalton Hooker*, n. 130. November, 1839,
spec. flor. et fruct.) In umbrosis ins. *Boa Vista*, (*Brunner*,
in herb. nostro spec. vegetum procerum.)
182. Solanum *fuscatum*, Jacq. *Coll.* 1. *p.* 51.—Ic. Jacq. *Ic.
rar. t.* 42.
HAB. In ins. *S. Jacobi*, valle *S. Antonii*, (*J. Dalton Hooker*,
n. 117. Nov. 1839, spec. fructifera,) ibid. *Chr. Smith*, in
herb. Mus. Brit. (cl. *J. D. Hooker.*)

XL. SCROFULARINEÆ, *Juss.*

§. *Campylantheæ*, Webb, *in Ann. Sc. Nat.* 3ème *Sér.* 3. *p.* 33.
Phyt. Can. 3. *p.* 125.

Character tribus reform.

Calyx 5-fidus; laciniis æstivatione imbricatis, subæqualibus,
2 superioribus paululum majoribus. *Corollæ* tubus infundibuliformis, laciniis planis, 2 anticis minoribus, æstivatione
interioribus. *Stamina* 2, postica anticorum vestigio nullo,
declinata, antheris arcte appressis, loculis acutis, confluentibus. *Stylus* clavatus, integer. *Capsula* coriacea, septicide bifida, valvis mox bipartitis quadrifida, columna pla-
centifera tota libera. *Semina* plurima, plana, campylo-
tropa, chalaza hiloque approximatis; testa reticulata,
appressa, in alam periphericam producta. *Embryo* peri-
phericus vel hippocrepidoideus.—*Frutices regionis* Maca-
ronesiacæ *desiccatione nigrescentes;* foliis *crebris sparsis,
superioribus alternis, sæpe crassis.* Flores *spicati, bracteati,
pedicellis basi bibracteolatis.—Campylantheas* nec *Salpiglot-
tideis* uti nunc in lucidissima coordinatione reconstituuntur,
nec *Gerardeis* ipsis convenientes in tribulum iterum suam
reducere maluimus.
183. Campylanthus *Benthami*, Webb. Campylanthus sal-
soloides, *ejusd. Phyt. Can.* 3. *p.* 126. quoad plantam Gor-
goneam non Roth.
Var. *a. glaber*, foliis filiformibus glabris, calycis laciniis glabellis ciliatis. Campylanthus glaber, *Benth. in DC. Prod.*

M 2

10. *p.* 508 *et* 596. Eranthemum salsoides, *Chr. Smith, l. c.*
p. 251. *herb. Mus. Brit.! ex cl. J. Dalton Hooker.*—Ic.
Tab. XV.

Var. β. *hirsutus,* foliis planis oblongis, superioribus filiformibus cum ramulis hirsutissimis, calycis laciniis puberulis
demum glabrescentibus. HAB. In rupestribus totius montis *Gurdo,* ins. *S. Nicolai,*
(*Forbes,* n. 9. var. *a*; n. 17. planta junior, foliis hirsutis,
var. floribus albis, sine num., die 30 Martii, 1822, sp. flor.
et fruct.) Ad apicem montis abrupti vallis *S. Dominici,*
ins. *S. Jacobi,* ad alt. 1200 ad 2000 ped. (*J. Dalton Hooker,*
n. 128. Junio, 1839, sp. var. *a.* florida et fruct.) In ins.
S. Antonii, (*Th. Vogel,* n. 35. b. sp. procera fructifera.) In
montibus ins. *S. Vincentii,* ad alt. 800 ped. "*Frutex* parvus. *Caulis* digiti crassitie ramis pluribus (sæpe dependentibus) depressis. *Folia* crassiuscula, subcarnosa, fructus
juglandis olent. *Flores* secundi. *Corollæ* tubus luteus;
limbus violaceus. *Stamina* medio corollæ tubo inserta,
loculo altero superiore, altero inferiore. *Stigma* perforatum,
ovarium multiovulatum." (*Th. Vogel,* n. 72. Junio, 1841,
spec. macra. var. β. florida et fructifera.)
Frutex duriusculus; ramis tenuibus, albidis, foliorum cicatricibus crebris nodulosis. *Folia* sparsa, stirpis junioris
(sive formæ β.) plana, oblongo-spathulata, acuta, basi in
petiolum attenuata, dense glanduloso-tomentosa : var. *a.*
filiformia, elongata, glabra, basi tantum et in axillis parce
pubescentia. *Ramorum* apices in forma β. hirsutissimi,
sed jam ad spicas floriferas glabrescunt. *Spicæ* rectius-
culæ, 3 vel plures aggregatæ, foliosæ, secundæ. *Bractea*
filiformis, bracteolis basalibus vel sub dimidio pedicelli
filiformis gracillimi nutantis sitis. *Calyx* junior subglabra-
tus, laciniis 2 inferioribus longioribus recurvis, glanduloso-
pubescens vel glaber, laciniis lanceolatis acutis margine in
varietatibus ambabus ciliatis, corollæ tubo subduplo brevioribus. *Corolla* ex cl. *Forbes,* (in var. *a.* aliquando alba,)
cœrulea, secundum cl. *Th. Vogel* in var. β. limbus violaceus, tubus luteus, ex cl. *J. Dalton Hooker,* (var. *a.*)

carnea, glabra, tubo crassiusculo æquali, medio circiter superne flexo, laciniis ovato-lanceolatis acutis. *Stamina* 2, sub medio tubi inserta; antheris oblongo-ovatis vel subreniformibus, glanduloso-pubescentibus, loculis junioribus basi acutiusculis, divaricatis, superne confluentibus. *Capsula* elongato-ovata, glabra, nervosa, nitidiuscula, apice rotundata, subcrenata, calyce persistente longiora. *Semina* rotundata, vel subreniformia, nigra, ala membranacea alba cincta.

Differt a *C. salsoloide*, Roth, foliis junioribus sæpe planis, hirsutissimis var. *a.* tenuioribus, spicis rectis, calycis laciniis lanceolatis, ciliatis glabris vel glabrescentibus, corollæ glabræ tubo crassiore, laciniis lanceolatis, antheris longioribus, acutioribus, capsula obtusiore calyce longiore, seminibus nigris.

TAB. XVI. *Fig.* 1. flower; *f.* 2. anthers; *f.* 3. germen; *f.* 4. capsule :—all *magnified.*

184. Celsia *betonicæfolia*, Desf. *Fl. Atl.* 2. *p.* 58. Benth. *in DC. Prodr.* 10. *p.* 245. Chr. Smith, *l. c. p.* 251. Celsia arborescens, *ejusd. in herb. Mus. Brit.*

HAB. Ad apicem montis abrupti, alt. 2000 ped. in valle *S. Dominici*, ins. *S. Jacobi*, (*J. Dalton Hooker*, n. 128. November, 1839, spec. glabra vegeta fructifera.) In dimidio superiore *Montes Verede*, ins. *S. Vincentii*, (*Th. Vogel*, n. 82. Junio, 1841, spec. deusta glutinoso-tomentosa florida et fructifera.)

185. Linaria *dichondræfolia*, Benth.; ramis rectiusculis rigidis albido-tomentosis, foliis utrinque dense pubescentibus albidis demum desicc. nigrescentibus inferioribus rhomboideo-cordatis superioribus rotundato-cordatis vel basi truncatis, petiolo tenui, pedunculis filiformibus decurvis petiolo longioribus, calycis laciniis lineari-lanceolatis linearibusque acutis tomentosis, corolla calyce subduplo longiore pubescente labio sup. 2-lobo inferiore elongato laciniis ovatis obtusis, calcare brevi recurvo, capsula ovata calyce sublongiore pubescente duriuscula ab apice usque ad medium valvatim dehiscente.—Linaria dichondræfolia, *Benth. in DC. Prodr.* 10. *p.* 270.

HAB. In declivibus umbrosis, ins. *S. Nicolai*, (*Forbes*, n. 16. die 27 Martii, 1822, sp. fructifera et florida.) In ins. *St. Vincentii*, (*Th. Vogel*, n. 13, 15, et 16, Junio, 1841, sp. fruct.)

186. Linaria *Brunneri*, Benth. *in DC. Prodr.* 10. *p.* 270. L. alsinæfolia, *Brunn. Ergebn. p.* 84. *non Spreng.*

Var. *a. vera;* ramis elongatis rectiusculis rigidulis, pilis densis longis patulis pilosis, foliis nigricantibus lanato-hirtis demum glabris, inferioribus rotundatis acutis basi subcordatis aliquando subsagittatis petiolis gracillimis, superioribus ovatis lanceolatisque basi cordatis vel rotundatis obsolete crenatis, pedunculis filiformibus, calycis laciniis oblongo-lanceolatis acutis hirtis, corolla calyce duplo longiore, labio superiore bifido, inferiore latiusculo 3-fido, laciniis ovatis, calcare brevi sacciformi leviter incurvo, capsula ovato-rotundata duriuscula pilosa calyce longiore ab apice usque ad medium valvatim dehiscente, seminibus oblongorotundatis, angulatis tuberculatis.

Var. *β. parietariæfolia;* ramis filiformibus pilis deflexis densiuscule pilosis, foliis tenuibus viridibus desiccatione subnigrescentibus pubescentibus demum glabris ciliatis inferioribus rhomboideo-rotundatis acutis basi cordatis sæpe auriculatis vel subsagittatis, superioribus ovato-rotundatis basi truncatis vel subcordatis eroso-crenatis, summis integris, calycis laciniis linearibus acutis, corolla calyce subduplo longiore, labio superiore 2-fido, inferiore 3-fido angusto elongato laciniis lanceolatis, calcare brevi angusto uncinatim recurvo, capsula magis chartacea.

HAB. In declivibus umbrosis et in rupestribus ad sinum *Tarrafal*, ins. *S. Antonii*, (*Forbes*, n. 22. d. 22 April. 1822, et n. 15. d. 2 April. sp. fl. et fruct.) Vulgaris præsertim in rupibus ins. *S. Jacobi*, (*J. Dalton Hooker*, n. 126. Nov. 1839, sp. fl. et fr.) In ins. *S. Vincentii*, (*Th. Vogel*, n. 9, 10, 11, 12 et 14. Junio, 1839, sp. fl. et fruct.) In ins. *S. Antonii*, Th. *Vogel*, (p. 26. sp. flor. et fruct.) In ins. *S. Jacobi*, (*Darwin.*) In insulis *Salis* et *S. Jacobi*, (*Brunner*, in herb. nostro.)

187. Scrofularia *arguta*, *Hort. Kew. p.* 302. *Phyt. Can.* 3.

p. 131. *Auch. exsicc.* 5057. S. peregrina, var.? *Wydl.*
Scrof. p. 28. *Walp. Repert.* 3. *p.* 106.
188. Doratanthera *linearis*, Benth. *in DC. Prodr.* 10. *p.* 347.
Anticharis Arabica, *Hochst. in Sched. Kotsch. Pl. Nub.*
exsicc. n. 90 ! *non Endl.*
HAB. In ins. *S. Antonii, (Th. Vogel,* n. 53. Junio, 1841, spec.
unicum adustum fructiferum.)

XLI. OROBANCHEÆ.

189. Phelypæa *Brunneri,* Webb ; scapo simplicissimo longissimo, florum squamis bractealibus oblongis navicula-
ribus in apicem longe attenuatis laceris acutis calycem
excedentibus, bracteolis linearibus, calycis dentibus lanceolatis acutis, corollis magnis luteis arcuato-tubulosis,
fauce ampliata, lobis subæqualibus acutis, genitalibus sub-
exsertis, filamentis basi cum corollæ tubo pilosis, antheris
sagittatis parce pilosis. Phelypæa lutea, *Brunner ! Ergebn.*
p. 100, *pro parte, non Desf.*
HAB. In herb. ins. Cap. Vir. (*Mus. reg. Par.*) et in ins. *Salis,*
(*Brunner.*)
The above description is taken principally from Brunner's
specimen. With this plant he sent at the same time what
he believed perhaps to be a younger specimen of the same,
but which is quite distinct, perhaps from N. Boro in Senegal,
of which we subjoin a description.*

* Phelypæa *Hesperugo,* Webb ; scapo simplicissimo juniore obtuso bractearum appendicibus exsertis comato, bracteis calyce longioribus—basi
lineari-lanceolatis apice in appendicem crassum linearem protensis,
bracteolis sublinearibus, calyce fisso lobo axili lineari posteriore basin
corollæ amplectente 4-dentato, dentibus lanceolatis margine subscariosis
acutis, corolla tubulosa tubo elongato calycem bis vel ter excedente
leviter arcuato cylindraceo fauce augusta vix dilatata glaberrima, lobis
superioribus lanceolatis basi ovatis cucullatis, lateralibus lanceolatis
acutis, inferiore longiore angustiore lingulata acuta, staminibus cum
corollæ tubi basi glabris, antheris sagittatis calvis, stylo glaberrimo,
stigmate lato cyathiformi. *Phelypæa lutea,* Brunn. ! l. c. pro parte ex
specimine ab indefesso inventore cum planta anteriore misso : non
Desf.

XLII. ACANTHACEÆ, *Juss.*

190. Dicliptera *verticillaris*, Juss.; *in Ann. Mus.* 9. *p.* 268, (ex cl. synon. præter Lamck. omnibus, ex scheda Jussiæana autographa herb. Desf.) Justicia verticillaris, *Lamck. Ill. p.* 40 (ex autopsia facta herb. Lamck. a celeb. a L. de Jussieu, non Linn. fil. nec Vahl, excl. syn. et patria Prom. Bon. Sp.)
HAB. In valle *S. Dominici* ins. *S. Jacobi* (*J. D. Hooker,* n. 122. Nov. 1839, spec. florida et fruct.)
It appears that Lamarck confounded a specimen of our plant, which he had received from Sierra Leone, with the *Justicia verticillaris*, Linn. fil. On this erroneous *J. verticillaris* the illustrious Jussieu founded his *Dicliptera verticillaris,* as is evident from the following note, in his own handwriting, attached to a specimen of our present plant from the West Indies in the herb. Desf., verbatim as follows: " *Justicia verticillaris,* Lam. ill., sic in herb. Lam., specimen ex Sierra Leone.—*Dicliptera verticillaris,* J." As the true *J. verticillaris,* L. fil. belongs to another genus, the name equally applicable to our species may remain. But should it be thought necessary to change it, the plant might be called *D. Jussiaei.*

191. Peristrophe *bicalyculata*, Nees ab Esenb. Dianthera bicalyculata, *Retz. Act. Holm.* 1775. *p.* 297. Dianthera Malabarica, *Linn. fil. Suppl. p.* 85. (ex cl. Syn. Rheed.) Dianthera paniculata, *Forsk. Fl. Æg. Arab. p.* 7. Justicia ligulata, *Lamck. Ill. p.* 40. *Cav. Ic.* 1. *p.* 52. Justicia Malabarica, *Chr. Smith, herb. Mus. Brit.* (ex cl. *J. D. Hooker.*)— Ic. Retz, *t. c. t.* 9. Lamck. *Ill. t.* 12. *f.* 2. Cav. *l. c. t.* 71.
HAB. In vallibus ins. *S. Jacobi* non infrequens (*J. D. Hooker,* n. 171. Nov. 1839, spec. flor. et fruct.)

192. Dicliptera *umbellata*, Juss. *l. c.* Justicia umbellata, *Vahl, Enum.* 1. *p.* 111.
HAB. In arvis ins. *Brava,* (*Brunner,* Ergebn. p. 60.)
The fragments of two other species of *Acanthaceæ* are

found in the herb. ins. Cap. Vir. (*Mus. reg. Par.*) but not in a state to be described.

XLIII. Primulaceæ, *Juss.*

193. Samolus *Valerandi*, Linn. *Sp. Pl.* 1. *p.* 243.—Ic. Gærtn. 1. *t.* 30. Lamck. *Ill.* 2. *t.* 101. *Fl. Dan. t.* 198. *Engl. Bot. t.* 703. Schkuhr, *Handb. t.* 40. Hab. In ins. *S. Vincentii* aquosis montanis ab alt. 500 ped. (*Th. Vogel,* n. 83. Junio, 1841. spec. flor. et fruct.) in ins. *S. Jacobi* (*Darwin.*)

194. Anagallis *cœrulea*, Schreb. *Spic. Fl. lips. p.* 5. Chr. Smith, *l. c. p.* 252 !—Ic. *Engl. Bot. t.* 1823. Hab. In ins. *S. Jacobi,* (*Chr. Smith,* in herb. Mus. Brit. ex cl. *J. D. Hooker.*)

XLIV. Sapoteæ, *Juss.*

195. Sapota *marginata*, Dne.; ramulis glabratis novellis tomentosis, foliis obovatis obtusis supra glabris subtus petiolisque tomentosis marginatis coriaceis dein glabratis, floribus axillaribus paucis glabris, pedicellis petiolum superantibus, foliolis calycinis rotundatis glabris, corollæ laciniis calycem parum superantibus rotundatis ciliolatis, squamis ligulatis obtusis, filamentis dilatatis, ovario superne piloso 6-loculare, loculis uniovulatis. (*Decaisne,* MSS. nobiscum benevole comm.)—Ic. nostra Tab. XIII. Hab. Hujusce arboris individua duo tantum viva ad ped. 20 alt. accedentia, scopulo ad apicem montis abrupti alt. circiter 2000 ped. vallis *S. Dominici* protecta, in ins. *S. Jacobi.* Fructus junior ramique succo albo lactescente scatent. Flores pauci. (*J. D. Hooker,* n. 114. Nov. 1839. sp. fructus juniores et florem unicum gerentia.) Tab. XIII. *Fig.* 1. flower; *f.* 2. portion of corolla, stamens and scales; *f.* 3. hair case to it.

XLV. Plumbagineæ, *Juss.*

196. Plumbago *occidentalis*, Sweet, *Hort. sub. p.* 428. Plumbago Zeylanica, var. *Hornem. Hort. hafn.* 1. *p.* 190. Plumbago Zeylanica, β. *Rœm. et Schult. Syst.* 4. *p.* 4.

HAB. In collibus vallis *S. Dominici*, ins. *S. Jacobi* haud frequens, flores candidi (*J. D. Hooker*, n. 181. Nov. 1839. spec. flor. et fruct.)

197. Plumbago *scandens*, Linn. *Sp. Pl. p.* 215.—Ic. Sloane, *Hist. of Jam.* 1. *t.* 133. *f.* 1. Jacq. *Stirp. Am. t.* 13. HAB. In herb. ins. Cap. Vir. (*Mus. reg. Par.*)

198. Statice *Brunneri*, Webb; foliis rosulatis rotundatis obcordatisque in petiolum attenuatis, scapo erecto aphyllo, ramulis secundis elongatis ramis abortivis pungentibus adnatis creberrime papillatis, bracteis exterioribus rotundato-ovatis interioribus oblongis vel rotundato-oblongis dorso hirsuto-pubescentibus, calycis tubo hirsuto. *Statice pectinata*, Brunner, Ergebn. non Hort. Kew.
HAB. In ins. *Salis* lapidosis (*Brunner*, in schedis herb. nostri.)

This plant is certainly nearly allied to *S. pectinata*, Hort. Kew. ; its inflorescence, however, is very different, its abortive branchlets recalling, though in a slighter degree, the *S. articulata*, Lois. The papillated branches and the broad bracteal scales give it something the look of *S. pruinosa*, Del., in place of the light appearance of *S. pectinata*, Hort. Kew.

199. Statice *Jovi-barba*, Webb; caule lignoso brevi, foliis dense imbricato-rosulatis oblongo-spathulatis margine ad apicem undulatis basi attenuatis infimis breviter petiolatis amplexicaulibus coriaceis glabris, scapo e foliorum rosula protruso gracili ancipite glabro apice subulato, spiculis secundis eleganter recurvis, bracteis exterioribus ovatis acutis interioribus elongato-lanceolatis acutis glabris, bracteola hyalina 1-nervi obliqua, calyce profunde 5-partito, laciniis lineari-lanceolatis acutis, tubo gracili 5-costato glabro, corolla zygopetala versus apicem campanulata breviter 5-fida, antheris ovatis papillatis, stylis apice clavatis stigmatoso-papillosis.
HAB. Copiosa in rupibus montis *Verede* ab alt. 1500 ped. usque ad apicem nec non in aliis montibus ins. *S. Vincentii* (*Th. Vogel*, n. 30. Junio, 1841, spec. florida.)

XLVI. Plantagineæ, *Juss.*

200. Plantago *major*, Linn. *Sp. Pl. p.* 163.—Ic. *Fl. Dan.* 461. *Engl. Bot. t.* 1558. Schkuhr, *Handb. t.* 23. Hab. In aquosis Montis *Verede*, ins. *S. Vincentii* ad alt. 1500 ped. et in ins. *S. Antonii* (*Th. Vogel*, n. 28 et 34. Junio, 1841, sp. flor. et fruct.)

201. Plantago *Psyllium*, Linn. *Sp. Pl. p.* 167.—Ic. Sibth. *Fl. Græc. t.* 149. Hab. In herb. ins. Cap. Vir. (*Mus. reg. Par.*)

XLVII. Nyctagineæ, *Juss.*

202. Boerhaavia *erecta*, Vahl, *Enum.* 1. *p.* 284.—Ic. Jacq. *Hort. Vind.* 1. *t.* 5. 6. Hab. In vallibus arenosis ins. *S. Vincentii*, boreæ et Favonio conversis (*Th. Vogel*, n. 21. Junio, 1841. sp. mancum sine flore et fructu sed huc referendum.)

203. Boerhaavia *paniculata*, Lamck. *Ill.* 1. *p.* 10. In *Boerhaaviarum* cognitione incerti multum adest; Œdipo suo egent nec hic synonymiano ullam adducere ausus sim. Planta nostra eadem est certo ac species Lamarckiana indicata, ex specimine herbarii Desf. ab ipso cum Lamarckii planta collato. A *Boerhaavia erecta*, cui fructu glaber folia punctata foliis impunctatis et fructu pilis glandulosis vestito differt. Hinc videtur Poiretium (*Encycl.* 5. *p.* 53) has plantas mutuo confudisse. Plantæ nostræ valde affinis est *B. procumbens*, Roxb., sed hujusce fructus magis elongatus pilis albidis vix viscosis hirtus, ex spec. herb. Lambert. Hab. In ins. *S. Antonii*: caules plures procumbentes, suffrutescentes, 3-pedales, flores rubri (*Th. Vogel*, n. 24. Junio, 1841. sp. fl. et fruct.)

204. Boerhaavia *dichotoma*, Vahl, *Enum.* 1. *p.* 290. Valeriana scandens, *Forsk. Fl. Æg. Ar. p.* 12. Hab. In herb. ins. Cap. Vir. (*Mus. reg. Par.*)

205. Boerhaavia *repens*, Linn. *Sp. Pl. p.* 5? Del. *Fl. d'Eg.*

p. 2. Cent. des pl. d'Afr. p. 93. *Vis. Fl. d'Eg. et Nub. p.* 4.
Boerhaavia vulvarifolia, *Poir. Encycl.* 5. *p.* 55. Boerhaavia
suberosa, *Chr. Smith, l. c. p.* 249. *Herb. Mus. Brit.* (ex cl.
J. D. Hooker.)—Ic. Del. *l. c. t.* 3. *f.* 1.
HAB. In ins. *S. Jacobi* prope Porto Praya locis umbrosis
vulgaris et in rupestribus (*J. D. Hooker,* n. 167. Nov.
1839, spec. flor. et fructif.)

XLVIII. AMARANTACEÆ, *Juss.*

206. Alternanthera *sessilis,* R. Br. *Prodr.* 1. *p.* 417. Ille-
cebrum sessile, *Linn. Mant. p.* 345.—Ic. Pluk. *Phytogr.*
t. 133. *f.* 2. Burm. *Zeyl. t.* 4. Rumph. *Amb.* 6. *t.* 15. *f.* 1.
HAB. In ins. *S. Jacobi* (*Darwin.*)
207. Achyranthes *argentea,* Willd. *Sp. Pl.* 1. *p.* 1191.
Achyranthes aspera, *a.* Sicula, *Linn. Webb, Phytog. Can.*
p. 194. Achyranthes virgata, *Poir !* *Encycl. Suppl.* 2. *p.* 10.
—Ic. Bocc. *Pl. Par. t.* 9. Lamck. *Ill. t.* 168. *f.* 1. Sibth.
Fl. Græc. t. 244.
HAB. In ins. *S. Jacobi* vulgatissima (*J. D. Hooker,* n. 170.
Nov. 1839. spec. fructifera, gracilia, caule albido, foliis
tenuibus pilosis vix appressis nec sericeis.)
208. Achyranthes *aspera,* Willd. *Sp. Pl.* 1. *p.* 1191. Achyranthes aspera, *β.* indica, *Linn. Sp. Pl. p.* 295. Achyranthes
obtusifolia, *Lamck. Encycl.* 1. *p.* 545. Achyranthes crispa,
Poir ! *Encycl. Suppl.* 2. *p.* 10.—Ic. Pluk. *Phytogr. t.* 10.
f. 4. Burm. *Zeylan. t.* 5. *f.* 3.
HAB. In collibus, ins. *S. Jacobi* (*J. D. Hooker,* n. 169. Nov.
1839. spec. flor. et fruct.) Ibid. (*Brunner,* l. c.)
209. Ærua *Javanica,* Juss. *Gen. p.* 88. Ærua, *Forsk. Fl. Æg.*
Ar. p. 170. Illecebrum Javanicum, *Linn. Syst. ed. Murr.*
p. 266. Achyranthes, *Chr. Smith, in herb. Mus. Brit.* (ex
cl. *J. D. Hooker.*)
This is a very variable plant, the forms it assumes may be
thus characterised.
Var. a. Forskalii, foliis oblongis vel ovatis obtusis, spica

crassiuscula, floribus majoribus, perigonii laciniis lanceolatis acutis. — Huc spectat varietas latifolia Vahl et
Æ. Ægyptiaca, Gmel. — Habitat in Ægypto, Senegalia et
ins. Gorgoneis.

Var. β. Bovei, foliis lineari-lanceolatis utrinque albescen-
tibus, spica tenuiore, floribus minoribus rotundatis spisse
lanatis, perigonii laciniis ovatis sæpe obtusis. — Habitat
in Arabiæ monte Sinai (*Bové*), Kenné in desertis, (*Sieber*.)
The flowers of our specimens, as well as of those collected
by Perrottet, Heudelot and Brunner in Senegal, are somewhat less woolly than those of the Egyptian plant, and the
leaves are always oblong; but I can perceive no specific difference.

HAB. In rupestribus sinus *Tarrafal* ins. *S. Antonii* (*Forbes,*
n. 25). In ins. *S. Jacobi* prope *Portum Praya* (*J. Dalton
Hooker,* n. 107. Nov. 1839.) In ins. *S. Vincentii* rupestribus (*Th. Vogel,* n. 79.)

210. Lestiboudesia *trigyna,* R. Br. Celosia trigyna, *Linn.
Mant. p.* 212. — Ic. Jacq. *Hort. Vindob.* 3. *t.* 15.

HAB. In valle *S. Dominici,* ins. *S. Jacobi* (*J. D. Hooker,*
n. 108. Junio, 1839, spec. flor. et fruct.)

211. Amaranthus *gracilis,* Desf., *Tabl. de l'Ecole Bot. ed.* 1.
(1804) *p.* 43. *Poir. Encycl. Suppl.* 1. *p.* 312. *Webb, Phyt.
Can.* 3. *p.* 191. Amaranthus viridis, *Linn. Sp. Pl.* 2. *p.* 1407.
quoad syn. Sloanii et Pisonis. Chenop. caudatum, *Jacq.
Coll.* 2. *p.* 325. Amaranthus oleraceus, *Lamck. Encycl.* 1.
p. 116.

HAB. In valle *S. Dominici,* ins. *S. Jacobi,* (*J. D. Hooker,*
n. 109. Nov. 1839, spec. fructifera.) In eadem ins. (*Darwin*) in ins. *S. Antonii* usque ad apicem montis *Verede*
(*Th. Vogel,* n. 12. 13 et 87. Junio, 1841, spec. fructifera et
flor.)

212. Amaranthus *spinosus,* Linn. *Sp. Pl. p.* 1407. — Ic. Herm.
H. Lugd. Bot. t. 33.

HAB. In ins. *Boa Vista* (*Brunner,* l. c.)

XLIX. Chenopodieæ, DC.

213. Chenopodium *murale*, Linn. *Sp. Pl. p.* 318. Moq·
Tandon, *Monog. p.* 32.—Ic. Curt. *Fl. Lond. t.* 20. *Engl.
Bot. t.* 1722.
Hab. In ins. *S. Jacobi*, (*J. Dalton Hooker*, n. 110. Nov. 1839.
spec. florida et fruct.) In ead. ins. (*C. Darwin.*) In ins.
S. Vincentii, (*Th. Vogel*, n. 86. Junio, 1841, sp. fructifera
sicca.)
214. Ambrina *ambrosioides*, Spach, *Suites à Buff.* 6. *p.* 297.
Moq. Tandon, *Monog. p.* 40. Chenopodium ambrosioides,
Linn. Sp. Pl. p. 320.—Ic. Barrel. *t.* 1185.
Hab. In ins. *S. Jacobi*, (*J. Dalton Hooker*, n. 111. Nov.
1849, spec. florida.) In eadem ins. (*C. Darwin.*)
215. Suæda maritima, Moq. Tandon, *Monogr. p.* 127. Chenopodium maritimum, *Linn. Sp. Pl. p.* 321.—Ic. *Fl. Dan.
t.* 489. *Engl. Bot. t.* 633.
Hab. In ins. *S. Antonii*, (*Th. Vogel*, sine num. Jun. 1841.
spec. florida et quædam fruct.)

L. Polygoneæ, *Juss.*

216. Persicaria *serrulata*, Webb et Moq. Tandon, *Phytog.
Can.* 3. 219. Polygonum serrulatum, *LaGasc. Nov. Gen.
Sp. p.* 14.
Hab. In valle *S. Dominici*, ins. *S. Jacobi*, (*J. Dalton Hooker*,
n. 104. Nov. 1839, spec. fructiferum.) In eadem insula,
(*C. Darwin.*) " *Rumicem* quendam quem maximum vocat
in ins. *S. Jacobi*, ad rivulos vallis *Pico* vidit *Brunner.*"
(*Ergebn.* p. 110.)

LI. Euphorbiaceæ, *Juss.*

217. Dalechampia *Senegalensis*, A. L.; caule volubili pubescente substriato, stipulis ovato-lanceolatis hirsutis acutis
integerrimis, foliis fere ad basin 3-lobis, lobis lanceolatis
acutis margine argute denticulatis, subtus grosse nervoso-

reticulatis pube spissa utrinque tomentosis inferioribus
sæpe lobulatis, involucri foliolis rotundato-ovatis apice
breviter 3-lobato-dentatis dentibus ovato-lanceolatis denticulatis utrinque tomentosis grosse 5-nerviis, appendiculis
stipulaceis 4 lanceolatis acutis tomentosis, florum masculorum involucello longe pedunculato cyathiformi truncato
grosse crenulato, floribus masculis pedicellatis, pedicello
flore longiore glanduloso basi bracteato, bractea lanceolata
acuta, antherarum subsessilium fasciculo longe stipitato
calycis lacinias lanceolatas excedente, florum fem. involucelli diphylli foliolis late ovatis fimbriatis, floris intermedii longius pedicellati calycis 9-laciniati lateralium 6-
laciniatorum laciniis lanceolatis pinnatis, pinnis glanduloso-
hirsutis, ovario orbiculari-depresso cum stylo glanduloso-
hirto, stigmate obtuso obsolete triangulari, florum pedicellis
cum calycis laciniis papposis longe accretis, capsula orbi-
culari-trigona depressa hirta.

Nostræ valde affinis est *Dalechampia papposa*, Endl. (Atakt.)
sed foliis dentatis aliisque notis distincta, sed affinior
Dalechampia quædam Nubica quam ad Montem Arasch-
Kool legit sedulissimus Kotschy (d. 30 Sept. 1839) quam-
que sub num. 84 fautoribus suis misit associatio itin.
adjecto in Schedis nomine *D. Cordofana*, Hochst., sed
nullibi a cl. viro descriptam invenio. Differt involucro
longiore florum femineorum (quos unicos vidi) involucello
latiore calycibus multo magis hirsuto-papposis ovarioque
strigosiore sed ulterius examinanda et forsan vix diversa.

HAB. In ins. *S. Jacobi,* (*Darwin,*) et in herb. ins. Prom. Vir.
(*Mus. reg. Par.*) Specimina sua habuit celeb. Jussiæus
ab Adansonio et a cl. Geoffroy, anno 1788.

218. Phyllanthus *scabrellus,* Webb; caule annuo diffuso,
ramulis debilibus 4-gonis angulis pilis dentiformibus scabris, stipulis apice filiformibus basi lanceolatis, foliis brevissime petiolatis ovatis vel suborbiculatis pallidis subtus
scabrellis mox glabris, margine obsolete denticulatis, nervis
evanidis, floribus in axillis foliorum solitariis erectiusculis,
calycis laciniis sæpe 5 ovatis vel lanceolato-ovatis late

scariosis medio dorso tantum herbaceis, ovario disco
obsolete subcrenato stipato, capsula complanata glaberrima
lucida. — Phyllanthus, (n. 987), *Schimper exsicc. Arab.
Associationis itin.* 1837, in arena prope Djeddam, et hujus
videtur esse var. glabrescens. Phyllanthus, (n. 89), *Kotsch.
It. Nub. a Monte Arasch-Kool.* HAB. In ins. *S. Jacobi*, (*J.
Dalton Hooker*, n. 105. spec.
fructiferum.) In ins. *S. Antonii*, (*Th. Vogel*, n. 17. Junio,
1841, spec. macrum adustum.) Extat quoque in herb.
Mus. reg. Par. cum priore in Senegambia ab indefessis
Perrottet et Leprieur lectus.
219. Phyllanthus *Thonningii*, Schum.! *Beskriv. Af. Guin. Pl.
p.* 192. Phyllanthus virgatus, *Vahl MSS. in herb. Juss.!*
caule lignoso, ramis virgatis 4-gonis glabris, stipulis basi
dilatatis apice acutis, foliis breviter petiolatis oblongo-
lanceolatis apice latioribus apiculatis basi subattenuatis
margine minutissime denticulatis subtus subglaucescentibus nervosis, floribus in axillis foliorum solitariis pendulis, calycis laciniis sæpissime 6 apice orbiculatis basi
attenuatis glabris margine scarioso angusto cinctis, ovario
glandulis 5-6 oblongis stipato, capsula pomiformi glabra.
HAB. In ins. *S. Jacobi*, (*J. Dalton Hooker*, n. 103 et 107.
Nov. 1839, specimina floribus femineis et fructibus onusta.)
220. Euphorbia *Chamæsyce*, Linn. *Sp. Pl. p.* 652.—Ic. Clus.
Hist. 2. p. 187.
HAB. In vallibus arenosis ins. *S. Vincentii* sub *Tamarice*,
(*Th. Vogel*, n. 57. Junio, 1841, spec. fruct.)
221. Euphorbia *Forskälii*, Gay, *in Phytog. Can. 3. ined.*
Euphorbia thymifolia, *Forsk. Fl. Æg.-Ar. p.* 94. *Del.! Fl.
Æg. Ill. p.* 63. non alior.
HAB. In ins. *S. Jacobi*, (*J. Dalton Hooker*, n. 105. Nov.
1839, sp. fl. et fruct.) In eadem ins. (*Darwin.*) In herb.
Mus. reg. Par. spec. florida et fructifera. Frequens in
ins. *S. Antonii*, (*Th. Vogel*, n. 17.) et in rupibus *S. Vincentii*, ad alt. circiter 500 ped. (*Th. Vogel*, n. 18. Junio, 1841,
spec. fruct.)
222. Euphorbia *hypericifolia*, Linn. var. *pusilla*, glaberrima,

prostrata, foliis integerrimis vel obsolete serrulatis, capsulis glabris.

Hab. In herb. ins. Cap. Vir. (*Mus. reg. Par.*) specimen unicum, ubi vulgo ex scheda vocatur " *Terbina.*"

223. Euphorbia *Brasiliensis*, Lamck, *Encycl. vol.* 2. *p.* 423. Euphorbia hypericifolia, *Linn. var. (excl. Gay, in schedis herb. nostr.)*

Hab. In herb. Cap. Vir. (*Mus. reg. Par.*) forma gracilis, foliis parvis vix serrulatis, capsulis hirtulis.

224. Euphorbia *scordifolia*, Jacq. *Coll. p.* 113. Willd. *Sp. Pl. vol.* 2. *p.* 896. Euphorbia tomentosa, *Pers. Enchir.* 2. *p.* 13.—Ic. Jacq. *Ic. rar.* 9. *t.* 476.

Hab. In herb. Cap. Vir. (*Mus. reg. Par.*) spec. florida.

225. Euphorbia *Tuckeyana*, Steud. *Nomencl. Bot. p.* 615. (*nomen sine descript.*) E. arborea, *Chr. Smith, in Tuck. Voy. p.* 251, (*nomen sine descript.*) *Herb. Mus. Brit.! (excl. J. D. Hooker*); arborescens, ramis crassis fuscis cicatricibus fol. not., foliis ad apicem ramorum oblongis vel oblongo-lanceolatis obtusiusculis vel subacuminatis margine membranaceis obsolete sinuatis, basi in petiolum attenuatis vel subsessilibus glabris floralibus amplis obcordato- vel tetragono-ovatis, involucri campanulato-tubulosi limbo 4-fido laciniis oblongis apice bifidis glandulis quadratis ad angulos brevissime 2-dentatis vel sublunatis bidentatis, staminum bracteis (perianthii masculi rudimentis) basi inter se connatis, laciniis filiformibus pectinato-dentatis, ovario ovato, capsula triquetro-pomiformi vel elongato-ovato, semine fusco marmorato, epistomio mediocre pileato.

Hab. Per totam ins. *S. Vincentii* communis ab alt. 200 ped. usque ad montium cacumina et alt. 2500 ped. frutex sæpius 2-3-pedalis sed aliquando arbor 6-pedalis, rami glabri, sub apice foliis circa 20 vestiti. (*Th. Vogel,* n. 122. Junio, 1841, spec. flor. et fruct.) nec non in alia scheda specimina vidimus non aliter diversa sed brevioribus involucri squamis apice lunatis et conspicue dentatis. Ad basin montis cujusdam abrupti vallis *S. Dominici,* ins. *S. Jacobi,*

N

178   SPICILEGIA GORGONEA.

frutex 5-8-pedalis, succo lacteo scatens; larvam perpulchram
Sphingis Euphorbiæ? quæ foliis hujus speciei vescebatur
observavi, (*J. Dalton Hooker*, n. 115. Nov. 1839, spec.
fructifera.)   In herb. Mus. reg. Par. specimen floridum.
Christian Smith says of this species in his journal, l. c.
p. 243 : " I found at last an *Euphorbia*, which bore so near a
resemblance to *E. piscatoria*, as scarcely to be distinguished
from it." Again, p. 27, in Tuckey's Journal, he says : " At
the height of about 1600 feet I at length found the hills and
small valleys covered with large bushes of a *Tithymalus* resembling *E. piscatoria*, but the identity difficult to be established."

226. Ricinus *communis*, Linn. ; *Sp. Pl. p.* 1430.—Ic. Lob.
Hist. p. 392. Blackw. *herb. t.* 148. Turp. *Fl. Méd. t.* 298.
Schkuhr, *Handb. t.* 312. *Nees ab Es. Gen. germ.* 2. *t.* 38.
HAB. In ins. *S. Jacobi, (J. Dalton Hooker*, n. 74. Nov. 1839,
spec. fruct.)

227. Dalechampia *Cordafana*, Hochst. *in Kotsch. exsicc. It.
Nub.* n. 84! Dalechampia inedita Senegalensis, *A. de Juss.
Euphorb. p.* 56. D. tripartita, *R. Br. in Salt, Abyss.?*
HAB. In ins. *S. Jacobi, (Darwin*, n. 287 et 288, sp. flor. et
fruct.)

Our specimens of this singular plant would resemble entirely the *D. papposa*, Endl. (*Atakt. t.* 20 *et* 21.) were its
leaves not always dentate. I can see no difference whatever
between this and Kotschy's plant : it is therefore a native of
either Ethiopia.

LII. MOREÆ, *Endl.*

228. Ficus *Lichtensteinii*, Link, *Enum. Hort. Ber.* 2. *p.* 451.
HAB. In vallibus ins. *S. Nicolai, (Forbes*, n. 20. d. 29 Martii,
1822, sp. fructifera.)

The young fruit of our species is turbinate, but when ripe
lemon-shaped, or nearly round (globuliform), about the size
of a small playing marble. The plant corresponds perfectly
with *F. Lichtensteinii* formerly cultivated at the Jardin du
Roi at Paris, and which probably came from Berlin, as it is

named thus without doubt by Desfontaines in his herbarium.
Sprengel unites the species with *Ficus Capensis*, Thunb.

LIII. URTICEÆ, *Juss.*

229. Forskåhlia *procridifolia*, Webb; ramis lignosis pilis patulis strigosis, foliis lanceolatis acutissimis basi attenuatis scabris supra viridibus subtus cinereo-tomentosis subtriplinerviis nervis ascendentibus margine argute serrato-dentatis dentibus spinosis, bracteis lanceolatis scariosis, involucro amplo turbinato tubo parce piloso, laciniis oblongis acutis nudis, antheris rotundatis, stylis pilosissimis longe exsertis, fructu elliptico subconvexo basi in pedicellum brevem attenuato.

230. Forskåhlia *candida*, Chr. Smith! l. c. et *herb. Mus. Brit.* (ex cl. *J. D. Hooker*) non *Linn. fil.*

This plant is nearly allied to *Forsk. tenacissima*, L., from which it differs in the shape and sharper teeth of its leaves, its larger and more decidedly funnel-shaped involucrum, by its styles much longer and more hairy, and by the shape of its fruit. HAB. In petrosis et in declivibus aridis vallium ins. *S. Nicolai (Forbes,* n. 19 et 39. d. 27 et 29 Martii, 1822, sp. fructifera.) Circa *Portum Praya*, ins. *S. Jacobi (J. Dalton Hooker,* n. 113.) Specimina vix dum florida. In ins. *S. Antonii (Th. Vogel,* n. 51.) et in ins. *S. Vincentii* frutex dumosus sæpe 2-pedalis (*Th. Vogel,* n. 71. Junio, 1841, spec. fructifera et florida.)

231. Forskåhlia *viridis*, Ehrenb.; caule basi lignoso, ramis pilis ascendentibus hirtis vel glabrescentibus, foliis ovatolanceolatis basi attenuatis subscabris utrinque viridibus vel tomento cinereo sordide albidis triplinerviis nervis divaricatis margine vix revolutis crenato-dentatis dentibus sæpe irregularibus vel subobsoletis, involucro turbinato apice dilatato tubo ad angulos piloso, laciniis late ovatis foliaceis obtusis vel obtusiusculis, stylo breviusculo pubescente, fructu ovali. Forskåhlia viridis, *Ehrenb. ex hort. Berol. Desf. Cat. Hort. Par. ed. 3. p. 347.*

HAB. In vallibus altioribus circa *Portum Praya*, ins. *S. Ja-*

N 2

cobi, (*J. D. Hooker,* n. 113. Nov. 1839, spec. vix dum
florida.) In herb. *Mus. reg. Par.* spec. florida.

LIV. ORCHIDEÆ, *Juss.*

232. Habenaria *Petromedusa,* Webb; petalis 3 exterioribus
ovato-lanceolatis 3-nerviis acutis, 2 interioribus profunde
2-partitis exteriorum fere longitudine lacinia superiore
oblonga paullulum breviore, inferiore setaceo, labello petalis exterioribus parum longiore usque ad medium 3-fido
laciniis filiformibus medio breviore lateralibus divergentibus, calcare germine breviore filiformi basi gracillimo
apicem versus sub latiore supra labellum incurvo, anthera
rotundata incurva subtus in cornua 2 horizontalia producta appendicibus 2 ipsa excedentibus (antheris abortivis)
horizontaliter porrectis apice deflexis crassis glandulosis
subtensa, germine gracili in pedicellum longum desinente.
HAB. In herb. ins. Cap. Vir. (*Mus. reg. Par.*)
Radix ...... *Caulis* sesquipedalis, erectus, gracilis, foliosus.
Folia ovato-lanceolata, divergentia, 4 poll. longa, 1½ poll.
lata.   *Spica* tenuis, sub-3-pollicaris, floribus circiter 15
laxis.   *Bracteæ* lineares, germine 4 lin. long. metiente
multo breviores.   *Petala* exteriora circiter 1½ lin. longa,
¾ lin. lata.

This may perhaps be identical with the species described
by Swartz in Persoon's Synopsis as *Orchis ichneumonea*
(*Habenaria ichneumonea,* Lindl.) found by Afzelius at Sierra
Leone; but it is not possible to decide this from his short
description.

Two other Orchideous plants, apparently of the Epiden-
drous tribe, occur in the Portuguese collection, but unfortunately without flower or fruit.

LV. ASPARAGINEÆ, *Juss.*

233. Asparagus *scoparius,* Lowe, *Prim. Mad. p.* 11. Asparagus plocamoides, *Webb, in Steud. Nomencl. p.* 150. A.
stipularis, *Brunn. Ergebn. p.* 20 ?
HAB. Fruticulus facie et magnitudine *A. officinalis,* Linn.

Ad apices monticulorum circa *Porto Praya* et montis abrupti in valle *S. Dominici*, ins. *S. Jacobi :* cortex argenteus, (*J. Dalton Hooker*, n. 10. 2 Nov. 1839, spec. sine flore et fructu, pedicellis solis superstitibus.)

This species is common in Teneriffe, on dry rocks, in the lower, or African region, where it acquires the height of 8 feet, or more. It differs from *A. officinalis* in having very numerous fasciculated flowers.

Asparagus.....?

A parte media usque ad apicem montis *Verede*, ins. *S. Vincentii :* frutescens, ramis longis divaricatis dependentibus reclinatis, (*Th. Vogel*, n. 4. Junio, 1841, spec. foliis fere destituta sine fl. et fruct.)

The young branches of this plant are green and striated, like those of *A. amarus*, but the leaves are more aciculated and the bracteæ subtended by a stronger and longer prickle.

There exists in the herb. Mus. Par. a single specimen of a species of *Smilax* very nearly allied to *S. Canariensis*, Willd., but evidently differing from it.

LVI. JUNCEÆ, *Juss.*

234. Juncus *acutus*, Lamck. *Encycl.* 3. *p.* 268. Juncus acutus, *a. Linn. Sp. Pl. p.* 463.—Ic. *Engl. Bot.* 1614.

HAB. In aquosis sub apice montis *Verede*, ins. *S. Vincentii.* (*Th. Vogel*, n. 42. Junio, 1841, spec. fructifera.)

LVII. COMMELYNEÆ, *R. Br.*

235. Commelyna *canescens*, Vahl, *Enum.* 2. *p.* 73. Rœm. et Schult. *Syst.* 1. *p.* 535. Mant. 1. *p.* 338. Kunth, *Enum.* 4. *p.* 50.

HAB. In valle *S. Dominici* ins. *S. Jacobi* (*J. Dalton Hooker*, n. 101. Nov. 1839, spec. flor.)

LVIII. NAIADEÆ, *Juss.*

236. Potamogeton *pusillus*, L. *Sp. Pl. p.* 184. Smith, *Fl. Brit. p.* 95.—Ic. *Engl. Bot. t.* 215. *El. Dan. t.* 1451. Nees ab Esenb. *fluvial. t.* 4. *f.* 9. *et seq.*

HAB. In ins. *S. Jacobi* (*Darwin*, spec. flor. et fructifera.)

182 SPICILEGIA GORGONEA.

LIX. Cyperaceæ, *Juss.*

237. Cyperus *alopecuroides,* Rottb. *Descr. et Ic. p.* 38. *t.* 8. *f.* 2. Kunth, *Enum.* 2. *p.* 19.

Hab. In valle *S. Dominici* ins. *S. Jacobi* Cyperum hunc unicum stirpemque aquaticam unicam inveni (*J. D. Hooker,* n. 100. Nov. 1839, spec. fructifera.)

238. Cyperus *mucronatus,* Rottb. *β. albidus,* Vahl, *Enum.* 2. *p.* 348. Kunth, *Enum. p.* 17. Cyperus lævigatus, *Linn. Mant. p.* 179. *Rottb. Gram. p.* 19. C. lateralis, *Forsk. Fl. Æg.- Arab. p.* 13. Cyperus monostachyus, *Link in Buch, Beschr. Can. Ins. p.* 138.—Ic. Rottb. *l. c. t.* 16. *f.* 1.

Hab. In monte *Verede* ins. *S. Vincentii* ad alt. 500 et 600 ped. (*Th. Vogel,* n. 113) et ad rivulos ins. *S. Antonii (id.* n. 64, 75 et 99. Junio, 1841, spec. flor. et fruct.)

239. Cyperus *Ægyptiacus,* Glox. *Obs. p.* 20. Kunth, *Enum.* 2. *p.* 48. Schœnus mucronatus, *Linn. Sp. Pl. p.* 63. Scheuchz. *Gram. p.* 367.—Ic. Glox. *l. c. t.* 3. Scheuchz. *l. c. t.* 8. *f.* 1. Mert. et Koch, *Deutschl. Fl.* 1. *p.* 450.

Hab. In herb. ins. Cap. Vir. *Mus. reg. Par.*

240. Cyprus *articulatus,* Linn. *Sp. Pl. p.* 66. Kunth, *Enum.* 2. *p.* 53.—Ic. Sloan. *Hist. Jam.* 1. *t.* 81. *f.* 1.

Hab. In rivulis ins. *S. Nicolai,* die 27 Martii, 1822, (*Forbes,* n. 38. spec. floriferum.)

241. Cladium *Mariscus,* R. Br. *Prodr. p.* 236. Schœnus Mariscus, *Linn. Sp. Pl. p.* 62. Cladium Germanicum, *Schrad. Fl. Germ.* 1. *p.* 75.—Ic. Schrad. *l. c. t.* 5. *f.* 7. *Fl. Dan. t.* 1202. *Engl. Bot. t.* 950.

Hab. Ab austrum ins. *S. Vincentii* in palude ad alt. circiter 500 ped. (*Th. Vogel,* n. 26. Junio, 1841, spec. flor.)

LX. Gramineæ,* *Juss.*

242. Setaria *verticillata,* Palis. de Beauv. *Agrost. p.* 51. *Parl.*

* Graminum Viridensium nomenclationem et descriptiones benevolentiæ debemus cl. atque amicissimi agrostographi Philippi Parlatore, Florentini.

Fl. Pal. 1. *p.* 37. *et in Webb Phyt. Can. ined.* Panicum
verticillatum, *Lin. Sp. Pl.* 82. *Desf. Fl. Atl.* 1. *p.* 57.—
Ic. *Hort. Gram. Aust.* 2. *p.* 11. *tab.* 13.

HAB. In ins. *S. Jacobi* (*J. Dalton Hooker,* n. 97. November,
1839.) In ins. *S. Antonii* (*Th. Vogel,* n. 59. Junio, 1841.)
OBS. Variat foliis omnino glabris (specim. ex S. Jacobi) vel
pilosis (specim. ex ins. S. Antonii.)

243. Pennisetum *cenchroides,* Rich. *in Pers. Syn.* 1. *p.* 72.
Kunth, *Enum. pl.* 1. *p.* 162. Parl. *Pl. Novæ, p.* 42. *et in*
Webb, *Phyt. Canar. ined. et in Fl. Pal.* 1. *p.* 34. Pennisetum
distylum, *Guss. Index Semin. H. R. Bocc. anno* 1826. *p.* 8.
et Fl. sicc. Prodr. Suppl. 1. *p.* 12. *et Syn. Fl. sicc.* 1. *p.* 115.
Parl. Fl. Panorm. 1. *p.* 71. *Bert. Fl. Ital.* 1. *p.* 593.
Cenchrus ciliaris, *Linn.! mant.* 320. *Desf.! Fl. Atl.* 2.
p. 387 *ex eorum herbariis.*—Ic. *Giesecke Ic. t.* 23. ex Kunth,
Phyt. Can. Ic. ined.

HAB. In insula *S. Antonii* (*Th. Vogel,* Junio, 1841.)
De identitate *Penniseti distyli,* Guss. cum *Cenchro ciliari,*
Linn. confer quod scripsi in *Plant. Nov.* et in *Fl. Palerm.*

244. Pennisetum *lanuginosum,* Hochst. *in Flora Botan. Zeit.*
ann. 1844. *p.* 252. *var. a.*

HAB. Graminum species copiosissima pabulum optimum et
semper viride præbet cum reliqua calore usta. (*J. Dalton*
Hooker, n. 95. November, 1839.)

In specimine ex insula Sancti Jacobi, quod possideo, culmus
superne non basi ramosus et folia villosula, sed spica, involucrum et spicularum partes cum descriptione et specimine Hochstetteriano ad amussim conveniunt.

245. Pennisetum *myurus,* Parl.; panicula spiciformi densiflora elongata subulata, spiculis solitariis, involucri setis
exterioribus paucis, interioribus basi plumoso sublanatis,
earum altera spiculos subduplo superante, culmis apice
ramosis, vaginisque glaberrimis, foliis late linearibus superne
scabris.

HAB. In insula *S. Jacobi* copiosa (*J. Dalton Hooker,* n. 92.
November, 1839.)

Radix deest. . . . *Culmus* erectus, teres, striatus, glaberrimus,

184 SPICILEGIA GORGONEA.

apice ramosus, vestitus. *Vaginæ* laxæ, striatæ, glabræ, internodio fere dimidio breviores. *Ligulæ* loco setarum series. *Folia* late linearia, 2-3 lineas lata, acuminata, striata, plana, in pagina superiore et margine scabra. *Panicula* spiciformis, terminalis, densiflora, elongata, 2-3½ poll. longa, subulata, subgracilis, multiflora, rufescens. *Rachis* subflexuosa, angulata, ad angulos sub lente ciliatoscabros denticulata. *Spiculæ* approximatæ, alternæ, sessiles, solitariæ. *Involucri* setæ exteriores paucæ, breves, inæquales, scabræ, interiores basi plumoso-lanatæ, longitudine valde inæquales, maxima parte spicula longiores, altera omnium longior, spicula solitaria intra involucrum breviter pedicellata, basi lanugine alba tecta, biflora. *Valvæ calycinæ* valde inæquales, inferior spicula tertia parte minor, ovata, acuta, enervis, superior spiculam longitudine superans, oblongo-lanceolata, apice acuminata et mucronato-aristata, hyalino-albida, puberula, quinquenervis, nervis obsolete purpurascentibus, a basi ad apicem protractis, nervo carinali puberulo et apicem versus ciliato-scabro. *Flosculus* inferior neuter, ejus valvæ inæquales, inferior major oblonga, obtusa, apice lacera, hyalino-albida, puberula, quinquenervis, carina ciliolato-scabra; superior lineari-lanceolata, acuta, apice aristulata, margine ciliolata. *Flosculi* superioris hermaphroditi valvæ cartilagineæ nitidæ, læves, inferior superiorem angustiorem amplectens. *Stamina* 3. *Styli* 2, stigmatibus exsertis, plumosis, purpurascentibus.

A *Penniseto lanuginoso* precipue differt panicula subgracili, densiflora, magis elongata et subulata, spiculis approximatis minoribus, solitariis, haud ternis, setis interioribus brevioribus nec dense lanatis, valva calycina superiore apice mucronato-aristata, aliisque notis.

246. Pennisetum *ciliatum*, Parl. ; panicula spiciformi densiflora elongata, spiculis solitariis, flosculo inferiore neutro, involucri setis exterioribus paucis brevibus setaceis, interioribus setaceis basi plumosis, omnibus spicula longioribus, altera longissima, culmo ramoso ad nodos pubescenti-

tomentoso, foliis linearibus, supra scabris, basin versos longiuscule villoso-ciliatis.

HAB. In insula *S. Vincentii* ad dimidium montis *Verede* (*Th. Vogel*, n. 112. Junio, 1841.)

Radix fibrosa, fibris crassiusculis, rufescentibus, subpubescentibus. *Culmus* erectus et subascendens, 1½-2 pedalis, teres, striatus, glaber, ad nodos pubescenti-tomentosus, totus vestitus, ramosus. *Vaginæ* laxiusculæ, striatæ, glabræ, internodio subduplo longiores. *Folia* linearia, 1½-2 lin. lata, acuminata, elongata, striata, plana, subtus levia, margine et pagina superiore scabra, basin versus longiuscule villoso-ciliata. *Ligula* semilinearis, ciliato-lacera. *Panicula* spiciformis, densiflora, 1½-3 pollices longa, subgracilis, acutiuscula, basi vagina folii supremi involucrata, pallida (ex sicco). *Rachis* glabra, denticulata. *Spiculæ* solitariæ, parvæ, alternæ, sessiles, biflorae. *Involucri* setæ exteriores paucæ, setaceæ, sub lente scabriusculæ, spicula breviores vel subæquales, setæ interiores setaceæ, sub vitro scabriusculæ, basi plumosæ, longitudine valde inæquales, inferior minor, lineari-oblonga, acuta, hyalino-membranacea, enervis, margine apicem versus sub lente subciliolato, superior spicula longitudine subæqualis, oblongo-lanceolata, acuta, submucronata, hyalino-membranacea, puberula, quinquenervis, nervo carinali scabriusculo. *Flosculi neutri* valde inæquales, valvula inferior major oblongo-lanceolata, apice obtuse subemarginata, in medio emarginaturæ mucronata, hyalino-membranacea, puberula, obsolete subquinquenervis, nervo carinali apicem versus sub lente scabriusculo, superior minor et angustior, acuta, margine scabriusculo. *Flosculi superioris* hermaphroditi valvæ cartilagineæ, nitidæ, inferiore majore superiorem amplectente. *Styli* connati, stigmatibus plumosis, elongatis, fuscis.

247. Digitaria *setigera*, Roth, *Nov. Sp.* 37. Rœm. et Schultes, *Syst. Veg.* 2. *p.* 474.

HAB. In insula *S. Jacobi* (*J. D. Hooker*, n. 87. November, 1839.)

248. Panicum *Daltoni*, Parl.; spicis simplicibus oblongis

obtusis, spiculis muticis, racheos ramis margine superne
sub spicularum insertione setosis, foliis glabris planis
margine cartilagineo eleganter serrulato-scabris, culmo
ramoso vaginisque glabris.
HAB. In insula *S. Jacobi* (*J. Dalton Hooker*, n. 83. Nov.
1839.

Radix fibrosa, fibris albidis, villosulis. *Culmus* ascendens,
ramosus, pedalis et ultra, glaber, levis, striatus, superne
aliquo tactu nudus.   *Vaginæ* compressæ, striatæ, laxæ,
glabræ. *Ligula* nulla.   *Folia* late linearia, 2-2½ lin. lata,
acuminata, plana, striata, glabra, levia, margine cartilagineo
eleganter serrulato-scabra. *Panicula* terminalis, subgracilis,
2-3 pollicaris, subsecunda, fere ut in *Panico Colono*, Linn.
Rachis flexuosa, striata, externe convexo-obtusangula, in-
terne ad spiculos excipiendos canaliculata, glabra, margine
precipue scabra.   Ejus rami alterni spicarum axem effor-
mantes, flexuosi, subcompressi, margine ciliato-scabri, et sub
spicularum insertione 1-2 setas albas, rigidas, spiculam subæquantes gerentes.   *Spicæ* alternæ, simplices, erectæ, rachi
subapproximatæ, oblongæ, obtusæ, superiores breviores e
viridi-purpurascentes.   *Spiculæ* ovatæ, unilaterales, muticæ.
Valvæ calycinæ valde inæquales, inferior minima, cordato-ovata, acuta, trinervis, nervis viridibus, scabris,
superior major, ovato-concava, quinquenervis, nervis validis, viridibus, scabris. *Flosculi hermaphroditi* valvæ cartilagineæ, glabræ, nitidæ, inferior major ovata, acuta, superiorem angustiorem longitudinaliterque striatam amplectens. *Flosculi neutri* valvæ corollinæ inæquales, inferior
major ovata, acuta, quinquenervis, nervis scabris, viridibus,
superior hyalino-membranacea, minor, obtusa.
HAB. Affine *Panico Colono*, Linn., sed foliis margine cartilagineo serrulato-scabris, spicis brevioribus, ramisque racheos
setosis, aliisque notis satis superque distinctum.   In herbario Musæi Britannici planta hæc ex India Orientali extat
sub nomine *Panici Cruris-Corvi* ut Panicum Crus-Corvi ex
herbario Linneano inspectione mihi a Panico Crure-Galli
vix diversum videtur et sicut hujus varietatem teneo.

249. Panicum *rachitrichum*, Hochst. *in Flor. Bot. Zeit.* 1844, *p.* 254.

HAB. In insula *S. Jacobi*, vallibus circa oppidum *Porto Praya (J. D. Hooker,* n. 83, November, 1839.)

250. Paspalum *scrobiculatum,* Linn. *Mant.* 1. *p.* 29. Rœm. et Schultes, *Syst. Veg.* 2. *p.* 296. Kunth, *Enum. Pl.* 1 *p.* 53. HAB. In insula *S. Jacobi (J. D. Hooker.*)

251. Aristida *Adscensionis,* Linn. *Sp. Pl.* 121. Kunth, *Enum. Pl.* 1. *p.* 160. Parl. *Pl. Novæ, p.* 44. *et in* Webb, *Phyt. Canar, ined. et Fl. Palerm.* 1 *p.* 59.—Ic. Aristida cœrulescens, *Desf. Fl. Atl. t.* 21. *f.* 2.

HAB. In ins *S. Jacobi,* circa *Porto Praya,* copiosa. Gramen fragile, culmi dirupti et aristæ glomos efformant qui cruribus ambulantium adhærent lædantque (*J. D. Hooker,* n. 91. November, 1839.)

252. Agrostis *stolonifera,* L.! *Sp. Pl.* 93. Guss. *Fl. Sic. Prodr.* 1. *p.* 57. Parl. *Fl. Palerm.* 1. *p.* 66. Agrostis verticillata, *Vill. Hist. des Pl. du Dauph.* 2. *p.* 74. Guss. *Fl. Sic. Syn.* 1. *p.* 134. Bert. *Fl. Ital.* 1 *p.* 408.—Ic. Trin. *Sp. Gram. fasc.* 3.

HAB. In insula *S. Vincentii (Th. Vogel,* Junio, 1841.) Confer observationes nostras in *Fl. Pal. l. c.*

253. Sporobolus *insularis,* Parl.; panicula spiciformi cylindracea conferta continua, ramis adpresso-erectis superioribus brevissimis, spiculis oblongis, valvis calycinis acutiusculis inferiore paleis breviore superiore subæquali, foliis linearibus subulatis rigidis glabris, ligula brevissima truncata ad angulos pilorum fasciculo munita, vaginis glabris margine superiore piloso-ciliatis, culmo ramoso rigido tereti.

HAB. In insula *S. Vincentii* in altioribus montis *Verede* (*Th. Vogel,* n. 97. Junio, 1841.)

Radix .... fibris albidis, villoso-lanatis, *Culmi* cæspitosi, erecti, subpedales, teretes, striati, glabri, penitus vestiti. *Folia* linearia, subulata, acuminata, rigida, striata, glabra, margine scabriuscula. *Vaginæ* striatæ, glabræ, margine superne villoso-ciliatæ, vagina suprema paniculæ basin involucrante. *Ligula* brevissima, truncata, ad extremitates pilorum fasciculo munita. *Panicula* spiciformis, cylindracea, conferta, continua, 4-5 pollices longa, ejus rami adpresso-erecti, superiores brevissimi. *Rachis* valde striata, glabra. *Spiculæ* parvæ, pallide flavæ, oblongæ, unifloræ. *Valvæ* calycinæ acutiusculæ, oblongæ, carinatæ, carina sub lente scabriuscula, albo-membranaceæ, glabræ, longitudine inæquales, inferior corolla brevior, superior subæqualis. *Valvæ* corollinæ albo-membranaceæ, glabræ, acutiusculæ, inferior uninervis, subcarinata, superior bicarinata. *Sporobolo Capensi*, quem ex Capite Bonæ Spei possideo, affinis, sed differt culmo tereti, ligulis vaginisque ad margines superne pilosis, panicula minus elongata, haud basi interrupta, valvis calycinis acutiusculis non acutatis.

254. Cynodon *Dactylon*, Pers. *Syn.* 1. *p.* 85. Parl. *Fl. Panorm. p.* 124. *et in* Webb, *Phyt. Canar. ined. et Fl. Pal.* 1. *p.* 88. Ten. *Fl. Nap.* 3. *p.* 70. Bert. *Fl. Ital.* 1. *p.* 112. Panicum Dactylon, *Linn. Sp. Pl.* 85. *Sibth. et Smith, Fl. Græc. Prodr.* 1. *p.* 40. Paspalon Dactylon, *DC. Fl. Franc.* 3. *p.* 15. Digitaria Dactylon, *Scop. Fl. Carn.* 1. *p.* 52. *Vis. Fl. Dalm.* 1. *p.* 54. Cynodon linearis, *Willd. Enum.* 30. Panicum lineare, *Burm. Ind.* 26. *t.* 10.—Ic. Sibth. et Smith, *Fl. Græc. t.* 60. *Host, Gram. Austr.* 2. *p.* 15. *t.* 18. HAB. In insula *S. Antonii* (*Th. Vogel*, Junio, 1841.)

255. Eleusine *Indica*, Gærtn. *l. c.* Kunth, *l. c.* Humb. et Kunth, *Nov. Gen. p.* 165. Parl. *in Webb, Phyt. Canar. ined.* Cynosurus Indicus, *Linn., Willd. l. c.*—Ic. Trin. *Ic. t.* 71. HAB. In insula *S. Antonii* (*Th. Vogel*, n. 62. Junio, 1841.)

256. Eragrostis *pulchella*, Parl.; panicula spiciformi laxiuscula stricta, ramis brevibus erecto-patentibus, spiculis 4-6-floris, valvis calycinis acutis corollinaque inferiore trinervi apice obtusiuscula cum acumine, carina sub lente ciliolato-scabris, valva corollina superiore longe pectinato-ciliata, culmo brevi gracili, foliis planis glabris ligula vaginisque margine longe piloso-ciliatis. HAB. In insula *S. Jacobi*, valle *S. Dominici* (*J. D. Hooker*, n. 81. November, 1839.)

Radix fibrosa, fibris tenuibus, albidis, pubescentibus. *Culmus* gracilis, teres, striatus, glaber, simplex, erectus, 4-5-pollicaris, totus vestitus. *Folia* linearia, 1-1½ lineam lata, 2 pollices fere longa, acuminata, plana, glabra, levia, margine scabriuscula, striata. *Vaginæ* laxæ, internodio longiores, striatæ, margine piloso-ciliatæ. *Ligulæ* loco pilorum series, ad margines longiorum. *Panicula* spiciformis, laxiuscula, stricta, subbipollicaris, basi subinterrupta. *Rachis* glabra, teres, ejus rami filiformes, breves, erecto-patentes vel subadpressi, sub lente scabriusculi. *Spiculæ* 4-6-floræ,ovatæ,parvæ. *Valvæ calycinæ* 2,subæquales,membranaceæ,albidæ,concavæ, carinatæ, carina viridi sub vitro ciliato-scabra, apice acuta, flosculo proximo breviores. *Valvæ corollinæ* membranaceæ, albidæ, parum inæquales, inferior major, lanceolato-oblonga, obtusiuscula cum mucronulo, trinervis, nervis viridibus, nervo carinali sub lente ciliolato-scabro: superior ovata, obtusa, bicarinata, carinis pilis longis pectinato-ciliatis.

257. Saccharum *Teneriffæ*, Lin. *fil. Suppl. p.* 106. Sibth. et Smith, *Fl. Græc. p.* 39. *t.* 53. Biv. *Sic. Pl. manip.* 4. *p.* 5. *t.* 1. Guss. *Fl. Sic. Prodr.* 1. *p.* 137. *et Syn. Fl. Sic.* 1. *p.* 159. Bert. *Fl. Ital.* 1. *p.* 328. Parl. *in Webb, Phyt. Can. ined.* Panicum Teneriffæ, *Brown, Prodr.* 1. *p.* 39. *Kunth, Enum. Pl.* 1. *p.* 98. Panicum villosum, *Presl, Cyp. et Gram. Sic. p.* 18. Tricholæna micrantha, *Schrad. in Schultes, Syst. Veg. Mant.* 2. *p.* 163.—Ic. Biv. *l. c.* Sibth. et Smith, *l. c.*

HAB. In ins. *S. Jacobi* rupestribus (*J. D. Hooker,* n. 84. Junio, 1839.) In ins. *S. Vincentii* (*Th. Vogel,* n. 98. Junio, 1841.) In ins. *S. Antonii* (*id.*)

258. Andropogon *hirtum*, Linn. *Sp. Pl.* 1482. Sibth. et Smith, *Fl. Græc. Prodr.* 1. *p.* 48. Desf. *Fl. Atl.* 2. *p.* 378. Parl. *Fl. Palerm.* 1. *p.* 269.—Ic. Host, *Gram. Austr.* 4. *t.* 1.

HAB. In insula *S. Vincentii* rupibus excelsioribus (*Th. Vogel,* n. 108. Junio, 1839.)

259. Andropogon *foveolatum*, Delil. *Fl. Ægypt.* 16. *t.* 8. *f.* 2. Kunth, *Enum. Pl. p.* 490.—Ic. Delil. *l. c.*

HAB. In insula *S. Antonii* (*Th. Vogel.*)

260. Andropogon, *sp.?*

HAB. In ins. *S. Antonii* (*Th. Vogel*, n. 636. Junio, 1839.)

261. Andropogon, *sp.?*

HAB. In ins. *S. Jagi* (*Th. Vogel*, n. 63.)

Mala specimina habeo unde harum duarum plantarum diagnosin facere vel descriptionem adumbrare haud possum.

262. Heteropogon *contortum*, Rœm. et Schultes, *Syst. Veg. 2. p.* 836. Nees ab Esenb. *in Linnæa, p.* 284. Andropogon contortum, *L. Sp. Pl.* 1480. *Brown, Prodr.* 1. *p.* 401.

HAB. In insula *S. Jacobi* montosis, ad altitudinem 800 pedum cujus inter culmos crassos siccosque errant seque abscondunt (*J. D. Hooker*, n. 89. November, 1839.)

263. Heteropogon, *sp.?*

HAB. In insula *S. Jacobi* ad apicem collium (*J. D. Hooker*, n. 90. November, 1839.)

Stirpem nec genus ob speciminis imperfectione determinare exacte nequeo.

MONACHYRON, *Parl.* Gen. Nov. *Andropogoneis* aff.?

Spiculæ trifloræ, flosculo inferiore neutro, univalvi, medio hermaphrodito, superioreque masculo, bivalvi. *Valva calycina* unica, a floribus remota, linearis, membranacea, mutica, flosculis valde brevior. *Flosculi inferioris* valva corollina unica, concavo-carinata, apice bifida, e sinu aristata, arista setacea, subulata, recta, arefactione subflexuosa. *Flosculi medii* valvæ corollinæ 2, subæquales, membranaceæ, muticæ, concavo-carinatæ, apice obtuse bifidæ. *Stamina* 3. *Styli* breves. *Stigmata* aspergilliformia? *Caryopsis* . . . .? *Flosculi superioris* valvæ corollinæ 2, subæquales, inferior valvam corollinam flosculi inferioris æmulans, superior bicarinata, apice subbifida.

Genus ab spicularum structura, quæ mea sententia difficillisime explicanda, certe singularis et ab omnibus Graminum generibus omnino distinctum.—Nomen ex gluma calycina unica μονος una et ἄχυρον gluma desumpsi. An valva corollina flosculis inferioris ut valvam alteram calycinam sit habenda et spiculæ bifloræ unde genus ad *Avenaceas* sit referendum?

264. Monachyron *villosum*, Parl.

HAB. In insula *S. Jacobi* (*J. D. Hooker*.)

Radix fibrosa, fibris tenuibus, puberulis. *Culmi* erecti, sub-pedales, graciles, basi geniculati, ramosi, teretes, superne aliquo tractu nudi ibique pube horizontali alba velutini. *Folia* anguste linearia, 1-1½ lin. lata, acuminata, longiuscula, striata, inferne levia, superne pilis brevibus adpressis sub lente munita. *Vaginæ* laxæ, valide striatæ, pilosæ, pilis brevibus velutinis, aliis longioribus crassioribus inter-mixtis. *Ligulæ* loco pilorum series. *Panicula* ramosa, 2-2½ poll. longa, acutiuscula, ramis subgeminis, ramo altero breviore, capillaribus, flexuosis. *Spiculæ* solitariæ, tri-floræ, flosculo inferiore neutro univalvi, intermedio her maphrodito bivalvi, superiore masculo bivalvi, pedicel-latæ, pedicellis apice incrassatis villosisque. *Valva ca-lycina* unica bracteæformis, membranacea, linearis, obtusa, basi piloso-barbata, apice sub vitro scabra. *Flosculi infe-rioris* valva corollina unica, membranacea, oblongo-lanceo-lata, apice acute bifida, exquisite quinquenervis, nervo carinali in aristam setaceam, subulatam, scabram, arefac-tione flexuosam, valva ipsa subæqualem terminato, dorso undique pilis albis vestita. *Flosculi medii* valvæ corollinæ 2, flosculo inferiore dimidio breviores, inter se subæquales, membranaceæ, concavo-carinatæ, carina a medio ad apicem glandulis nigris minutis præsertim in valva superiore munitæ, apice obtuse vel truncato-bifidæ, valva inferior latiuscula, quinquenervis, valvam inferiorem genitaliaque amplectens. *Stamina* 3, antheris linearibus, luteis, utrin-que emarginatis, grandibus. *Styli* breves. *Stigmata* as-pergilliformia ex apice floris egredientia. *Caryopsidem* non vidi. *Flosculi superioris* valvæ corollinæ inæquales, inferior valvam corollinam flosculi inferioris æmulans sed paullo angustior, superior brevior, linearis, bicarinata, carinis apicem versus longiuscule ciliato-villosis, apice sub-bifida. *Stamina* 1-2.

192   SPICILEGIA GORGONEA.

LXI. Equisetaceæ.

265. Equisetum *pallidum*, Bory! *Exp. de Morée, Bot. p.* 282.
Equisetum ephedroides, *ejusd. Fl. du Pelop. p.* 66.—Ic.
Bory, *Expéd. de Mor. Bot. t.* 35. *Fl. du Pelop. t.* 37.
Hab. In herb. ins. Cap. Vir. (*Mus. reg. Par.*)

LXII. Filices.

266. Ophioglossum *reticulatum*, Linn. *Sp. Pl. p.* 1518.
Ophioglossum cordatum *et* reticulatum, *Plum. fil. p.* 141. —
Ic. Plum. *l. c. t.* 164. Lamck, *Ill. t.* 864. *f.* 2.
Hab. In herb. ins. Cap. Vir. (*Mus. reg. Par.*)
267. Cystopteris *odorata*, Presl, *Tent. p.* 93. Aspidium odo-
ratum, *Bory! in Willd. Sp. Pl.* 5. *p.* 286.
Hab. In declivibus siccis umbros. ins. *S. Nicolai* (die
29 Martii, 1822. *Forbes*, n. 23. spec. sorophora.) Ad
occidentem et meridiem ins. *S. Vincentii* ad alt. 600 ped.
(*Th. Vogel*) apicem versus monticuli cujusdam in valle
S. Dominici ins. *S. Jacobi* (*J. D. Hooker.*)
268. Pteris *ensifolia*, Desf. *Atl.* 2. *p.* 401. Bory, *Fl. Pel.
p.* 68. Pteris longifolia, *Guss. Syn. Fl. Sic.* 2. *pars* 2.
p. 657, *an Linn.?* a qua secundum cl. Bory. l. c. differt.—
Ic. Alp. *Exot. p.* 66. Bocc. *Mus. t.* 46. Bory, *l. c. t.* 39.
Hab. In declivibus umbrosis humidis ins. *S. Nicolai* (*Forbes,
p.* 24. d. 30, Mart. 1822.) In ins. *S. Antonii* (*Th. Vogel*).
269. Adiantum *Capillus Veneris*, Linn. *Sp. Pl. p.* 1558.—Ic.
Matth. (*Valgris* 1565) *p.* 1201. Cam. *epit. p.* 924. *Engl.
Bot. t.* 320. Turp. *Fl. Méd. t.* 94.
Hab. In ins. *S. Antonii* et *S. Nicolai* (*Forbes* n. 14 et 25.)
270. Adiantum *Capillus Gorgonis*, Webb; stipite tereti læte
purpurascente pilis mollibus subpaleaceis fulvis hirtulo,
rachide apice nudiuscula sæpe radicante, fronde pinnata,
pinnis remotiusculis subsessilibus subflabellato-3-angula-
ribus sulcato-lineatis pilosis, margine inferiore a basi leviter
incurvo, interiore a rachide dimoto subincurvo, margine
superiore profunde 3-4-laciniato, sinubus latis divaricatis

laciniis oblongis apice latioribus rotundatis, crenatis, indu-
siis latis quadratis hirtis.

HAB. In muris *Sacchari* agrorum villium insulæ *S. Nicolai*
(*Forbes*, n. 21. d. 30 Martii, 1822.) In sylvis *Euphorbiæ*
et alibi passim in ins. *S. Nicolai* (*Forbes*, 3. d. 30 Martii,
1822) ad apicem Montis *Verede*, ins. *S. Vincentii* (*Th.
Vogel*, n. 5. Junio, 1841.) In herb. ins. Cap. Vir. (*Mus.
reg. Par.*)

Though this species comes near the *Adiantum caudatum*,
Linn., and *A. hirsutum*, Bory, the shape and the incisures of
the leaves, as well as of the indusia, are very distinct. It
may perhaps be the plant called *A. incisum* by Forskähl,
(*Fl. Æg. Arab. p.* 187) but it does not agree with the short
description he gives of that species.

272. Asplenium *polydactylon*, Webb; stipitis robusti per
totam longitudinem paleacei nervo medio tereti marginibus
revolutis, fronde ad apicem stipitis a basi digitato-radiata
sexies dichotoma, dichotomia summa brevi æquali pinnis
linearibus latiusculis planiusculis apice 2-3-dentatis, dentibus 2-apiculatis.

HAB. In herb. ins. Cap. Vir. (*Mus. reg. Par.*)

A nostra *B. radiatum*, Presl, quod verum quoque *Asplenium*
nec *Blechnum*, differt stipitibus debilibus filiformibus basi
nudis, fronde quater raro quinties dichotoma, pinnis angustis
sæpe longitudine inæqualibus, apice acutis vel irregulariter
ad latera dentatis.

Our comparison has been established with a plant brought
by Commerson from the Mauritius in the herbarium of Desfontaines, and ticketed " *Asplenium radiatum*, Sp. Pl. 308,"
in the handwriting of Willdenow, which agrees perfectly with
the figure of Vahl. To which, or whether to either of these,
the synonymy of Forskåhl should be appended, appears doubtful, unless Vahl had himself seen the plant of that author.

273. Asplenium *palmatum*, Lamck. *Encycl.* 2. 302. Swartz,
Syn. p. 75.—Ic. Pluk. *Phytogr. t.* 287. *f.* 4.

HAB. Ad arborum radices in declivibus umbrosis ins. *S. Nicolai*, (*Forbes*, d. 3 Martii, 1822.)

O

194   SPICILEGIA GORGONEA.

274. Asplenium *Canariense*, Willd. *Sp. Pl.* 5. *p.* 336. A. gemi-
naria, *Bory, Iles Fort. p.* 313.
HAB. In ins. *S. Vincentii (Th. Vogel.)* In herb. ins. Cap.
Vir. (*Mus. reg. Par.*)
275. Aspidium *molle*, Swartz, *Syn. p.* 49. Willd. *Sp. Pl.* 5.
p. 246. Link, *Fil. Spec. p.* 100. Brunner, *Ergebn. p.* 20.—
Ic. Jacq. *Ic. Rar.* 3. *t.* 640. Schkuhr, *Crypt. Germ. t.* 34. *b.*
HAB. In umbrosis humidis ins. *S. Nicolai (Forbes,* n. 22.
die 30 Martii, 1822.)
276. Aspidium *elongatum,* Swartz, *Syn. p.* 55. Polypodium
elongatum, *Hort. Kew.* 3. *p.* 465. Nephrodium elongatum,
Lowe Nov. Fl. Mad. p. 5.
HAB. In declivibus umbrosis ins. *S. Nicolai (Forbes,* n. 27.
die 29 Martii, 1822.)
A species of *Nephrodium* occurs in the herbarium of the
Cape de Verd islands of the Paris Museum, but without sori,
and scarcely therefore determinable.
277. Nothochlæna *Maranta,* R. Br. *Prod. p.* 146. N. subcordata,
Desv. *Journ.* 3. *p.* 92. *Encycl. Suppl.* 4. *p.* 110. Acrostichum
Maranta, *Linn. Sp. Pl. p.* 1527. A. subcordatum, *Cav. Anal.
de Cienc. Nat.* 4. *p.* 97. A. Canariense, *Willd. Sp. Pl.* 5.
p. 121 ! (cx spco. herb. Desf. a viro celeberrimo notato.)
Pteris Canariensis, *Presl, Syn. p.* 145. Ic. Schkuhr, *Crypt.
Gewachst. t.* 4.
HAB. In herb. ins. Cap. Vir. (*Mus. reg. Par.*)
278. Nothochlæna *lanuginosa,* Desv. *Encycl. Suppl.* 4. *p.* 110.
Acrostichum velleum, *Hort. Kew. ed.* 1. 3. *p.* 457. Acros-
tichum lanuginosum, *Desf. Fl. Atl.* 2. *p.* 400.—Ic. Desf.
l. c. t. 256. Schkuhr, *Crypt. Germ.* 1. *t.* 1.
HAB. In herb. ins. Cap. Vir. (*Mus. reg. Par.*)

LXII. MUSCI.*

279. Notarisia *crispata,* Montag. *in Webb et Berth. Fl. Can.
Sect. ult. p.* 14. N. Capensis, *Hampe in Linnæa.* Ortho-

* I am indebted to my learned collaborator and excellent friend,
Dr. Montagne, for the following list of the few Cellular plants contained in the collections from the Cape de Verd Islands.—P.B.W.

trichum? crispatum, *Hook. et Grev. Edinb. Journ. of Sc.*
1. *p.* 115. *Coll.* n. 119.

HAB. In summo Monte *Verede* ad altitudinem 400 hexap.
supra mare lecta, ins. *S. Vincentii,* (*Th. Vogel.*)
280. Macromitrium, sp?
Specimina sterilia, cum priori mixta inveni. (*Th. Vogel*).
M. Nepalensi quoad habitum simile at foliorum forma
diversum, vix vero determinandum.

LXIII. HEPATICÆ.

281. Marchantia *papillata,* Raddi.—Nees ab Esenb. *Hep. Eur.*
4. *p.* 109. *Syn. Hepat. p.* 258.
HAB. In herb. ins. Cap. Vir. (*Mus. reg. Par.*)
282. Frullania *nervosa,* Montag. *l. c. p.* 55. Nees, Lindg.
et Gottsche, *Syn. Hepat. p.* 439.
HAB. Cum prioribus et eisdem immixta.
OBS. Specimen, licet singulum adsit, hancce speciem in
plantis Canariensibus olim describendis conditam genui-
nam esse pulchre et abunde confirmans.

LXIV. FUNGI.

283. Coniothecium *anisosporum,* Montag. MS.; acervulis
minutis hemisphæricis confluentibus erumpentibus atris,
sporis conglomeratis magnitudine variis fusco-atris e glo-
boso-oblongis angulatis impellucidis, episporio papuloso.
HAB. Ad corticem arborum. (*Th. Vogel.*)
Pustulæ minutæ, vix $\frac{1}{5}$ millim. diametro metientes, aterrimæ,
sphæriam quamdam mentientes, sparsæ aut tandum con-
fluentes. *Sporæ* sub epidermide candicante conglobatæ
et conglutinatæ, quoad magnitudinem multum variantes,
$\frac{3}{200}$ ad $\frac{5}{200}$ millim. crassæ, globosæ, oblongæ, angulatæ,
opacæ, episporio papuloso vestitæ, corticem haud cons-
purcantes.
Ab omnibus speciebus hucusque cognitis hæc episporio papuloso differre videtur.

LXV. Lichenes, *Linn.*

284. Evernia *flavicans*, Fries, *Lich. Eur.* Borrera, *Ach.*
Hab. In ins. *S. Vincentii* lecta, (*Th. Vogel,* n. 117 et 118.)
285. Ramalina *usneoides,* Montag. *Crypt. Bras. nec non in Explor. Scient. Alger. cum icone.* Alectoria usneoides, *Ach.*
Hab. In arboribus *S. Vincentii.*
286. Ramalina *scopulorum,* Ach.
Hab. In rupibus basalticis ins. *S. Vincentii* ad altitud. 250 hexapod. supra mare lecta. (*Th. Vogel,* n. 115.)
287. Roccella *tinctoria,* Ach.
Hab. Ad rupes cum priori (Collect. n. 116.)
288. Parmelia *leucomela,* Fries, *Lich. Eur.* Borrera, *Ach.*
Hab. In herb. ins. Cap. Vir. (*Mus. reg. Par.*)

LXVI. Algæ, *Linn.*

281. Sargassum *fissifolium,* Ag. *Syst. Alg.* 1. *p.* 303. Montag. *l. c. p.* 132 cum descriptione. Fucus fissifolius, *Mert. in Mém. Mus. Hist. Nat. Paris, t.* 5. *p.* 180.
Hab. In mari Atlantico prope Cap. Vir. (*J. D. Hooker*).
290. Cystoseira *Abies-marina,* Ag. *Spec. Alg.* 1. *p.* 54. Montag. *l. c. p.* 137.
Hab. In littore Ins. Cap. Vir. (*J. D. Hooker.*)
291. Galaxaura *umbellata,* Lamx.
Hbb. In oris insulæ *S. Vincentii* inter maris rejectamenta, (*Th. Vogel.*)
292. Galaxaura *fragilis,* Lamx.
Hab. In littore ins. *S. Antonii,* cum priore rejecta.
293. Liagora *decussata,* Montag. MS.; fronde calce incrustata filiformi tereti virgato-ramosa, ramis oppositis, ramulis decussatis subulatis erectis apice nudo vinoso-lilacinis.
Hab. Ins. *S. Vincentii,* ad oras rejecta. (Legit cl. *Forbes.*)
Alga maxime spectabilis et inter omnes hujusce generis . species pulcherrima. *Frons* basi scuto parvulo rupibus affixa, albo-incrustata, filiformis, setam porcinam, si denudata, crassa, cum crusta vero sat crassa et fragilissima

qua obducta est pennam passerinam adæquans, sensim attenuata, spithamæa, a basi ramosa. *Rami* ut plurimum oppositi, longissimi, virgati, ramulis instructi brevibus, bilinearibus, decussatis! subulatis, erectis, inferne ut et frons calce porosa incrustatis albis, apice nudo lilacinis vel hoc colore quem Galli vocant *lie de vin* tinctis. Totius frondis nec non ramorum primariorum circumscriptio generalis lanceolata, ita ut planta taxiformis dici potuisset. *Structura* omnino generis. *Sporarum* glomeruli inter fila moniliformia horizontaliter irradiantia nidulantes et in ramulis laterales, sphærici, $\frac{6}{100}$ millim. crassi. *Sporæ* minimæ, obovatæ, inter paraphyses e nucleo centrali quoquo versus irradiantes, dichotomas, articulatas obviæ.

Species, si qua, genuina nec cum ulla alia confundenda.

293. Alsidium *triangulare*, I. Agarah, *in Linnæa* 15, *p.* 28.

HAB. In ins. Cap. Vir. (*Forbes.*)

FLORA NIGRITIANA;

OR, A

CATALOGUE OF THE PLANTS

OF THE

RIVER NIGER, THE ISLAND OF FERNANDO PO,

AND ADJACENT PARTS OF

WESTERN TROPICAL AFRICA;

FROM THE COLLECTIONS OF

DR. TH. VOGEL:

TO WHICH ARE ADDED THOSE OF

MR. G. DON AND OTHER TRAVELLERS:

BY

JOSEPH DALTON HOOKER, M.D., R.N., F.R.S., L.S., &c.

AND

GEORGE BENTHAM, ESQ.

PREFACE

THE following pages contain a complete enumeration of all the plants collected by the late Dr. Vogel and his companions at Sierra Leone and other parts of the Guinea Coast, on the River Niger and its branches, and in the Island of Fernando Po, and of those gathered by Mr. George Don, chiefly at Sierra Leone and in the Island of St. Thomas, when collecting for the Horticultural Society. It had been wished to include also all other collections from the same country to which access could be had, but it was found that the work would thus have been extended beyond the prescribed limits, both as to bulk and time; it has therefore been confined to the above-mentioned enumeration, with the occasional addition of a few others of peculiar interest, from the Hookerian Herbarium. At the same time, all other species hitherto published from Western Tropical Africa, including Senegambia, are mentioned by name, and wherever unpublished species have been found in the Herbaria of Sir W. Hooker or of Mr. Bentham, they have been alluded to under their respective genera, and the geographical range of each species is given as far as known wherever they extend beyond the region here in question. The geographical botanist will thus find a Flora as complete for his purpose as the materials could furnish.

The manuscript, as far as the *Leguminosæ* inclusive, was prepared, with such exceptions as are specially indicated by notes, by Dr. Joseph Dalton Hooker previously to his departure for India. This manuscript has been revised by Mr. Bentham, who has prepared for press the remainder of the natural Orders. In this, however, he has been guided by a considerable number of analyses made by Dr. Hooker, and some memoranda made by Dr. Planchon in the Hookerian Herbarium.

FLORA NIGRITIANA.

I. RANUNCULACEÆ.

1. Clematis *grandiflora,* DC. *Syst. Veg.* 1. *p.* 151.—C. chlorantha, *Lindl. Bot. Reg. t.* 1234.—Sierra Leone, *Don.*

The west coast of Africa offers a striking instance of the scarcity of *Ranunculaceæ* in all hot, damp and low climates; for there it is represented by a solitary genus, and of this genus one other species only is known as an inhabitant, the *C. hirsuta,* Guill. and Perr., from the Cape de Verd. It is probable that, as in Tropical Asia, the genus affects hilly localities, for no species occur in any of the collections made on the flat and humid coasts south of Sierra Leone.

II. DILLENIACEÆ.

1. Tetracera *scabra,* Hook. fil. ; ramis pedunculis foliisque subter præcipue scaberulis, his petiolatis obovatis obtusis basi angustatis super glaberrimis albo-punctatis subter nervis pilosis, marginibus integerrimis tenuiter recurvis, paniculis racemosis multifloris, pedicellis villosis, sepalis rotundatis pubescentibus ciliatis.—Nun River, *Vogel;* Niger, below Abòh, *Ansell.*

Frutex volubilis. *Rami* asperi. *Folia* 3-5 unc. longa, 1½-2½ lata, super (siccitate) saturate fusca subnitida, subter pallida ; juniora subspathulata utrinque scaberula ; seniora punctulata, punctis demum concavis. *Sepala* extus rubescentia. *Petala, staminaque* albida. *Ovaria* in omnibus quos examinavimus floribus tria, quorum duo semper abortiri videntur. *Ovula* pauca. *T. Senegalensi* proxima, differt ramis foliisque subter asperis, floribus minoribus, sepalisque rotundatis et pubescentibus.

In the presence of several species of *Tetracera*, Tropical Western Africa partakes of the botanical features both of the east coast of South America and Intertropical Asia. The species are confined to the hottest, dampest and most shaded forests of all these regions. All of the African species are more nearly allied to those of the New than of the Old World.

2. Tetracera *Senegalensis*, DC. *Prodr. v.* 1. *p.* 68. *Guill. et Perr. Fl. Seneg. p.* 2.—T. obovata, *DC. et Guill. et Perr. l. c.*—Senegambia, Sierra Leone and the Bight of Benin, *Afzelius, Don,* &c.

Apparently a very common species north of the Equator, but not occurring in the collections of the Niger Expedition.

3. Tetracera *alnifolia*, Willd.; DC. *Prodr. v.* 1. *p.* 68.—Senegambia, Sierra Leone and Guinea, *Smeathman, Afzelius, Don,* &c.

The only other species of this order known as W. African, is *T. rugosa*, Guill. and Perr., a name unnecessarily changed by Steudel to *T. Guillemini*, on the supposition that there was already a *T. rugosa*, H.B.K., which we cannot find to have been anywhere described.

III. ANONACEÆ.

1. Anona *muricata*, Linn. (Sour-sop).—Sierra Leone (cult.), *Vogel.*

2. Anona *squamosa*, Linn. (Sweet-sop).—Sierra Leone (cult.), *Vogel.*

A plant now equally and universally diffused over the tropics of both the New and Old World, and to which St. Hilaire, with some shew of probability, assigns the E. Indies as its native place. It is particularly abundant about Portuguese colonies, and its Brazilian name of "Ata" is considered by that author as all but conclusive in favour of its Asiatic origin. On the other

hand, it is unknown on the continent of Asia except as a cultivated plant, and there universally regarded as a colonist from
the W. Indies. It is not, however, the less valued on account
of its foreign origin, for we read of it as, in seasons of famine,
proving the staff of life to the Hindoos, whilst its acrid seeds
are used to clean vermin from the body, for which purpose they
are powdered and mixed with the flour of Cajana or *Ervum
Lens.*

3. Anona *cherimolia*, Mill. ; DC. *Prodr. v.* 1. *p.* 85.—A. tri-
petala, *Ait.* (Cherimolia.)—Cape de Verds and W. Africa,
(cult.)

A very abundant plant, and quite naturalized in the Cape de
Verd Islands. In a dried state it is with difficulty distinguished
from the preceding, and chiefly by the more pubescent leaves.
This plant is an undoubted native of the New World.

4. Anona *palustris*, L. *Hook. Bot. Mag. t.* 4226, (*Alligator
apple.*)—Grand Bassa Cove, *Vogel.*

Not alluded to by Vogel as in a state of cultivation, though
we cannot doubt that such was the case. It is a native of the
West Indies, and common along the Brazilian coasts. Its fruit
is hardly edible, and certainly not palatable. Hitherto it has
not been introduced into Asia, though it is one of the few
species of this genus that has flowered and ripened its fruit in
England.

5. Anoua *Senegalensis*, Pers. DC. *l. c. Guill. et Perr. Fl.
p.* 5. *Deless., Ic. Sel.* 1. *p.* 23, *t.* 86.—A. arenaria, *Schum.
et Thonn. Beskr. p.* 257.—Senegambia, SierraLeone, Bight of
Benin, Congo River, *Perrottet, Afzelius, Don, Christ. Smith.*

Doubts have been thrown upon the Asiatic and even African
origin of any of the species of *Anona,* from the fact of none
having been hitherto found in Asia, and from those species
which that continent shares with Africa being assuredly natives
of the New World. The number of stations, however, assigned
to *A. Senegalensis,* from between localities so widely apart as
the Congo and Sierra Leone, is in favour of the genus being
African, as is the fact that neither this nor *A. glauca* and *chry-*

socarpa have been found in any part of America. We are
inclined to cite *Anona* as an instance of the greater affinity
existing between W. Africa and E. America, than the latter
shares with any other part of the Old World.

6. Anona *glauca,* Thonn. et Schum. *Beskr. p.* 259. *Guill. et
Perr. Fl. Seneg.* 1. *p.* 5.—Senegambia and Cape Coast Castle,
Brunner and *Brass* (in Mus. Brit.)

A seventh species is *A. chrysocarpa.* Guill. et Perr.

It is singular that the *Anona reticulata* (Bullock's heart,
of the West Indies) is nowhere mentioned as cultivated in
Tropical Africa; it is a far from unpalatable fruit, and very
abundantly spread over the East and West Indies.

The *Monodora myristica,* Dun. (Calabash nutmeg), does not
exist in any of our W. African collections, but is probably a native
of the continent of Africa. It is always quoted as an inhabitant
of Jamaica, where, according to Mc. Fadyen, (Flora of Jamaica,
p. 12), but one tree of it exists, and where the generally-received
opinion is, that it was introduced from the continent of South
America. Mr. Brown, on the other hand, argues for its African
origin, and the probability of its having been carried by the
negroes to Jamaica.

1. Hablitzia *Æthiopica,* Alph. DC. *Mem. Anon. p.* 31.—West
coast of Africa, from Senegambia to the Bight of Benin.

It was with considerable anxiety that this, the " Guinea" or
" Malaghata pepper," was sought for, but in vain, amongst the
collections of the Niger Expedition, for it is a plant of which
we know but little botanically; though its seeds were an article
of export for upwards of two centuries, and were once highly
prized as a condiment, it is now never seen and seldom
heard of, except by the curious.

So important an article of commerce was it, that the name of
" Grain Coast" was given to a long tract of land in the Bight of
Benin, and the establishment of the towns of Grand Bassa and
Cape Palmas was due to its importance. Up to the close of the
18th century, the Guinea pepper was in great request; when
the still more aromatic and pungent grain of the Eastern

Archipelago drove the milder condiment from the table and market.

It was a plant very early known to the Arab physicians; Serapim calls it "fulful alsuaden," that is, Pepper of the Black people, whence our name of Æthiopian pepper. The French, "Grain de Zelim," is derived from the Arabic name of "Azelim," given to it by Avicenna.

Alphonse De Candolle cites this genus as confined to Western Africa and the Western Indies. The only other African species is *H. undulata*, A. DC., (*Xylopia undulata*, Pal. Beauv.) from Benin.

1. Cœlocline *parviflora*, A. DC. *l. c.* Uvaria parviflora, *A. Rich. in Fl. Seneg. p.* 9, *t.* 3, *f.* 1.—Senegambia, *Perrottet;* Quorra, *Vogel.*

Flores plerumque solitarii, axillares, ¼-⅓ unc. longi. *Sepala* 3, late ovata, acuta. *Petala* linearia, sepalis quintuplo longiora. *Ovaria* 4.

A very similar species, or probably variety of this, from the Congo River, (*Christ. Smith*), has the leaves narrower and sharper at the base; whilst a third, also in the British Museum, and gathered by Smeathman, has villous and hairy ramuli, with longer and still more acuminate leaves. The flowers and fruit of all are very similar. To these three species, as many other W. African ones may be added; *C. acutiflora*, A. DC., *C. polycarpa*, A. DC., and *C. oxypetala*, A. DC., all from Sierra Leone.

1. Artabotrys *macrophylla*, Hook. fil.; glabra, foliis amplis late ovatis v. elliptico-oblongis utrinque rotundatis et apice abrupte acuminatis subcoriaceis super nitidis, pedunculis oppositifoliis lateralibusve uncinatis ramosis multifloris, pedicellis brevissimis crassis, sepalis e basi lata acuminatis, petalis ovatolanceolatis calyce duplo longioribus.—Fernando Po, *Vogel.*

Arbor parva, apice ramosa. *Rami* diam. pennæ anserinæ, teretes; cortice atro, striato. *Folia* brevissime petiolata, fere pedalia, 8-10 unc. lata, subcoriacea, super splendentia, subter pallida, opaca, costa muricata, venis prominulis. *Panicula* 2 unc. longa, lignosa; pedunculo primario caule continuo et ejusdem

diametro, statim ramoso ; ramo altero uncinato-recurvo, ramulis
brevibus multifloris bracteatis pubescentibus ; bracteis ovatis v.
obovatis, extus velutinis. *Pedicelli* vix 2 lin. longi, infra florem
incrassati ; floribus pro planta parvis. *Sepala* crassa, patentia,
3-4 lin. longa, basi 3 lin. lata, tumida, extus velutina, in acumen
lineare recurvum producta. *Petala* 6, coriacea, inter se
subæqualia, extus velutina, basi dilatata et intus lamina carnosa aucta, supra laminam paullo constricta ed divergentia.
Stamina numerosa, multiseriata. *Ovaria* plurima.

The specimens are imperfect, but the inflorescence as well as
the flowers, in so far as we have been able to ascertain their
structure, sufficiently indicate their close affinity to the genus
Artabotrys, of which the species hitherto described are all from
the East Indian Archipelago, or from the south-eastern parts of
the Asiatic continent.

1. Uvaria ? *Vogelii,* Hook. fil. ; glabra, foliis breviter petiolatis
anguste obovali-oblongis breviter acuminatis margine subun-
dulatis basi rotundato-subcordatis subter glaucis, pedicellis
unifloris solitariis geminisve, calyce obtuse repando-trilobo,
petalis ovatis obtusis crassis exterioribus latis calyce duplo
longioribus interioribus oblongis minoribus, staminibus glandulosis, carpellis (v. carpellorum articulis ?) breviter stipitatis
oblongis monospermis.   (Tab. XVII.)—On the Quorra, at
" Sterling," *Vogel.*

Rami validi ; *ramuli* horizontales, patentes, dein erecti, elongati
et hinc inde semel in spiram torti, unde verosimiliter scandentes, epidermide atra, punctis albis conspersa. *Folia* submembranacea, 3-6 unc. longa. 1½ unc. lata, petiolo 2 lin.
longo, super in sicco nitidula, subter pallidiora glaucescentia
v. rubescentia. *Pedicelli* semiunciam longi, fructiferi in
parte inferiore denudata ramulorum siti, ad axillas foliorum
delapsorum. *Flores* quorum fragmenta tantum adsunt, parvi.
Calycis lobi brevissimi, obtusi. *Petala* exteriora, (forte nondum perfecte accreta), 1½ lin. longa, crassa, glabra, æstivatione verosimiliter anguste imbricata. *Stamina* pauca, (sub-
definita ?) ; filamentum breve, crassum ; connectivum filamento subæquale et sicut illius apex, glandulis conspersum,

subquadratus ; loculi oblongi marginales. *Carpella* (v. carpellorum moniliformium articuli inferiores ?) circa 6, 3½-4 lin.
longa, siccitate nigra, lævia et glabra, oblonga, obtusa cum
mucrone parvo, sed ex speciminibus haud patet si mucro e
styli reliquiis superest, vel stipitem indicat articulorum superiorum arbortientium seu delapsorum. *Pericarpium* tenuiter
carnosum, semini arcte adhærens. *Semen* in carpello (seu
articulo) unicum, loculum arcte implens, exarillatum, raphe
completa percursum. *Testa* tenuiter coriacea, integumentum
interius membranaceum, cum exteriore conferruminatum, intus
productum in plicas numerosas transversales parallelas, cum
illas albuminis alternantes et juxta raphin plica angusta
verticali inter se connexas. *Albumen* corneum, ruminatum,
laminas format horizontales numerosas cum plicis integumenti
alternantes, et irregulariter inter se connexas, lamina verticali
fere continua raphi opposita ad peripheriam tamen haud
attingente ; lamina altera verticali cum priore ad angulam
rectam disposita hinc inde laminas duo v. plures connectente.
PLATE XVII. *Fig.* 1. flower ; *f.* 2. one of the outer petals ;
f. 3. stamen ; *f.* 4. vertical, and *f.* 5. transverse section of
the carpel and seed, (in which, however, by an error of the
artist, the vertical plates of the albumen are represented as
continuous with the pericarp) ; *f.* 6. portion of the surface
of the connectivum showing the glands ; *all magnified, especially the last.**

* The sketch made by Dr. Hooker of the only tolerably complete
flower that he could find, shows that the number of stamens is much
fewer than in most *Uvariæ :* this circumstance, together with the
form of the anthers and the apparently monospermous carpels, induced
Dr. Planchon to suggest that this plant should constitute a distinct
genus, under the name of *Clethrospermum,* allied to *Oxandra.* The
state of the flower examined was such, however, that it was not possible to ascertain whether the number of stamens was really definite,
nor yet to investigate the structure of the pistils ; and although the
carpels look as if they were complete and constantly monospermous,
yet precisely the same appearance is often assumed by the moniliferous
fruits of some *Uvariæ,* when reduced by accident or by abortion to a single
articulation, and it is therefore impossible, without further materials,

2. Uvaria *gracilis*, Hook. fil.; glabra, ramulis gracilibus ultimis pubescentibus, foliis breviter petiolatis submembranaceis obovato-lanceolatis longe et obtuse acuminatis basi
subangustatis et juxta petiolum obscure cordatis subter pallide glaucis venis rubris, pedunculis axillaribus solitariis,
sepalis patulis obovatis obtusis, carpellis glaberrimis breve
cylindraceis lævibus subglaucescentibus longe stipitatis (an
nunc moniliformibus?) toro parvo capitato insertis monospermis.—Sierra Leone, *Don.*

Rami crassit. pennæ corvinæ, parce ramulosi; cortice cinereo,
striato, nunc albo-punctato. *Folia* 3 unc. longa, 1¼ lata,
ima basi emarginata v. cordata, supra medium gradatim
latiora, deinde angustata, apice subobtusa v. acuminata, super
pallide viridia vix nitentia, subter alba, glauca; petiolo 2
lin. longo. *Pedicellus* fructus uncialis. *Lobi calycini* ¼ unc.
longi, coriacei, persistentes. *Torus* parvus, 1½ lin. lat., apice
planus. *Carpella* parva, patentia, ¾ unc. longa, pedicello
æquilongo suffulta, utrinque obtusa, apiculata.

Specimens rather imperfect, but belonging to a very distinct
species. Some of the carpels are distinctly monospermous, while
others appear to be the lowest loculus of a moniliform carpel.
Seeds very aromatic.

3. Uvaria *globosa*, Hook. fil.; ramis gracilibus, ramulis velutinis, foliis breviter petiolatis oblongo-ellipticis lanceolatisve
basi rotundatis apice angustatis utrinque nitidis ad venas
subter præcipue pubescentibus marginibus tenuiter recurvis,
floribus axillaribus solitariis v. binis brevissime pedicellatis

to give any character to distinguish this species from *Uvaria*, as generally
extended to include *Unona*, or even from those species of true *Uvariæ*,
which Blume includes in his group *Ambiguæ*. The vertical laminæ of the
seed are by mistake described by Dr. Planchon as folds of the integument almost meeting in the axis; when in fact they are the continuous
portions of the albumen itself, by which the horizontal plates are more
or less connected together. They are continuous with each other, occasionally forming a cross in the centre, and extend nearly to the circumference, the most complete of them being opposite to the raphis, from
whence a narrow vertical fold of the integument projects into a slight
furrow in the albumen.—(G. B.)

velutinis, carpellis 3-4 breviter stipitatis globosis dense ferrugineo-pubescentibus toro capitato insertis, seminibus biseriatis.—Accra ; *Vogel.*

Rami graciles, teretes, ramulosi, crassitie pennæ anatinæ ; cortice atro tenuiter striato albo-punctato; ramulis patentibus,
ascendentibus, pube rufa velutinis. *Folia* 2-4-uncialia, $\frac{3}{4}$-1
unc. lata, forma varia, pleraque lanceolata, rarius ovalia v.
oblonga, e basi semper rotundata, ad $\frac{3}{4}$ longit. sensim latiora,
deinde angustata, summo apice acuminata v. obtusa, utrinque
siccitate luride rufo-fusca; petiolo 1 lin. longo. *Pedunculi*
1-2-flori, 1 lin. longi, validi, velutini. *Flores* parvi, extus
dense sericei, pilis rufis nitidis. *Sepala* late ovato-triangu-
laria, basi connexa, 2 lin. longa. *Petala* (in flore manco
observata) exteriora late ovata obtusa, calyce paullo longiora, æstivatione verosimiliter imbricata, interiora haud visa.
Stamina haud numerosa, lineari-clavata; antheræ filamento
æquilongæ, connectivo incrassato, loculis linearibus lateralibus.
Carpella pauca, singula 3-4 lin. diametro, stipite linea breviori, rufo-velutina, vertice obtusissima v. depressa et notata
cicatrice styli. *Semina* 4-7, horizontalia, laminis endocarpii
separata, exarillata, mutua pressione variis modis verticaliter
compressa, testa nitida coriacea, integumento interiore more
plerumque *Anonacearum* intra plicas albuminis producto,
costa verticali interiore vix in laminam producta.

This species is evidently congener with several of the Eastern
Uvariæ retained in the genus by those who confine it within the
narrowest limits, although the stamens appear to be less numerous
than they usually are. It is very nearly allied to, if not identical
with an undescribed Cape Coast specimen, which has rather
broader, less acuminated leaves, almost cordate at the base, and
somewhat larger flowers.

Four other West African species of *Uvaria* are enumerated ;
two by DC., *U. macrocarpa* and *U. ovata*; and two by Schumacher, *U. cordata* and *U. cylindrica,* which may belong to
some of the above or following species. There are further, four
undescribed in the British Museum, two of them from Sierra
Leone, one from the Congo River, and the fourth from Cape

Coast Castle, making in all eleven species of this or closely allied
genera. The difficulty, however, of ascertaining even the
generic characters is very great, without the presence of very
good specimens, both of the flower and fruit; the relation
between the carpological and floral characters not having yet
been sufficiently made out by the monographists who have
studied the *Anonaceæ*.*

Another West African genus of *Anonaceæ, Hexalobus*, A.D.C.
including two species, completes the Order as existing on that
coast, which thus enumerates upwards of twenty native species,
a very large proportion for a Flora so little known, and so
defective in number of species. In the predominance of
Anonaceæ, this Flora resembles that of the islands of the Indian
Archipelago, to which the whole coast is related more markedly
in its botany, than to the continent of America.

IV. MENISPERMACEÆ.†

Gen. Nov. JATEORHIZA —Fl. dioici.—Masc. *Sepala* 6, ovata,
biseriata, exteriora paulo minora, æstivatione imbricata. *Petala*
6, ovata, sepalis breviora, apice truncata, lateribus introflexis
stamina tegentibus. *Stamina* 6, petalis opposita: *filamenta*
crassa, apice arcte refracta, et in connectivum amplum car-

---

* The æstivation of the corolla especially has been little attended to,
and is likely to afford valuable auxiliary characters. In most general
works, as in Endlicher's "Genera" and Lindley's "Vegetable Kingdom,"
the petals are said to be valvate in each series; and although, in the most
recent work on the subject, Martius and Endlicher's "Flora Brasiliensis,"
the imbricate æstivation of the petals of *Duguetia* is noticed, yet even in
that work the valvate æstivation is included, as well in the ordinal character, as in the generic character of *Guatteria ;* whereas in most, if not
all species of true *Guatteriæ, Uvariæ, Unonæ*, and some others, the petals
will be found more or less to overlay each other in the bud, as readily
indicated by the rounded form of their apex. In *Anona*, and all others where
they are truly valvate, that arrangement naturally occasions them to
terminate in a point, at least in the young state.—(G. B.)

† The MS. of this order has been entirely drawn up by Mr. Miers,
from whose able pen we may shortly hope for a complete monograph,
where the species, here only alluded to, will be fully described.

nosum terminata: *antheræ* extrorsæ, dorso affixæ, 4-lobæ, 2-valvæ, rima transversali hiantes. *Ovaria* rudimentaria 3, centralia, punctiformia.—FL. FŒM. *Sepala* ut in masc. *Petala* 6, cuneato-obovata, crassiuscula, apice emarginata, lateribus introflexis stamina volventibus. *Stamina* sterilia 6, petalis dimidio breviora: *filamenta* tenuiora, compressa, lobo rotundato apiculata. *Ovaria* 3, libera, erecta, oblonga, gibba, extus dense glanduloso-pilosa, supra gynophoram sub-3-gonam imposita, 1-locularia, ovulo unico funiculo brevi angulo interno supra medium appenso. *Stylus* brevis, crassus, subexcentricus. *Stigma* 3-partitum, laciniis 2-3-fidis, reflexis. *Drupæ* 3, abortu pauciores, ovatæ, carnosæ, 1-spermæ. *Nux* ovata, dorso convexa, tuberculata, pilis fibrillosis densissime plexis induta, ventre lævis, concava. *Semen* loculo conforme, meniscoideum. *Embryo* intra albumen carnosum quasi 2-laminare fere rectus, lamello exteriori simplici, tenui, interiori crassiori, et in rugis plurimis transversalibus profunde ruminatis, *testa* tenui in plicis insinuata, *cotyledonibus* membranaceo-foliaceis, spathulato-oblongis, lateraliter divaricatis, et in locellis sejunctis utrinque positis, *radicula* supera, brevi, tereti, ad apicem spectanti, centrifuga.—Suffrutices Africæ tropicæ *debiles, volubiles, setis rigidis, vel pilis setosis vestitæ;* folia *alterna, magna, petiolata, cordata, rotundata, palmatim 3-5-7-loba;* racemi *axillares, elongati,* pedicellis *laxis, 3-7-floris,* floribus *vagis, pro ordine majusculis, bracteatis, sessilibus,* bracteis *longissime setoso-ciliatis.*

Jateorhiza *strigosa,* Miers. Cocculus? macranthus, *Hook. fil. in Hook. Ic. Pl.* 759 (TAB. *nostr.* XVIII); foliis rotundato-3-lobatis, basi profunde inciso-cordatis, lobis 3-angularibus acutis mucronatis auriculis basalibus rotundatis, marginibus parallelis fere approximatis, submembranaceis reticulatis supra nitidis subtus pallidioribus 7-nervibus nervis utrinque setoso-strigosis, setis adpressis rigidis rufulis longiusculis, margine dense setoso-ciliatis; petiolo striato auriculis basalibus duplo longiore arctc setoso-strigoso; racemo axillari.—Clarence Cove, Fernando Po, *Vogel;* Congo, *Tuckey* in Herb. Mus. Brit.

This is very distinct from the two other well-known species ;*
and I have retained for it the more appropriate specific name of
strigosa, previously given by me to the Congo specimen : this
point, however, would certainly have been ceded in favour
of the distinguished author of the Flora Antarctica, did not
the name of *macrantha* convey a very incorrect idea of the
species, for its flowers, when expanded, are scarcely more than
2 lines in diameter, and although large for the order, they are
not greater than those of *J. palmata,* or the flowers of some
other genera, and are assuredly diminutive when compared
with those of plants in general. The leaves, from the insertion of
the petiole, to the summit of the middle lobe, are $7\frac{1}{4}$ inches long,
the lateral lobes measure 6 inches, the depth of the basal lobes
is $2\frac{1}{4}$ inches, so that the total length is $9\frac{1}{2}$ inches, their extreme
breadth being 9 inches, and the length of the petiole $5\frac{1}{2}$ inches.
PLATE XVIII. *Fig.* 1. Masc. flower, forced open ; *f.* 2. three of
the six stamens ; *f.* 3. female flower ; *f.* 4. one of the petals
of the female flower, with three sterile filaments ; *f.* 5. ovaria
and styles.

1. Cissampelos *Vogelii,* Miers ; ramulis demum glabris ; foliis
♂ palatis, ♀ subpeltatis, cordatis deltoideo-obovatis apice
obtusiusculis emarginatis sinu mucronatis, supra sparse pubescentibus, subtus griseo-glaucis et pubescentibus 5-7-nervibus,
petiolo tomentoso sæpe refracto limbo fere æquilongo; racemis ♂ ternis petiolo 4-plo brevioribus, ♀ axillaribus solitariis
rarius binis scorpioideo-flexuosis gracilibus folio longioribus,
floribus pedicellatis 7-9 fasciculatis folio reconditis, sepalo
oblongo extus piloso, petalo obcordato minimo ovarioque
glabris.—On the Quorra River ; *Vogel.*

A very distinct species : the leaves of the male plant are
larger than those of the female, being $2\frac{1}{2}$ inches long, including
the basal lobes, and $2\frac{1}{4}$ inches from the petiole at the sinus, they
are $2\frac{1}{8}$ inches broad, the petiole being $1\frac{1}{2}$ to 2 inches long, and
its insertion half a line within the margin. The female inflo-

* 1. Jateorhiza palmata, *Miers,* (Cocculus palmatus, DC. Hook. Bot.
Mag. tab. 2970-2971) ;—and 2. J. Columba, *Miers,* C. palmatus, Wall. Cat.
n. 4953, (in hort. Bot. Calcuttæ cult.)

rescence consists of axillary, slender, lax, scorpioid spikes, $2\frac{1}{2}$ inches long, with very small bracteiform mucronate leaflets, each enclasping about 7 minute pubescent flowers.

2. Cissampelos *comata*, Miers ; foliis suborbicularibus cordatis apice emarginatis mucronatis, petiolo sub-brevi, racemulis 3-4 in axillis junioribus fasciculatis dichotome divisis, pedicellis capillaceis, petalo glabro margine crenato, anthera 8-loba.—On the Quorra ; *Vogel?*

The specimen consists only of a floriferous branchlet about 10 inches long, with a single basal leaf of the parent stem, about $1\frac{1}{4}$ in diameter, on a short petiole only $\frac{1}{4}$ of an inch ; the axils of the floriferous branch exhibit leaflets of similar form, the lower ones being half an inch diameter and expanded, diminishing upwards to the size of 2 lines ; there are generally 3 capillary racemules about 1 inch long ; the sepals are pilose outside, the petals quite glabrous.—Is it the female of *C. Vogelii?*—but the shape of its leaves does not accord with the supposition.

The other Tropical W. African species of *Cissampelos* are *C. Owariensis*, Beauv., from Cape Coast and Oware, *C. mucronata*, A. Rich., extending across from Senegambia to Abyssinia and the Island of Bourbon, and three unpublished species, of which two from Congo, are in the British Museum, and one from Senegambia in the Hookerian herbarium.

The *Cocculus Cebatha*, DC., (which includes *C. Leæba, epibaterium* and *ellipticus*, DC.), mentioned under the first of these names above, (Spicil. Gorgon, p. 97), extending from Senegambia, and Cape Verd, to Egypt, Abyssinia and Arabia ; and *Tinospora Bakis*, Miers, or *Cocculus Bakis*, A. Rich., found also both in Senegambia and Abyssinia and a new species of my genus *Holopeira*, (founded on *Cocculus villosa*, DC., and its allies), complete the list of Tropical W. African *Menispermaceæ*, (J. Miers.)

V. NYMPHÆACEÆ.

1. Nymphæa *cærulea*, Savigny ; DC. *Prodr.* 1. *p.* 114.—Senegambia, *Perrottet* ; Cape Coast, *Don.*

216 FLORA NIGRITIANA.

The limited accommodation under which Dr. Vogel suffered, probably prevented him from preserving the species of this genus, which appears to abound in Western Tropical Africa. Three other species inhabit Senegambia, *N. rufescens, micrantha* and *abbreviata* (all of Guill. and Perr.); whilst a fourth abounds along the coast, *N. dentata*, Schum. and Thonn. (*N. Lotus*, Pal. de Beauv.); and two additional ones, *N. maculata* and *N. Guineensis*, Schum. and Thonn, have been described from Guinea.

VI. PAPAVERACEÆ.

The widely diffused *Argemone Mexicana* is included by Guillemin and Perrottet in the Flora of Senegambia.

VII. CRUCIFERÆ.

An Order as impatient of hot, low and humid climates as are the *Ranunculaceæ*. One species of *Nasturtium* is enumerated by De Candolle as a West African plant. *Cruciferæ* in general appear to be in a great measure represented by the following order.

VIII. CAPPARIDEÆ.

1. Ritchiea *erecta*, Hook. fil.; (TAB. XIX et XX.) fruticosa, erecta, ramosa, ramis verrucosis, foliis patulis longe petiolatis 3-foliolatis, foliolis oblongo-lanceolatis breviter acuminatis basi angustatis integerrimis, racemo terminali multifloro, sepalis lanceolatis acuminatis, petalis lineari-ligulatis, staminibus numerosis calyce longioribus, antheris parvis.—Fernando Po, *Vogel*.

Frutex glaberrimus. *Petioli* teretes, 3-4-unciales, stricti. *Foliola* petiolo æquilonga vel longiora, nunc 6-8 unc. longa, breviter petiolata. *Racemus* terminalis, 2-3-uncialis, cicatricosus; pedicellis ⅓ unc. longis, erectis, basi utrinque bracteolatis. *Sepala* ½-pollicaria, acuminata, marginibus puberulis. *Petala* anguste linearia filamentaque albida. *Antheræ* parvæ, nigrescentes.

A very handsome species, and quite distinct from the following in the erect, branching and not climbing habit, the much longer petioles, smaller flowers and differently shaped leaflets and petals.

2. Ritchiea *fragrans,* Br. App. Clapperton, p. 225.—Sierra
Leone, *Afzelius, Don.*

From Brown's observations in the paper referred to, it appears
that there are other African species of the same genus known to
him.

1. Capparis *linearifolia,* Hook. fil.; glaberrima, caule gracili
scandente? tereti parce ramoso, stipulis aculeiformibus minutis vix recurvis, foliis breviter petiolatis lineari-oblongis
muticis subcoriaceis arcte conduplicatis integerrimis aveniis,
pedunculis multifloris axillaribus solitariis folio æquilongis
brevioribusve patentibus interdum ramosis, floribus parvis
corymbosis pedicellatis, alabastris globosis, sepalis ovalibus
concavis, petalis calycem vix superantibus obovatis, staminibus circa 15, petalis æquilongis toro brevissimo insertis.—
Sierra Leone, *Forbes.*

Caulis crassitie pennæ corvinæ, subflexuosus; internodiis $\frac{1}{2}$-1
unc. longis. *Petioli* 2 lin. longi. *Folia* 2-pollicaria, $\frac{1}{2}$ unc.
lata, siccitate viridia. *Pedunculi* patentes, in axillis foliorum
omnium solitarii, horizontales, teretes, glaberrimi. *Flores*
5-8 apicem versus pedunculi, circiter 4 lin. diametro. *Antheræ* majusculæ.—An planta dioica?
This is a remarkably distinct species.

2. Capparis *erythrocarpa, Isert. Berl. Nat.* 9, *p.* 339, *t.* 9. DC.
Prodr. 1. *p.* 246.—*An* C. Afzelii, *DC. l. c.?*—Accra, *Vogel;*
Guinea.

The descriptions in DC. Prod. do not serve to distinguish
C. erythrocarpa from *C. Afzelii,* though the latter agrees best
with Vogel's plant, simply from its meagreness. The apices of
the leaves are obtuse in this specimen.

There are several other published W. African species of *Cap-
paris,* some of which extend over a remarkably wide range,
viz.: *C. polymorpha,* A. Rich. from Senegambia and Abyssinia;
C. corymbosa, Lam., from Senegambia and Sennaar; *C. tomen-
tosa,* Lam., from Senegal and Delagoa Bay (*Forbes*); *C. puberula,*
DC., from Cape Coast, (*Brass*); and Gambia, (*Don*); *C. Brassii,*
DC., and *C. fascicularis,* DC., both from Cape Coast (*Brass*);
C. Thonningii, Schum. and *C. reflexa,* Schum., both from

Guinea; but possibly these may be the same as some of those previously named, though very imperfectly described.

1. Mærua *Currori,* Hook. fil.; glaberrima, caule tereti, cortice pallido albo-punctato, foliis ad apices ramulorum sparsis breviter petiolatis oblongo-obovatis obtusis apiculatis basi angustatis integerrimis aveniis subcoriaceis, corymbis axillaribus terminalibusque paucifloris, pedicellis flore brevioribus, calyce basi anguste-cylindraceo, corona petaloidea subbipartita.— Elephants' Bay, *Dr. Curror.*

Rami crassitie pennæ anserinæ, hinc inde tumidi; cortice lævi, pallide rufa, punctis albidis aspersa; ramulis cicatricosis. *Folia* uncialia, plana, siccitate subcoriacea, (an carnosula?), obtusa v. emarginata, apicula acuta, costa vix prominula. *Pedicelli* 3 lin. longi. *Alabastra* ½-uncialia. *Calycis tubus* segmentis acutis longior.

A very curious species, but the specimens are far from good. There are four other W. African species of *Mærua,* viz. *M. Angolensis* and *M. Senegalensis,* from Angola and Senegal, *M. rigida,* Br., common to Senegal and Central Africa, and an undescribed species found both in Sennaar and Senegal.

1. Cleome *pentaphylla,* Linn. *Sp. Pl.* 938.—Sierra Leone, *Vogel,* and elsewhere along the coast, abundant.

2. Cleome *Guineensis,* Hook. fil.; caule erecto simplici v. e basi ramoso sparse piloso folioso, foliis 3-foliolatis, floralibus breve petiolatis, foliolis ellipticis utrinque acutis integerrimis pilosiusculis ciliatis, pedicellis axillaribus gracillimis glandulosopilosis, sepalis lineari-lanceolatis, petalis anguste spathulatis, staminibus 6 toro brevi insertis, capsulis stipitatis linearibus, stylo gracili, valvis glaberrimis reticulatim venosis, seminibus rufis biseriatis orbiculatis profunde transversim sulcatis.— *Gynandropsis triphylla,* DC. *in part.*?—Sierra Leone, Cape Coast, Grand Bassa and Stirling, *Vogel;* Senegal and Guinea.

Herba bipedalis. *Caulis* sulcatus, superne hispidulus. *Folia* inter se conformia; foliolis ¾-pollicaribus, petiolo æquilongis, latitudine variis. *Flores* 4 lin. longi, petalis pallide purpureis, ovario staminibusque inclusis. *Pedicellus* fructifer ⅓ unc.

longus. *Siliqua* suberecta, 1½ unc. longa, valvis planis ¼ unc. latis.

3. Cleome *foliosa*, Hook. fil. ; caule prostrato basi lignoso, ramis erectis foliosis petiolis pedunculisque dense glanduloso-pubescentibus, foliis 5-foliolatis supremis 3-foliolatis, foliolis brevissime petiolulatis late obovatis obtusis apiculatisve utrinque pubescentibus integerrimis subcoriaceis, floribus in axillis supremis pedunculatis, sepalis anguste lineari-lanceolatis glandulosis, petalis late obovato-spathulatis, toro brevissimo, staminibus 6 basi monadelphis, thecaphoro valido glanduloso, siliquis stipitatis linearibus in stylum crassiusculum obtusum angustatis.—Elephant's Bay, (South of the Line), *Dr. Curror.*

Caulis basi lignosus, cortice pallido. *Rami* pedales, stricti, parce ramosi, foliosi, teretes, striati. *Folia* inter se conformia, petiolis ⅓ unc. longis. *Foliola* petiolis æquilonga. *Pedunculi* ½ unc. longi. *Sepala* submembranacea, ¼-pollicaria. *Petala* (sicca) pallide rufo-purpurea, calyce duplo longiora. *Stamina* vix exserta. *Capsulæ* 2-pollicares, vix 2 lin. latæ, pedicello ⅓ unc. suffultæ ; valvis paulo concavis. *Semina* perplurima, immatura minuta, rufa.

A species closely resembling *Polanisia viscosa* (an abundant tropical, and even African plant, though not in Herb. Vogel), but differing in the obsolete torus and pedicellate capsules. Foliage resembling *C. pentaphylla.*

The other W. African species of *Cleome,* are *C. monophylla,* Linn., and *C. angustifolia,* Forsk., from Senegambia, and *C. acuta,* Schum., from Guinea.

There are also two other West African genera of *Capparideæ,* which, however, appear confined to the dryer and more desert latitudes of the North Tropic, and these range from Western Asia westward ; these are *Cadaba,* of which *C. farinosa* is common to Senegal, Sennaar and Arabia, and *Boscia,* containing two species, *B. Senegalensis,* A. Rich., and *B. angustifolia,* A. Rich, both Senegambian plants. The *Streblocarpus angustifolia* of Senegal, and *Cratæva Guineensis,* Schum. et Thonn., from Guinea, together with *Strœmia trifoliata,* Schum. et

Thonn. (probably the same plant as *Cadaba farinosa*, complete the list of W. African *Capparideæ* known; the *Calycandra pinnata* of A. Richard being a leguminous plant.

IX. FLACOURTIANEÆ.

1. Flacourtia *Vogelii*, Hook. fil.; arborea, dioica, ramis nunc spinescentibus, foliis alternis petiolatis elliptico-lanceolatis apice angustatis obtusis serratis nitidis transverse reticulatis, racemis axillaribus brevibus puberulis petiolo paulo longioribus 6-8-floris, floribus 4-meris, sepalis late ovatis obtusis ciliatis, stigmatibus 5-6.—Niger River, at Abôh, *Vogel*.

Rami validi, teretes, cortice rufo-brunneo, punctis albidis notato; ramulis sub lente puberulis. *Folia* 3-4 unc. longa, 1½ lata, basi subacuta, apice obtusa, obtuse serrata, utrinque nitida; petiolo ¼ unc. longo, glabro v. puberulo. *Pedunculi* erecti. *Fl.* ♀ pedicellati; pedicello 1-2 lin. longo calyce basi intruso. *Ovarium* sub 6-loculare. *Fructus* carnosus, ruber.

Closely resembling the Indian *F. sapida*, but the leaves are longer, narrower, more beautifully shining, reticulated and regularly serrated, and the female flowers are very different. There is a new species of this genus from Senegal in the Hookerian Herb.; and two others from Guinea, *F. edulis*, Schum., and *F. flavescens*, Willd., are described by Thonning.

1. Oncoba *glauca*, Hook. fil.—Ventenatia glauca, *Pal. de Beauv. Fl. Ow. et Ben. 1. p. 29. t. 17.*—Fernando Po, *Vogel;* Benin, *Pal. de Beauv.*

The identity of the genera *Oncoba*, Forsk., *Heptaca*, Lour., *Ventenatia*, Pal. de Beauv., *Lundia*, Thonn. et Schum., and *Xylotheca*, Hochst., is shown by Dr. Planchon in the sixth vol. of the London Journal of Botany, p. 295, where also he has given the distinctive characters of the several species. Among them, a second W. African species is the *O. spinosa*, Forsk., extending apparently from Senegambia to the Yemen.

2. Bixa *Orellana*, Linn.—Grand Bassa, *Vogel;* Sierra Leone and Accra, *Don.*

X. Violariæ.

1. Ionidium *thesiifolium,* DC. *Prod.* 1. *p.* 309.—I. rhabdo-
spermum, *Hochst. in Herb. Un. Itin.*—Senegal, *Heudelot ;*
Accra, *Vogel, Don.*
Var. *β. chenopodiifolia,* Guill. et Perr.—Senegal, *Heudelot ;*
Cape Coast, *Don.* Also a native of Upper Egypt.
2. Ionidium *suffruticosum,* Ging. *in DC. Prod.* 1. *p.* 311.—
Accra, *Vogel, Don.*

A protean plant, and probably only a large var. of the preceding, owing its stature to the more humid atmosphere of the
Bight of Benin. *I. linifolium,* DC., is probably another variety,
as well as *I. enneaspermum,* Vent., and *Viola Guineensis* and
V. lanceifolia, Schum. et Thonn. The species has a very wide
range, extending through Nubia and Upper Egypt to Eastern
Tropical Africa, Madagascar, East India and Tropical Australia.

Some American species are hardly to be distinguished from
the African, except by their smooth, not striated seeds.* The
two Oware species of *Ceranthera,* described by Palisot de
Beauvois, and two other Senegal ones (in Herb. Hook.) complete the small catalogue of known W. African *Violariæ.*

XI. Sauvagesiaceæ.

1. Sauvagesia *erecta,* Linn.—Sierra Leone, *Vogel, Don ;* Senegal,
and abundant throughout Tropical America.

XII. Polygaleæ.

1. Polygala *arenaria,* Willd. *Sp.* 3. *p.* 880.—On the Quorra,
Vogel and Ansell; Guinea.

Allied to the *P. Vahliana* (of East India), but the foliage is
very different. Flowers about 20, deflexed, densely imbricated
into a sort of strobilus.

* This appears, however, to be a constant, and therefore important
character.—(G. B.)

2. Polygala *nutans*, Hook. fil.; annua, pubescens, caule erecto parce ramoso, foliis oppositis patulis subsessilibus linearibus v. lineari-lanceolatis obtusis utrinque pilosis, racemis axillaribus erectis 8-10-floris foliis dimidio brevioribus, floribus imbricatis deflexo-nutantibus, sepalis ext. liberis int. late ellipticoovatis hirsutis ciliatisque capsula pilosa breviter oblonga profunde emarginata ⅓ longioribus, carina dorso longe cristata.— Accra, *Vogel.*

Caulis gracilis, 6-uncialis, patentim pubescens. *Folia* uncialia, 2-4 lin. lata. *Racemi* ½ unc. longi. *Flores* albidi v. virescentes, 2-3 lin. longi, alis obtusis.

Also an Abyssinian plant, and closely allied to *P. erioptera* and *P. arenaria.* The flowers are similar in all, but are not so densely imbricated in this as in the latter, from which it further differs in the narrower foliage; a variable character however.

3. Polygala *rarifolia*, DC. *Prod.* 1. *p.* 332.—Sierra Leone, *Don.*

Suffrutex bipedalis, glaberrima. *Rami* stricti, graciles, diam. pennæ corvinæ, virgati, nudi, profunde sulcati, cortice viridi. *Folia* per-pauca. *Racemi* axillares, sed e lapsu foliorum quasi ramei, unciales, stricti, erecti, multiflori. *Flores* lactei, majusculi, ¼ unc. longi, breviter pedicellati; sepalis parvis, concavis, 2 anticis coalitis, apicibus liberis, interioribus liberis, oblongis, subcoriaceis, valde concavis; carina cristata. *Stigma* petaloideum, concavum; bracteolis floralibus 3, minimis.

4. Polygala *Donii*, Hook. fil.; caule suberecto, gracili, puberulo, foliis plurimis alternis linearibus utrinque angustatis subacutis v. obtusis obscure puberulis 1-nervibus marginibus vix recurvis, racemis terminalibus gracilibus multifloris, pedicellis basi 3-bracteolatis, floribus majusculis nutantibus, sepalis ext. liberis, alis oblique obovato-rotundatis basi angustatis ciliatis capsula quadrata emarginata pilosa duplo longioribus, carina dorso cristata.—Sierra Leone? *Don.*

Caulis in exempl. solitario simplex, 5-uncialis, teres. *Folia* omnia conformia, erecta, pollicaria, 2 lin. lata, obscure puberula, subter uni-costata; internodiis ¼ unc. longis. *Racemi* 3,

terminales, intermedio 2-unciali 20-floro, lateralibus abbreviatis. *Florum pedicelli* gracillimi. *Alæ* 3 lin. longæ, ciliatæ.
Capsula subquadrata, angulis rotundatis.

Apparently a very distinct species, easily recognized by the
form of the *alæ.*

5. Polygala *Guineensis,* Willd. *Sp.* 3. *p.* 882.—P. multiflora,
Poir. Dict. 5. *p.* 497?—Accra, *Ansell.*

Herba bipedalis, di-trichotome ramosa. *Rami* gracillimi, erecti,
virgati. *Folia* 1½-2 unc. longa, filiformia, vix ¼ lin. lata,
subflexuosa, erecta, viridia. *Racemi* terminales, erecti, 4-5
unc. longi. *Flores* secundi, parvi, bracteolis setaceis valde
caducis suffulti, pedicellati, patentes v. nutantes, cœrulei,
pedicellis 1 lin. longis, subclavatis. *Sepala* 2 exteriora coalita,
apicibus liberis. *Alæ* obovatæ, glaberrimæ, concavæ, 1½ lin.
longæ. *Carina* longe cristata. *Capsula* obovato-quadrata,
glaberrima, apice biloba.

A remarkably distinct, and singularly slender species.

The remaining W. African species known to me are, 1 *P.
erioptera,* DC., a species common to Senegal, Egypt and Arabia :
it is probably identical with *P. triflora* and *linearifolia,* Roth,
(non auct.) in which the leaves are sharp; in all the carina is
crested, though *P. triflora* is placed erroneously in the group
"carina nuda." 2. *P. paniculata,* Lin. var. *Africana,* DC.,
from Senegal, a species widely diffused in America, some
specimens from the Plata River have the flowers as small as
the African, but white, and it is possibly further identical with
the *P. capillaris,* E. Meyer, from South Africa ; and, lastly,
3. *P. obtusata,* Guill. et Perr., from Senegal, which completes
the list of W. African *Polygalæ.*

1. Lophostylis *oblongifolia,* Hochst. *in Ratisb. Flora,* 1842, *n.*
15.—On the Quorra, *Vogel,* who describes it as a middle-
sized tree ; Senegal.

Dr. Vogel's specimen is imperfect, and in fruit only, but I
think referable to this species, which has a wide range, being
found also in Abyssinia. The propriety of separating this
from *Securidaca* is perhaps doubtful.

1. Carpolobia *lutea*, Don, *Gen. Syst.* 1. *p.* 370.—Sierra Leone, *Don;* Senegambia, *Heudelot.*
Rami graciles, virgati, puberuli, teretes, flexuosi, patentes. *Folia* alterna, subdisticha, breve petiolata, obovata, in acumen elongatum apice obtusum angustata, glaberrima, integerrima, submembranacea, reticulatim venosa, viridia, 2-4 unc. longa. *Racemi* axillares, solitarii, breves, 2-5-flori, infra folia orti et iis tecti, graciles, ¼ unc. longi, pubescentes. *Flores* erecti, ½ unc. longi, brevissime pedicellati; pedicellis basi bracteolatis. *Sepala* 5, ciliata; 3 longioribus ovato-lanceolatis, obtusis, cæteris oblongis. *Petala* calyce ter longiora, in tubum fissum basi coalita; 4 subæqualia, oblongo-spathulata, venosa; quinto carinæformi, duplo majore. *Stamina* 5; filamentis ad medium monadelphis. *Stylus* gracilis; stigmate parvo, capitato.

A very handsome plant, ill described by its discoverer. The calyx and corolla are both irregular, and formed of five pieces, of which one of the latter resembles the ala of a papilionaceous flower. This explains the structure of the flower of those *Polygalæ* in which two of the petals are wanting, or only represented (as in *P. Donii*) by a lobe at the base of the two smaller petals.

2. Carpolobia *alba*, Don, l. c.—Sierra Leone, *Don.*
The two other species described by Don are leguminous plants, which will be noticed hereafter under *Baphia* and *Bracteolaria.**

XIII. Droseraceæ.

There are two species of *Drosera*, from W. Tropical Africa, preserved in the British Museum, *D. Burmanni,* Vahl, and *D. Indica,* Linn. They are both from Sierra Leone (*Afzelius*), and are also natives of India. To the *D. Indica,* Dr. Plan-

* See my observations and generic character of *Carpolobia,* in Hook. Journ. Bot. v. 4. p. 104.—(G. B.)

chon refers also the *D. minor* of Thonn. and Schum. from
Guinea.

XIV. FRANKENIACEÆ.

The *Frankenia pulverulenta* extends as far south as Senegal.

XV. CARYOPHYLLEÆ.

1. Mollugo *nudicaulis,* Lam.—M. bellidifolia, *Ser. in DC.
Prodr.* 1. *p.* 391.—Cape Coast, *Don;* on the Quorra,
Vogel; Senegal.
Common to the W. Indian Islands, and also to Egypt, Mada-
gascar and the East Indies. Three other species, *M. cervina,* Ser., *M. verticillata,* Linn.,
and *M. denticulata,* Guill. et Perr., are W. African.

XVI. LINEÆ.

1. Hugonia *Planchoni,* Hook. fil. (Tab. XXVII.); ramis
petiolisque ferrugineo-pubescentibus, foliis lanceolato-oblongis
cuspidatis utrinque acutis remotiuscule serrulatis glabris
nitidis rigide chartaceis pulchre reticulato-venosis, stipulis
bracteisque pinnatipartitis laciniis subulatis, cymis axillaribus
brevibus 3-5-floris, stylis staminibus longioribus—Sierra
Leone, *Afzelius, Vogel;* Accra, *Vogel.*
A most distinct and handsome species.
Frutex scandens. *Cirrhi* in parte inferiore ramulorum alterni,
nunc nulli. *Ramuli* angulati, ramique subteretes cicatricibus
stipularum albis notati. *Folia* sat conferta, alterna, 3-5½
poll. longa, 10-20 lin. lata, petiolo vix 2 lin. longo. *Stipulæ*
ob lacinias subulatas conspicuæ, bracteis conformes. *Sepa-
lorum* pars in alabastro externa subsericeo-rufescens, parte
tecta glabra nitida. *Petala* flava (ex *Vogel*) anguste v. latius-
cule cuneata, staminibus stylisque longiora. *Bacca* sicca, globosa, *Piso* subæqualis, mucronata, calyci subæqualis.*

* The above character and description were drawn up by Dr. Planchon.

Q

PLATE XXVII.—*Fig.* 1. flower, *slightly magnified; f.* 2. petal; *f.* 3. stamens and pistil; *f.* 4. fruit; *f.* 5. transverse section of the same ; *the three last magnified.*

A second West Tropical African species, the *H. Afzelii,* Br., from Sierra Leone, has been described by Planchon in the Lond. Journ. Bot. v. 7. p. 525.

XVII. MALVACEÆ.

1. Malachra *capitata,* Linn.—Congo River, *Christ. Smith.*
Another species, *M. hispida,* Guill. et Perr., is confined to Senegal.

1. Urena *lobata,* Linn.—U. diversifolia, *Schum. et Thonn. Beskr. p.* 308.—U. virgata, *Guill. et Perr.*? *Fl. Seneg. p.* 48.
A most abundant W. African plant, of which I think I can recognize the following varieties :

a. Caule griseo, pilis stellatis sparsis aspero, foliis integris 3-5-lobisve subter petioloque cano-pubescentibus, calyce hispido.
—Fernando Po, Nun and Quorra Rivers, *Vogel;* Sierra Leone, *Don.*
The common East and West Indian form of the plant.

β. Foliis super pilis sparsis fulvis obsitis subter canis vix pilosis, calyce dense hispido lobis elongatis.—On the Nun, *Vogel.*

γ. Undique pilis patentibus in caulem stellatis obsita, calyce dense hispido.—Fernando Po, *Vogel.*
Very similar to a Phillippine Island state of the plant.

δ. Foliis profunde 3-lobis, lobo intermedio lineari-elongato, lateralibus oblongis subacutis.—Cape Coast, *Vogel.*
Similar to a Brazilian variety.

ε. Foliis minoribus flaccidis pilis stellatis patulis pubescentibus, calyce hirsuto.—Senegambia, *Heudelot.*
Identical with cultivated specimens from the West Indies and indigenous East Indian ones.

Guillemin and Perrottet describe two other *Urenæ; U. obtusata* and *U. virgata,* whose claims to specific distinction appear doubtful to me.

1. Paritium *tiliaceum*, Linn. Hibiscus Guineensis, *DC.*, *Prod.*
5, *p.* 454, *non Don.*—Senegal, Sierra Leone, and Nun River,
Vogel.

2. Paritium *quinquelobum*, Hook. fil.; ramis puberulis, petiolis
elongatis pubescentibus pedunculis foliisque super punctis
minimis asperis, foliis late cordato-rotundatis coriaceis 3-cus-
pidatis v. dense 3-lobis suberoso-dentatis super puberulis
subter ad nervos stellatim pilosis, floribus in racemum sub-
spicatum axillarem folio æquilongum dispositis, involucelli
foliolis calyce brevioribus, calyce capsulam vix superante.—
Hibiscus quinquelobus, *G. Don, Gard. Dict.* 1. *p.* 482.—
Sierra Leone, *Don.*

Ramuli ultimi herbacei, teretes, crassitie pennæ anserinæ, striati,
pube sparsa simplici v. stellata. *Folia* 5 unc. longa, 7 lata,
basi profunde cordata, sinu lato v. contracto. *Pedunculus*
axillaris, strictus, 9 unc. longus, basi nudus, superne ramosus,
ramis abbreviatis. *Pedicelli* brevissimi, 2 lin. longi, basi
bracteolati, bracteolis nunc foliaceis. *Calyx* 5 lin. longus,
cano-pubescens; involucelli foliolis subulatis. *Corolla* (purpurea?) 1½ unc. longa.

There are three other W. African *Paritia : P. virgatum*,
Guill. et Perr., *P. sterculiæfolium*, G. et P., and a third undescribed species—all from Senegal.

1. Abelmoschus *esculentus*, Wight et Arn.—Hibiscus esculentus,
Linn.—W. Africa, (*cult.*)
2. Abelmoschus *moschatus*, Mœnch.—Hibiscus Abelmoschus,
Linn.—Grand Bassa, *Vogel ;* Senegal.
1. Hibiscus *vitifolius*, Linn. ; β. caule patentim piloso, foliis ca-
lyceque pilis fulvis stellatis.—H. strigosus, *Schum. et Thonn.*
Beskr. p. 314—Cape Coast, *Vogel.*
This differs from the usual East Indian state of the plant, in
the greater villosity and many stellate setæ, or hairs, which,
together with the toothing of the leaves, are very variable
characters. In some varieties, the whole plant is merely hoary.
2. Hibiscus *physaloides*, Guill. et Perr., *Fl. Seneg. p.* 52—H. adscendens, *G. Don, Gard. Dict.* 1. *p.* 482.—St. Thomas,

Don; Valley of St. Domingo, St. Yago, Cape de Verds, *J. D. Hooker.**

3. Hibiscus *asper,* Hook. fil.; caule erecto superne angulato suffruticoso tuberculis minimis sub-aculeato, foliis patentibus longe petiolatis petiolo scabrido palmatim 7-partitis utrinque asperulis, lobis linearibus acutis obtusisve integris v. sinuato-dentatis, floribus axillaribus solitariis breviter pedunculatis, involucelli laciniis subulatis calycem pilosum æquantibus.— Sierra Leone, *Miss Turner,* (in Herb. Hook.)

Statura, habitusque *H. cannabinæ,* sed major, caule inermi, foliisque 7-lobis. *Flores* 1½ unc. diam.

A very fine species. The petioles and laciniæ of the leaf are 3-5 inches long.

4. Hibiscus *Guineensis,* G. Don, *Gard. Dict.* 1. *p.* 481, *non DC.*; caule herbaceo inermi piloso (pilis furcatis), foliis sublonge petiolatis late rotundato-cordatis superioribus 3 cæteris 5-lobis obtuse serratis utrinque sed subter præcipue pilosis lobis acutis acuminatisve, pedunculis brevibus clavatis axillaribus v. in racemum dispositis hispido-pilosis, involucelli foliolis lineari-subulatis calycem capsulamque æquantibus, calycis segmentis appresse pilosis ovato-subulatis sinu obtuso, corolla ampla, capsula ovato-globosa acuminata hispido-strigosa, seminibus angulatis vix tuberculatis.—St. Thomas, *Don.*

The above description closely accords with Wight and Arnott's (Prodr. Flor. Pen. Ind. Or. p. 49) character of Wallich's *H. lunarifolius;* and indeed the only distinction I can trace between this and original specimens of the E. Indian plant is, that in the latter, the calyx and involucellum are considerably longer than the fruit, in *H. Guineensis* shorter than that organ.

The *Hibiscus cannabinus* of the E. Indies is also a native of Senegal, and a fifth species, from the same country, *H. verrucosus,* Guill. et Perr., is described in the Flora Senegambiæ.

5. Hibiscus *Surattensis,* Linn.—Accra and St. Thomas, *Don.*; Senegal. Also an E. Indian species.

* Omitted above, p. 107.

The *H. diversifolius* (*Hibiscus*, Jacq., *H. scaber*, Mich.), a
N. American species, is a native of Senegal, as is also *H. rostel-
latus*, Guill. et Perr.; and four species, *H. versicolor*, *H.
triumfettæfolius*, *H. congener*, and *H. obtusatus*, are described
by Schumacher from Guinea.

The old genus, *Hibiscus*, is a very important one in W. Africa,
there being, besides the above-enumerated species : one of *Bom-
bycella: Hibiscus clandestinus*, Cav., from Senegal; three of
Cremontia, all natives of Senegal : *H. Senegalensis*, Cav., *H. tubulosus*, Cav. (also an E. Indian plant), and *H. ribesiifolius*, Guill.
et Perr.; one sp. of *Fugosia : F. digitata*, Pers.; and three of
Pavonia : P. Zeylanica, Willd., *P. triloba* and *P. hirsuta*, Guill.
et Perr.

1. Gossypium *Barbadense*, L.—G. punctatum et G. prostratum,
Schum. et Thonn. Beskr., p. 310, 311.—Fernando Po and
Cape Palmas, *Vogel*, (near habitations).

Var.? *hirsutum*, ramis molliter et patentim pilosis, seminibus
obovatis fusco-castaneis glabriusculis, ubique v. basi gossypinis, gossypio albo.—Sierra Leone, *Miss Turner*, (in Herb.
Hook.) ; Accra, *Don.*

Dr. Vogel's specimens being in flower only, I cannot determine the variety to which it may belong. Don's specimens
differ from Miss Turner's only in the seeds being cottony all
over.

1. Wissadula *rostrata*.—Sida rostrata, *Schum. et Thonn., Beskr.
p.* 306—S. stellata, *G. Don, Gard. Dict.* 1. *p.* 499.—Abutilon
laxiflorum, *Guill. et Perr., Fl. Seneg. p.* 66.—A. parviflorum,
A. St. Hil. Fl. Bras. Merid. 1. *p.* 201.—Sida periplocifolia,
β. caribæa, *DC. Prodr.* 1. *p.* 468 —Abutilon periplocifolium,
Don, Webb, supra, p. 108.—Cape Coast, *Vogel, Thonning ;*
St. Thomas, *Don. ;* Senegal.

Var. β. foliis subrugosis.—Accra, *Vogel.*

A common West Indian plant, closely allied to *W. hirsuta*,
Presl, but apparently differing, as well in the absence of the
rigid hairs characteristic of that species, as by the fruit, of
which the carpels are more divergent at the apex, with longer
points. The Ceylonese plant, which probably originally suggested the specific name *periplocifolia*, appears to be quite dis-

tinct.   It has been called *W. Zeylanica* by Medik, the founder
of the genus *Wissadula*, but ought perhaps to retain the Linnæan specific name.
1. Abutilon *Asiaticum*, G. Don, (Sida, *Linn*.) Accra, *Vogel.;*
   Senegal.
   Common to both the East and West Indies.
2. Abutilon *Indicum*, G. Don, (Sida, *Linn.*)—an S. glauca,
   Cav. Ic. 1. *t.* 11 ?—S. grandiflora, *G. Don, Gard. Dict.* 1.
   p. 501.—St. Thomas, *Don.*
   Equally common in the Eastern and Western hemispheres.
   Five other species are recorded by Guill. et Perr. as natives of
   Senegal, viz. : *A. ramosum, tortuosum, macropodum, sparman-
   nioides*, and *fruticosum*, and the *Sida Guineensis*, Schum. et
   Thonn., is probably also an *Abutilon.*
1. Sida *rhombifolia*, Linn.—S. retusa, *G. Don, Gard. Dict.* 1.
   p. 492.—S. rugosa, *Schum. Thonn. Beskr.,p.* 304 ?—Senegal ;
   Nun River, and Ebu, *Vogel ;* St. Thomas, *Don.*
   The majority of the specimens of this plant which I have
   examined, have shortly bicuspidate carpells.   It is a native of
   both the E and W. Indies.
2. Sida *linifolia*, Cav.—S. linearifolia, *Schum. et Thonn. Beskr.*
   p. 303. —Senegal ; Sierra Leone, Cape Coast, Accra, and
   Quorra River, *Vogel, Don,* &c.
   A West Indian plant, varying much in stature and the breadth
   of the leaves.
3. Sida *cordifolia*, Linn.—S. althæifolia, *Sw.*—S. Africana, *Pal.*
   Beauv. 2. *p.* 87. *t.* 116.—S. decagyna, *Schum. et Thonn. Beskr.*
   p. 307 ?—Senegal to Benin, *Vogel, and others.*
   Var. foliis minoribus.   *S. cordifolia*, L. ?—St. Thomas and Accra, *Don.*
   A very common species in the warmer regions of both hemispheres.
4. Sida *urens*, Linn.—S. sessiliflora, *G. Don, Gard. Dict.* 1.
   p. 491, et S. debilis, *G. Don, l. c. ?*—Senegal, St. Thomas,
   Don.
   Also both an East and West Indian plant.   The West African
   specimens are very slender, and the carpels shortly 2-cus-
   pidate.

5. Sida *stipulata*, Cav. *diss.* 1. *p.* 22. *t.* 23. *f.* 10.—S. prostrata,
G. *Don*, *Gard. Dict.* 1. *p.* 490.—Sierra Leone, *Vogel*,
Don.

These specimens have slender aristæ to the carpels. The
species is a native of both the East and West Indies, and is considered by Dr. Planchon as a mere variety of *S. acuta.*

6. Sida *retusa*, Linn.—St. Thomas, *Don.*

The leaves of these specimens are either obtuse or acute, and
scarcely retuse, as is frequently the case with East Indian specimens, where it occurs as far east as the Phillippine Islands.
The carpels terminate in a subulate point. Dr. Planchon considers it a mere variety of *S. rhombifolia.*

7. Sida *acuta*, Burm., *Cav. diss.* 1. *p.* 15. *t.* 2. *f.* 3.—S. ovata,
G. *Don*, *Gard. Dict.* 1. *p.* 492.—Cape Coast, *Vogel;* St.
Thomas, *Don.*

Certainly identical with the East Indian species of this name.
Don's specimens have luxuriant foliage.

8. Sida *Vogelii*, Hook. fil.; gracilis, erecta, glaberrima, caule
virgatim ramoso, foliis petiolatis lineari-lanceolatis acutis
inæqualiter subacute serratis basi rotundatis, stipulis glaberrimis petiolum superantibus, pedunculis axillaribus solitariis
(rarius binis) unifloris gracilibus medio articulatis petiolis ter
longioribus, calyce glaberrimo segmentis acuminatis, carpellis
subsenis 2-rostratis.—Fernando Po, *Vogel.*

Suffrutex 2-3-pedalis, ramosus. *Caulis* crassitie pennæ corvinæ. *Stipulæ* majusculæ, petiolum gracilem ⅕ unc. longum
superantes. *Folia* utrinque glaberrima, 1½-2 unc. longa,
4-5 lin. lata, submembranacea. *Pedunculi* graciles, erecti.
Carpella dorso rugosa, rostris paulo divaricatis.

I believe this to be a very distinct species, more similar to the
N. American *S. Elliottii* than to any other. It differs from
S. spinosa and all its varieties in wanting the spine, and in the
long peduncles; from *S. stipulata*, (which it otherwise much
resembles), in the stipules; from *S. acuta*, by the same characters, and from *S. rhombifolia* in the smooth leaves, which are
not hoary.

There are four other W. African *Sidæ*, all from Senegal;

S. grewioides, Guill. et Perr., *S. spinosa*, L., *S. canescens*, Cav., and a possibly undescribed species.

The above, with two plants belonging to other genera of *Malvaceæ*, viz., *Lagunea ternata*, Cav., and *Bastardia angulata*, Guill. et Perr., complete the catalogue of W. African *Malvaceæ*.

XVIII. BOMBACEÆ.

1. Adansonia *digitata*, Linn.—Cape Verd to Congo, various travellers, *Vogel, Don*, &c.
2. Bombax *Buonopozense*, Pal. Beauv. *Fl. Ow. et Ben.* 2. *p.* 42. *t.* 83? glaberrimum, foliis palmatis longe petiolatis, foliolis anguste elliptico-lanceolatis membranaceis utrinque angustatis acuminatis apicem versus remote ciliato-serratis.— Station uncertain in *Vogel's* collection; Sierra Leone, *Miss Turner*.

The flowers of *B. Buonopozense* are contained in Herb. Hook., and to that plant the leaves preserved by Vogel may belong, though there are probably other W. African species of the genus.

3. Eriodendron *anfractuosum*, DC. *Prodr.* 1. *p.* 479. An *Bombax Guineense*, Schum. et Thonn. Beskr. p. 302. ?— Congo River, *Christ. Smith.* A common East Indian tree.

XIX. STERCULIACEÆ.

1. Sterculia *tragacantha*, Lindl., *Bot. Reg.*, *t.* 1353.—S. pubescens, *G. Don, Gard. Dict.* 1. p. 515.—Sierra Leone, *Don.*

One of Mr. Don's specimens is in fruit, and the following description may be added:

Carpella 2, globosa, breviter stipitata, subrostrata v. mucronata, reticulata, dense ferrugineo-pubescentia.

This species, like the East Indian *S. urens*, yields a gum resembling *Tragacanth*. The *S. pubescens* of Don's Herb. (not *Gard. Dict.*) is the following plant.

2. Sterculia *obovata*, Br. *in Pl. Jav. Rar. p.* 233.—Congo

River, *Christ. Smith ;* Sierra Leone, *Don ;* Senegambia, *Herb. Hook.*

Leaves smaller than in the foregoing, and more shortly petiolate.

The *S. tomentosa,* Guill. et Perr., is the only other W. African species.

1. Cola *acuminata,* Br. *Pl. Jav. Rar. p.* 237.

a. Foliis longe petiolatis anguste lineari-obovatis acuminatis. *Sterculia acuminata,* Pal. de Beauv.—St. Thomas, *Don.*

β. Foliis breviter petiolatis latioribus coriaceis.—*S. nitida,* Vent. ?—*S. macrocarpa,* Don.—*S. verticillata,* Schum. et Thonn.—*Lemania Bichy,* DC.—Fernando Po, *Vogel ;* St. Thomas and Sierra Leone, *Don.*

This is the well-known Cola-nut, of which the var. *β.* is the only one cultivated in the New World.

The *Sterculia cordifolia,* Cav., from Senegambia, is considered by Brown (Pl. Jav. Rar. p. 237) to be a second species of *Cola.* In the same work, a third plant, allied to *Cola,* but scarcely of the same genus, is alluded to as gathered in Sierra Leone by Afzelius, and three W. African species of *Courtenia* are enumerated : *C. Afzelii,* Br., from Sierra Leone and Congo ; *C. triloba,* Br., from Senegambia ; and *C. ? heterophylla,* Br. (*Sterculia heterophylla,* Pal. Beauv.), from Oware.

XX. Byttneriaceæ.

1. Waltheria *Indica,* Linn. An *W. Guineensis* et *W. Africana,* Schum. et Thonn. Beskr. p. 295, 296. ?

A very common W. African plant ; also abundant both in the East and West Indies.

A second species, closely allied to one from Brazil, is in Heudelot's Senegambian collection.

1. Melochia *corchorifolia,* Linn.—Polychlæna simplex et P. ramosa, *G. Don, Gard. Dict.* 1. *p.* 488.—Senegal, Quorra river and Cape Palmas, *Vogel ;* St. Thomas, *Don.*

Very variable in the size and breadth of the foliage.

Another, and probably new species of this genus, occurs in Senegambia.

XXI. TILIACEÆ.

1. Corchorus *tridens,* Linn.—Abundant along the West coast
of Africa, also in the East Indies.
2. Corchorus *acutangulus,* Lam. Dict.—C. alatus, *G. Don,*
Gard. Dict. 1. *p.* 542, 2. *p.* 104.—Senegal, Cape Coast, and
Quorra River, *Vogel;* St. Thomas, *Don.*
A frequent East and West Indian species.
3. Corchorus *olitorius,* Linn.—C. lanceolatus et C. longicarpus,
G. Don, l. c., p. 543.—Senegal, Quorra, *Vogel;* St. Thomas,
Don.
A fourth species, *C. brachycarpus,* Guill. et Perr., occurs in
Senegal. The three Guinea species, described by Schumacher
and Thonning, *C. angustifolius, C. polygonus,* and *C. muricatus,*
are probably the same as some of the preceding ones.
1. Triumfetta *rhomboidea,* Jacq. *Amer. p.* 147. *t.* 90.—Cape
Coast, Grand Bassa and Quorra River, *Vogel;* St. Thomas,
Don.
Certainly identical with the W. Indian plant.
Var. *β. glabriuscula.*—Bassa Cove, *Ansell.*
Var. *γ.* foliis omnibus brevi-petiolatis basi ovato-cuneatis.—Cape
Coast and Accra, *Don.*
Also a W. Indian variety, having all the leaves like the upper
ones of the first. All the varieties have membranous leaves, broad
and undivided, the lower abrupt, and not cuneate at the base.
2. Triumfetta *glandulosa,* Lam. *Dict.* 3. *p.* 421?—Quorra
River, *Vogel.*
Caulis glabriusculus. *Folia* submembranacea, super vix puberula, subter velutino-pubescentia, superiora basi cuneata,
inferiora basi latiora; omnia subcordata, obscure triloba.
Stamina 12 et plura. *Fructus* deest.
The plant agrees tolerably with Lamarck's description, except
the leaves being less velvety.
3. Triumfetta *velutina,* Vahl, *Symb.* 3. *p.* 62.—Accra, *Vogel.*
Differs (possibly not specifically) from *T. mollis,* in the more
cuneate base of the coriaceous leaves, which have shorter petioles,
and are more tomentose beneath; the stems, too, are more

shrubby, and the toothing of the leaves less decided. The young capsules are densely villous between the aculei, in which respect the plant differs from Vahl's description, but this character depends on age.

4. Triumfetta *trilocularis*, Roxb. *Fl. Ind. 2. p.* 462.—Nun River, *Vogel.*

Specimens very bad, but apparently identical with a Zanzibar species, called *T. semitriloba* by Bojer, but which differs from De Candolle's description of that plant in the leaves not being velvety beneath.

Caules erecti, rigidi. *Folia* late ovata, obscure lobata v. integra, basi ovato-cuneata, rigida, utrinque puberula, inæqualiter serrata, 1-1¼ unc. longa; petiolo 1½-pollicari. *Capsulæ* immaturæ globosæ, pubescentia, inter setas albida.

The *T. trilocularis,* Guill. et Perr., appears to be the *T. rhomboidea,* Jacq., judging from the imperfect specimen in Herb. Hook.

5. Triumfetta *angulata,* Lam, Dict. 3. p. 421?—Gambia, *Capt. Boteler,* (in Herb. Hook.)

Folia superiora sessilia, ovata v. ovato-lanceolata, nunc late rhomboidea, triloba, pubescentia et sericeo-pilosa, subter velutina. *Capsulæ* immaturæ globosæ, pedicellatæ, setosæ, tomentoque albido dense opertæ.

6. Triumfetta *eriophlebia,* Hook. fil.; caule erecto ramoso laxe patentim piloso villosiusculo, foliis petiolatis deflexis superioribus lanceolatis, inferioribus late ovatis sub 3-lobis, omnibus acuminatis basi rotundatis ad petiolum cordatis subgrosse irregulariter dentatis utrinque hirsutis pilis patulis, nervis subter dense albo-lanatis, floribus mediocribus axillaribus pedicellatis subracemosis, sepalis longe acuminatis, staminibus 10, ovario hispido, capsula immatura lanata setisque uncinatis tecta.—Fernando Po, *Vogel.*

Petioli ½ unc. longi. *Folia* 2½-pollicaria, submembranacea, pilis patulis fulvis vestita.

Complete specimens, both flowering and fruiting, are required of the *Triumfettæ* to determine the limits of the species; and such are rare in our Herbaria. The size of the flower and arming of the capsule probably afford better characters than the variable foliage; it should always be mentioned whether the capsules described are ripe or not. Wight and Arnott hint at the probability of the number of East Indian species being much exaggerated, for these authors remark, that characters drawn from the suppression of parts of the flower and shape of the leaves, are not to be depended upon.

7. Triumfetta, sp.?—Sierra Leone, *Don.*
Caulis lignosus, ramosus, pilis patulis stellatis velutino-pubescens. *Folia* breve petiolata, petiolo ¼ unc. longo, 1½-pollicaria, basi lata, ad petiolum cordata, utrinque dense velutinotomentosa, coriacea, triloba, inæqualiter serrata, segmentis infimis glandulosis. *Capsula* immatura uncinato-setosa, inter setas pubescens; matura globosa puberula, breviter setosa, 3-5-cocca; coccis 1-spermis; seminibus majusculis.

In a genusal ready involved in so much confusion, I unwillingly insert another doubtful species. The present is out of flower, but would appear very distinct from its congeners.

Two other species of this genus are mentioned and first described by Guillemin and Perrottet, *T. cordifolia* and *T. pentandra,* both from Senegal. Schumacher and Thonning have also described one from Guinea, under the name of *T. mollis;* a specimen from Senegal of Heudelot's is either this species or one closely allied to it, and another gathered by Don, at Sierra Leone, in leaf only, if not the *T. cordifolia,* may be distinct from all the foregoing.

1. Grewia *carpinifolia,* Juss. *Ann. Mus.* 2. *p.* 91. *t.* 51. *f.* 1.—Cape Coast, St. Thomas and Accra, *Vogel and Don.* Also a native of Senegal and Oware.

Grewia, as was to be expected, is a large W. African genus, whence the poverty of the Niger collections in one so conspicuous and easily collected is remarkable. Besides the above, six species are enumerated, chiefly however from Senegal, viz.:
1. *G. betulæfolia,* Juss., closely allied to, and probably identical with the *G. populifolia,* Vahl, and if so, a plant ranging from Senegal, through Arabia and Persia, to the Peninsula of India ;
2. *G. corylifolia,* Guill. et Perr. (*G. villosa,* Hort. Mal.) also

a native of Senegal, Nubia and the East Indies. 3. *G. bicolor*, Guill. et Perr., from Senegal. 4. *G. megalocarpa*, Pal. Beauv., from Benin. 5. *G. guazumæfolia*, Juss., from Senegal and East India. 6. *G. mollis*, Juss., a native of Senegal and Benin?
1. Omphacarpus *Africanus*, Hook. fil.; ramulis puberulis, foliis breve petiolatis ovatis acuminatis super glaberrimis nitidis subter puberulis integerrimis v. obsolete serratis, fructibus oblique obovato-cuneatis compressis.—Sierra Leone ; *Don.*
Frutex majusculus. *Rami* teretes ; cortice atro ; ramulis fusco-pubescentibus, viscosis ? *Folia* 2 unc. longa, subcoriacea. *Flores* paniculati. *Fructus* ¾ unc. longus.

Of this curious plant, whose only congener is a Borneo species described by Korthals, I have seen but very imperfect specimens. Except by the somewhat different shape of the fruit, the two species are hardly distinguishable. It affords a remarkable instance of the relation between the littoral Flora of W. Africa and that of the hot damp Malayan Archipelago, and which contrasts so strongly with the Flora of the drier northern parts of Tropical West Africa, as Senegal, Cape Verd, &c., the types of whose vegetations, and many of the species themselves, are prolonged eastward through Sennaar, Abyssinia, and Arabia, to the Peninsula of India.

GLYPHÆA, *Hook. fil.* (nov. gen.)

Calyx ad basim 5-partitus, laciniis oblongis, æstivatione valvatis, deciduis. *Petala* anguste lanceolata, sessilia, basi nuda. *Stamina* plurima, hypogyna ; *filamentis* gracilibus haud complanatis ; *antheris* basifixis immobilibus erectis linearibus, connectivi angusti productione brevissime apiculatis loculis 2 laterali-introrsis, apice rimula brevi poriformi introrsum dehiscentibus. *Ovarium* subsessile (gynophoro saltem haud conspicuo) in *stylum* apice acuto stigmaticum attenuatum, abortu (?) 3-loculare, *loculis* ad angulum internum superposite pauciovulatis, et inter ovula contracto-interruptis, inde in locella superposita uniovulata divisis. *Fructus* subcapsularis ? (fragmenta ejus tantum video) fusiformi-oblongus, verticaliter pluricostatus (10-costatus, *Hook. fil.*), *mesocarpio* crasso aride suberoso, *locellis* monospermis paucis (pro carpello singulo 2-3 uniseriatis), *endocarpio* cartilagineo subindehiscente limitatis ; *columella* in fructu forsann on sponte irregulariter fracto in fila soluta. *Semina* ad medium anguli interni locelli cujusve peritrope inserta, transverse late oblonga, anatropa. *Embryo* in axi albuminis rectus ; cotyledonibus semini conformibus, haud crassis, facie plana sibi invicem applicitis, radicula exserta lineari-oblonga versus hilum directa.—*Frutex* Africæ occidentalis tropicæ, facie et vegetationis *Grewiæ*, ramis virgatis. *Folia* alterna, disticha, petiolata, lanceolata, cuspidata, remote et inæqualiter repando-serrata v. denticulata, triplinervia, cæterum penninervia, rigide membranacea, glabriuscula. *Stipulæ* non visæ. *Umbellæ* 3-4-floræ, pedunculatæ, sæpius oppositifoliæ, nunc axillares, basi ebracteatæ, bracteolis ad basim pedicellorum caducis. *Flores* lutei.*

1. Glyphæa *grewioides*, Hook. fil. (TAB. XXII.)—Grewia lateriflora, *G. Don*, *Gard. Dict.* 1. *p.* 549.—Fernando Po, *Vogel*.
Ramuli, petioli, pedicellique pube parca stellata conspersi. *Folia* variant lanceolata v. late elliptica, basi subcordata v. obtusiuscula. *Pedunculi* umbellæ 1-2-pollicares. *Pedicelli* 5-12 lin. longi. *Fructus* erectus, 1¼ unc. longus, utrinque attenuatus, subacutus, axi fibroso percursus. *Semina* 2 lin. longa, lævia, pallide brunnea.

A very distinct genus, allied closely to the Javanese *Diplophractum*, Desf., (Mem. Mus. 5. p. 34. t. 1.) as well as to the true *Grewia*.

PLATE XXII. *Fig.* 1. expanded flower, *slightly magnified;* *f.* 2. ovary ; *f.* 3. fruit, *natural size; f.* 4. the same, cut across ; *f.* 5. longitudinal section of the same ; *f.* 6. seed ; *f.* 7. the same, cut through in the direction of the rhaphe.

1. Christiana *cordifolia*, Hook. fil., foliis petiolatis oblongo-ovatis obtusis basi cordatis 5-nervibus super glaberrimis subter molliter ferrugineo-pubescentibus obsolete 3-lobis integerrimis, corymbo terminali, pedunculis pedicellisque velutino-

---

* The above character is copied from that given by Dr. Planchon in the " Icones Plantarum," t. 760.

pubescentibus, calyce 3-lobo persistente, carpellis 2-4 brevissime stipitatis subglobosis velutino-tomentosis 1-locularibus
2-valvibus 1-spermis.—On the Quorra, opposite Stirling,
Vogel.

Arbor excelsa; trunco mediocri. *Rami* validi, teretes; cortice
pallido fibroso, punctis albidis notato. *Petioli* stricti, $2\frac{1}{2}$ unc.
longi, vix puberuli. *Folia* 8-10 unc. longa, 5-6 lata, oblongo-
cordiformia, obtusa, subobliqua, nunc obscure lobata, plana,
siccitate ferrugineo-fusca, membranacea. *Corymbus* 6 unc.
longus, amplus, pluries ramosus : ramis primariis validis, subelongatis; pedicellis breviusculis. *Calycis* lobi 2 lin. longi,
coriacei, obtusi. *Carpella* diametro *Pisi sativi,* molliter
tomentosa. *Testa* pulchre irrorata, e membranis 3 constans;
exteriore dense crustacea; intermedia tenuior, inter crus-
taceam et membranaceam, atra; interior tenuissima, albumini appressa. *Albumen* copiosum, carnosum. *Cotyledones*
maximæ, foliaceæ, venosæ; *plumula* minima; *radicula* teres
crassa.

A very handsome plant, agreeing well with *Christiana* in the
three-lobed calyx, but in the fruit and habit allied to *Brown-
lovia,* Roxb.

2. Christiana *Africana,* DC. *Prod.* 1. *p.* 516.—On the Congo
River, *Christ. Smith.*

1. Honckneya *ficifolia,* Willd. *in Ust. Del. ex DC. Prod.* 1.
p. 506.—Clappertonia, *Meissn. Gen. p.* 36 (28).—Grand
Bassa, *Vogel;* Sierra Leone, *Don.;* Senegal.

Sepala 5, 3 v. omnia glandula globosa apiculata instructa.
Petala late rotundata, breviter unguiculata, imbricata, convoluta, rubro-cœrulescentiaio, *(Vogel).* *Stamina* basi coalita,
plura incompleta, setiformia, 12 elongata, antherifera; antheris versatilibus, elongatis; loculis utrinque liberis. *Ovarium*
8-loculare setosum; *loculis* multiovulatis, ovulis 2-serialibus,
placentis axillaribus affixis. *Semina* parva, orbicularia, plano-
convexa. *Testa* e membranis 3 constans; exteriore membranacea; intermedia crustacea; interiore tenuissima. *Albumen*
carnosum. *Embryo* axillis, latitudine albuminis. *Cotyledones*
latissimæ, planæ; radicula crassa terete.

This outer coat of the *testa* is usually described as an *arillus*. The hairs of the fruit are described by Vogel as always of a reddish-brown colour on one side of it, and green on the other.

XXII. DIPTEROCARPEÆ.

1. Lophira *alata*, Banks.—*Guill. et Perr. Fl. Seneg. p.* 108. *t.* 24.—Sierra Leone, *Don.* ; Senegal.

A low shrub, 2 to 3 feet high, according to Don. The structure of the wood is highly curious.

XXIII. CLUSIACEÆ.

1. Pentadesma *butyracea*, G. Don, *Gard. Dict.* 1. *p.* 619.— Sierra Leone and St. Thomas, *Don.* (The Butter- and Tallow- tree of W. Africa.)

XXIV. TERNSTRŒMIACEÆ.

The true *Ternstrœmiaceæ*, now known to be so numerous in Tropical America and Asia, have no representative in West Tropical Africa since *Ventenatia* and *Cochlospermum* have been removed by Planchon, the one to *Bixineæ*, the other to the neighbourhood of *Geraniaceæ*. The following genus, however, and the others forming Planchon's group of *Ixionantheæ*, are so nearly related to *Ternstrœmiaceæ*, that it may be convenient to consider them merely as a tribe of that order.

1. Ochthocosmus *Africanus*,* Hook. fil. (TAB. XXIII.) rhachidi- bus et pedicellis exceptis glaberrimus, foliis alternis brevissime petiolatis oblongis sparsis cuspidatis, cuspide callis paucis subglandulosis secus marginem instructo, utrinque acutis margine leviter incrassato et revoluto integris v. subrepandis rigide chartaceis nitidis subtus pallidioribus pulchre et te- nuissime venosis, racemis axillaribus 1-3 folio brevioribus, pedicellis fasciculatis petala sub fructu æquantibus rhachidi- busque puberulis, petalis sub fructu induratis calyce plus

---

* This character is copied from that drawn up by Dr. Planchon, "Icones Plantarum," t. 773.

duplo longioribus, staminibus styloque exsertis.—Sierra Leone,
Don.
Although at first sight this interesting plant might appear to
differ generically from the original *O. Roraimæ*, Benth., from
Guyana, a more close inspection shows those differences to be
merely specific. The leaves, which have in both the same firm
texture and glossy surface, are here scattered on the branchlets,
instead of being collected rather densely towards their apex;
the inflorescence consists of racemose fascicles, not of a sub-
corymbose panicle; the petals become thicker, and might be
called almost woody, a character which, connected with all
others, marks out the affinity of both plants with the genus
Ixionanthus of Jack. (*Planchon.*)
PLATE XXIII. *Fig.* 1. flower, long after fecundation; *f.* 2.
petal, with two stamens; *f.* 3. pistil, with the disc and lower
parts of the filaments; *f.* 4. vertical section of the same;
f. 5. fruit in the persistent flower; *f.* 6. transverse section of
the same; *f.* 7. seed, with the arilliform production of the
exostome.

XXV. ERYTHROXYLEÆ.

1. Erythroxylon, possibly *E. ferrugineum*, Cav.?—From Vogel's
collection, without the precise station. The specimen is
evidently of this genus, but quite undeterminable as to species.
It does not appear to be the same as *E. emarginatum*, Schum.
et Thonn. from Guinea, the only species as yet published
from Western Africa.

XXVI. HYPERICINEÆ.

1. Psorospermum *ferrugineum*, Hook. fil.; caule tereti erecto,
foliis brevissime petiolatis elliptico-obovatis oblongisve obtusis
v. acutis adultis glabratis subter reticulatis immaturisque
utrinque ferrugineo-tomentosis, marginibus leviter revolutis.
pedunculis axillaribus sub-3-chotomis pedicellisque rufo-
tomentosis, sepalis obtusis pubescentibus fasciculos staminum
pentandros æquantibus, petalis intus subvillosis. — Sierra
Leone, *Don.*

R

242 FLORA NIGRITIANA.

Rami cortice pallide griseo rimoso tecti; ramulis ferrugineo-tomentosis. *Petioli* ½ unc. longi. *Folia* 1-2-uncialia, ¾-1 unc.
lata, super siccitate fusca, subnitida, subter pallidiora, reticulatim
venosa, pube stellata ferrugineo-subtomentosa. *Cymæ* sub
12-floræ; pedicellis post anthesin elongatis, ½ unc. longis.
Flores 3 lin. longi. *Androphora* ciliato-barbata. *Baccæ* immaturæ globosæ, acuminatæ, 5-loculares, loculis-3-vacuis,
cæteris 2-spermis; testa seminis scrobiculata.

Closely allied to the *P. febrifugum*, Spach, but the leaves are
smaller, and not white underneath, nor are the cymes dense.

2. Psorospermum *tenuifolium*, Hook. fil. (TAB. XXI.); glaberrimum, ramis teretibus, ramulis oppositis, foliis petiolatis
membranaceis ellipticis utrinque angustatis punctis opacis
notatis, pedunculis axillaribus, cymis multifloris, pedicellis
sepalisque subacutis glaberrimis, petalis intus villosis, andro-
phoris pentandris, ovario 5-loculari, loculis 2-ovulatis, bacca
globosa abortu 3-loculari, loculis 1-2-spermis, testa subcoriacea scrobiculata.—Nun River, *Vogel.*

Frutex 10-pedalis, ramosus. *Rami* teretes, cortice pallido striato;
ramulis gracilibus. *Petioli* 2-4 lin. longi. *Folia* 3-4-polli-
caria, 2-2½ unc. lata, sæpe inæquilateralia, glaberrima, apice
angustata, acuminata; super siccitate fusco-castanea, vix
nitida, subter pallidiora, utrinque punctis minimis sparsa,
costa prominula; penninervia, nervis parallelis. *Pedunculi*
½ unc. longi, deinde ramosi, pedicellis perplurimis, 1-2 lin.
longis. *Flores* parvi. *Petala* albido-lutescentia ex Vogel.
Baccæ siccitate nigræ, in vivo ex Vogel nigrescenti-purpu-
rascentes; 2-3 lin. diametro. *Testa* seminum coriaceo-car-
nosa, albumine nullo; cotyledonibus conduplicatis, sparse
nigro-punctatis.

In this and the following species the cotyledons are remarkably
conduplicate, a character possibly peculiar to the African species,
as may be also the delicately membranous nature of the perfectly
smooth leaf.

PLATE XXI. *Fig.* 1. flower; *f.* 2. pistillum; *f.* 3. fruit, *nat.
size; f.* 4. the same, *magnified; f.* 5. transverse section of
the same; *f.* 6. seed; *f.* 7. embryo.

3. Psorospermum *alternifolium*, Hook. fil. ; glaberrimum, ramis
validis teretibus, ramulis crassis elongatis glaucis, foliis petiolatis inferioribus alternis omnibus obovato-oblongis obtusis
acutisve valde coriaceis basi angustatis super nitidis subter
glaucis reticulatim venosis sparse nigro-punctatis, pedunculis
axillaribus lateralibusve laxe paniculatis, paniculis di-tri-cho-
tomis, sepalis 4-5 glaberrimis obtusis, petalis basi squamula
minima auctis apice inflexis acuminatis intus subvillosis, an-
drophoris barbatis 5-7-andris, ovario 5-loculari, loculis sub
2-ovulatis, bacca abortu 2-3-loculari, loculis 2-spermis, testa
scrobiculata.—Sierra Leone ; *Don.*

Frutex 2-pedalis. *Ramuli* diametri pennæ corvinæ, læves, glaberrimi. *Petioli* 4-6 lin. longi. *Folia* erecta, superiora
opposita, cætera alterna, 4-7 unc. longa, 3-4 lata, seniora
rigida, coriacea, super nitida, venosa, subter opaca, albido- v.
cœruleo-glauca, nigro-punctata. *Panicula* 3 unc. longa,
ramis gracilibus erectis. *Flores* 3 lin. longi. *Baccæ* parvæ,
vix 2 lin. latæ, calyce paulo aucto suffultæ. *Semina* omnino
ut in *P. tenuifolio.*

A noble species ; remarkable for the alternate lower leaves,
stout ramuli, the glaucous hue of the latter, as of the petioles,
and under sides of the leaves, the large panicles, small
berries, and conduplicate cotyledons.

The other species of this genus are *P. Senegalense,* Spach,
which is the *Hypericum Guineense,* Linn., (but perhaps not
Vismia Guineensis of Choisy), a Senegambian plant, and *P.
febrifugum,* Spach, fromAngola.

1. Haronga *paniculata,* Pers. *Syn.* 2. *p.* 91 (sub *Arungana*).—
Grand Bassa and Fernando Po, *Vogel;* Sierra Leone, *Don. ;*
Senegambia.

I cannot distinguish the W. African specimens from those of
the Mozambique, Mauritius and Madagascar. Vogel remarks
that it is a shrub, and Don a tree 20 ft. high.

1. Vismia *Leonensis,* Hook. fil. ; ramulis teretibus oppositis
velutino-pubescentibus, foliis petiolatis late elliptico-ovatis
oblongo-lanceolatisve utrinque angustatis membranaceis

super glabratis subter ferrugineo-tomentosis nigro-punctatis, cymis axillaribus paucifloris, pedicellis elongatis pedunculo æquilongis sepalisque obtusis pubescentibus, bacca globosa pulposa 5-loculari, loculis polyspermis, seminibus subcylindraceis.—Sierra Leone ; *Vogel*, (cult.), *Don*.

Rami graciles ; cortice pallide cinereo rimoso ; ramulis divaricatis. *Petioli* 2-3-pollicares. *Folia* inferiora parva, late elliptico-oblonga, 1-1½ unc. longa, superiora majora, oblongolanceolata, 4-pollicaria, in apicem acuminatam angustata, submembranacea, margine obscure undulata, super fusco-brunnea, opaca, subter pallidiora, flavo-fusca v. rufo-ferruginea, pube stellata subtomentosa, penninervia. *Pedunculi* graciles, 4 lin. longi, dichotome v. subumbellatim ramosi. *Baccæ* diametro *Pisi* majusculi. *Testa* crustacea pallide flavo-brunnea, leviter reticulata, albumine v. endopleura parca ; embryone tereti, cylindraceo.

The specimens are flowerless, but the structure of the fruit and seed so entirely accord with the S. American *Vismiæ*, that though perhaps the only extra-American species of the genus, I refer it to that with little hesitation.

It is to be borne in mind that it appears to abound at Sierra Leone, where Vogel distinctly says *it is in cultivation*.

The *Lancretia suffruticosa*, Del., a plant common to Senegal and Egypt, is the only other W. African *Hypericinea* known to me.

XXVII. MALPIGHIACEÆ.

1. Acridocarpus *plagiopterus*, Guill. et Perr. *Fl. Seneg*. 1. *p.* 123. t. 29.—Anomalopteris obovata, *G. Don, Gard. Dict*. 1. 642.—Sierra Leone, *Don.* ; Senegambia.

2. Acridocarpus *Smeathmanni*, Guill. et Perr. *l. c. p.* 124. Anomalopteris spicata, *Don, l. c.*—Sierra Leone, *Don*.

3. Acridocarpus *longifolius*, Hook. fil. ; glaberrimus, ramis gracilibus, foliis alternis breve petiolatis lineari-oblongis elongatis apice angustatis acuminatisque margine undulatis membranaceis subter vix glandulosis, petiolo biglanduloso, racemo

terminali breviusculo paucifloro, pedicellis gracillimis sepalisque puberulis, bracteolis subulatis eglandulosis.—Anomalopteris longifolia, *Don, l. c.*—St. Thomas, *Don.*
Frutex 8-pedalis. *Rami* graciles, crassitie pennæ anatinæ,
teretes ; cortice pallide cinereo ; ramulis rufescentibus. *Folia*
5-10 unc. longa, 2½-4 lata, oblongo-obovata v. lineari-obovata,
ramea obovato-lanceolata ; membranacea, utrinque glaberrima,
integerrima, in petiolum non angustata, superne in apicem
obtusum acutumve angustata, super fusco-viridia, reticulatim
venosa, subter pallide viridia, glandulis paucis notata ; petiolo
¼ unc. longo. *Racemus* terminalis, 4-pollicaris, pedunculo
pube ferruginea vestito ; pedicellis gracilibus, ½ unc. longis.
Flores flavi, ½ unc. lati. *Sepala* late oblonga, puberula,
unico basi glandula maxima depressa notato. *Petala* calyce
quadruplo longiore ; marginibus erosis.

This is Don's *Anomalopteris longifolia,* so called in his
collections, though differing from the insufficient and inaccurate
description in the "Gardener's Dictionary." It cannot, however, be the *A. Guineensis* of Adrien Jussieu, which that author
particularly describes as having very coriaceous leaves, whereas
those of this plant are even more membranous than in *A.
Smeathmanni.*

4. Acridocarpus *Guineensis,* Adr. Juss. *Malpigh. p.* 231 ; ramis
    validis supremis puberulis, foliis alternis breve petiolatis co-
    riaceis lineari-lanceolatis acuminatis integerrimis glaberrimis
    super lævibus subter reticulatim venosis, petiolo biglandu-
    loso ; racemis lateralibus, pedunculo valido, pedicellis sepalis-
    que oblongis pubescentibus, bracteolis subulatis, samaris
    glabris, ala gradatim dilatata apice oblique rotundata.—
    Fernando Po, *Vogel.*
Frutex sarmentosus, 4-5-pedalis ; ramis pendulis, apices versus
præcipue puberulis. *Folia* 6-10 unc. longa, 3-4 lata, coriacea,
super opaca lævia, subter pallidiora venis prominulis reticulata,
remote punctata, basi glandulis majusculis 2-3 instructa ;
petiolo incrassato, ⅓ unc. longo. *Racemi* 3-4 unc. longi,
erecti, multiflori ; pedicellis infimis curvatis, ¾ unc. longis.

Calycis laciniæ oblongæ, pubescentes. *Corolla* flava; petalis fere ⅓ unc. longis. *Ovarium* dense pubescens. *Samara* 2-pollicaris, glabra v. puberula, e basi dilatata, latior quam in *A. Smeathmanni* et *plagiopteride.*

5. Acridocarpus *corymbosus,* Hook. fil. (Tab. XXIV); ramis sparse tuberculatis, foliis breviter petiolatis coriaceis ellipticoovatis oblongisve acutis integerrimis glaberrimis super lævibus subter reticulatim venosis basi biglandulosis, racemis axillaribus terminalibusque corymbosis pubescentibus folio multoties brevioribus, pedunculo sursum incrassato, pedicellis elongatis gracilibus, floribus parvis.—Cape Coast Castle, *Vogel.*

Arbuscula. *Rami* diametro pennæ anatinæ; cortice fusco, tuberculis sparsis instructo. *Folia* 3 unc. longa, 1¾ lata, basi obtusa, glandulis (nisi ad apicem petioli) nullis, v. inconspicuis, super fusco-brunnea, subter pallidiora, opaca; petiolo 2 lin. longo. *Pedunculus* vix uncialis post anthesin subclavatus. *Inflorescentia* corymbosa, corymbo 1 unc. lato, multifloro; bracteolis parvis, subulatis; pedicellis gracilibus, ½ unc. longis. *Calycis* lobi rotundati, extus puberuli, unico basi glandula depressa instructo. *Flores* ¼ unc. lati.

Plate XXIV. *Fig.* 1. flower, without the petals; *f.* 2. stamen; *f.* 3. ovary; *all more or less magnified.*

The other W. African species of this genus are, *A. Cavanillesii,* Adr. Juss., from Sierra Leone; *A. Angolensis,* Adr. Juss., from Angola; and an undescribed Senegambian species in the Hookerian Herbarium, collected by Heudelot.

1. Heteropterys *Africana,* Adr. Juss., *Malp. p.* 202.—Sierra Leone, and Grand Bassa, *Vogel;* Senegambia.

2. Heteropterys *Jussieui,* Hook. fil.; foliis elliptico-oblongis lineari-lanceolatisve acuminatis coriaceis super lucidis planis bullatisve subter reticulatim venosis glaucis, paniculis terminalibus trichotome ramosis pubescentibus, calycis laciniis biglandulosis, samara circumscriptione semicirculari ala plana semini conforme.—Sierra, Leone, *Don, Vogel.*

There is little but the very different form of the samara to

distinguish this from *H. Africana*. Each carpel has a broad
semicircular wing, produced equally above, below, and out-
wardly.

The above are the only two extra-American representatives of
a genus numbering no less than eighty-two species, and afford
a striking example of the relation subsisting between the East
American and West African Floras.

1. Triaspis *odorata*, Adr. Juss. *in Deless. Ic.* 3. *p.* 21. *t.* 36.—
Fernando Po, *Vogel;* Guinea, *Thonning.*
The *T. flabellaria,* Adr. Juss., from Senegambia, concludes
the catalogue of W. African *Malpighiaceæ.*

XXVIII. SAPINDACEÆ.

1. Cardiospermum *Halicacabum,* Linn., var. hirsutum.—Frequent along the coast, from Cape Verd to the Niger River.
2. Cardiospermum *microcarpum,* H.B.K. *Nov. gen. et sp.* 5.
p. 104.—Senegambia; also in Vogel's collection, without the
precise station, but probably as frequent as *C. Halicacabum.*
A very distinct plant from the former, in the small, short,
broadly-triangular, trigonous capsules, depressed at the top,
although when in flower only it is difficult to distinguish it.
Both species vary in the greater or less degree of pubescence
of the stems, young leaves, and pods, or in their perfect smooth-
ness, yet it is probably the *C. Halicacabum* that Schumacher and
Thonning refer to as *C. hirsutum,* and that the *C. microcarpum* is their *C. glabrum.* Both species are so widely diffused
over Tropical America, the whole of Africa, the East Indies, and
the islands of the Pacific, that we have no other data to determine which is more particularly their native country, than this,
that America is the exclusive station for all other known species
of the genus. The *C. microcarpum* has been since published
by Miguel (Linnæa, p. 18. 359), under the name of *C. acuminatum,* and Cape-Verd specimens of it were included under *C. Halicacabum* in the first portion of this vol., p. 114.

1. Paullinia *pinnata*, Linn.—Senegal, *Sieber;* Sierra Leone,

Cape Coast, Fernando Po, St. Thomas, and Congo River,
Christ. Smith, Vogel, Don, &c.

An abundant W. African plant, variable in the size of the
foliage, from which *P. Senegalensis,* Juss., *P. uvata,* Schum. et
Thonn., and *P. Africana,* Don, do not appear to be distinct.
It is also frequent in the West Indies. The only other West
African species of this large American genus, and the only one
hitherto published as extra-American is the *P. sphærocarpa,*
Rich., from Guinea.

1. Schmidelia *Africana,* DC., *Prod.* 1. *p.* 610.—Ornitrophe
tristachyos, *Schum. et Thonn., Beskr. p.* 188?—Sierra Leone,
Abòh, Quorra, and Grand Bassa, *Vogel;* Senegal, *Sieber,* &c.

Apparently a very common species along the coast, and very
closely allied to the *S. Abyssinica,* Hochst., from Abyssinia, and
to the *S. melanocarpa* and *leucocarpa,* Presl. (*Rhus.,* E Mey. in
Pl. Drège MS.), from South Africa ; but all these appear to have
rather larger flowers. The *S. serrata* from E. India, and
S. Cominia from the West Indies, have the leaves always downy
underneath.

2. Schmidelia *hirtella,* Hook. fil.—S. monophylla, *Hook. fil.
in Ic. Pl. t.* 775 (TAB. XXV.) Ramis petiolis foliisque
subter pubescenti-pilosis, foliolo solitario membranaceo obovato-oblongo basi angustato apice abrupte longe acuminato
remote argute dentato, costa super puberula, racemis axillaribus
multifloris petiolo paulo longioribus, floribus parvis 4-meris,
petalis intus dense villoso-barbatis, glandulis (inter stamina et
petala) obcuneatis.—Fernando Po, *Vogel.*

Frutex. Ramuli graciles, subflexuosi, teretes, pubescenti-pilosi ;
cortice pallido. *Folia* alterna 1-foliolata. *Petiolus* ½-un-
cialis, erecto-patens. *Foliolum* amplum, 6-7 unc. longum,
3 latum, ad apices venularum dentatum, super luride virescens subter pallidius, ad axillas venarum lanuginosum. *Racemi* unciales ; pedicellis 1-2-floris. *Pedunculus* gracilis, pubescens. *Flores masculi* globosi, ½ lin. lati. *Sepala* 2 exteriora
minora et angustiora, interiora lata, concava. *Petala* obovata, unguiculata, sepalis multoties minora, intus barba de-

pendente aucta. *Squamulæ* carnosæ, petala æquantes, emarginatæ. *Stamina* 8; filamentis crassiusculis glaberrimis.
Fl. fœminei ignoti.

A very distinct species, though belonging to a unifoliolate
group, which is common to the W. Indies, Brazil, S. Africa,
and Ceylon. One of the S. African ones named by E. Meyer, in
Drège's collection, *Rhus monophylla*, has been rightly placed
by Presl in *Schmidelia*, under the name of *S. monophylla* in
his *Botanische Bemerkungen*, a work we had not seen when we
first gave the same name to the present species.
PLATE XXV. *Fig.* 1. bud; *f.* 2. flower without the calyx; *f.* 3.
petal; *all magnified.*

The *S. affinis*, Guill. et Perr., of Senegal, probably the same
as *Ornitrophe magica*, Schum. et Thonn. from Guinea, and *Ornitrophe thyrsoidea*, Schum. et Thonn., are the only other West
African congeners.

1. Sapindus *Senegalensis, Poir. Dict.* 6, *p.* 666.—S. Guineensis,
G. Don, Gard. Dict. 1. *p.* 666?—Senegal, *Brunner;* on the
Gambia, *Don.*

The *S. saponaria,* Linn., is said by Brunner to be cultivated
in Senegambia and at St. Yago (Cape de Verd), and to have
become almost wild in woods of the valley of San Domingo, in
the latter island.

1. Deinbollia *pinnata,* Schum. et Thonn. *Beskr. p.* 242.—Prostea
pinnata, *Camb. in Mem. Mus. Par.* 18. *p.* 39.—Guinea, in
Vogel's collection without the exact locality.

2. Deinbollia? *grandifolia,* Hook. fil.; glaberrima, ramis validis, petiolis elongatis teretibus lævibus basi incrassatis, foliolis
amplis alternis subremotis breve petiolulatis lineari-oblongis
lanceolatisve basi inæqualiter cuneatis rotundatis integerrimis,
paniculis axillaribus folio brevioribus ramosis, fructibus didymis subbaccatis.—Cape Palmas, *Vogel.*

Arbor 8-pedalis. *Rami* læves, crassitie digiti minoris. *Petiolus*
2-pedalis, teres, glaberrimus, cortice pallide fusco. *Foliola*
8 unc. ad pedalia, super leviter, subter profundius reticulata,
pallide viridia, subcoriacea, vix nitida; petiolulis 3 lin. longis.
Panicula fructifera pedalis, ramosa, erecta; ramis subangu-

latis pallide punctatis, sub lente puberulis. *Pedicelli* fructus lignosi, 3 lin. longi.

A very handsome species, but unfortunately neither in flower or fruit. The latter having fallen away, I have taken the brief description of that organ from a note on the ticket, by Dr. Vogel, which adds that it is lemon-colour. The leaflets are very large.

3. Deinbollia *insignis*, Hook. fil.; glabra, foliolis 13 v. ultra amplis alternis brevissime petiolulatis oblongo-ellipticis acuminatis undulatis basi obtusissimis, paniculis infrafoliaceis ad axillas foliorum delapsorum elongatis, sepalis extus puberulis, petalis glabris ciliatis, staminibus ciliatis.—Fernando Po, *Vogel.*

Truncus subarborescens, orgyalis, 2 poll. diametro, apice coronam fert foliorum bipedalium. Horum uncium adest nec perfectum, pars enim inferior deest; glaberrimum est, rhachis crassa obtuse subangulata. *Foliola* alterna, 8-10 poll. longa, 3-4 poll. lata, petiolulo crasso brevissimo v. vix ullo, utrinque viridia, nervis pinnatis et venulis reticulatis numerosis prominulis scabriuscula. *Paniculæ* semipedales ad pedales, infra comam e trunco ortæ, parum ramosæ, novellæ puberulæ. *Ramuli* breves, irregulariter subcymosæ. *Pedicelli* breves, superne incrassati, tomentosi. *Sepala* 5, orbiculata, concava, valde imbricata, parum inæqualia, 2½-3 lin. diametro. *Petala* 5, sepalis minora, orbiculata, intus basi squama lata plus minus ciliolata aucta. *Discus* carnosus, parum prominulus. *Stamina* circa 20; *filamenta* brevia, hirta; *antheræ* oblongosagittatæ, filamentis paullo longiores, dorso ciliatæ. *Ovarium* hirtellum, trilobum. *Stylus* brevis, trilobus. *Fructus* deest.

1. Blighia *sapida*, Kœn. *Ann. Bot.* 1806. *v.* 2. *p.* 571—Cupania edulis, *Schum. et Thonn., Beskr. p.* 190.—Cultivated at Frederiksgaue, *Vogel;* found wild on the plains of Guinea, according to *Thonning.*

LECANIODISCUS, Planch.* (nov. gen.)

Calyx 5-partitus. *Petala* 0. *Discus* calycis fundum occupans

---

* From Vogel's imperfect specimens, Dr. Hooker was unable to make

obscure 10-crenatus. *Stamina* 10, intra disci marginem inserta, antheris oblongis. *Ovarium* villosum, apice vix in stylum brevissimum attenuatum, intus triloculare. *Stigma* crassum, reflexo-trilobum. *Ovula* in loculis solitaria erecta. *Drupa* obovoidea v. globosa, styli reliquiis apicalibus, intus abortu unilocularis. *Semen* arillo mucoso involutum. *Embryo* rectus, cotyledonibus crassis conferruminatis, radicula parva.—*Frutex?* Africæ occidentalis tropicæ, habitu *Cupaniæ*. *Folia* impari-pinnata, foliolis oppositis v. alternis integerrimis. *Racemi* breves axillares, floribus secus rhachin fasciculatis, breviter pedicellatis.

1. Lecaniodiscus *cupanioides*, Planch. MS.—Accra, *Vogel ;* Sierra Leone, *Don ;* also Senegambia, *Heudelot,* in the Hookerian Herbarium.

Frutex v. arbor, ramulis sulcatis pubescenti-tomentosis. Foliorum *petiolus* communis 3-6-pollicaris, sulcatus, puberulus. *Foliola* 6-11, breve petiolulata, obovata obtusa v. brevissime obtuso-acuminata, basi breviter angustata, margine integerrima, at obscure undulata, submembranacea, supra siccitate fusco-brunnea, subtus pallidiora, glabra v. subtus ad costas puberula, penninervia, venulis inter nervos reticulatis. *Racemi* 1½-2 poll. longi, rhachide rufo-tomentello. *Bracteæ* parvæ, caducissimæ. *Pedicelli* solitarii v. 2-3-ni, puberuli, 2 lin. longi. *Calycis* laciniæ fere 2 lin. longæ, oblongæ, obtusæ, crassiusculæ, intus extusque pubescentes, per anthesin reflexæ. *Discus* 1¼ lin. diametro, plano-patellæformis, crenaturis vix conspicuis et fere usque ad marginem calycis fundo adnatus. *Stamina* glabra, ovario vix longiora. *Antheræ* filamento paullo breviores, biloculares, rimis longitudinalibus dehiscentes. *Ovarium* sessile, ovoideo-globosum, dense villosum. *Stylus* brevissimus in lobos latos crasse stigmatosos ovato-hippocrepideos deflexos brevissime divisus. *Drupa*

out the genus of this plant, and had referred it with doubt to *Cupania*, of which it has the habit. Dr. Planchon having since found a flowering specimen in Heudelot's collection, and ascertained that they belonged to an entirely new genus, I have drawn up the abbreviated character from the two specimens.—(G. B.)

semi-unciam longa, extus tomentosa et in vivo (ex Vogel) lutea, pericarpio tenuiter carnosa, endocarpio tenui. *Semen* cavitatem fructus implens, arillo mucoso (cujus rudimentum jam sub ovulo ante anthesin apparet) involutum.

The apetalous flowers, and the remains of the stigmate at the summit of the fruit, and not lateral, distinguish this genus from *Sapindus*, and bring it nearer to *Schleichera*, from which it is readily distinguished by the calyx, the anthers, and the embryo.

1. Dodonæa *viscosa*, Linn.—On the Gambia, *Don ;* Senegal, *Sieber*.

Don's specimens, in excellent fruit, are undoubtedly identical with the Jamaica plant, which Schlechtendahl, with reason, considers as the best entitled to retain the Linnæan specific name. Sieber's specimens, which the same botanist establishes as a distinct species, under the name of *D. Kohautiana*, appear to be only a slight variety, partly accidental, from the manner in which the resinous exudation has dried, so as to give them a scaly appearance. *D. repanda*, of Schum. and Thonn., from Guinea, is evidently closely allied, but may be distinct.

The *Erioglossum cauliflorum*, Guill. et Perr., from Senegambia, referred by Arnott to the *Cupania canescens*, Pers., is the only other W. African *Sapindacea* known to me besides the following, considered by Planchon as forming with *Melianthus* and *Bersama*, a distinct Order, *Meliantheæ*, but which undoubtedly bears considerable affinity to, if it be not a mere tribe of, *Sapindaceæ*.

1. Natalia *paullinioides*, Planch. (Tab. XXIX) ; foliis cum impari 7-10-jugis, foliolis oppositis v. passim alternis petiolatis lanceolatis breviter cuspidatis utrinque acutis glabriusculis (nervis subtus tantum pilosulis) remote serrulatis serraturis incurvis, supra siccitate nigrescentibus subtus pallidis, racemo oppositifolio pedunculato plurifloro, bracteis parvis subulatis, pedicellis calyce brevioribus v. eum subæquantibus, petalorum lamina lineari-oblonga cristulis parvulis basi ornata v. nuda, ungue pro parte sericeo-albido, stylo inferne piloso staminibusque exsertis.—Sierra Leone, *Vogel*.

Frutex (verosimiliter scandens) facie *Paulliniæ*. *Ramuli* pe-

tiolique communes rhachidesque racemi sulcati et pube de-
tersibili primum hinc inde sparsi, demum glabrati. *Stipulæ*
in unam extra-axillarem brevem, ovatam, dorso sericeam concretæ. *Racemus* 7-pollicaris, inferne nudus, medio cicatricibus pedicellorum notatus, apice confertiflorus. *Flores* illis
Æsculi Hippocastani minores, leviter irregulares. *Calyx*
profunde 4-fidus, lacinia infera (antica) apice bidentata (e 2
constante). *Petala* 5, æstivatione imbricata, infimo emarginaturæ laciniæ infimæ calycis respondente, inde sepalis 2
connatis alterno, cæteris angustiore. *Stamina* 4. *Filamenta* basi dilatata, duorum petalorum infimorum connata, 2
lateralium libera. *Glandula* carnosa, brevis, sepalo postico
opposita, propter stamina externa. *Ovarium* 4-loculare pilis
rufis vestitum. *Stigma* pyramidato-truncatum. (*Planchon*).
PLATE XXIX. *Fig.* 1. bud, side view; *f.* 2. flower; *f.* 3. the
same, with only the stamens, pistil, gland, lower petal and
one of the posterior petals; *f.* 4. stamens, gland, and pistil,
back view; *f.* 5. gland; *f.* 6. ovary, vertical section; *f.* 7.
summit of the style; *all more or less magnified.*

XXIX. MELIACEÆ.*

1. Turræa *Vogelii,* Hook. fil.; ramulis pubescentibus, foliis
elliptico-ovatis acuminatis integerrimis ad venas pubescentibus, floribus pedicellatis pentameris, calyce brevissime dentato, tubi staminei dentibus 10 setaceis demum patentibus,
ovario 12-loculari, styli parte inflata apice tantum stigmatosa
petala non excedente.—Sea Coast, Fernando Po, *Vogel.*
Frutex (fide Vog.) ramosissimus, ramis elongatis sarmentosis,
pube brevi subvelutina obtectis. *Folia* 4-6 poll. longa, 2-3
poll. lata, acumine obtusiusculo, margine integerrima subun-
dulata, inæquilatera et basi sæpius obliqua, ima basi obtusa et
rotundata v. brevissime aculata, membranacea, penninervia et

* The characters and descriptions in this and the two following Orders
are drawn up by myself from Dr. Hooker's memoranda, as well as from
my own examination of the specimens. (G. B.)

reticulata, costis et venis primariis præsertim subter pube brevi tomentellis, petiolo circa 3 lin. longo pubescente; folia novella venis dense hirtis flavicant. *Pedunculi* in axillis superioribus bipollicares, apice umbellam ferunt pluri- (8-10?) floram. *Pedicelli* tomento minimo canescentes, 4-5 lin. longi, basi bracteis parvis sericeis confertis stipati. *Calyx* cupuliformis, tomentellus, lineam longus. *Petala* 10-11 lin. longa, lineari-oblonga, basi longe angustata, extus in sicco vix tomento tenuissimo leviter canescentia, in vivo fide Vogel intus albida et patentia. *Tubus stamineus* dimidio petalorum longior, tenuiter cylindracea, intus infra apicem pilosulus, cæterum glaber; dentes apicales setacei, antheris vix breviores, superne integri v. bifidi et papilloso-serrulati, in alabastro erecti et inter antheras stylo appressi, per anthesin reflexo-patentes.* *Antheræ* ad apicem tubi subsessiles, oblongo-lineares, connectivo in apiculam uncinato-inflexam producto. *Stylus* usque ad apicem antherarum tenuis, dein inflatus, oblongo-linearis, glandula stigmatosa crassa subintegra depressa coronatus. *Ovarii* loculi 12 vidi in flores paucos quos examinavi.

2. Turræa *propinqua*, Hook. fil.; ramulis glabriusculis, foliis elliptico-oblongis obtusis v. vix acuminatis integerrimis (v. apice lobatis?) basi angustatis glabris v. vix ad venas minute puberulis, floribus pedicellatis pentameris, calyce brevissime dento; tubi staminei dentibus 20 setaceis demum patentibus, ovario 12-loculari, styli parte inflata apice tantum stigmatosa, petala superante.—St. Thomas, *Don*.

Rami quam in præcedente tenuiores, nonnisi juniores tenuissime puberuli. *Folia* minora et angustiora, majora in specimine vix tripollicaria, semelque in specimen folium apice lobatum occurrit. *Pedunculi* pollicares, pedicellis in umbella semi-

---

* The arrangement of these curious appendages in the bud, as well as the texture of the upper portion of them, seem to indicate that they are destined, as well as the hooks on the top of the anthers, to perform some function at the time of fecundation, perhaps analogous to that of the *collecting hairs* found in so many plants at the same period on the style or other contiguous parts.

pollicaribus 6-10-nis bracteolis paucis parvis basi stipatis. *Flores* minores quam in *T. Vogelii,* albi, fragrantissimi (ex Don). *Petala* semipollicaria crassiuscula, extus siccitate canescentia. *Calyces* et genitalia *T. Vogelii* in omnibus nisi stylo ratione calycis longiore et dentibus tubi staminei (in flore unico a me examinato) 20 nec 10. *Ovarii* loculi 12. The *T. heterophylla,* Sm., the only species mentioned by Bennett in his review of the genus as from West Tropical Africa, is different from either of the above. Don has likewise published (Gard. Dict. 1. p. 678) a *T. quercifolia* from Sierra Leone, which may possibly be the same as *T. lobata,* since published and figured by Lindley (Bot. Reg. 1844, t. 4.) from the some country. This belongs to Bennett's first division, although Rœmer, on geographical grounds, places it in his genus *Rutæa,* founded on Bennett's second division.

1. Melia *Azederach,* Linn.—M. augustifolia, *Schum. Thonn. Beskr.,p.* 212.—Sierra Leone, *Don,* probably cultivated.

1. Trichilia *emetica,* Vahl, *Symb.* 1. *p.* 31.—Goniostephanus tomentosus, *Fenzl, Flora,* 1844, *p.* 312.—Elkaja emetica, *Forst. Rœm. Syn. Mon. Hesperid. p.* 116.—Sierra Leone, *Herb. Hook.;* Senegambia, *Heudelot.* Also a native of Nubia and Arabia.

The *T. Prieuriana,* A. Juss., from Senegambia (to which *T. Ruppeliana,* Fresen., from Abyssinia, appears closely allied) and an undescribed one from Senegambia are the only other West African species known to me, but Dr. Planchon suggests that the *Limonia? monadelpha,* Schum. et Thonn. Beskr. p. 217. is probably also a *Trichilia.*

1. Carapa *Guineensis,* Sweet, Hort. Brit. et in A. Juss. Mem. Mel. p. 90.—C. Touloucouna, *Guill. et Perr. Fl. Seneg.* 1. *p.* 128.—Touloucouna gigantea, *Rœm. Syn. Mon. Hesperid.* 1. *p.* 123.—Senegal, Sierra Leone, *Don ?* This tree produces an oil employed in making soap for anointing the body, as is the case with the *C. Guianensis,* of which species Ad. Juss., and others who have examined the two plants, suspect that the African one may be a mere variety.

The number of parts of the flowers, although usually different in he two, is expressly stated by Jussieu not to be so constant in *C. Guianensis* as to warrant the making use of it as a specific distiction, still less as a generic one, as proposed by Rœmer, apparently without re-examination of specimens.

1. Khaya *Senegalensis,* Guill. et Perr. *Fl. Seneg.* 1. *p.* 130. *t.* 32.—Senegal, *Brunner ;* Sierra Leone, *Don.*

Wood like Mahogany, and very useful for various purposes. The *Ekebergia Senegalensis* is the only other West African plant of this order known to me. (G. B.)

XXX. AURANTIACEÆ.

1. Glycosmis? *Africana,* Hook. fil. ; foliolis solitariis oblongo-ellipticis breviter acuminatis margine recurvis coriaceis, drupis obovoideo-oblongis (abortu ?) monospermis.—St. Thomas, *Don.*

Specimen unicum fructiferum, formis unifoliolatis latifoliis *G. citrifoliæ (Limonia parviflora,* Sims, *L. citrifolia,* Willd. et DC. non Roxb. quæ *Glycosmis? citrifolia,* W. et Arn. et Rœm. Syn) simile, sed baccarum forma certe diversa. *Petioli* teretes, 3-4 lin. longi, apice articulati. *Foliolum* 4-6 poll. longum, 2-2½ poll. latum, coriaceum, nitidum, pellucido-punctatum, venulis a costa divergentibus tenuibus crebris parallelis reticulatisque. *Inflorescentia* omnino *G. citrifoliæ. Flores* desunt sed ex calycis vestigiis pentameri videntur. *Drupæ* 4-5 lin. longæ, stigmate subsessili disciformi coronatæ, pericarpio carnoso cellulis oleiferis numerosis, endocarpio membranaceo ; pulpa in sicco nulla apparet. *Semen* unicum, ex apicem pendulum, cavitatem implens ; testa rigide membranacea ; embryo ad hilum spectans ; cotyledones crassæ, carnosæ, basi integræ, cellulis oleiferis numerosis ; radicula brevissima, plumula minima.

1. Claussena *anisata,* Hook. fil.—Amyris anisata, *Wild. Spec.* 2. *p.* 337.—Fagarastrum anisatum, *G. Don, Gard. Dict.* 2. *p.* 87.—Cape Coast, *Vogel.*

The genus *Amyris* had already been restricted to American plants by Kunth and others; Wight and Arnott showed that the Indian ones at least belonged to *Aurantiaceæ*, and chiefly to *Claussena*. Don, however, in establishing the genus *Fagarastrum* for the African species, retained them among *Terebinthaceæ*, without alluding to *Aurantiaceæ*. Presl has since perceived the affinity to the latter order of the South African species, but without comparing it either with *Claussena* or with Willdenow's *A. anisata*, (from which it is separated by characters so slight as possibly not even to be specific), created a new genus under the name of *Myaris*. Both African species, *C. anisata* and *C. inæqualis*, are very near in habit and character to the *C. Willldenowii*, but differ, not only in the form of the leaves and other minor points, but also in what at first might appear more important, that the ovules are usually, especially in *C. inæqualis*, collateral and not superposed. It must be observed, however, that even in the Indian species the ovules are collaterally inserted, although from the form of the cell they place themselves one above the other as they are developped, the placenta becoming slightly elongated into an umbilical cord. In *C. anisata* and *inæqualis*, the number, form and size of the cells is variable; they are usually small, and according to that form, the ovules at the time of flowering lie either more or less superposed, or absolutely side by side, especially in *C. inæqualis*.

1. Citrus *aurantium*, Linn.—C. articulata, *Willd. in Spreng. Syst. 3. p.* 334?—Cape Palmas and Isle of S. Antonio, *Vogel*, (probably cultivated.)

The only other W. African *Aurantiacea* published is *Citrus paniculata*, Schum. and Thonn., from Guinea, which, however, from the character given, can scarcely be a true *Citrus*.—(G. B.)

XXXI. OLACINEÆ.

1. Heisteria *parvifolia*, Sm. *in Rees' Cycl. v.* 17; ramulis angulatis, foliis ovatis oblongisve acuminatis coriaceis, calycis fructiferi profunde lobati lobis subcordato-ovatis acutiusculis

s

v. obtusis post fructus delapsos patentibus sinubus reflexis.—
Sierra Leone, *Whitfield;* Grand Bassa and Fernando Po,
Vogel; Senegal.

The specimen described by Sir J. Smith was a small-leaved
one, hence his name is not very appropriate although there do
not appear to be sufficient grounds for changing it. The species
is so closely allied to a common, although hitherto undescribed
Brasilian species,* that one is almost tempted to consider it as a
mere variety. The leaves are, however, usually rather smaller,
narrower and thinner, and the divisions of the enlarged calyx
not so blunt, with the *sinus* more reflexed. Both, however, may
be mere forms of *H. cauliflora,* Sm.†

Some specimens, in leaf only, gathered by Vogel in the woods
of Fernando Po, and stated by him to be those of a shrub
bearing a bitter fruit called *Kola,* of which the seeds are chewed
by the natives, are conjectured by Dr. Planchon to belong to a
new species of *Heisteria,* but there is no evidence to confirm
the supposition, and some remains of flower-stalks seem to show
an inflorescence very different from that of *Heisteria.*

1. Strombosia ? *grandifolia,* Hook. fil.; foliis amplis obovali- v.
elliptico-oblongis acuminatis, floribus axillaribus congestis breviter pedicellatis, calycis brevissime adhærentis limbo profunde
5-fido, ovario sublibero triovulato, stigmate obsolete trilobo.—
Fernando Po, *Vogel.*

Frutex arborescens (ex *Vog.*) *Rami* teretes, læves, graciles, uti
tota planta glaberrimi. *Folia* 5-7 poll. longa, 3-4 poll. lata,
petiolo semipollicari, apice in acumen subsemipollicare producta, margine integerrima v. undulato-sinuata, basi rotundata
v. cuneato-acuta, submembranacea, nitida, costa valida et

---

* *H. Raddiana,* (Benth MS.) ramulis subangulatis, foliis ovatis oblongisve obtusis v. breviter acuminatis crasso-coriaceis nitidis, calycis fructiferi profunde lobati lobis ovato-orbicularibus obtusissimis post fructus
delapsos patentibus, sinubus subreflexis.—Rio Janeiro, *Raddi, Gardner,*
5379 and 5378 ?

† To this I should refer, besides Cayenne specimens, *Hostmann's* n.
194, from Surinam, and *Gardner's* n. 2516, 2787, and 5974, from Tropical
Brasil.

nervis primariis pinnatis subtus prominulis percursa, et venulis transversis creberrimis arcuato-subparallelis reticulata. *Nodi* floriferi axillares, flores plures parvos inconspicuos breviter pedicellatos ferunt. *Pedicelli* apice incrassati, carnosuli, cum toro et calycis tubo continui. *Calyx* minimus, subcampanulatus, tubo brevissimo fere omnino adhærente, lobis brevibus ovatis obtusis. *Petala* 5, oblonga, æstivatione valvata, apice inflexa, basi inter se cohærentia, superne intus villosa. *Filamenta* petalis opposita, numero iis æqualia et alte adnata. *Antheræ* ovato-oblongæ. *Ovarium* crasso-carnosum apice pulvinato-depressum et (pressione petalorum) 5-angulatum, prope basi intus excavatum in loculos tres spurios apice confluentes. *Ovula* 3, e parte uniloculare pendula. *Stylus* brevissimus, apice crasse stigmatosus et obsolete trilobus. *Fructus* non vidi.

There is little doubt, even without having seen the fruit, that this plant is referable to Blume's genus, *Strombosia,* as characterized by Gardner, although in the Ceylonese plant, and probably also in the Javanese, the ovary is pentamerous, and the calyx adheres rather higher up. Dr. Gardner is evidently correct in his views of the affinities of the genus, although we can scarcely agree with him and Blume in describing the ovary as immersed in a fleshy disk. The thick fleshy mass at the base of the style appears to us to be perfectly continuous with, and to form part of the ovary itself, which in most *Olacineæ* is very fleshy and thick compared with the ovuliferous cavity.

RHAPHIOSTYLIS, *Planch.* (nov. gen.)

Flores hermaphroditi. *Calyx* parvus, liber, 5-partitus. *Petala* 5. *Stamina* totidem, iis alterna, sterilia nulla. *Ovarium* uniloculare, ovulis 2, hinc ab apice pendulis collateralibus. *Stylus* excentricus, basi postice gibbus, gibbo sulcato. *Fructus* . . . . *Inflorescentia* axillaris.—*Frutices* Africæ tropicis glabri. *Folia* alterna integerrima perennantia. *Flores* in nodos axillares plures pedicellati, iis *Apodytis* similes.

1. Rhaphiostylis *Beninensis,* Planch. MSS.—Apodytes Beni-

nensis, *Hook. fil. in Ic. Pl. t.* 778. *et tab. nostr.* 28.—
Cape Palmas, *Vogel.*

Frutex (ex Vog.) glaberrimus. *Ramuli* teretes v. striati. *Folia*
petiolata, ovali- v. elliptico-oblonga, breviter et obtuse acuminata, integerrima, margine recurvo, basi obtuse angustata,
rigide membranacea v. subcoriacea, costa venisque paucis primariis validis, rete venarum tenui; 2-4 poll. longa, ¾-1½ poll.
lata, petiolo bilineari canaliculato. *Pedicelli* in nodos axillares per 10-12 aggregati, tenues, erecti, uniflori, 2-3 lin.
longi, ima basi minute bracteolati. *Alabastra* oblonga, 3 lin.
longa. *Calyx* minutus, fere ad basin 5-partitus, laciniis ovato-
triangularibus acutis basi leviter imbricantibus. *Petala* linearia, (teste Vog. albo-viridia), apice uncinato-inflexa, intus
glaberrima *Filamenta* tenuissime ciliata, a basi ultra medium
concavo-dilatata, superne filiformia. *Antheræ* oblongæ, polline trigono. *Ovarium* sessile, ovatum, compressum, glabrum,
obliquum. *Stylus* excentricus, filiformis, incurvus, basi postice
dilatatus in gibbum superne sulcatum, sulco fere ad medium
styli obscure continuo; styli apex clavato-stigmatosus obliquus.

The characters derived from the flower are so nearly those of
Apodytes, that in the absence of the fruit, Dr. Hooker had
described it as a new species of that genus. Dr. Planchon has,
however, named it as a new genus, and in this I should be disposed to agree with him, chiefly on account of the inflorescence,
which in *Olacineæ* appears very constant. The character derived from the ovary and style is also remarkable, and forms
a positive distinction from those of *Apodytes.* The fruit remains
unknown.

PLATE XXVIII. *Fig.* 1. flower, before expansion; *f.* 2. stamen;
f. 3. ovary and calyx; *f.* 4. vertical section of the ovary;
f. 5. transverse section of the same; *f.* 6. ovule, which should
have been drawn in the inverse position, as in *f.* 4.

A fine specimen, also in flower only, from Heudelot's Senegambian collection, is considered by Dr. Planchon as a second
species, but I do not see any character to distinguish it by.
Dr. Hooker observes that in *Rhaphiolepis* the base of the corolla

(so called) is inserted on the apex of the pedicel, but not immediately within the calyx, whose true nature is probably that of an involucre, as suggested by Brown.

Olacineæ, considering the smallness of the Order, rather abound in Western Tropical Africa, there being besides the above, four other species known belonging to other genera, viz. : *Ximenia Americana,* L., common to both the New and the Old World ; *Groutia celtidifolia,* Guill. et Perr., a Senegal plant closely allied to an Abyssinian congener ; *Icacina Senegalensis,* Juss., and another undescribed species of *Icacina,* gathered by Heudelot in Senegambia.—(G. B.)

XXXII. VINIFERÆ.

1. Cissus *cæsia,* Afz. *in DC. Prod.* 1. *p.* 628 ; caule glaberrimo terete glauco, foliis petiolatis late cordatis obscure angulatis subacutis ciliato-denticulatis super glabratis subter puberulis reticulatim venosis, pedunculis gracilibus superne pedicellisque elongatis puberulis paucifloris, floribus parvis, drupis late obovatis. — Guinea, *Afzelius;* Sierra Leone, *Vogel, Don,* (Country Grapes.)

Rami crassitie pennæ olorinæ, pulchre glauci. *Folia* 5 unc. longa, super pube fulva sub lente subtilissime conspersa, luride viridi-fusca, subter pube tenui grisea ornata ; petiolo puberulo, sub ¾ unc. longo. *Cirrhi* validi, bi-multifidi. *Pedunculus* 2-pollicaris, di- trichotomus, gracilis, pedicellis ¼ unc. longis, floribus triplo longioribus; alabastris breviter cylindraceis, obtusis ; petalis 4, solutis.

This differs from *C. rufescens* in the longer and more slender peduncles, in the fewer, smaller and less crowded flowers, and in the glaucous stems. The nerves of the leaf are not rufescent in any of Vogel's specimens : those of Don's Herb. are between the Senegalese and Dr. Vogel's in this character.

2. Cissus *argute,* Hook. fil. ; glaberrima, caule subgracili obscure tetragono basi subpolygono, foliis sublonge petiolatis ovatis acuminatis basi profunde cordatis argute serratis dentibus erectis, stipulis late ovatis, cirrhis gracilibus, pedun-

culis petiolo subæquilongis plerisque trichotomis puberulis,
cymis subumbellatis 8-10-floris.—On the Quorra, at Ibu,
Vogel.

β. Foliis paulo majoribus magis angustatis, floribus majoribus,
petalis non cohærentibus.—On the Quorra, *Vogel.*

Rami fusco-virides, crassitie pennæ anserinæ. *Folia* utrinque
glaberrima, siccitate rugulosa et crispata, 2-3 unc. (in β. 4 unc.)
longa; 2½ lata, basi profunde cordata, pleraque late ovata,
abrupte acuminata, superiora angustiora; petiolo ½-¾ longit.
folii. *Stipulæ* late ovatæ. *Pedunculi* subgraciles, puberuli;
ramis pedicellisque pube fulva tectis. *Calyx* cyathiformis;
margine integro. *Petala* 1 lin. longa, apice cohærentia v.
soluta; stamina basi glandulis aucta.

Easily to be recognized by the crisp, smooth, and sharply-
toothed leaves.

3. Cissus *uvifera*, Afz.? *DC. Prod.* 1. *p.* 628; glaberrima,
caule valido obscure tetragono, foliis sublonge petiolatis coriaceis late ovatis acuminatis basi obtuse cordatis retusisve
remote serratis subter nervosis, cirrhis validis, baccis longe
pedicellatis globosis 1-locularibus 1-spermis, petalis 4 apice
cohærentibus.—An *C. populnea*, Guill. et Perr. *Fl. Seneg. p.*
134?—Sierra Leone and Fernando Po, *Vogel.*

Caules validi, diam. pennæ olorinæ, 4-goni v. obscure polygoni,
angulis siccitate nunc tuberculatis, striatis sulcatisve. *Folia*
3-5 unc. longa, 2-3½ lata, basi truncata v. cordata, sinu latissimo; petiolo ½-2-pollicari subpeltatim affixo. *Cymæ* paucifloræ? ramis puberulis; pedicellis ½-uncialibus.

Specimens very imperfect, and differing from the descriptions
of *C. uvifera* in the leaves not being entire.

4. Cissus *petiolata*, Hook. fil.; glaberrima, caule suberoso obtuse tetragono striato glaberrimo, foliis longissime petiolatis
pallide viridibus subcoriaceis opacis late ovatis obtusis basi
5-nerviis latissime cordatis obscure sinuato-denticulatis, pedunculis ramisque cymæ elongatis dichotomis paucifloris, stylo
elongato, baccis majusculis oblongis.—Aguapim, *Vogel.*

Caules siccitate pallide flavi, fragiles, suberosi, striati, ramulique

profunde 4-sulcati, angulis obtusis. *Folia* 3-4 unc. longa,
ovata, lobis basi rotundatis, utrinque fusco-viridia, opaca,
nervis non prominulis ; petiolo foliis longiore, gracili ; stipulis caducis. *Racemi* petiolo subæquilongi, trichotome ramosi ; ramis glaberrimis, gracilibus, fructiferis divaricatis.
Baccæ paucæ, virides, ½ unc. longæ.

A most distinct species, though in an imperfect state. It is
allied to an Abyssinian plant, and also to the Cape *C. fragilis*,
E. Mey., but the remarkable length of the petioles and pedicels
will at once distinguish this.

5. Cissus *producta*, Afz.? *DC. Prod.* 1. *p.* 629—Sierra Leone,
Don.

Folia integra, lanceolata, acuminata, obscure serrata, glaberrima,
basi rotundata, baccis obovatis.

Specimens too imperfect for determination.

6. Cissus *glaucophylla*, Hook. fil.; caule erecto? glaberrimo
tereti lævi subglauco, foliis longe petiolatis late ovatis acuminatis profunde cordatis lobis rotundatis integerrimis v. sinuato-
dentatis coriaceis super lævibus adultis cœruleo-glaucis,
nervis subter prominulis obscure puberulis, stipulis late ovato-
rotundatis, cirrhis nullis, paniculis terminalibus trichotomis
multifloris, petalis 4 cohærentibus.—Fernando Po, *Vogel.*

Ramuli obscure 4-goni. *Folia* late ovato-cordata v. suborbi-
culata, subpeltatim petiolata, 3-5 unc. longa, 2-3½ lata, in
acumen elongatum producta, super lævia, juniora atro-fusca,
seniora pulchre glauca, subter castanea, opaca ; petiolo folio
breviore. *Racemi* 2-3-unciales, compositi, subcymosi, multiflori, pedicellis ramisque puberulis. *Flores* umbellati, parvi,
1 lin. longi. *Calyx* cyathiformis, depressus. *Petala* 4, ca-
lyptratim cohærentia, basi calyce latiora. *Stamina* 4. *Stylus*
gracilis, breviusculus.

7. Cissus *tetraptera*, Hook. fil.; glaberrima, ramis crassis carnosis tetrapteris ad nodos constrictis striatis, foliis breve
petiolatis late reniformi-rotundatis profunde cordatis 5-lobis
carnosis utrinque sub lente (e rhaphidibus perplurimis) creberrime punctato-striatis subargute serratis, cirrhis crassis
elongatis, pedunculo terminali brevi apice umbellato, pedicellis

elongatis, floribus majusculis 4-petalis.—Elephant's Bay, (S. of the Line), *Dr. Curror.*

Rami crassitie digiti minoris, striati, valde carnosi, tetraquetri, angulis alatis, alis undulatis, pallide virides. *Stipulæ* transverse elongatæ, breves. *Petiolus* 3-4-uncialis. *Folia* 2 unc. lata, crassa, carnosa, siccitate viridia, subtiliter reticulata, subpellucida, fasciculis rhaphidium valde conspicuis. *Pedunculus* oppositifolius, pollicaris, striatus, ¼ unc. diametro, apice 5-radiatus. *Rami* ¾ unc. longi, umbellulam sub 7-florem gerentes; pedicellis 4 lin. longis. *Flores* majusculi. *Calyx* brevis. *Petala* breviter ovato-oblonga; staminibus 4; stylo cylindraceo stigmate simplici.

Possibly a young branch of *C. Currori,* but in that plant I find no trace of an alate stem; the panicle is different, as are the toothed lobes of the leaves. The raphides are in both so conspicuous as to cause a projection of the cuticle over the crystals, and give the semblance to the whole plant of being pubescent.

8. Cissus *Leonensis,* Hook. fil.; caule robusto tereti puberulo et setoso, foliis late orbiculatis cordatis palmatim 5-lobis super pubescentibus subter rufo-lanatis, lobis ovato-oblongis acuminatis argute serratis, cirrhis multifidis, panicula vage decomposite ramosa, ramis alternis, corolla pentapetata calyptræformi. —Sierra Leone, *Vogel.*

Caulis herbaceus, crassus, teres, fuscus, pubescens, setisque patentibus sparsis instructus. *Folia* 8 unc. lata, submembranacea, late cordata, supra medium lobata, super luride fusca, sub lente subarachnoidea et puberula, subter lana tenui rufa subappressa instructa; nervis 5, validis, radiantibus, pubescentibus; petiolo 4 unc. longo, pubescenti; stipulis deciduis. *Panicula* brevis, pubescens, 2 unc. longus; pedicellis brevissimis. *Flores* parvi, fere lineam longi, globosi. *Calyx* cyathiformis, margine integro submembranaceo. *Petala* breviter ovata, crassiuscula, apice arcte cohærentia. *Stamina* 3. *Ovarium* depressum, pentagonum, angulis sulcatis, filamenta foventibus; stylo brevi, conico, crasso, truncato, 10-sulcato, apice depresso v. subinfundibuliformi.

This should probably be referred to *Vitis*, from the decomposed panicle, with always alternate branches, characters which would afford better characters for distinguishing these genera than those now in use.

9. Cissus *Currori*, Hook. fil.; glaberrima, foliis amplis 3-foliolatis, petiolo valido, foliolis petiolulatis ovatis obtusis basi cordatis grosse et obtuse subduplicato-crenatis carnosis punctis prominulis (raphidibus) notatis, stipulis ovatis acutis, panicula effusa, pedunculo elongato ramis dichotomis divaricatis, floribus majusculis, petalis 4 non cohærentibus.—Elephant's Bay, (S. of the Line), *Dr. Curror.*

Species omnium e sectione *trifoliolata* ornatissimus, arborescens, ramosus, carnosus. *Rami* crassitie digitis majoris, profunde striati, glabrati. *Stipulæ* ramo angustiores, ½ unc. longæ. *Petiolus* 3-4-uncialis, striatus. *Foliola* 6 unc. longa, 4 lata, plana, siccitate pallide flavo-viridia, pellucida, basi cordata, sinu angusto, petiolulo folioli intermedii fere unciali. *Pedunculi* axillares v. terminales, 3-4 unc. longi, erecti, stricti, petiolo graciliores, dichotome ramosi; ramis divaricatis, pluries divisis; pedicellis brevibus crassis. *Cirrhi* nulli? *Flores* ⅓ unc. longi. *Calyx* parvus, cyathiformis. *Corolla* calyce latior; petalis breviter ovato-oblongis, obtusis. *Stamina* 4? *Ovarium* depressum, latum, 4-gonum, profunde 4-sulcatum, stylo valido, subelongato, stigmate simplici. *Bacca* junior ovoidea.

A noble species, to which I have attached the name of its lamented discoverer. It is described by him as a much branched and very succulent tree.

10. Cissus *Ibuensis*, Hook. fil.; parce pubescens, caule gracili tereti obscure angulato apice subtomentoso, foliis breve petiolatis 3-foliolatis, foliolis petiolulatis elliptico-ovatis ovato-lanceolatisve acuminatis argute serrato-dentatis utrinque sed subter præcipue puberulis, cirrhis filiformibus divisis, pedunculis elongatis alterne ramosis pubescentibus, ramis pedicellisque brevissimis, floribus parvis, petalis 4 apice demum liberis, obovatis.—Ibu and Nun River, *Vogel.*

Species gracilis, scandens. *Caules* crassitie pennæ corvinæ,

266 FLORA NIGRITIANA.

pube tenui superne densiore sparsa. *Stipulæ* parvæ, ovatæ.
Petiolus uncialis. *Foliola* patula, intermedio paulo longiore,
1½-2 unc. longa, ½-¾ lata, basi rotundata, utrinque fusca,
opaca, petiolulo ¼-½ unc. longo. *Pedunculus* 2-4-uncialis,
tenuis, ramis alternis divaricatis. *Calyx* breviter cyathiformis.
Petala 4, ovata, apice non cohærentia. *Stamina* 4. *Stylus*
brevissimus. Inflorescence imperfect in these specimens. Closely allied to
Vitis carnosa, Wall., but the whole plant is less hairy.

11. Cissus *tenuicaulis*, Hook. fil.; caule gracili striato parce
piloso, foliis longe petiolatis 5-foliolatis, foliolis petiolulatis intermedio majore lateralibus geminis lanceolatis acuminatis
basi rotundatis grosse serratis membranaceis utrinque pilosis,
cirrhis elongatis gracillimis, racemis folio æquilongis, fructiferis dichotome ramosis, ovario disco carnoso immerso,
baccis late pyriformibus.—Sierra Leone, *Vogel.*
Caulis crassitie pennæ passerinæ, glabratus v. superne precipue
parce pilosus, obscure striatus. *Stipulæ* late ovatæ, obtusæ.
Petiolus gracilis, glaberrimis, 2-pollicaris. *Foliola* 1½-2 unc.
longa, super fusco-viridia, subter pallidiora, utrinque pilis
albidis sparsa, petiolulo subhispido-pubescente. *Flores* minimi. *Calyx* breviter cyathiformis. *Petala* 4, apice cohærentia. *Stamina* 4. *Discus* urceolaris. *Stylus* crassus, brevis;
stigmate capitato. *Baccæ* sub 3-lin. longæ.
Except in the pubescent leaves, this hardly differs from
C. Japonica. It is also very near the E. Indian *C. capreolata*,
which is however a densely pubescent plant.

12. Cissus *membranacea*, Hook. fil.; glaberrima, flaccida, caule
gracillimo tereti striato, foliis petiolatis trifoliolatis, foliolis
petiolulatis lateralibus longioribus ovatis ovato-lanceolatisve
acuminatis basi valde inæqualibus serrato-dentatis membranaceis, cirrhis gracillimis, pedunculis trichotomis paucifloris
fructiferis elongatis petiolo æquilongis, floribus minimis. —
Among *Vogel's* plants, without the precise station.
Caulis diametr. pennæ passerinæ. *Petioli* 2-3 unc. longi; stipulis parvis, membranaceis, late ovatis, obtusis. *Foliola* 2-3
unc. longa, 1-1½ lata, subpellucida, summa obscure puberula,

intermedio plerumque basi obtuso v. in petiolulum angustato, lateralibus geminis basi valde inæqualibus, latere exteriore deorsum angustato, interiore rotundato, v. in lobum producto; petiolulo intermedio elongato, nunc pollicari. *Flores* ut in *C. tenuicauli.*

The oblique bases of the more regularly and conspicuously serrated leaflets, and smaller flowering panicle, will at once distinguish this from the *C. Japonicus.*

13. Cissus *Vogelii,* Hook. fil.; setoso-pubescens, caule herbaceo tereti crassiusculo profunde striato, stipulis orbiculari-ovatis acuminatis, petiolis pubescentibus elongatis 5-foliolatis, foliolis obovato-lanceolatis in petiolulum angustatis acuminatis dentatis membranaceis super glaberrimis subter ad nervos præcipue pubescentibus, panicula effusa axillari ampla alterne et dichotome ramosa, ramis multifloris, floribus pedicellatis cylindraceis pubescentibus, petalis linearibus apice fornicatis dorso setis glanduloso-capitatis ornatis.—Fernando Po, on the sea shore, *Vogel.*

Caules prostrati, ramosissimi, sarmentosi. *Rami* herbacei, pallide flavi, profunde sulcati, pubescentes et setis sparsis ornati. *Stipulæ* majusculæ, late ovato-rotundatæ, acuminatæ. *Petioli* 4-6 unc. longi, graciles, pubescentes. *Foliola* omnia plana, 2-3 unc. longa, 1-1½ lata, basi in petiolulum ¼ unc. longum pubescentem angustata, super glabrata v. glaberrima, subter puberula, nervis discoloribus rufo-pubescentibus. *Panicula* composita 6-8 unc. lata ; pedunculo stricto 4 unc. longo, pubescente, setoso et striato, ad axillas bracteolato, bracteolis oblongis ligulatisve, pedicellis inæquilongis. *Calyx* breviter cyathiformis, pubescens, sub 4-lobus. *Petala* erecta, pubescentia, linearia. *Ovarium* oblongum, profunde 4-sulcatum ; stylo elongato ; stigmate simplici.

In the unusual form of the flower this is related to the *C. cymosa,* but the stipitate glands, or glandular hairs of the petals, form a prominent and beautiful diagnostic character, and the whole plant is much less pubescent.

It is allied to the Abyssinian *C. mollis,* Steud., but the leaflets are smaller and more delicate, less pubescent and broader; the stipules smaller; panicles much larger, and of a different form. All these three species have the four or five glands of the disc firmly cohering with the ovarium, which thus appears deeply 4-grooved.

14. Cissus *cymosa,* Schum. et Thonn. *Beskr. p.* 82.—Guinea, *Thonning;* Accra, *Vogel.*

Flores clavati.　*Petala* 4, erecta, apice fornicata, cucullata, dorso gibboso-incrassata, pubescentia.

The predominance of this genus in Western Africa is highly indicative of its humid atmosphere and jungly coast. Four other species are described as inhabiting the same country : *C. rufescens,* Guill. et Perr., of Senegambia (very closely allied to *C. cæsia*); *C. quadrangularis,* Wall. (*C. triandrus,* Schum.), a plant also common to Arabia and the continent of India; *C. gracilis,* Guill. et Perr., and *C. bifida,* Schum. et Thonn.

1. Leea *Guineensis,* Don.—Sierra Leone, Cape Palmas, St. Thomas, and Fernando Po, *Vogel, Don.*

XXXIII. Cochlospermeæ.*

1. Cochlospermum *Planchoni,* Hook. fil. ; caule subarborescente, ramis puberulis striatis foliosis, ramulis petiolis foliisque subter subvelutino-tomentosis, foliis late orbiculari-reniformibus profunde cordatis 5-lobis, lobis rotundatis obtusis

---

* *See* Lond. Journ. Bot. 6, p. 294, where Dr. Planchon has, with great sagacity, pointed out the affinities of *Cochlospermum* and *Amoreuxia,* and I perfectly agree with him in their separation from *Ternstrœmiaceæ,* although I cannot subscribe to all his speculations on the grouping of this and several of the following Orders. Their real relative positions appear to me to be far from being, as yet, satisfactorily ascertained, I therefore leave them, for the present, nearly in the order in which De Candolle had placed them, however convinced I am that several of the smaller groups might be advantageously united as tribes of larger Orders.—(G. B.)

obscure sinuato-dentatis, floribus in ramulos ultimos axillaribus, sepalis 5 inæqualibus rotundatis pubescentibus 2 exterioribus minoribus.—Quorra River, in savannahs, *Vogel.*
Arbuscula 6-pedalis. *Rami* crassitie pennæ olorinæ, pube grisea. *Petioli* ¾-1 unc. longi. *Folia* coriacea, 2½-3 unc. longa, 3-4 lata, super atro-fusca, (siccitate) nitida, subter pube densa grisea; venis primariis palmatim radiatis, venulis obscuris. *Alabastra* ½ unc. longa. *Flores* lutei.
This and the *C. tinctorium,* A. Rich., are the only W. African species known to me.

XXXIV. GERANIACEÆ.

Though S. Africa may be considered as the head-quarters of *Geraniaceæ,* and the N. shores of the same continent are not deficient in species, yet one species only exists within the Tropic, the *Monsonia Senegalensis,* Guill. et Perr.

XXXV. OXALIDEÆ.

1. Biophytum *sensitivum,* D.C. *Prod.* 1, *p.* 690.—On the Quorra at Attah, *Vogel.*
An abundant E. and W. Indian plant.
It is remarkable that no species of *Oxalis,* not even the elsewhere ubiquitous *O. corniculata,* or *O. stricta,* appears to be found on the West Intertropical African coast.
A species, apparently of *Averrhoa,* is in the Hookerian Herbarium, collected in Senegambia by Heudelot.

XXXVI. ZYGOPHYLLEÆ.

1. Kallstrœmia *minor,* Hook. fil.; subsericeo-pilosa pubescensve, foliolis 3-jugis oblique ovato-oblongis obtusis v. mucronulatis, pedicellis petiolo brevioribus, floribus parvis, coccis dorso muricatis 1-locularibus 1-spermis.—Tribulus pubescens, *G. Don, Gard. Dict.* 1, *p.* 669.—Cape Coast, *Don, Vogel.*
A. *Tribulo cistoidi* differt capsula 10-cocca, a *Kallstrœmia maxima* statura, foliolisque paucijugis.

I have separated this from *K. maxima*, on the grounds of its constantly smaller size, and the few leaflets.

1. Tribulus *cistoides*, Linn.—Sierra Leone? *Don.*

This is one of the very few plants common to the West Indies and Pacific Islands, being found in Oahu. It varies much in the size of the fruit, which in these W. African specimens is particularly large.

The *Tribulus terrestris* is a native both of Senegal and Guinea.

1. Zygophyllum *simplex*, Linn.—Benguela, *Dr. Curror.*

A plant common to the shores of the Red Sea, the banks of the Nile, and the Cape de Verd Islands, but not that I am aware of to any other part of the W. coast of Africa, except Benguela.

Fagonia Arabica, a native of Arabia, as well as of N. Eastern Africa, is also a Senegal plant.

1. Balanites *Ægyptiaca*,* Del. *Fl. Æg. p. 77, t. 28, f. 1.*— Senegal, Sierra Leone, *Whitfield.*

Also a native of Abyssinia, and a variety of it, by some considered as a distinct species, extends to the dry plains of Bengal.

XXXVII. ZANTHOXYLEÆ.

1. Zanthoxylum *rubescens*, Planch. in Herb. Hook.; ramis aculeatis, foliolis circa 11 suboppositis ovali-oblongis longe acuminatis basi acutis crebre pellucido-punctatis petiolo aculeato supra canaliculato, floribus diclinis, masculis paniculatis parvis tetrameris.—Cape Coast, *Vogel.*

Frutex orgyalis, ramulis rubentibus, aculeis validis conicis recte reflexis v. recurvis. *Foliorum petiolus* 8-9-pollicaris. *Foliola* 2-3 poll. longa, pollicem lata, glabra, membranacea. *Panicula* ampla, bis terve racemoso-ramosa, ebracteata. *Flores* in

* Brown removes *Balanites* from *Zygophylleæ*, but as I am not aware that he has published his views of its real affinities, and as, at any rate, it is in some measure related to *Zygophylleæ*, I have left it here at the end of the Order.—(G. B.)

specimine omnes abortu masculi, parvi, albidi, per 2-3 e tuberculis secus ramos sessilibus v. pedunculatis orti, pedicello lineam longo fulti. *Sepala* 4, minima, orbiculata. *Petala* 4, ovato-oblonga, lineam longa. *Stamina* 4, petalis subæquilonga. *Ovarii* rudimentum carnosum. This agrees in so many respects with the description given in the Flora Senegambiæ of the *Z. Leprieurii*, which is drawn up from imperfect female plants, that our plant might be taken for the male of the same species, were it not that the number of parts of the flower appear to be constantly quaternary, not quinary.

2. Zanthoxylum? a very bad specimen, affording no materials to distinguish it from the American *Z. pterota.*—Cape Palmas, *Ansell.*

The *Z. Senegalensis,* D.C., and the above-mentioned *Z. Leprieurii,* Guill. et Perr., both from Senegal, are the only other W. African species known; the *Z. polygamum* of Schum. and Thonn. being probably the same as the *Z. Senegalensis.*

XXXVIII. Simarubeæ.

1. Brucea *paniculata,* Lam. *Dict.* 1, *p.* 472.—Sierra Leone, *Don.*

The only other W. African species of this Order known, is the *Hannoa undulata,* Planch. (*Simaba? undulata,* Guill. et Perr.) from Senegal.

XXXIX. Ochnaceæ.

1. Ochna *dubia,* Guill. et Perr. *Fl. Seneg.* 1. *p.* 137. *t.* 35.—Sierra Leone, *Don.*

Another W. African *Ochna* has been published, the *O. multiflora,* DC., from Sierra Leone.

1. Gomphia *glaberrima,* Pal. Beauv. *Fl. Ow. et Ben.* 2. *p.* 22. *t.* 71.—Oware, *Beauvois;* Sierra Leone, *Whitfield, Don.*

The slender panicles, and smaller, broad, nearly globose carpels, are the best characteristics of this species. The nerves,

which are defined on the leaf, form a more or less oblique angle
with the costa.

2. Gomphia *reticulata*, Pal. Beauv. *l. c. t.* 72.—Benin, *Beauvois;* Sierra Leone, *Forbes, Vogel.*

It is not easy to distinguish this from *G. glaberrima*, without
the flowers. The whole plant is more robust, much darker in
colour when dry, the leaves have stronger and better defined
nerves, forming a right-angle with the costa. The foliage
varies much in breadth, and is more or less (but never sharply)
serrated. Judging from the buds, the flowers are smaller.
Panicles simple or branched in both. Neither Beauvois
figures or descriptions assure me that these two are the plants
he describes, nor that the latter are distinct from one another.

3. Gomphia *Vogelii*, Hook. fil. ; foliis oblongo-lanceolatis acuminatis basi in petiolum brevem angustatis obscure sinuato-
dentatis coriaceis utrinque lucidis venis e costa ascendentibus, panicula robusta ramis infimis elongatis ascendentibus,
calyce fructifero majusculo, carpellis 3-4 globosis calyce inclusis lobis dimidio brevioribus.—Grand Bassa, *Vogel.*

Frutex ramosus. *Rami* teretes ; cortice pallide brunneo. *Folia*
4-6 unc. longa, 1½-2 lata, siccitate utrinque pallide flavo-
brunnea, costa valida; venis primariis ½-¾ unc. distantibus,
prominulis; venulis creberrimis, parallelis. *Panicula* 3 unc.
longa ; pedicellis calyce æquilongis longioribusve. *Calyx*
fructifer majusculus ; lobis non reflexis, suberectis, subcoriaceis, fere ½ unc. longis. *Carpella* magnitudine grani
piperis.

This I distinguish from *G. reticulata* by its more coriaceous leaves, remote ascending veins, which run obliquely, and
are united by venules of extreme tenuity and regularity.
The much larger, broader, calycine segments distinguish it
from *G. glaberrima*.

4. Gomphia *flava*, Schum. et Thonn. *Beskr. p.* 216. (fide *Planch.*)
—G. *macrocarpa*, Hook. fil. MSS. *Planch. in Lond. Journ.
Bot. v.* 6, *p.* 2 ; foliis anguste elliptico-oblongis acuminatis
in petiolum angustatis argute serratis coriaceis planis reticulatim venosis, venis primariis subremotis ascendentibus pani-

cula basi ramosa, carpellis calyce reflexo longioribus teretibus
utrinque obtusis.—*Fernando Po, Vogel.*
Rami teretes. *Folia* subnitida, 6-pollicaria, 2 unc.
lata, super fusco-viridia, subter pallidiora, utrinque reticulatim venosa;
venis e costa ascendentibus, deinde margine parallelis, remotis, vix prominulis; venulis transversis conspicuis reticulata; supra basin ad apicem serrata; petiolo 1¼ unc. longo.
Panicula terminalis; florifera 4 unc. longa; fructifera elongata; pedicellis ¼-⅓ unc. longis. *Calycis* lobi pedicellum
æquantes v. breviores, post anthesin reflexi. *Carpella* 1-3,
⅓ unc. longa, teretia, breviter cylindracea, utrinque rotundata.

Readily to be distinguished by the sharply serrated leaves,
which are very coriaceous, and especially by the large cylindrical carpels. The veins and venules are not distant from one
another, as in the following, the former ascend from the costa,
and on approaching the margin run parallel to it for a considerable distance.

5. Gomphia *Turneræ*, Hook. fil.; foliis elliptico-oblongis v.
anguste lineari-lanceolatis in apicem longe acuminatis et in
petiolum angustatis, valde coriaceis, obscure crenatis integerrimisve, utrinque lucidis lævibus, venis inconspicuis, panicula elongata ramosa, ramis patentibus gracilibus, floribus
ternis subfasciculatisve, pedicellis calyci æquilongis.—Sierra
Leone, *Miss Turner, Don, Vogel.*
Rami teretes, subgraciles; cortice pallido. *Folia* elongata,
utrinque fusco-castanea v. viridia, nitida, subter pallidiora, 4-6
unc. longa, 1¼-2 lata, in petiolum ¼ unc. longum angustata,
apice in acumen gracile producta; costa valida; venis primariis subremotis, ascendentibus; venulis valde inconspicuis;
folia juniora margine obscure crenulata. *Panicula* 6-8 unc.
longa, ramis elongatis gracilibus. *Flores* bini, terni v. subfasciculati, nutantes, ¾ unc. lati; pedicello ¼-⅓ unc. longo.
Petala late obovata, orbiculata, subunguiculata, intense
lutea.

The flowers of this plant entirely resemble those figured by
Beauvois as *G. reticulata,* from which the compound panicle,

T

very narrow, coriaceous, and plane (never undulate) leaves, at once distinguish the present species.

6. Gomphia *calophylla,* Hook. fil. ; foliis obovato-lanceolatis basi gradatim angustatis, abrupte acuminatis rarius apice angustatis marginibus undulatis venis parallelis confertissimis creberrime striatis subnitidis, racemo laterali foliis subæquilongo v. breviore, pedunculo compresso v. ancipiti, pedicellis gracilibus, laciniis calycinis post anthesin patulis, carpellis subglobosis.—Sierra Leone, *Don, Vogel;* Cape Coast and Fernando Po, *Vogel.*

Rami teretes ; cortice cinereo, ramulis compressis. *Folia* 5-7 unc. longa 1½-2½ lata, basi ad petiolum brevem cuneata, deinde gradatim dilatata, apicem versus rotundata v. angustata et acuminata, margine undulata v. subcrispato-incrassata, venis transversis perplurimis parallelis ; stipulis brevibus, ovato-triangularibus. *Racemi* basi nudi, supra medium densiflori ; pedunculo compresso ancipiti ; pedicellis solitariis binis ternisve gracilibus ¼ unc. longis. *Calycis* laciniæ post anthesin patentes, lineari-oblongæ. *Carpella* parva, globosa, laciniis calycinis breviora.

A very handsome species, of which the flower is unknown to me. The nervation of the foliage exactly resembles that of *Elvasia Hostmannia,* Planch.

7. Gomphia *affinis,* Hook. fil. ; foliis oblongo-lanceolatis acuminatis basi angustatis submembranaceis nitidis majoribus undulato-crispatis integerrimis, nervis parallelis confertissimis transversis, panicula terminali, ramis gracilibus angulatis, pedicellis subelongatis, carpellis calyce longioribus globosis.— Fernando Po, *Vogel.*

Rami teretes ; cortice pallido subrugoso striato. *Folia* apices versus ramulorum, 3-4 unc. longa, 1-1½ lata, submembranacea, utrinque nitida. *Panicula* 2-3 unc. longa ; ramis strictis paucifloris gracilibus ; pedicellis fructiferis ⅔ unc. longis, gradatim incrassatis. *Calycis* laciniæ parvæ, pedicellis ⅓ breviores.

Allied to the *G. calophylla* in the nervation of the leaves, which are, however, smaller, more membranous and glossy, and narrower above the middle. The inflorescence, too, is paniculate, not racemose, the peduncle angular and not so compressed, the calycine segments smaller, and the carpels larger.

XL. Rhamneæ.

1. Zizyphus *Baclei*, DC. *Prod.* 2. *p.* 20.—Guill. et Perr. *Fl. Seneg. t.* 37.—Attah and Quorra, *Vogel;* Senegal. *Z. orthacantha*, DC., which is possibly a variety of *Z. jujuba*, is a native of Senegambia. The true *Z. jujuba* is found at Mozambique, and thence eastward through the Peninsula of India to the Indian Archipelago.

1. Ventilago *denticulata*, Willd., *DC. Prod.* 2. *p.* 38.—V. maderaspatana β, *W. et Arn. Prod. Fl. Pen. Ind. Or.* 1. *p.* 164.—Celastrus diffusus, *G. Don, Gard. Dict.* 2. *p.* 6.— St. Thomas, *Don.*

The disc of the flower is smooth, or only very slightly hairy, in other respects I am unable to distinguish these specimens from some of the forms from the Indian Peninsula. I have not, however, seen the fruit.

XLI. Chailletiaceæ.*

1. Chailletia *toxicaria*, Don, *DC. Prod.* 2. *p.* 57; foliis petiolatis ovato-oblongis v. oblongo-sublanceolatis obtuse acuminatis basi acutis rotundatisve subcoriaceis glabris, cymulis contractis raro foliiferis in pedunculo axillari vix ramoso solitariis paucisve, petalis bifidis, stylo breviter trifido, drupis canescentibus.—Sierra Leone, *Don, Vogel;* and, apparently the same species, Senegambia, *Heudelot.*

Frutex dumosus, inflorescentia partibusque novellis canescentibus, cæterum glaber. *Folia* 3-4-pollicaria, sæpe obliqua, acumine brevi obtuso v. retuso, margine sæpius undulata,

* By G. Bentham.

T 2

adulta utrinque glabra, supra opaca, subtus siccitate sub-
rubentia, ad axillas venarum foveolata, reticulato-venulosa,
petiolo crassiusculo 3-4 lin. longo fulta. *Stipulæ* minutæ,
caducæ. *Pedunculi* in axillis superioribus solitarii, nunc
breves cymulam unicam ferentes, nunc 1-3-pollicares, cymulis
pluribus sessilibus v. breviter pedicellatis, nudis v. bractea
foliacea fultis; cymula infima sæpe ex ima basi pedunculi
orta; omnes in glomerulum contractæ v. rarius leviter evolutæ, cano-tomentosæ. *Pedicelli* florentes vix lineam longi,
fructiferi longiores, incrassati. *Sepala* 5, ovata, extus tomentosa, fere 1½ lin. longa, æstivatione imbricata. *Petala* oblongo-linearia, calyce paullo longiora, apice breviter bifida,
extus puberula, intus glabra et linea elevata a sinu loborum
decurrente carinata. *Stamina* petalis æquilonga. *Squamæ*
hypogynæ petalis oppositæ, breves, emarginatæ, tomentosæ,
inter se liberæ sed continuæ. *Ovarium* dense tomentosum,
conicum, triloculare. *Styli* glabri fere ad apicem coaliti v.
rarius demum ad medium soluti. *Ovula* in loculis gemina.
Drupa ovoidea v. subglobosa, pollicem longa, obtusa v. acuminata, extus tomentosa, abortu monosperma v. rarius disperma.

One of G. Don's Sierra Leone specimens has narrower leaves,
and may possibly be the plant described by D. Don under the
name of *C. erecta.* If so, it would appear not to be specifically distinct from *C. toxicaria.* The form of the fruit is very
variable in the dry specimens, owing perhaps to its being
gathered at different stages of maturity.

2. Chailletia *affinis*, Planch. *in Herb. Hook.*; foliis longiuscule
petiolatis ovali-ellipticis obovatisve obtuse acuminatis basi
acutis rotundatisve subcoriaceis glabris, cymis laxiusculis
in pedunculo axillari libero v. petiolo adnato solitariis paucisve,
drupa glabrata.—Fernando Po, *Vogel.*

Closely resembling *C. toxicaria* in the colour, venation and
consistence of the leaves; it appears, however, to differ in the
longer petioles, broader leaves; looser inflorescence and smooth
fruits, and from some remains of petals and stamens, the flowers
appear to have been larger. There are, however, neither per-

fect flowers nor mature fruits to admit of determining whether
it be really specifically distinct, or a mere variety of *C. toxicaria.*

3. Chailletia *subcordata,* Hook. fil.; foliis breviter petiolatis
late-ovatis vix acuminatis basi subcordatis glabris v. ad costas
puberulis, cymis in pedunculo brevi axillari libero solitariis
multifloris, ramulis evolutis, petalis profunde bifidis, stylo
breviter trifido.—Fernando Po, *Vogel.*

Frutex ramosus, orgyalis et altior, ramulis tomentellis. *Folia*
3-4 poll. longa, 2½-3 poll. lata, nunc obtusissima, nunc
acumine brevi acutiusculo terminata, margine integerrima v.
obsolete sinuata, basi late rotundata obtusissima v. ad petiolum
sæpius cordato-emarginata, novella tomentella, adulta glabrata,
rigide membranacea v. subcoriacea, venis primariis subtus
puberulis, axillis venarum haud foveolatis, sed glandulæ adsunt scutelliformes hinc inde per paginam inferiorem sparsæ.
Petioli 2 lin. longi. *Stipulæ* angustæ, acutæ, petiolo breviores. *Pedunculi* petiolo longiores. *Cymæ* juniores densæ,
mox dichotome evolutæ, ramulis vulgo 4 demum semipollicaribus. *Pedicelli* vix semilineam longi, bracteola parva subtensi, sub flore articulati et infra articulationem post flores
delapsos persistunt. *Flores* ut videtur exsiccatione cadusis-
simi, alabastra juniora tantum in speciminibus supersunt,
globosa, tomentosa, vix lineam diametro. His *sepala* æstivatione imbricata; *petala* brevia, lata et fere bipartita; ovarium et stylus *C. toxicariæ.*

4. Chailletia *oblonga,* Hook. fil.; foliis petiolatis oblongis acuminatis basi acutis ramulisque glabris, cymis laxis in pedunculo brevi libero axillaribus v. terminalibus subpaniculatis,
pedicellis calyce sublongioribus, sepalis lanceolatis, petalis
calyce dimidio longioribus stamina subæquantibus, stylo elongato apice breviter bifido, drupa obovali-oblonga tomentosa.
—Fernando Po, *Vogel, Ansell.*

Arbor, ramulis tenuibus foliisque glabris v. novellis vix tomentellis. *Folia* ramulorum florentium 2-3 poll. longa, 1-1½
poll. lata; inferiora tamen et ramorum sterilium duplo ma-

jora; etiam in sicco virentia, apice in acumen breve latum
producta, basi acuta, petiolo 1-2-lineari fulta; foveolæ paginæ
inferioris omnino deesse videntur. *Stipulæ* minutæ. *Cymæ*
leviter tomentellæ, graciles et laxe dichotomæ, folio tamen multo
breviores. *Bracteæ* minutæ v. obsoletæ. *Pedicelli* 2-2½ lin.
longi, supra medium articulati. *Sepala* angusta, 1½ lin. longa,
extus tomentosa, æstivatione valde imbricata. *Petala* glabra,
anguste linearia, fere 3 lin. longa, ad duas tertias integra,
dein biloba, lobis vix divergentibus, sinu acuto intus carinato-
prominente. *Stamina* petalis vix longiora. *Glandulæ* hypogynæ basi brevissime connatæ. *Ovarium* breve, tomentosum,
biloculare. *Stylus* staminibus longior, glaber v. basi leviter
pubescens, apice breviter bifidus. *Drupa* ultrapollicaris,
fulvo-tomentosa, sæpius bilocularis, disperma. *Seminis* testa
membranacea; cotyledones crassæ, carnosæ; radicula supera,
brevissima.

5. Chailletia *floribunda*, Planch. *in Hook. Ic. t.* 792. (Tab. XXX);
ramulis cinereo-tomentellis, foliis petiolatis amplis ovali-
oblongis basi acutis glabris, cymis amplis multifloris in pedunculo brevi libero axillaribus, pedicellis brevissimis, sepalis
oblongis, petalis calyce plus dimidio longioribus quam stamina
brevioribus, stylo elongato apice breviter bifido, drupa obovoidea tomentosa.—Fernando Po, *Vogel.*

Ramuli tomento diu persistente cinerei, crassiores ac in præcedente. *Folia* 6-9 poll. longa, 2½-4 poll. lata, acumine obtuso
v. acuto interdum brevissimo, margine integerrima v. obsolete
sinuata, siccitate fusco-rubentia, ad costas pilis raris puberula,
cæterum glaberrima, petiolo 4-6 lin. longo fulta. *Stipulæ*
parvæ, deciduæ. *Cymæ* pluries dichotomæ, usque ad 3 poll. diametro, nunc pedunculo ima basi bifido geminæ videntur, nunc
pedunculo communi brevi fultæ. *Bracteolæ* minutæ. *Pedicelli*
vix semilineares, articulati. *Sepala* linea paullo longiora, obtusiuscula, extus tomentosa, distincte imbricata. *Petala* et genitalia *C. oblongæ*, sed stamina longiora. *Drupa* etiam pariter
dense tomentosa sed latior et brevior, maturam tamen non vidi.

Plate XXX. *Fig.* 1. flower; *f.* 2. petal, with the hypogynous scale opposed to it; this scale is however represented too narrow at the base, and too evidently connected with the petal; *f.* 3. anther, back view : *all magnified.*

The *Rhamnus paniculatus,* Schum. et Thonn.; which De Candolle had, without examination, referred doubtfully to *Ceanothus,* under the name of *C.? Guineensis,* (Prod. 2. p. 30), is evidently, from Thonning's detailed description since published, a *Chailletia* nearly allied to *C. toxicaria,* and if really a distinct species, should receive the name of *C. paniculata.* A seventh species, as yet unpublished, is among Heudelot's Senegambian plants, which gives to Tropical Africa nearly half the total number of species now known of this small Order, whose affinities with *Hippocrateaceæ* and *Celastrineæ* become more and more evident as the species are better known. The valvate calyx, mentioned among the distinctive characters by Lindley (Veg. Kingd. p. 583), is a mistake; all the species known to me have it imbricated in æstivation, as originally described by De Candolle; so much so, that one division of the calyx is usually entirely concealed by the others in the bud. The characteristic disc of *Celastrineæ,* which in *Hippocrateaceæ* is united with the filaments in a fleshy mass, is represented among *Chailletiaceæ* by the hypogynous glands, which are sometimes slightly connected, so as to form a real disc, only differing from that of several *Celastrineæ* by being more deeply lobed. The three groups might indeed be considered, without inconvenience, as three tribes of one natural Order.—(G. B.)

XLII. HIPPOCRATEACEÆ.

1. Hippocratea *rotundifolia,* Hook. fil.; caule tereti scandente? cortice lævi, foliis petiolatis late oblongis rotundatisve obtusis v. subacuminatis rugosis utrinque reticulatis coriaceis, paniculis axillaribus terminalibusque dichotomis ramis erectis elongatis, petalis rotundatis concavis, disco depresso concavo, antheris 4-lobatis extrorsum dehiscentibus.—Sierra Leone, *Don.*

Rami cortice griseo-fusco lævi tecti. *Folia* opposita, 4-5 unc.

longa, $3\frac{1}{2}$-$4\frac{1}{2}$ lata, margine undulata, pallide flavo-viridia, utrinque opaca, consimilia; venis prominulis reticulata, rugulosa, subcoriacea. *Panicula* folio longior; ramis strictis, ad axillas compressis, gracilibus. *Flores* lutei.

Very near a West Indian species common to Demerara and St. Vincents, but the leaves are much broader, not marked with raised dots, more rugulose and opaque on the upper surface. There are five other W. African species of this genus, viz.: *H. Richardiana*, Guill. et Perr., from Senegal; *H. Indica*, Willd., common to Senegambia, the East Indies, and probaby Madagascar; *H. paniculata*, Vahl, ranging from Sierra Leone to Senegal; *H. macrophylla*, Vahl, from Sierra Leone; and *H. velutina*, Afz., from Guinea.

1. Salacia *prinoides*, DC. *Prod.* 1. *p.* 571; ramis teretibus sparse pustulatis, ramulis compressis, foliis (inferioribus suboppositis) petiolatis valde coriaceis late ellipticis utrinque obtusis integerrimis v. obscure sinuato-dentatis opacis super luride virescentibus subter pallidioribus nervis divaricatis, pedicellis axillaribus solitariis paucisve aggregatis petiolo æquilongis, lobis calycinis brevibus obtusis, petalis late oblongis obtusis, disco elevato, filamentis ovario æquilongis, antheris sub-urceolatis.—Grand Bassa, *Vogel.*

Frutex? glaberrimus. *Rami* cortice atro-castaneo lævi obscure pustulato tecti. *Folia* 3-4 unc. longa, $1\frac{1}{2}$-2 lata, suprema opposita, inferiora approximata, sed vere alterna, super vix nitida, subter pallidiora, venis inconspicuis reticulata; petiolo $\frac{1}{4}$ unc. longo. *Pedicelli* validi, erecti, 1-flori, infra florem incrassati. *Flores* flavo-virides, $\frac{1}{4}$ unc. diametr. *Calycis* lobi lati, coriacei, orbiculati. *Petala* calyce ter longiora, obtusa, fusco-striata. *Discus* erectus, subelongatus. *Filamenta* compressa, ligulata, recurva; *antheris* rubris, filamento bis latioribus, tranverse elongatis, 1-locularibus, rima lata superne hiantibus.

I am unable to distinguish this from the *Salacia prinoides* of Malacca, but have given a detailed description of the African specimens, with which more copious ones of the Indian species than I have had access to, should be compared.

2. Salacia *Senegalensis*, DC. *Prod.* 1. *p.* 570.—Sierra Leone
and Accra, *Vogel;* Senegal.
Specimens of a similar Senegambian plant, collected by
Brunner, are possibly a different species.

3. Salacia *affinis*, Hook. fil.; ramis teretibus, cortice pallide
rufo verrucis 4-lobis pallidis dense consperso, foliis suboppositis alternisque petiolatis elliptico-oblongis acuminatis
basi angustatis subintegerrimis coriaceis super luridis subnitidis, subter pallidioribus, venis reticulatis, pedicellis axillaribus fasciculatis unifloris gracilibus petiolum superantibus, alabastris cylindraceis, lobis calycinis late rotundatis, petalis late
oblongis obtusis apicibus incurvis, disco conico, filamentis
planis elongatis.—Sierra Leone, *Whitfield.*
Rami diametro pennæ corvinæ. *Cortex* pallide ruber, undique
verrucis parvis flavo-fuscis 4-sectis sparsa. *Folia* 3 unc.
longa, 1⅓ lata, versus apicem obscure serrata; petiolo ⅓ unc.
longo. *Flores* axillis in superioribus perplurimi, ¼ unc. lati,
pedicellis fere ½ uncialibus.
Very closely allied to *S. Senegalensis,* but the leaves are very
obscurely serrated, the flowers larger, petals broader, and the
warts on the stem larger and more numerous.

4. Salacia *cornifolia*, Hook. fil.; ramis teretibus, cortice griseo
verruculato, ramulis lævibus glaberrimis, foliis oppositis petiolatis ellipticis utrinque angustatis apice obtusis rarius elliptico-lanceolatis longe acuminatis coriaceis super nitidis subter
pallidioribus reticulatis margine tenuiter recurvo obscure
sinuato-serrato, pedicellis 2-3 axillaribus validis 1-floris petiolo longioribus, alabastris globosis, disco depresso annulari,
filamentis brevissimis, antheris rima hippocrepiformi extrorsum hiantibus vix 2-locularibus, baccis parvis pyriformibus 3-
locularibus.—Sierra Leone, *Vogel.*
Rami sparse verruculati; ramulis oppositis, patentibus, compressis. *Folia* 3-4 unc. longa, 1½ lata, forma varia, utrinque
angustata, apice plerumque obtusa, petiolo ⅓ unc. longo.
Flores ¼ unc. diam.; pedicellis ½ pollicaribus, superne incrassatis. *Stamina* fere ut in *Hippocratea,* sed rima vere dorsalis.

282   FLORA NIGRITIANA.

Fructus ⅓ unc. longus, obscure trigonus, siccitate fusco-
viridis.

In the structure of the flower, the depressed disc, short
filaments, and especially in the apparently all but 1-celled
anthers, this approaches *Hippocratea*, from which genus, however, the inflorescence and fruit remove it.

5. Salacia *pyriformis*, Walp. *Rep.* 1. *p.* 402 ; ramis præcipue ad
petiolum compressis, cortice atro-fusco lævi, foliis petiolatis oppositis alternisque subcoriaceis oblongis utrinque rotundatis nunc subacuminatis plerumque integerrimis super nitidis subter pallidioribus opacis reticulatis, pedicellis plurimis
axillaribus 1-floris petiolo æquilongis, alabastris globosis,
fructu pyriformi obtuse trigono.—Calypso pyriformis, *G.
Don in Gard. Dict.* 1. *p.* 629.—Mountains of Sierra Leone,
Don ; Senegambia.

Frutex scandens ?. *Rami* elongati, teretes. *Folia* 4-5 unc.
longa, 2-2½ lata, siccitate super fusca, nitida, subter pallidiora, subferruginea ; venis divaricatis ; petiolo ½-⅔ unc.
longo. *Flores* ¼ unc. diametro, petalis oblongis obtusis, disco
conico, filamentis basi latissimis. *Fructus* (fid. *Don*) magnitudine *Pyri* "Bergamot" dicti, edulis.

This is one of the few edible fruits of Western Africa, of which
an account, drawn up by Sabine, appears in the Hort. Soc. Trans.
v. 5. p. 459. under the name of *Tonsella pyriformis*.

6. Salacia *elongata*, Hook. fil. ; ramis validis obscure tuberculatis v. granulatis, ramulis lævibus ad axillas foliorum compressis, foliis petiolatis oblongo-obovatis elongatis basi angustatis apice rotundatis v. subacuminatis obscure sinuato-
serratis integerrimisve super nitidis subter pallidioribus opacis
coriaceis, pedunculis axillaribus fructiferis petiolo longioribus.—St. Thomas, *Don*.

Exemplar miserrimum, priori affine, differt foliis magis coriaceis
angustioribus basi non rotundatis, venisque foliorum multo
minus a costa media divergentibus. *Alabastra* globosa.

This is evidently a distinct species from the former, although
it appears to have been confounded with it in Don's Herbarium.

7. Salacia *erecta*, Walp., *Rep.* 1, *p.* 402; ramis teretibus superne subangulatis, cortice subgranulato, ramulis compressis angulatis, foliis oppositis breve petiolatis ellipticis elliptico-lanceolatisve utrinque angustatis regulariter serratis super subnitidis, pedicellis axillaribus, fructu ovato-cordato obtuse trigono.—Calypso erecta, *G. Don, Gard. Dict.* 1. *p.* 629.—Sierra Leone, *Don.*

Frutex habitu *Theæ viridis. Folia* 2½ unc. longa, 1 lata, subcoriacea, basi angustata, sed in petiolum non desinentia, apicem versus obtuse acuminata, super lævia, subter venis prominulis reticulata. *Fructus* ¾ unc. longus. Specimens rather imperfect.

8. Salacia *debilis*, Walp. *Rep.* 1. *p.* 402; caule tereti striato lævi, foliis breve petiolatis subcoriaceis ovalibus v. ellipticooblongis utrinque subangustatis vix acutis obscure serratis, floribus parvis axillaribus fasciculatis, pedicellis gracillimis erectis.—Calypso debilis, *G. Don, Gard. Dict.* 1. *p.* 629.— Senegambia, *Heudelot ;* Sierra Leone, *Don.*

Rami cortice fusco tecti. *Folia* sæpius elliptica, 2-2½ unc. longa, ½-1½ lata, super lævia, subnitida, subter pallidiora, venis prominulis reticulata. *Pedicelli* ½-1 unc. longi, filiformes, stricti, erecti, uniflores. *Flores* patentes, 1 lin. lati. *Lobi calycini* rotundati, ciliati. *Petala* patula, lineari-oblonga, obtusa. *Stamina* 3; filamentis brevissimis; antheris late subreniformibus, emarginatis, transverse dehiscentibus, unilocularibus.

Easily to be recognized by the small flower and very slender pedicels. The habit is that of an *Elæodendron.*

9. Salacia? *rufescens*, Hook. fil. ; ramis teretibus lævibus striatis, ramulis creberrime flavo-punctatis, foliis breve petiolatis submembranaceis elliptico-oblongis v. oblongo-lanceolatis utrinque angustatis obtusis siccitate subcrispatis rufescentibus sinuato-dentatis, pedicellis solitariis binisve axillaribus 1-floris petiolum superantibus, alabastris globosis, lobis calycinis petalisque? rotundatis, antheris transverse elongatis.—Sierra Leone, *Vogel.*

Rami cortice fusco tecti. *Folia* patula, 2-3 unc. longa, 1-1½

lata, in apicem obtusum angustata, basi in petiolum 1-2 lin.
longum desinentia, opaca, subter pallidiora; venis inconspicuis. *Pedicelli* ½ unc. longi. *Flores* 1½ lin. lati. *Calycis* lobi concavi, late rotundati. *Petala* brevissime unguiculata. *Discus* amplus, planus. *Stamina* filamentis brevissimis;
antheris transversee longatis, cylindraceis; polline trigono.

The solitary specimen differs very much from any other
species, though wanting any more striking character than
the rufescent colour. Having seen no fruit, the genus is
perhaps doubtful. The form of the anther is nearly that of
Hippocratea, but the inflorescence is very different.

The only other described West African *Salacia*, the *S. Africana*, DC. (*Tonsella*, Willd.) from Guinea, may possibly be the
same as some one of the preceding species.

XLIII. CELASTRINEÆ.*

1. Celastrus (Catha) *Senegalensis*, Lam. DC. Prod. 2. *p.* 8.—
On the Gambia, *Don;* Senegal.

The *C. coriaceus*, Guill. et Perr., also from Senegal, and
the *C. lancifolius*, Schum. et Thonn., from Guinea, are the
only other W. Tropical African species, and belong likewise to
the section or genus *Catha*, well distinguished in most cases
from the true *Celastri* by the axillary inflorescence, short style
and thin incomplete arillus, although the numerous South
African species have not yet been sufficiently examined to ascertain the real value of these characters. Presl has farther separated some species to form his two genera *Encentrus* and *Polyacanthus*, to the former of which the *C. coriaceus* might be
referred, were the chief character, the two-celled ovary and
capsule, constant, but though the number of cells be indeed
generally two, I have occasionally found three. So again, in
Polyacanthus stenophyllus, neither the supposed quaternary
parts of the flower, nor the unilocular capsule are by any means
constant in one and the same individual, and both genera must

* By G. Bentham.

necessarily be reunited with *Catha*, whether the latter be retained
as a separate genus or as a section of *Celastrus.*—(G. B.)

XLIV. TEREBINTHACEÆ.*

1. Canarium? *edule*, Hook. fil.—Pachylobus edulis, *G. Don,*
Gard. Dict. 2. *p.* 89.—St. Thomas, *Don.*
   The fruit described by Don, on which he founded the genus
Pachylobus, does not appear to differ from that of *Canarium;*
and the foliage, of which alone there is a specimen, is very like
that of some species of that genus. It differs from that of
C. commune in the midrib of the leaflets being hispid underneath. The flower is unknown.
   The only other W. Tropical African plant known of the tribe
of *Bursereæ* is the *Balsamodendron Africanum*, Arn., (*Heude-
lotia*, A. Rich.) from Senegal.
1. Spondias *lutea*, Linn.—S. aurantiaca, *Schum. et Thonn.
Beskr. p.* 225?—Sierra Leone, cultivated, *Vogel.*
2. Spondias *dubia*, A. Rich. *Fl. Seneg.* 1. *p.* 153.—Grand Bassa,
Vogel; St. Thomas, *Don;* Senegal, *Sieber.*
3. Spondias? *Zanzee*, G. Don, *Gard. Dict.* 2. *p.* 79.—Fernando Po? *Vogel;* St. Thomas, *Don.*
   This species, and another from Senegambia closely allied to it,
(*S. microcarpa*, A. Rich.), differ from *Spondias* in their polygamous and tetramerous flowers, with the petals much more
decidedly imbricate, and when better known will probably be
found to form with *Harpophyllum caffrum*, Bernh. (or *Spondias
caffra*, Meissn.), a distinct section of *Spondias*, or possibly a good
separate genus. The specimens of *S. Zanzee* are very imperfect,
and insufficient to afford any decided character to distinguish it
from *S. caffra;* the young fruit is like that of the true *Spondias*,
but small.
   Another W. African species, *Spondias Birrea*, A. Rich., from
Senegal and Abyssinia, has since been established as a distinct
genus by Hochstetter, under the name of *Sclerocarya.*

* By G. Bentham.

1. Odina *Oghigee,* Hook. fil.—Spondias Oghigee, *G. Don, Gard. Dict.* 2. *p.* 79.—Sierra Leone, *Don;* Grand Bassa, *Vogel,* who states that the bark is converted into powder by the natives and mixed with other substances to form a paint for the face. This species appears to be so very near to the East Indian *O. Wodier,* that the imperfect specimens of the collection afford no positive character to distinguish it. They are perhaps more perfectly glabrous than the East Indian ones, and the leaflets rather less numerous. The ovary is, as in *O. Wodier,* unilocular, with one ovule suspended from the top of the cavity, and is crowned by four short styles, each of which is truncated and apparently stigmatic at the apex. There are two other W. African species, *O. acida* and *O. velutina,* both from Senegal, and published by A. Richard, under the new generic name of *Lannea,* but since correctly referred to *Odina* by Endlicher.

1. Sorindeia *heterophylla,* Hook. fil. ; foliis simplicibus pinnatisve foliolisque oblongis coriaceis glabris, venulis anastomosantibus convergentibus in venulas spurias versus axillas venarum reversas, paniculis axillaribus laxis, floribus masculis 10-15-andris.—Sapindus simplicifolius, *G. Don, Gard. Dict.* 1. *p.* 666.—Sierra Leone, *Don.*

Habitus, glabrities, inflorescentia et flores masculi omnino *S. Madagascariensis;* flores fœminei et fructus desunt. *Foliola* speciminis alterius omnia simplicia, inferiora 8 poll. longa, 4 poll. lata, petiolo ultrapollicari, superiora 3 poll. longa, 2 poll. lata, petiolo 3-4-lineari; omnia brevissime et obtuse acuminata, basi acutiuscula et æqualia, nisi folio infimo maximo cui basis hinc valde dilatatur; *paniculæ* in axillis superioribus folio multo longiores, paucifloræ. In speciminibus cæteris pariter floriferis folia omnia pinnata; his foliola 3-7, foliis simplicibus similia sed sæpius angustiora et basi plus minus inæqualia. *Calyces* ut in *S. Madagascariensi* breves, brevissime 5-dentati. *Petala* 4, æstivatione valvata. *Stamina* 10-15, disco planiusculo inordinate inserta, pleraque tamen marginalia.

The singular venation of the leaflets and leaves of this plant,

may be traced in a slight degree, and very irregularly, in the
leaflets of the *Sorindeia Madagascariensis*, and more distinctly
in the simple leaves of a species from Penang (Wallich, n.
8505) and Malacca (Griffith) apparently referable to the same genus,
as well as in the *Dupuisia juglandifolia* mentioned below, but
not in any other terebinthaceous plant I am acquainted with.
A species of *Sorindeia*, from Congo, is alluded to by Brown,
and has been named *S. Africana* by De Candolle, but being as
yet undescribed, I have no means of judging whether it be
different from the above *S. heterophylla* or not.

1. Dupuisia? *longifolia*, Hook. fil. ; foliolis 15-18 anguste oblongis coriaceis glabris supra glaucis subtus elevato-penni-
nervibus rete venularum inconspicua, panicula mascula ampla
floribunda ferrugineo-tomentella, disco staminifero rufo-hirto.
—Sierra Leone, on the borders of marshes, *Vogel.*

Arbor excelsa. *Foliorum* petiolus cum rhachi bipedalis et longior.
Foliola 6-8 poll. longa, 1½-2 poll. lata, inferiora sublanceo-
lata, basi rotundato-cuneata, apice breviter acuminata; ultima subæqualiter oblonga basi longe acutata ; omnia subtus
siccitate leviter rubro-fusca; petioluli breves, crassi. *Paniculæ* semipedales ad pedales, e ramis ortæ ad axillas foliorum
delapsorum, pyramidato-ramosissimæ, tomento minuto ferru-
gineæ. *Flores* in specimine omnes masculi, brevissime pedicellati, subfasciculati, ebracteati, nutantes, duplo fere majores
iis *Sorindeiæ Madagascariensis.* *Calyces* breves, lati, ferruginei, dentibus 5 brevibus distantibus. *Petala* 1¼ lin.
longa, crassa, glabra, æstivatione valvata, per anthesin patentia. *Discus* planiusculus dense hirsutus. *Stamina* 5,
petalis breviora ; antheræ filamento longiores.

The original species, *Dupuisia juglandifolia*, A. Rich., from
Senegal, differs from the above by its broader leaflets, of a
thinner texture, with the venation of *Sorindeia*, by its smaller
flowers and smoother panicle, &c. Both are evidently nearly
allied to *Sorindeia*, and satisfactory specimens are wanting to
ascertain with certainty whether the two genera ought or not
to be united. In the mean time the number of the stamens,

equal to that of the petals, not double that or more, will serve
to characterize *Dupuisia*.

1. Anacardium *occidentale*, Linn.—St. Thomas, *Don;* Fernando Po, *Vogel*.

The remaining *Terebinthaceæ* mentioned as inhabiting W.
Tropical Africa are the *Rhus villosa*, Linn., a Cape plant inserted
in the Flora Senegambiæ as found at Cape Verd, and a species
of *Anaphrenium*, E. Mey., or *Heeria*, Meissn., from Senegal,
distributed amongst Sieber's plants under the name of *Vitex
terna*. It appears to be identical with a Cape plant occurring
in some old collections, as well as in Drège's, and since published by Bernhardi, under the name of *Anaphrenium mucronatum*, and by Presl under that of *Rhus salicifolia;* it is also
scarcely distinct from the Abyssinian *Anaphrenium Abyssinicum*, Hochst., or *Ozoroa insignis*, Delile, (G. B.)

XLV. CONNARACEÆ.*

There has been considerable confusion in the circumscription
of the genera of this small Order, owing to De Candolle's having
overlooked the fact that Gærtner's *Omphalobium* was founded
on the fruit of Linnæus' *Connarus monocarpus*, the original
species of both genera being evidently one and the same plant,
Gærtner's name must consequently be entirely suppressed, and
the chief character of the three genera, so well defined by
Brown, and which still include the whole Order, would stand
thus:

1. *Connarus*. Calyx imbricatus. Ovarium et stylus 1 (rarissime 2 ?). Ovula sutura ventrali affixa. Capsula stipitata. Semen exalbuminosum.—*Omphalobium*, Gærtn.

2. *Rourea*.† Calyx imbricatus. Ovaria et styli 5. Ovula e

* By G. Bentham.
† Wight and Arnott, taking the same view of the limitation of the
genera, have (Prodr. 1. p. 143) by mistake described the calyx of *Connarus monocarpus* as glabrous, which has misled others as to the identity
of this typical species. The calyx is clothed with a short rusty down in

basi ovarii erecta. Capsulæ (abortu sæpe solitariæ) sessiles.
Semen exalbuminosum. — *Connarus,* DC. ; *Byrsocarpus,*
Schum. et Thonn. ; *Anisostemon,* Turczan.
3. *Cnestis.* Calyx valvatus. Ovaria et styli 5. Ovula e basi
ovarii erecta. Capsulæ (abortu sæpe solitariæ) sessiles. Semen
albuminosum.

Since Brown published his observations in his celebrated
Appendix to Tuckey's Congo, as little has been added to our
knowledge of the affinities of the Order as to that of the genera
themselves. Their intimate connection with *Averrhoa* and the
Oxalideæ on the one hand, and with *Copaifera* and allied
Leguminosæ on the other, so clearly pointed out by him, has
only been further confirmed. Planchon has indeed proposed
the uniting *Oxalideæ, Connaraceæ* and *Leguminosæ* into one
group, or even Order, but for this there appear to be no better
grounds than there would be for uniting *Rubiaceæ* with *Scrophularineæ, Gentianeæ* and *Apocyneæ,* on account of the intermediate *Loganiaceæ.* Arnott, and latterly Lindley, in order to
obviate the anomaly of placing an Order where the stamens are
almost always very distinctly hypogynous, in a group of professedly perigynous Orders, have removed *Connaraceæ* from
their usual place next to *Leguminosæ* to the neighbourhood of
Oxalideæ and *Zanthoxyleæ.* That may be their best situation,
but as the real distinction between hypogynous and perigynous
insertion of the stamens, important as it usually is, is not yet
correctly understood, and as in this respect *Leguminosæ* themselves are variable, I have preferred leaving *Connaraceæ* in their
old place. Several species, both of *Rourea* and of *Cnestis,* have
very evident stipules ; almost all *Leguminosæ* of the tribe of
Cynometreæ have the flowers nearly or quite as regular as in
Connaraceæ ; and in *Copaifera,* besides the arillus, we may
observe the radicle at a considerable distance from the hilum on
the back of the seed near its base, although not so far as in
Connarus, where it is also on the back of the seed, but near its
summit.

this as well as in the greater number of true *Connari,* as mentioned by
Linnæus in his generic character.

U

1. Connarus *Africanus,* Lam. *Dict.* 2. *p.* 95.—Omphalobium
Africanum, *DC. Prod.* 2. *p.* 85.—Sierra Leone, *Don.*
  Two other W. African species of *Connarus* have been published : *C. floribundus,* Schum. et Thonn. (*Omphalobium Thonningii,* DC.) from Guinea, and *O. Smeathmanni,* DC., from
Sierra Leone.
1. Rourea *coccinea,* Hook. fil. ; glabra, foliolis 7-11 (parvis)
oblique ovali-ellipticis orbiculatisve obtussissimis retusisve
membranaceis v. demum coriaceis reticulatis, cymis laxe 3-5-
floris, staminibus stylos duplo superantibus.—Byrsocarpus
coccinea, *Schum. et Thonn. Beskr. p.* 226.—Cape Palmas,
Accra and on the Quorra, *Vogel;* Senegambia, *Heudelot;*
Guinea, *Thonning.*
  The leaflets vary much in number, breadth and consistence,
but these differences appear to depend much on age and on the
vigour of the shoots ; they seldom attain an inch in length.
Small stipules may often be observed on the young sterile shoots.
The flowers are precisely those of the true *Rourea,* the stamens
most decidedly hypogynous, the ovaries hairy, the pod smooth,
of a bright scarlet when fresh, and straighter than in most
Roureæ, but not otherwise differing from the generic type.
  Schumacher and Thonning describe a second species of
Rourea from Guinea, under the name of *Byrsocarpus punicea,*
and *Omphalobium villosum,* DC., from Sierra Leone and Senegambia, is a third. To the latter species may probably be
referred the *Cnestis obliqua,* Pal. de Beauv., from Oware. I
have a fourth unpublished species, gathered by Captain Middleton at Grand Bassa, remarkable in the calyx, which is only very
shortly divided into five imbricately æstivated lobes, or rather
teeth, and in the stamens less distinctly hypogynous, although
in the absence of fruit it cannot be generically distinguished
from *Rourea.*
1. Cnestis *corniculata,* Lam. *Dict.* 2. *p.* 23 ?—Grand Bassa,
Vogel.
Frutex, ex Vog., arborescens. *Ramuli* et foliorum juniorum
  petioli communes ferrugineo-villosi, demum glabrati et verruculosi. *Stipulæ* parvæ, rigidæ, acutissimæ. *Foliola* 7-9,

ovata v. ovato-lanceolata, acuminata, basi rotundata, 1-1½-pollicaria, coriacea, venulosa, juniora ad costam subtus hirsuta, adulta glabrata. *Paniculæ* floriferæ laxæ, graciles, foliis longiores. *Bracteæ* ad ramificationes setaceæ, villosæ. *Pedicelli* ultimi 2-3 lin. longi, capillares, infra medium articulati. *Calyx* lineam longus, sepalis lanceolatis acutis æstivatione valvatis, extus pallide roseis. *Petala* calyci æquilonga, lanceolata, extus puberula, alba. *Stamina* hypogyna, basi vix connata, calyce breviora, alterna alternis breviora. *Carpella* 5, sessilia, pubescentia, in stylos breves desinentia. *Ovula* 2, a basi ovarii erecta.

The other W. Tropical African species of *Cnestis* are *C. ferruginea*, DC., from Sierra Leone, and *C. pinnata*, Pal. de Beauv., from Oware : Brown mentions also several new species as being contained in C. Smith's Congo collection.—(G. B.)

XLVI. LEGUMINOSÆ.*

1. Crotalaria *genistifolia*, Schum. et Thonn. *Beskr. p.* 335 ; *Benth. Enum. Leg. in Lond. Journ. Bot.* 2. *p.* 479.—Guinea, *Isert* ; Accra, *Vogel*.

2. Crotalaria *Vogelii*, Benth. *l. c. p.* 561.—On the Quorra, at Stirling, *Vogel*.

3. Crotalaria *ononoides*, Benth. *l. c. p.* 572.—Sierra Leone, *Don*.

4. Crotalaria *ochroleuca*, G. Don, *Gard. Dict.* 2. *p.* 138 ; erecta? ramis virgatis ramulis petiolisque tenuiter subsericeo-pilosis, stipulis minutis, foliolis elongatis lineari-lanceolatis supra glabris subtus appresso-puberulis, racemis terminalibus subelongatis plurifloris, calycis dentibus tubo subtriplo brevioribus, vexillo late elliptico acuminato, carinæ rostro falcato, legumine subsessili oblongo polyspermo.—St. Thomas, *Don*. Habitu, statura et magnitudine florum *C. brevidenti*, Benth., simillima, differt vexillo acuminato, carina falcata et legumine majore, ampliore.

In habit and foliage this species may also be compared with *C. lanceolata*, E. Mey., from South Africa, which however has

---

* By J. D. Hooker and G. Bentham.

much smaller flowers; the acuminated standard is like that of
C. macrocarpa. There is also a *C. pallida,* Ait., from "Africa,"
which is only known by a very short phrase, as applicable to
this as to several other species.

5. Crotalaria *falcata,* Vahl, *Benth. l. c. p.* 585.—C. obovata, *G.
Don, Gard. Dict.* 2. *p.* 138.—Bassa Cove, *Ansell;* Cape
Palmas and Cape Coast, *Vogel;* Accra, *Don.*

6. Crotalaria *striata,* DC., *Benth. l. c. p.* 586.—Senegal, *Perrottet;* St. Thomas, *Don.* This is found in many parts of
Africa, South Asia, and in the West Indies, but in some of
the stations whence it has been sent it is probably cultivated
or introduced.

7. Crotalaria *incana,* Linn., *Benth. l. c. p.* 587.—Senegal, *Perrottet;* Sierra Leone, *Don.*—An American species perhaps
introduced only to the Old World.

8. Crotalaria *Goreensis,* Guill. et Perr., *Benth. l. c. p.* 589.—
Senegal, *Perrottet;* Gambia, *Captain Boteler, Heudelot;*
Accra, *Thonning, Don, Vogel.* Also a native of Upper Egypt.

9. Crotalaria *lotifolia,* Linn. *DC. Prod.* 2. *p.* 134.—Cape Coast,
Vogel.—A West Indian species, from which Vogel's imperfect
specimen does not appear to differ.

10. Crotalaria, *sp. n.?* allied to some of the S. American shrubby
trifoliolate species, but the specimen is insufficient for determination.—Sierra Leone, *Don.*

This is, next to *Indigofera,* one of the most numerous dicoty-
ledoneous genera in W. Tropical Africa, as besides the above
ten, no less than sixteen other species are known to inhabit that
region, viz.: 1. *C. arenaria,* Benth. 2. *C. glauca,* Willd. 3. *C.
Leprieurii,* Guill. et Perr. 4. *C. calycina,* Schranck, (a common
S. Asiatic species.) 5. *C. Perrottetii,* DC. 6. *C. gracilis,* Walp.
7. *C. ebenoides,* Walp. 8. *C. iodina,* Benth. 9. *C. macrocalyx,*
Benth. 10. An unpublished species from Senegambia allied to
C. medicaginea, Lam. 11. *C. sphærocarpa,* Perrot. 12. *C. polycarpa,* Benth. 13. *C. Senegalensis,* Bacle, (also found in Nubia).
14. *C. lathyroides,* Guill. et Perr. 15. *C. podocarpa,* DC. (also
found in Cordofan), and 16. *C. cylindrocarpa,* DC.

One species of *Lupinus, L. termis,* Forsk., is cited in the

Flora Senegambiæ, besides which the only other W. African
species known of the tribe of *Genisteæ*, is the *Xerocarpus hirsutus*, Guill. et Perr., from Senegal and Cordofan.
Of the tribe of *Trifolieæ*, no species has as yet to my knowledge been found in W. Tropical Africa, although a few have
been gathered in the eastern portion of that continent within
the Tropics as well as in North and South Africa.

ACANTHONOTUS, Benth. *Indigoferæ* sp., Auct. *Char. gen. ref.*

Calyx profunde quinquefidus, laciniis lanceolato-subulatis. *Petala* breve unguiculata, inter se calycique æquilonga ; vexillum
obovatum, alæ lineari-oblongæ liberæ, carina recta, edentula,
segmentis fere a basi connatis. *Stamina* diadelpha, antheræ
late didymæ, connectivo in acumen producto basi dorso piloso.
Ovarium breviter stipitatum, lineare, obliquum, in stylum va-
lidum uncinatum productum. *Stigma* subsimplex. *Legumen*
indehiscens, falcatum, latiusculum, compressum, echinatum,
hispido-setosum, dorso bicarinatum. *Semen* solitarium.

1. Acanthonotus *echinatus*, Benth.—Indigofera echinata, *Willd.
et Auct.*—Senegambia, *Heudelot ;* on the Quorra, at Attah,
Vogel.
A widely distributed plant, common to West Africa, Ceylon,
and the Peninsula of India. The leaves vary much in form,
from broadly obovate and very blunt, to oblong and rather
acute at both extremities, or elliptical ; all these forms may be
seen in the W. African specimens, although the Indian ones are
rather more constantly obovate.

1. Indigofera *bracteolata*, Guill. et Perr. *Fl. Seneg.* 1. *p.* 176.
—Senegambia ; on the Quorra, *Vogel.*
2. Indigofera *enneaphylla*, Linn.—S. Thomas, *Don ;* Cape Palmas
and Grand Bassa, *Vogel.*
3. Indigofera sp. very near *I. tetrasperma*, Vahl, and *I. paniculata*, Perr., but possibly distinct. There is only a single
specimen in Vogel's collection without any precise locality.
It appears to be an erect annual, four or five feet high, much
branched in the upper part, the upper leaves are simple and
like those of *I. tetrasperma*, but the lower leaves are wanting,

and there is no fruit, so that it cannot be accurately described.

4. Indigofera *Nigritana*, Hook. fil.; caule tereti erecto ramoso, ramulis gracilibus glabratis, stipulis lineari-subulatis, foliolis lineari-obovatis subacutis strigillosis, caulinis 2-3-jugis, rameis ultimis floralibusque simplicibus, pedicellis axillaribus filiformibus unifloris, calycis strigosi 5-partiti laciniis lanceolato-subulatis subæqualibus, vexillo oblongo calyce bis longiore, ovario biovulato, legumine appresse piloso breviter oblongo compresso dispermo.—On the Quorra, *Vogel.*

Caulis herbaceus, 2-3-pedalis, superne pyramidatim ramosus, ramulis strictis sulcatis, ultimis filiformibus. *Petioli* graciles, ½-¾ poll. longi. *Foliola* ramea 3 lin. longa, floralia minora, ultima bracteæformia. *Pedunculi* ¼-¾ poll. longi, apice curvati v. geniculati. *Flores* 1½ lin. longi. *Calyx* glandulosus. *Legumen* fere 2 lin. longum, atrofuscum, seminibus suborbiculatis.

5. Indigofera *endecaphylla*, Jacq. *DC. Prod. 2. p.* 228.—I. anceps, *Vahl*, DC. *l. c.*—I. Schimperiana, *Hochst. Pl. Abyss.* —Grand Bassa and Cape Palmas, *Vogel;* found also in Senegal and Guinea, in Abyssinia and East Tropical Africa, southward to Port Natal.

6. Indigofera *simplicifolia*, Lam. *DC. Prod. 2. p.* 222.—On the Quorra, at Attah, *Vogel;* Sierra Leone, *Smeathmann.*

These are luxuriant specimens, above three feet high, though apparently an annual, the stems are nearly simple, as described by Lamarck, and certainly not very much branched, as appears to have been the case with the specimen seen by De Candolle. The larger leaves are five inches long and half an inch broad, but the upper ones are scarcely above an inch and a half long as stated by Lamarck. Vogel's specimens are not in fruit, but the ovary is long and slender, with numerous ovules.

7. Indigofera *dendroides*, Jacq. *DC. Prod. 2. p.* 227.—Savannahs of the Quorra, *Vogel;* Senegal and Guinea.

8. Indigofera *hirsuta*, Linn. *DC. Prod. 2. p.* 228.—I. astragalina, *DC. l. c.*—I. ferruginea, *Schum. et Thonn. Beskr. p.* 370.— I. fusca, *G. Don, Gard. Dict. 2. p.* 211.—Cape Coast, and on the Quorra *Vogel;* St. Thomas, *Don;* Senegal and Guinea.

A widely diffused species, extending from Tropical Africa
through the whole of Southern Asia to the Philippine Islands
and North Australia, and varying considerably in most of these
localities.

9. Indigofera *Anil*, Linn.—I. uncinata, *G. Don, Gard. Dict.* 2. *p.*
208.—Sierra Leone and on the Quorra, *Don, Vogel;* cultivated.

The name *Anil,* given by Linnæus to the West African and
American indigo, is derived from a Hindostanee term applied to
the Indian indigos, and especially to the *I. tinctoria,* and signi-
fying *blue.* Both the *I. tinctoria* and *I. argentea* are also cultivated in West Tropical Africa.

Besides the above nine species, the same region possesses at
least twenty-seven other species of *Indigofera,* viz. : 1. *I. (Ame-
carpus) Senegalensis,* Lam. (*I. tenella,* Schum. et Thonn., and
Brissonia trapezicarpa, Desv.) from Senegal and Nubia ; 2. *I.
diphylla,* Vent., from Senegal ; 3. *I. Perrottetii,* Guill. et Perr.,
from Senegal ; 4. *I. oligosperma,* DC. (*I. glutinosa,* Schum. et
Thonn.) Senegal, Guinea and Nubia ; 5. *I. macrocalyx,* Guill. et
Perr., Senegal ; 6. *I. nigricans,* Vahl, (*I. elegans,* Schum. et
Thonn.) Guinea ; 7. *I. pulchra,* Vahl, Senegal and Nubia ;
8. *I. procera,* Schum. et Thonn., Guinea ; 9. *I. trichopoda,*
Guill. et Perr., Senegal ; 10. *I. tetrasperma,* Schum. et Thonn.,
(nec Vahl ex Webb, supra p. 121), Guinea ; 11. *I. paniculata,*
Pers., Sierra Leone ; 12. *I. viscosa,* Lam. (*I. glutinosa,* Guill. et
Perr., and possibly *I. lateritia,* Willd.), Senegal and Guinea, and
thence across Tropical Africa to the East Indian Peninsula ;
13. *I. sessiliflora,* DC., Senegal ; 14. *I. linearis,* DC., Senegal
and Cape Verd Isles ; 15. *I. subulata,* Vahl., (*I. Thonningii,*
Schum.) Guinea ; 16. *I. pilosa,* Poir. (*I. Guineensis,* Schum. et
Thonn.), Guinea ; 17. *I. aspera,* Guill. et Perr., Senegal and
Cordofan ; 18. *I. Prieureana,* Guill. et Perr., Senegal ; 19. *I.
lasiantha,* Desv., Angola ; 22. *I. macrophylla,* Schum. et Thonn.,
Guinea ; 23. *I. secundiflora,* Poir., Guinea, and four unpublished
species from Senegal. The *I. ornithopodiodes* of Schum. and
Thonn., cultivated in Guinea, appears to be the *I. tinctoria.*

Of the allied genus, *Cyamopsis,* there is one Senegambian
species, *C. Senegalensis,* Guill. et Perr., also found in Nubia.

296   FLORA NIGRITIANA.

The widely diffused genus, *Psoralea*, has as yet no representative in West Tropical Africa, although one species at least has been found in Nubia. Of the two known species of *Requienia*, one is from Senegal, *R. obcordata*, DC.

1. Tephrosia *Vogelii*, Hook. fil.; fruticosa, ramulis pedunculisque velutino-tomentosis, stipulis lanceolato-subulatis caducis, foliolis 6-10-jugis lineari-oblongis obtusis emarginatisve apiculatis subtus præcipue adpresse sericeis, floribus amplis, bracteis late ovatis acuminatis, calycis lati dentibus late oblongis obtusis, vexillo emarginato dorso sericeo, alis cum carina basi connatis, filamento superiore basi apiceque libero, legumine breviter falcato lineari-oblongo valde compresso dense fulvo-villoso.—On the Quorra, and Fernando Po, *Vogel.*

Frutex arborescens, 8-10-pedalis, ramis suberectis, ramulis sulcatis, pube conferta. *Stipulæ* 2 lin. longæ. *Petioli* pubentes, 6-8 poll. longi, pube fulva dense obtecti. *Foliola* 1½-2½-pollicaria, ½-¾ poll. lata, inferne subangustata, supra grisea sericeo-pubescentia, subtus dense albo-sericea, venis prominulis oblique parallelis striata, costa ferruginea. *Racemi* terminales, validi, erecti, pluriflori, subpyramidati. *Bracteæ* caducæ. *Flores* magnitudine *Pisi sativi*. *Calyx* late hemisphæricus, supra medium 5-fidus. *Corolla* purpurea; vexillum latius quam longum, erectum, patens, dorso argenteo-sericeum, basi albo-maculatum, breviter unguiculatam; alæ late dolabriformes, obtusæ, transverse rugosæ. *Ovarium* lineare, velutinum. *Legumen* 5 poll. latum, planum, sub-18-spermum.

A very handsome species of the same group as *I. toxicaria*, and which, like that species in the W. Indies, is cultivated by the natives and used for poisoning fish.

2. Tephrosia *densiflora*, Hook. fil.; fruticosa, ramulis pedunculisque pube tenui dense obtectis, stipulis lineari-subulatis, foliolis obovato-oblongis late emarginatis obcordatisve subtus præcipue pubescenti-sericeis, racemis terminalibus, bracteis anguste lanceolatis, calycis villosuli dentibus 4 brevissimis quinto subulato revoluto, vexillo sericeo, alis liberis, legumine leviter arcuato late lineari dense villoso.—Patteh, *Vogel.*

Frutex ramosus. *Folia* 3-5 poll. longa; foliola 1¼-1½ poll. longa,

6-9 lin. lata, supra pube tenui appressa, subtus sericea, costa ferruginea. *Racemi* sub-4-pollicares. *Bracteæ* extus sericeo-tomentosæ, intus glabræ. *Pedicelli* 6 lin. longi, villosuli. *Calyx* 1½ lin. longus, latior quam longus. *Vexillum* fere ½ poll. longum, latissime orbiculatum, dorso dense sericeo-villosum, brevissime unguiculatum, apice emarginatum ; alæ basi tranverse rugosæ. *Filamentum* superius basi apiceque liberum. *Legumen* pendulum, 3-4 poll. longum, 5 lin. latum, planum, apice oblique et abrupte acuminatum, undique dense villosum.

Allied to *T. Vogelii*, but less hairy, with much smaller flowers, narrower pods, broader and more emarginate leaflets, and very different calyx and bracts. It is said by Vogel to be a cultivated species.

3. Tephrosia *Ansellii*, Hook. fil. ; caule erecto basi lignoso parce ramoso, ramis pedunculisque laxe patentim pilosis, stipulis subulato-filiformibus, foliolis sub-7-jugis lineari-oblongis obtusis vix emarginatis, subtus laxe et molliter sericeo-pilosis, racemis terminalibus strictis elongatis multifloris, bracteis lineari-subulatis, calycis dentibus 4 brevissimis inferiore longiore, vexillo sericeo, legumine late lineari-oblongo subarcuato compresso parce piloso marginibus villoso-ciliatis.

—Savannahs on the Quorra at Stirling, *Ansell, Vogel.*

Suffrutex erectus, 3-pedalis, simpliciusculus. *Caulis* strictus, griseus v. pallide rufescens, villis mollibus laxis. *Stipulæ* 5 lin. longæ, persistentes. *Petioli* 4-7 poll. longi, pube pallide fulva. *Foliola* 1½-2 poll. longa, 3-5 lin. lata, basi vix angustata, apice retusa, pilis albidis marginata. *Racemi* semipedales ad pedales, floribus secus axem geminis. *Bracteæ* stipulis consimiles, pedicellos duplo triplove superantes. *Calyx* vix lineam longus, hemisphæricus, dentibus 2 superioribus connatis, cæteris subulatis, inferiore longiore. *Vexillum* late elliptico-oblongum, emarginatum, unguiculatum, dorso pilis fusco-brunneis dense sericeum ; alæ basi cum carina connatæ, transverse rugosæ. *Filamentum* superius basi et apice liberum. *Ovarium* dense sericeum. *Legumen* 1½-1¾ poll. longum, 4 lin. latum, plano-compressum, basi apiceque an-

gustatum, sub-7-spermum, faciebus pilosis, marginibus incrassatis villoso-ciliatis.

4. Tephrosia *elongata,* Hook. fil.; suffruticosa, erecta, caule
tereti ramulisque cinereo-pubescentibus, stipulis oblongo-lanceolatis, petiolis elongatis, foliolis sub-12-jugis anguste lineari-
oblongis obtusis apiculatis supra glabratis subtus appresse
sericeis, racemis longissimis multifloris, bracteis ovato-oblongis,
pedicellis brevibus, vexillo calyce breviter 5-dentato sextuplo
longiore dense velutino, legumine leviter arcuato lineari compresso velutino.—On the Quorra, *Vogel.*
Suffrutex tripedalis. *Caulis* strictus, ramis raris valde elongatis
laxe foliatis. *Folia* spithamea, suberecta, petiolo velutino-
pubescente, stipulis 1½ lin. longis extus sericeis. *Foliola* sesquipollicaria, 2½ lin. lata, vix basi angustata, costa subtus
ferruginea. *Racemi* interdum bipedales, parte florifera per
anthesin 4-unciali, fructifera pedali. *Flores* sæpius gemini,
erecto-patentes, purpurei, 9 lin. longi, pedicello 2-lineari
bracteaque sericeis. *Calyx* brevissimus, dente inferiore subulato, 2 summis fere ad apicem connatis. *Vexillum* sessile,
late obcordatum, dorso dense fulvo-sericeum; alæ basi carinæ vix cohærentes, medio transverse rugulosæ. *Filamentum* supremum basi et apice liberum. *Legumen* erecto-
patens v. horizontale, acuminatum, 3 poll. longum, 2½ lin.
latum, plano-compressum, marginibus incrassatis, sub-14-
spermum.

5. Tephrosia *fasciculata,* Hook. fil.; fruticosa, ramis validis
teretibus pubescentibus, petiolis elongatis, stipulis ovato-lanceolatis, foliolis 6-9-jugis linearibus emarginato-bilobis supra
glabratis subtus albo-sericeis, racemis terminalibus erectis
brevibus, pedicellis brevissimis, floribus erectis fasciculatis,
legumine lineari compresso velutino.—On the Quorra, *Vogel.*
Frutex erectus, orgyalis. *Stipulæ* infimæ foliaceæ, 3-4 lin.
longæ. *Petiolus* 6-8-pollicaris. *Foliola* 2 poll. longa, 3 lin.
lata. *Racemi* ut videtur densiflori, soli fructiferi mihi
noti. *Flores* delapsi iis *T. elongatæ* similes. *Legumina* fasciculata, patula, 2¼ poll. longa, ¼ poll. lata, sub-11-sperma,

faciebus pilis flavo-fuscis pubescentibus, marginibus incrassatis. Most nearly allied to *T. elongata*, but the leaves are longer, the stipules very different, and the pod broader and fewerseeded.

6. Tephrosia *flexuosa*, G. Don, *Gard. Dict.* 2. *p.* 232 ; caule erecto sublignoso flexuoso superne ramulisque pubescentipilosis, stipulis filiformi-subulatis, foliolis 4-5-jugis linearilanceolatis subacutis supra pubescentibus subtus argenteosericeis, racemis terminalibus abbreviatis sericeo-villosis, floribus confertis, bracteis subulatis, calyce brevissimo, legumine late-lineari subtoruloso sub-8-spermo.—St. Thomas, *Don*.

Fruticulus 1-1½-pedalis, caulibus sæpe angulatim flexuosis basi denudatis stipulis pollicaribus solis persistentibus. *Petioli* molliter flavo-pubescentes. *Foliola* supra grisea, subtus pilis rufescentibus tecta. *Racemus* florens sesquipollicaris, fructifer parum elongatus. *Legumen* suberectum, vix arcuatum, faciebus torulosis sericeo-pubescentibus, marginibus incrassatis fulvo-tomentosis. *Semina* late oblonga, testa nitida, nigra, flavo-irrorata.

7. Tephrosia *elegans*, Schum. et Thonn. *Beskr. p.* 376.—Savannahs on the Quorra, *Vogel;* Guinea.

Simillima *T. brevipedi*, e Guiana ; differt foliolis plerisque bijugis, floribus minoribus, calycis laciniis latioribus petalisque obtusioribus.

8. Tephrosia *pulchella*, Hook. fil. ; caule erecto simplici basi lignoso superne subfastigiatim ramoso, ramis petiolisque gracilibus adpresse puberulis, foliolis 4-5-jugis anguste linearicuneatis retusis emarginatisve supra glabratis subtus argenteosericeis, floribus axillaribus solitariis parvis brevissime pedicellatis, calycis segmentis subæqualibus subulatis, legumine erecto dense villoso.—On the Quorra, at Stirling, *Vogel*.

Suffrutex gracilis, tripedalis, ramulis erectis virgatis. *Stipulæ* minimæ. *Petiolus* 4-9 lin. longus. *Foliola* sensim e basi angustata, ½-¾ poll. longa, ¾-1 lin. lata, inferiora raro sesquipollicaria, supra obscure griseo-puberula subtus argentea costa ferruginea. *Flores* parvi, pedicello 1 lin. longo bracteam sub-

ulatam excedente. *Calyx* dense sericeus, ad medium 5-fidus.
Vexillum orbiculari-oblongum, unguiculatum ; alæ cum carina
vix adhærentes, medio obscure tranversim reticulatæ. *Ovarium* dense sericeo-pilosum, sub-12-ovulatum.
9. Tephrosia *linearis,* Pers. *Syn.* 2. *p.* 330.—Senegal ; Accra,
Vogel; also a native of Cordofan.

The genus *Tephrosia* seems peculiarly abundant towards the
northern or drier regions of West Tropical African Flora, all
but one of the following additional known species being found
in Senegambia, viz. : 1. *T. bracteolata,* Guill. et Perr., and 2.
T. platycarpa, Guill. et Perr., both from Senegal ; 3. *T. humilis,*
Guill. et Perr., from Senegal and Cape Verd ; 4. *T. apollinea,*
DC., from Senegal ? Nubia and Arabia ; 5. *T. leptostachya,* DC.,
from Senegal and Nubia ; 6. *T. lineata,* Schum. et Thonn.
(perhaps the same as *T. leptostachya,* DC., and *T. purpurea,*
Pers.) from Guinea ; 7. *T. gracilipes,* Guill. et Perr., from Senegal ; 8. *T. uniflora,* Pers. (to this may possibly belong the
Senegambian *T. apollinea,* as well as the *T. anthylloides,* Hochst. ;
the true *Galega apollinea,* Delille, being probably a very different
species), Senegal and Nubia ; 9. *T. lathyroides,* Guill. et Perr.,
from Cape Verd ; 10. *T. hirsuta,* Schum. et Thonn. (perhaps,
as well as *T. lathyroides,* the same as *T. uniflora*) from Guinea ;
11. *T. digitata,* DC., from Senegal ; and one, or perhaps two,
unpublished species from Senegambia.

The *T. toxicaria,* presumed by Tussaud to have been introduced into the West Indies from West Africa, does not appear
in any of the collections from the latter country, where other
species are cultivated for the same purpose of poisoning fish.
It may also be stated, as further evidence that *T. toxicaria* is
really American, not African, that the American Continent possesses other indigenous species much nearer allied to that one
than to any African ones.

1. Sesbania *aculeata,* Pers.—Accra, *Vogel;* Senegambia, and
   widely diffused over Africa and the East Indies, and introduced into the West Indies.
2. Sesbania *Ægyptiaca,* Pers.—*S. punctata,* DC.—Sierra Leone
   and Quorra River, at Attah, *Vogel;* Senegambia, and, like

the preceding species, diffused over Africa and East India as far as the Philippine Islands.

Three other Senegambian species have been described, *S. leptocarpa*, DC., *S. pachycarpa*, DC. and *S. pubescens*, DC., of which two at least, if not all three, are also found in Nubia.

1. Agati *grandiflora*, Desv., *DC. Prod.* 2. *p.* 266.—Sierra Leone, *Don.*—Introduced from E. India, where also it appears to be generally, if not always, more or less in a state of cultivation.

No other genus of true *Galegeæ*, nor yet any of the large tribes of *Astragaleæ* or *Viceæ*, have as yet been found in West Tropical Africa, although a few have been gathered in Nubia.

1. Stylosanthes *Guineensis*, Schum. et Thonn. *Beskr. p.* 351, *et G. Don, Gard. Dict.* 2. *p.* 281 ?—S. erecta, *Pal. de Beauv. Fl. Ow. et Ben.* 2. *p.* 28. *t.* 77.—*Vog. Linnæa*, 12. *p.* 68.—Senegambia to Guinea ; Cape Coast Castle, Grand Bassa and Nun River, *Vogel;* Whydah, *Don.*

Certainly very closely allied to the West Indian *S. procumbens*, and perhaps only a luxuriant variety. It is usually a much larger plant, thickly covering large patches of ground, and rising to the height of a foot and a half, but not really erect, on which account it may be better to adopt Schumacher's name than Palisot de Beauvois's. The flowers are usually more numerous than in *S. procumbens*, but not always so.

The East Indian *S. mucronata*, Willd., slightly differing from the preceding, has also been found in Senegal.

2. Stylosanthes *viscosa*, Sw., *Vog. Linnæa*, 12. *p.* 66.—Sierra Leone, *Don.* The specimens are precisely similar to the common American form.

1. Arachis *hypogæa*, Linn.—Abundantly cultivated in West Tropical Africa. In addition to Mr. Brown's observations (App. Cong.) on the probable country and migrations of this plant, it may be stated that the discovery of several other species of the same genus in Brazil is additional evidence of the American origin of the *A. hypogæa*.

1. Zornia *diphylla*, Pers., var. *glochidiata.*—Z. glochidiata,

Reich. DC. Prod. 2. *p.* 316.—Z. biarticulata, *G. Don, Gard. Dict.* 2. *p.* 288.—Accra, *Vogel, Don ;* Senegambia.

The distinctions between the common American, Asiatic and African forms of this widely diffused plant prove to be very inconstant, as already observed by Vogel ; and it is now no longer possible to view them as forming more than one species, found in almost every hot country visited by botanists.

A new species of *Geissaspis,* remarkable for its entire, not ciliated bracts, is amongst Heudelot's Senegambian plants.

The same Senegambian collection includes the *Herminiera elaphroxylon,* Guill. et Perr., which must be removed to *Hedy-sareæ,* the pod being certainly articulated when quite ripe, and an allied plant, which appears to be the same as the Abyssinian *Acrotaphros bibracteata,* Steud.

1. Ormocarpum *verrucosum,* Pal. de Beauv. *Fl. Ow. et Ben.* 1. *p.* 96. *t.* 58.—Grand Bassa, in marshy places and in maritime sands on the Nun River, and in the island of Fernando Po, *Vogel;* Oware.

2. Ormocarpum *coronilloides,* G. Don, *Gard. Dict.* 2. *p.* 279 ; foliis impari-pinnatis, foliolis multijugis oblongis utrinque obtusis mucronulatis, leguminis articulatis striatis glabris v. pilis glanduliferis raris instructis.—Rathkea glabra, *Schum. et Thonn. Beskr. p.* 355 ;—var. *a.* petiolis pedicellisque glanduloso-puberulis, St. Thomas, *Don ;*—var. *β.* petiolis pedicellisque glabrioribus, on the Niger, *Mc William.*

Very similar to a Philippine Island plant, which appears to be the *Æschynomene colutoides* of A. Richard, and which as well as our plant, may be mere varieties of the East Indian *O. sennoides,* the degree of glandular hairiness of the inflorescence and pods being evidently very variable.

1. Æschynomene *aspera,* Linn. *DC. Prod.* 2. *p.* 320.—Nun and Quorra Rivers, *Vogel;* Accra, *Don.* These specimens differ but slightly from the common East Indian form in the bracts and calyxes being more acuminated and smooth, but do not appear to be specifically distinct.

The *Æ. Indica,* Linn., which probably includes *Æ. macro-*

poda, DC., *Æ. quadrata*, Schum. et Thonn., and *Æ. sensitiva*, Pal. de Beauv., a common East Indian plant, is found in Senegal and Guinea, extendiug also to Cordofan ; and the American *Æ. sensitiva*, Sw., as well as an apparently new species, are in Heudelot's Senegambian collection.

Sprengel, under the name of *Smithia spicata*, has described a presumed Senegambian plant, which must remain a puzzle until it has been seen by some more accurate botanist. No true *Smithia* has as yet been found to our knowledge in Western Africa, athough one species (*Kotschya Africana*, Endl.) is a native of Upper Egypt.

1. Uraria *picta*, Desv., *DC. Prod. 2. p.* 324.—Cape Palmas and Quorra River, at Addanda, *Vogel;* St. Thomas, *Don.*— A common East Indian species, also recorded as a native of Guinea by Schumacher and Thonning, whose plant is erroneously referred by Walpers to *Desmodium*.

1. Desmodium (Pleurolobium) *oxybracteatum*, DC. *Prod. 2. p.* 334?—D. grande, *E. Mey. Comm. Pl. Afr. Austr. p.* 124. —D. paleaceum, *Guill. et Perr. Fl. Seneg. p.* 209.—Abòh, *Vogel;* Senegambia; also S. East Africa, Madagascar and Mauritius. Vogel's specimens closely resemble the eastern ones, Heudelot's are rather less vigorous, but all appear to belong to one species, varying very much in the hairiness of the fruit, as is the case with so many *Desmodia*.

2. Desmodium (Chalarium) *latifolium*, DC. *Prod. 2. p.* 328.— D. lasiocarpum, *DC. l. c.*—Hedysarum deltoideum, *Schum. et Thonn. Beskr. p.* 361.—Accra and Quorra River, *Vogel;* St. Thomas, *Don ;* Senegambia and Guinea. A very common East Indian plant extending as far as the Philippine Islands.

3. Desmodium (Heteroloma) *Mauritianum*, DC. *Prod. 2. p.* 334.—H. fruticulosum, *Schum. et Thonn. Beskr. p.* 363.— D. linearifolium, D. ramosissimum et D. tenue? *G. Don, Gard. Dict. 2. p.* 294.—St. Thomas and Sierra Leone, *Don ;* Grand Bassa, Cape Palmas, &c., *Vogel.*—A Mauritius plant, extending probably all across the African continent.

4. Desmodium (Heteroloma) *incanum*, DC. *Prod. 2. p.* 332.— D. sparsiflorum, *G. Don, Gard. Dict. 2. p.* 294.—Sierra

Leone and St. Thomas, *Don*; Fernando Po, *Vogel.*—A
common American species, said to be found also in the
Mauritius

5 Desmodium (Heteroloma) *oxalidifolium*, G. Don, *Gard. Dict.*
2. *p.* 295.—St. Thomas, *Don;* sandy shores of the Nun
River, near the sea, *Vogel.*

Caules e basi radicante ½-2-pedales, prostrati, graciles, apicibus
adscendentibus, parce subsericeo-pilosi. *Stipulæ* 3 lin. longæ,
persistentes, lanceolatæ, setaceo-acuminatæ. *Petioli* ½-1 poll.
longi. *Foliola* late obovato-orbiculata, 1-1½-pollicaria, membranacea, pallide viridea, supra sparse puberula. *Pedunculi*
terminales, elongati, pauciflori, pedicellis subgeminis filiformibus semipollicaribus. *Bracteæ* lanceolatæ setaceo-acuminatæ. *Flores* 2 lin. longi. *Calyx* pubescens, profunde partitus, segmentis subulatis. *Legumen* pollicare, articulis
4-5 longioribus quam latis, hinc fere rectis, illinc convexis,
faciebus planis glochidiato-pubescentibus.

Apparently allied to *D. cæspititium*, DC., and closely resembling a Javanese species.

Besides the above, there are three described *Desmodia* from
W. Tropical Africa, viz.: *D. lanceolatum*, Schum. et Thonn.,
from Guinea, *D. ovalifolium*, Guill. et Perr., and *D. terminale*,
Guill. et Perr. (referred by Webb to the common W. Indian
D. tortuosum), both from Senegambia, and a fourth apparently
new one in Heudelot's collection.

1. Nicholsonia *reptans*, Meissn. *Linnæa*, 21. *p.* 260.—Desmodium triflorum, *DC. Prod.* 2. *p.* 334.—D. Bullamense,
G. Don, Gard. Dict. 2. *p.* 294.—Hedysarum granulatum,
Schum. et Thonn. Beskr. p. 362.—Accra, *Vogel.*—A very
common weed in hot damp climates within the Tropics, both
of the New and the Old World. Meissner is quite right in
transferring it from *Desmodium* to *Nicholsonia*, to which
genus belong four or five other E. Indian and Mauritius
species published as *Desmodia.*

1. Alysicarpus *vaginalis*, DC. *Prod.* 2. *p.* 353.—Stirling, at the
confluence, *Ansell;* also Senegal, Cordofan, and very abundant in East India, as far as the Philippine Islands.

A second species, *A. rugosus*, DC., to which probably belong also *Hedysarum rugosum* and *H. ovalifolium*, Schum. et Thonn., is a native of Senegal and Guinea, extending to Cordofan.

Of the genus *Abrus*, which connects *Vicieæ* with *Phaseoleæ*, one species, the common *A. precatorius*, is included in the Senegambian Flora. It is found abundantly within the Tropics, both in the New and the Old World, but in many places evidently introduced.

1. Centrosema *decumbens*, Mart.—*Benth. in Ann. Mus. Vind. 2. p.* 120.—Cape Coast, *Vogel.*—A common South American plant, ranging from the West Indies to South Brazil. The genus is also, with this exception, exclusively American.

Clitoria Ternatea, a common East Indian plant, introduced into many parts of America, is found also in Senegambia, and over a great portion of Tropical Africa.

1. Glycine *labialis*, Linn.—*Wight et Arn. Prod. Penins. Ind. Or. 1. p.* 200.—St. Thomas, *Don.*—A common Tropical plant, both in the New and the Old World.

The *G. Senegalensis*, DC., is generally supposed to be a mere variety of *G. parviflora*, Lam., (which latter is correctly referred to *G. labialis*); but some specimens in Heudelot's Senegambian collection, agreeing well with De Candolle's character, are certainly distinct. Besides the longer racemes, and longer and smoother pods, the calyx is essentially different, being divided into four instead of five teeth, and each tooth is broader. Hochstetter's *Kennedya Arabica*, from Cordofan, appears to be the same as the Senegambian plant.

1. Johnia *Willdenowii*, Hook. fil.—Glycine hedysaroides, *Willd. Spec. 3. p.* 1060.—Accra, *Don*; in *Vogel's* collection without the precise locality.

This species, well described by Thonning under Willdenow's name, is certainly congener to the *Johnia Wightii*, Arn., and to the *Bujacia anonychia*, E. Mey., which is scarcely specifically distinct from the Indian plant. In all I find the stamens monadelphous, and all bearing anthers, the upper stamen being free only at the base. In the *Johnia vestita*, Arn., however, the upper stamen is entirely free. All these small genera, allied to *Glycine*, require a general revision.

Another Guinea plant, the *Glycine biflora*, Schum. et Thonn., may possibly be a *Johnia*, but is at present insufficiently described to determine the genus.

1. Dioclea (Pachylobium) *reflexa*, Hook. fil.; ramulis petiolisque patentim pilosis, foliolis ovatis brevissime acuminatis tenuiter coriaceis supra glabris subtus hirsutis, inflorescentia rufo-tomentosa, floribus confertis brevissime pedicellatis, bracteis lanceolatis persistentibus reflexis, calycis campanulati lacinia infima tubo vix breviore, carina rostrata alis breviore, legumine ovato-oblongo planiusculo rufo-villoso.—Dolichos coriaceus, *Grah. in Wall. Cat. Herb. Ind. n.* 5562.—Cape Palmas, near the sea-coast on the Quorra, and Fernando Po, *Vogel.*

Frutex scandens, ramis teretibus glabratis ad nodos ramulisque patentim pilosis. *Stipulæ* medio affixæ, lanceolato-subulatæ, reflexæ. *Petioli* 3 poll. longi, pilis flaccidis patentibus fulvis obsiti, stipellis setaceis. *Foliola* 3-5 poll. longa, 2-3 lata, breve petiolulata, reticulato-venosa, supra nitida, subtus pallidiora. *Pedunculi* bipedales, inferne glaberrimi, supra medium pubescentes et dense florigeri. *Bracteæ* semipollicares, ligulatæ, acuminatæ, rufo-velutinæ, recurvæ. *Flores* per 2-4 ad quemquam nodum brevissime pedicellati, splendide rubro-purpurei. *Calyx* rufescenti-sericeus, 4 lin. longus, basi bibracteolatus, tubo late ovato-hemisphærico, ore 5-lobo, lobis lateralibus late oblongis obtusis, summis paullo brevioribus, infimo angustiore longiore. *Vexillum* glaberrimum, calyce bis longius, ungue late lineari, lamina latissime obcordata; alæ vexillo paullo longiores, obovato-quadratæ, basi auricula reflexa appendiculatæ; carina coriacea. *Ovarium* villosissimum. *Legumen* planum, crasso-coriaceum, in specimine adhuc immaturum sed jam 5 poll. longum, 2 poll. latum, sutura seminifera valde incrassata.

A species, as far as hitherto known, only found in East India, and some other parts of the Old World, although belonging to an otherwise exclusively American genus. It closely resembles the

Brazilian *D. violacea*, whose violet-coloured bracts are straight
and erect, whilst in the West African one they are always reflexed,
as in another yet unpublished Brazilian species, (Gardner, n.
2117), in which however the flowers are very different.
Two species of *Canavalia* are mentioned as inhabiting W.
Tropical Africa, the *C. obtusifolia*, DC., common on the
sea-coast of Senegal and Guinea, and of East Tropical Africa,
Asia and America, and *C. gladiata*, DC., from Guinea, a
commonly cultivated East Indian species. The first of these
is described by Schumacher and Thonning under the name of
Dolichos obovatus ; and that species is followed by a *D. ovali-
folius*, which is unknown to us, but judging from the expression "cætera uti in præcedenti," it may be another *Canavalia*.

1. Mucuna *urens*, DC., *Prod.* 2. *p.* 405.—Fernando Po, *Vogel;*
Accra, *Don;* also Guinea.—Apparently identical with the
West Indian plant figured by Plumier.

2. Mucuna *flagellipes*, Vogel, MS. ; caule foliisque glabris,
foliolis oblique ovatis subcordatis apice abrupte acuminatis
membranaceis nitidis, pedunculis longissimis apice racemosis,
bracteis late ovato-cymbæformibus sericeis, calycis late hemi-
sphærici dentibus 3 æquilongis obtusis.—On the banks of the
Niger, most abundant, *Vogel.*

Caulis lignosus, scandens, arbores altissimas superans (fide Vogel).
Ramuli teretes. *Petioli* 3-4 poll. longi, sulcati. *Foliola* 5
poll. longa, 3 poll. lata, basi valde oblique rotundata v. plerumque plus minusve cordata, apice in acumen 4-5 lin. longum
obtusum producta, petiolulis 2 lin. longis. *Pedunculi* 3-6
pedes longi, funiformes, nudi, apicibus floriferis incrassatis pubescentibus, parte florifera tripollicari geniculatim flexuosa,
nodis florigeris 12-16. *Bracteæ* 1¼ poll. longæ, latissimæ,
pilis fulvis pungentibus obsitæ, tridentatæ. *Vexillum* 1¼
poll. longum, viridi-lutescens ; alæ obtusæ, vexillo æquilongæ,
basi lineis sericeis percursæ, cum carina breviore subacuta
connatæ.

1. Erythrina *Vogelii*, Hook. fil. ; inermis ? subglabra, caule sulcato, foliolis ovato-oblongis oblongisve obtusis coriaceis reti-

culatim venosis, racemis terminalibus strictis multifloris, calyce spathaceo vix puberulo apice reflexo obscure tridentato, vexillo calyce multo breviore, alis calycem æquantibus, carina triente longiore.—Fernando Po, *Voyel;* Accra, *Don.* *Caulis* lignosus, cortice albido. *Petioli* 2-3-pollicares, infra petiolulos 1 lin. longos glanduligeri. *Foliola* 4 poll. longa, lateralibus minoribus, supra viridia, subtus pallidiora. *Pedunculi* 8-12-pollicares, stricti, multiflori, dense pubescentes. *Flores* 1¼ poll. longi, solitarii v. gemini, brevissime pedicellati. *Calyx* 4 lin. longus, coriaceus. *Vexillum* paullo curvatum. *Legumen* deest.

Vogel cites this as a medicinal plant. Don's specimens are hardly determinable: their flowers appear rather larger than Vogel's.

2. Erythrina *Senegalensis,* DC. *Prod.* 2. *p.* 413.—E. Guineensis, *G. Don, Gard. Dict.* 2. *p.* 371.—Sierra Leone, *Don.* Senegambia to Guinea, extending, according to A. Richard, as far as Abyssinia.

1. Phaseolus *lunatus,* Linn.—Fernando Po, *Vogel.*—A plant extensively cultivated in Tropical countries, especially in Asia and Africa.

The *P. vulgaris* is enumerated by Schumacher and Thonning as being in cultivation in Guinea, and Guillemin and Perrottet have described another species, *P. Senegalensis,* as a native of Senegambia.

1. Vigna *oblonga,* Benth. *Bot. Sulph. p.* 86.—Sandy banks of the Nun River, near the sea, and Fernando Po, *Vogel.*

An American sea-coast plant, very near *V. glabra,* but the leaflets are always remarkably blunt, besides some differences in the flowers.

2. Vigna *multiflora,* Hook. fil.; pilosula v. glabra, stipulis breviter auriculatis, foliolis membranaceis ovato-rhombeis, pedunculis folio longioribus supra medium multifloris, pedicellis calyce subbrevioribus, calycis late campanulati dentibus tubo brevioribus supremo latissimo integro lateralibus obtusis infima angustiore, carina nuda erostri, leguminibus glabris leviter falcatis.—Fernando Po, on the sea-coast, *Vogel.*

Herba volubilis, *V. gracili* affinis, sed omnibus partibus major.
Rami hinc inde pilis reflexo-patentibus hirti v. omnino glabrati. *Stipulæ* 1-2 lin. longæ, late lanceolatæ, acutæ, striatæ,
basi in auriculam brevem acutam infra insertionem productæ,
glabræ v. piloso-ciliatæ. *Petioli* infra foliola 2-3 poll. longi,
inter foliola semipollicares, hinc inde præsertim prope basin
patentim pilosi. *Stipellæ* parvæ, obtusæ, fere glanduliformes.
Petioluli 1-1½ lin. longi, villosi. *Foliola* 1½-3-pollicaria ;
terminale late rhombeum, lateralia valde inæquilatera, basi
truncata v. subcordata, omnia apice breviter acuminata, utrinque viridia, margine ciliata, ad costas pilis nonnullis utrinque
hispidula et in pagina superiore pilis raris conspersa. *Pedunculi* infra flores 3-5-pollicares, parte florida 1½-2-polli-
cari. *Pedicelli* ad quemquam nodum gemini v. rarius 3-4-ni,
per anthesin calyce sæpius breviores, fructiferi paullulum elongati. *Bracteæ* et bracteolæ parvæ, oblongo-lineares. *Calyx*
fere *V. glabræ* nisi dentibus latioribus multo obtusioribus.
Petala glabra, iis *V. glabræ* subsimilia. *Legumen* 15-18 lin.
longum, vix 2 lin. latum.

Like the *V. gracilis,* this species cannot be generically separated from the *V. glabra* of Savi, which is much nearer to the
true *Dolichos* than some of the following. Unfortunately, the
very indifferent specimens existing in herbaria, scarcely admit
as yet of any exact limitation of this and the allied genera.

3. Vigna ? sp., apparently near *V. multiflora,* but the specimen
is too bad to determine.—Near the town of St. Ann de Chiaves,
Don.

4. Vigna *unguiculata,* Walp. *Rep.* 1. *p.* 779.—Dolichos ungui-
culatus, *Linn. et Auct.*—Cape Coast, *Vogel.*—Extensively
cultivated in W. Tropical Africa, as it is in other parts of
Africa, Asia and America.

5. Vigna *linearifolia,* Hook. fil.; caule volubili scabro v. piloso,
stipulis lanceolatis acuminatis pilosis, petiolis hispidis, foliolis
longe lineari-lanceolatis hispidulis transverse reticulatis, pedunculis pilosis apice 2-3-floris, calycis laciniis 5 longe lan-
ceolato-setaceis, legumine densissime velutino-villoso.— Savannahs of the Quorra, *Vogel.*

Caules pluri-pedales, inter gramina volubiles, rigiduli, teretes, inferne glabrati, superne scabri et pilosi, internodiis elongatis. *Stipulæ* 3-4 lin. longæ, basi in auriculam brevem obtusissimam adnatam infra insertionem productæ. *Petioli* 1½-2-pollicares, petiolulique pilis rigidulis hispidi. *Foliola* rigidula, 6-8 poll. longa, vix semipollicem lata, utrinque scabrida et pilosula, penninervia et lineis parallelis crebris transversis pulchre reticulata. *Pedunculi* oppositifolii, validi, 1-1½ poll. longi, apice incrassati. *Flores* majusculi, sessiles v. brevissime pedicellati. *Calyx* semipollicaris, tubuloso-campanulatus, tubo brevi basi obtuso, laciniis plus duplo tubi longitudine, subulato-acuminatis. *Vexillum* late obcordatum, breviter unguiculatum, lamina basi utrinque hamata, extus lutea, intus purpurascens ; alæ purpureæ, carinæ cohærentes, basi hinc calcaratæ. *Stigma* laterale vix productum. *Legumen* 1¾ poll. longum, 2 lin. latum, sub-10-spermum, pilis rigidis atropurpureis vestitum.

6. Vigna *reticulata,* Hook. fil.; caule volubili superne foliis pedunculisque setosis, stipulis ovato-lanceolatis acuminatis, foliolis anguste ovato-lanceolatis acutis creberrime transversim reticulatis, pedunculis apice sub-2-floris, calycis laciniis 5 subulato-setaceis, legumine velutino-tomentoso pilis fulvis consperso.—Savannahs at Accra, *Vogel.*

V. linearifoliæ simillima, sed duplo major, pilis fulvis rigidioribus, foliolis stipulisque latioribus, legumine fere 2½-pollicare, indumento diverso.

The beautiful tranverse reticulations on the leaflets of this and the preceding species distinguish them from any of their congeners.

6. Vigna *Nigritia,* Hook. fil.; caule volubili sparse piloso, stipulis ovato-lanceolatis basi in auriculas 2 productis, petiolis hispidis, foliolis ovato-lanceolatis utrinque pilosis, pedunculis pilosis folio longioribus apice multifloris, calycis glabri brevius campanulati dentibus 5 obtusis, legumine glabrato apice hamato-mucronato.—On the Quorra, *Vogel.*

Caulis pallide flavo-fuscus, scaberulus, superne pilosus. *Stipulæ* 2½ lin. longæ, striatæ, glabriusculæ, ciliatæ, auriculis infra insertionem brevibus acutis. *Petioti* pollicares. *Foliola* 2 poll. longa, ½-2½ lata, basi rotundata truncata v. cuneata, trinervia, reticulatim venosa, pallide viridia, lateralia obliqua. *Pedunculi* 2-4-pollicares, validi, sulcati, pilis retrorsis. *Flores* brevissime pedicellati. *Calyx* 1 lin. longus, glaberrimus. *Vexillum* 4 lin. longum, latissimum, recurvum, pallide roseum, ungue brevissimo, lamina basi utrinque auriculata; alæ intensius coloratæ, basi hinc auriculatæ. *Legumen* sub lente minutissime puberulum, leviter curvatum, 1½ poll. longum. *Semina* oblonga, rufo-fusca.

A most distinct species, of which the specimen is very imperfect.

8. Vigna *Thonningii*, Hook. fil.—Plectrotropis hirsuta, *Schum. et Thonn. Beskr. p.* 339.—Cape Coast and Fernando Po, *Vogel.*

This answers very well to Thonning's description of the plant he gathered at Aguapim. It comes very near to the American *V. carinalis*, Benth. *Bot. Sulph.*, and like that species, the *V. angustifolia*, and some other African and Asiatic species, is remarkable for the much-curved oblique keel with a lateral spur on one side only, on which character Schumacher and Thonning founded their genus *Plectrotropis*. Although they can scarcely be admitted to the generic rank thus accorded, they will probably be found to constitute a good sectional group in the now extensive genus *Vigna*.

Besides the above eight species, West Tropical Africa possesses at least three others, viz.: *V. gracilis*, Hook. fil. (*Dolichos*, Guill. et Perr.), from Senegambia; *V. Nilotica*, Hook. fil. (*Dolichos*, Delile), from Senegambia, Nubia and Egypt; and *V. angustifolia*, Hook. fil. (*Dolichos*, Vahl, *Plectrotropis*, Schum. et Thonn.), from Senegambia and Guinea.

Of the genus *Dolichos*, although it be essentially African, no true representative appears to have been found within the West Tropical region, the *D. nervosus*, Schum. et Thonn., being probably the *Lablab vulgaris*, Savi, which is common over a great part of Africa, and exists in Senegambia and Guinea, either wild or cultivated, as well as *Pachyrrhizus angulatus*, *Rich.*,

Voandzeia subterranea, Dup. Thou., and *Cajanus Indicus,* Spr.,
to which last must probably be referred the *Cytisus Guineensis,*
Schum. et Thonn.—A Senegambian plant, supposed to be a
Psophocarpus, but of which only the fruit and foliage are
known, has been published by Desvaux under the name of
P. palustris, and by Guillemin and Perrottet under that of
P. palmettorum. It appears, however, to be at least as nearly
allied to the Brazilian *Diesingia* as to the Asiatic *Psophocarpus.*

1. Cyanospermum *calycinum,* Hook. fil.—Rhynchosia calycina,
   Guill. et Perr. Fl. Seneg. 1. p. 214.—Sierra Leone, *Don,*
   Vogel.

Although the caruncle of the seed is extremely small in this
species, yet the general habit, the calyx and corolla are those
of *Cyanospermum* rather than *Rhynchosia,* and it has also the
peculiar blue seed of the former genus. The constricted pod
occurs also in *Rhynchosia phaseoloides,* which is in every other
respect a true *Rhynchosia.*

1. Rhynchosia *Memnonia,* DC. *Prod.* 2. p. 386.—St. Thomas,
   Don; Senegal, Nubia, and Upper Egypt.
2. Rhynchosia *debilis,* Hook. fil.; prostrata v. volubilis, pubescens, stipulis parvis ovato-lanceolatis acuminatis subulatisve,
   foliolis membranaceis rhombeo-orbicularibus abrupte acuminatis lateralibus inæquilateris, racemis axillaribus densifloris
   petiolo multo brevioribus, bracteolis lineari-lanceolatis pedicello longioribus, calycis laciniis lanceolato-subulatis infima
   elongata, legumine patentim piloso rufo.—Dolichos debilis,
   Don, MSS.—Glycine macrophylla, *Schum. et Thonn. Beskr.*
   p. 348 ?—St. Thomas, *Don.*

Caulis gracilis 2-3-pedalis. *Stipulæ* deciduæ, 2 lin. longæ,
striatæ. *Petioli* 2-pollicares, graciles, canaliculati. *Foliolum* terminale 2½ poll. longum, lateralia minora, omnia utrinque puberula et subtus glandulis miminis creberrime punctata. *Racemi* 1-1½-pollicares, densiflori. *Bracteolæ* stipulis
angustiores, sæpius rufescentes et rigiduli ; pedicelli suberecti,
1-2 lin. longi. *Calyx* 3 lin. longus, anguste campanulatus,
pubescens et glandulosus, laciniis subulatis tubo longioribus,

inferiore cæteris triente majore. *Vexillum* calycem superans, oblongum, basi biauriculatum ; alæ angustæ, vexillo breviores, carinæ breviori cohærentes.

Closely allied to the East Indian *R. densiflora*, it differs chiefly in its much smaller calyces. The other W. Tropical African species are : 1. *R. minima*, DC. from Senegambia and Guinea, a common Tropical plant in both hemispheres ; 2. *R. caribæa*, DC., a West Indian plant, found also in Senegal and Guinea, unless the species alluded to under that name in African Floras, be rather the *R. Memnonia;* 3. *R. faginea*, Guill. et Perr., from Senegambia, a species which should perhaps be transferred to *Eriosema* or *Arcyphyllum;* and 4. *R. argentea*, Desv., from Angola, which is entirely unknown to me.

1. Eriosema *glomeratum*, Guill. et Perr. *Fl. Seneg.* 1. *p.* 216, (sub. *Rhynchosia*).—Sierra Leone, *Don;* Cape Palmas and Quorra River, *Vogel;* Senegal and Guinea.

β. *minor*, ramis laxe villosis, foliolis vix pollicaribus.—Sierra Leone, *Don.*

2. Eriosema *spicatum*, Hook. fil. ; molliter rufo-pubescens v. glabrescens, caule suberecto, stipulis liberis lanceolatis vix acuminatis, foliolis elliptico-ovatis obtusis, pedunculis folio pluries longioribus apice racemum spiciformem ferentibus, floribus reflexis, bracteis minimis, calyce obtuse 5-dentato, legumine oblongo-obovato rufo-tomentoso.—Sierra Leone, *Don ;* Senegal.

Fruticulus pedalis, parce ramosus. *Stipulæ* 2 lin. longæ. *Petioli* pollicares. *Foliola* rigidula, sesquipollicaria, minute glanduloso-punctata, reticulato-venosa, nervis venulisque subtus prominulis, lateralia terminali minora. *Pedunculi* semipedales, stricti ; parte florifera 1-2-pollicari. *Flores* brevissime pedicellati, arcte reflexi, 4 lin. longi. *Calyx* breviter campanulatus, 1 lin. longus, puberulus, sub lente glandulosus. *Vexillum* oblongo-obcordatum, lamina basi biauriculata et biappendiculata, ungue brevi ; alæ carinam subæquantes. *Legumen* fere semipollicem longum, 2 lin. latum, tomento molli rufo subsericeum.

3. Eriosema *podostachyum,* Hook. fil. ; caule erecto piloso v. glabrato, stipulis liberis lanceolatis acuminatis, foliolis ovatis acutis v. obtusiusculis puberulis v. supra glabratis, pedunculis folio pluries longioribus puberulis apice racemum spiciformem ferentibus, floribus reflexis, bracteis lanceolato-subulatis, calycis 5-fidi dentibus late ovatis acutis, legumine patentim sericeo-piloso.—Grand Bassa, *Vogel, Ansell.*

Caulis tripedalis, strictus, basi perennis, rufo-fuscus, superne præcipue patentim et retrorsum pilosus. *Stipulæ* 4 lin. longæ. *Petioli* pollicares, sæpius patentim pilosi et rufo-pubescentes. *Foliola* bipollicaria, viridia, subtus pallidiora, ad nervos puberula, punctis glandulosis minutis confertis. *Pedunculi* 8-10-pollicares, stricti, pilosi v. glabrati. *Flores* iis *E. spicati* similes, vexillo alisque apice rubris. *Legumen* late oblongum, semipollicare, pilis patentissimis subsericeis villosum. *Semina* nitida, flavo nigroque irrorata, hilo elongato.

Near *E. spicatum,* but a much larger plant, more or less clothed with spreading hairs, longer peduncles, larger stipules and bracteæ, and more pointed teeth to the calyx.

The *E. cajanoides,* (Rhynchosia *Guill. et Perr.*) from Senegambia, is the only other species of the genus known to be a native of W. Tropical Africa.

Don has also described a *Flemingia Guineensis,* from Guinea, but there do not appear any specimens in his collection.

1. Ecastaphyllum *Brownei,* Pers.—DC. *Prod. 2. p.* 241.— Grand Bassa and Nun River, *Vogel;* Senegal and Guinea. —A plant widely diffused over Tropical America, from the West Indies to Brazil.

1. Dalbergia *saxatilis,* Hook. fil. ; inflorescentia excepta glabra, foliolis 4-jugis oblongis utrinque rotundatis emarginatis venoso-reticulatis terminali obovato, paniculis folio ter brevioribus pilosiusculis, bracteolis parvis lineari-oblongis, calycis late campanulati glabrati vix striati dentibus lateralibus brevibus obtusis inferiore elongato, legumine elliptico-lineari graciliter stipitato.—Sierra Leone, *Don;* Senegambia, *Heudelot.*

Ramuli nitidi, obscure striati. *Petioli* 3-4-pollicares, graciles.

Foliola pollicaria, subtus glaucescentia. *Calyx* basi rotundatus, medio constrictus. *Stamina* ut in congeneribus. *Ovula* 2-3. *Legumen* 4-4½ poll. longum, 1-1¼ poll. latum, breviter et laxe reticulatum, membranaceum, albidum, stipite subpiloso 4 lin. longo. *Semen* in parte centrali suberosa unicum.

There are two other specimens in a very imperfect state, which are either varieties of the preceding, or distinct species, differing apparently in the calyx as follows :

β. *Donii,* calycis glabriusculi basi obtusi dentibus lateralibus longioribus obtusis.—Sierra Leone? *Don.*

γ. *Ansellii,* calycis puberuli basi acutiusculi dentibus lateralibus brevioribus acutiusculis.—Cape Palmas, *Ansell.*

2. Dalbergia *pubescens,* Hook. fil. ; rufo-pubescens, foliolis 5-jugis oblongis v. obovato-oblongis utrinque obtusis basi sæpe inæqualibus, racemis axillaribus terminalibusque folio brevioribus, bracteolis parvis, calycis late campanulati velutini dentibus latis acutis, ovario hirsuto.—Sierra Leone, *Don ;* Senegambia, *Heudelot.*

Rami lignosi, cortice fusco, ramulis obscure angulatis. *Petioli* 3-4 poll. longi, stricti. *Foliola* 1-2-pollicaria, coriacea, obtusa v. emarginata, basi rotundata v. rarius superiora angustata, supra pilis sparsis puberula, subtus pilis rufis densius vestita. *Racemi* 1-1½-pollicares. *Pedicelli* 1½-lineares. *Calyx* 1 lin. longus, latiusculus, 4-dentatus, dente superiore latiore emarginato, lateralibus latis brevibus subacutis. *Vexillum* late oblongum, emarginatum, ungue gracili laminæ æquilongo ; alæ lineari-oblongæ obtusæ basi hinc hamato-auriculatæ ; carina bis latior. *Filamentorum* tubus more generis bipartitus, antheris basifixis didymis. *Ovarium* longe stipitatum, pedicello piloso.

Two other species of *Dalbergia* occur in Senegambia, the *D. melanoxylon,* Guill. et Perr., and an apparently undescribed one in Heudelot's collection.

1. Drepanocarpus *lunatus,* Meyer.—*DC. Prod. 2. p.* 420.— Grand Bassa and Nun River, *Vogel ;* Senegal and Guinea.

1. Pterocarpus *esculentus,* Schum. et Thonn. *Beskr. p.* 330.

—Quorra River, at Abòh, *Vogel*, who remarks that the fruit is eaten by the negroes, and that the tree forms a thicket along one side of the creek. It is found also in Senegal and in Guinea. The Senegambian collections contain also the *P. erinaceus*, Poir.; the *P. lucens*, Lepr., and a fourth apparently undescribed species.

OSTRYOCARPUS, *Hook. fil.* (nov. gen.)

Calyx breviter campanulatus, basi contractus, ore obscure 5-dentato. *Vexillum* late rhomboideum, ungue brevissimo lato, glabrum, recurvum. *Alæ* vexillum æquantes, oblongo-cultriformes, basi hinc auriculatæ, ungue gracili. *Carinæ* petala alis conformia. *Stamina* 10, diadelpha, antheris ovatis. *Ovarium* sessile, sericeo-pubescens, triovulatum, stylo glabriusculo filiforme apice stigmatoso. *Legumen* orbiculare, plano-compressum, coriaceum, margine seminifero incrassato canaliculato, abortu monospermum. *Frutex* sarmentosus, foliis impari-pinnatis, floribus paniculatis.

1. Ostryocarpus *riparius*, Hook. fil.—Sierra Leone and Fer-Po, *Vogel;* Sierra Leone, *Don;* Senegambia, *Heudelot.*
Rami lignosi, cortice pallido verrucoso, uti folia glaberrimi. *Petioli* stricti, teretes, supra sulcati, 4-8 poll. longi. *Foliola* bijuga cum impari, opposita, 6-8-pollicaria, coriacea, petiolulata, oblonga, utrinque rotundata v. subacuta, reticulatim venosa. *Paniculæ* axillares terminalesque, petiolo æquilongæ v. longiores, ramosæ, ramis pubescenti-sericeis multifloris. *Flores* flavidi, in ramulis propriis glomerati, pedicellis ½ lin. longis. *Calyx* pubescens 1½ lin. longus, bracteolis parvis rotundatis appressis. *Petala* 3 lin. longa. *Legumen* in sicco atrum, 2 poll. longum, ut videtur indehiscens, valvis planis obscure reticulatis. *Semen* immaturum 1 poll. longum, oblongum, in medio legumine situm.

This plant has the habit of some of the smaller-flowered *Lonchocarpi*, but the stamens are distinctly diadelphous, and the pod is remarkable.

1. Lonchocarpus *Formosianus*, DC. *Prod. 2. p.* 260.—Robinia

argentiflora, *Schum. et Thonn. Beskr. p.* 352.—St. Thomas, *Don ;* on the Quorra, Cape Palmas, Grand Bassa, &c., *Vogel.* From Senegal to Guinea, and thence to the east coast, where it is said to be planted; and if, as is probable, this be but a mere variety of the *R. sericeus,* DC., it is an American species not uncommon both in the West Indies and in Brazil.

Among the numerous species which have been collected under *Lonchocarpus,* but very few have as yet been described in fruit, and in those few so many differences are observable in the pod, that almost each has been proposed as the type of a separate genus. The present species, although one of those on which Kunth's *Lonchocarpus* was originally founded, has not the membranous wingless pod of *L. latifolius* and *macrophyllus,* which he described, nor the coriaceous wingless one of *Sphinctolobium virgilioides,* nor yet the curiously winged one of *Neuroscapha Guilleminiana,* but one in some measure intermediate between those of the two latter. With the habit and flowers of *Neuroscapha,* and a pod in general form like that of the same genus, the broad wings are replaced by a slightly prominent longitudinal nerve on each side of the seminal suture. It is, therefore, safer for the present to leave the genus *Lonchocarpus* entire, until a sufficient number of fruits shall have been observed, to show whether these slight modifications correspond with any real differences in habit or in flower.

2. Lonchocarpus ? *macrostachyus,* Hook. fil.; ramis foliisque glabris, foliolis 5-7 oblongis obtuse acuminatis basi rotundatis coriaceis, panicula ampla multiflora, ramulis pedicellisque brevibus velutinis, calycis velutini dentibus 5 brevibus, bracteolis late ovatis, vexillo glaberrimo.—On the Quorra, at Ibaddi, *Vogel.*

Rami ramulique lignosi, teretes, cortice pallide fusco verrucis pallidis notato. *Petioli* 5-8 poll. longi, stricti. *Foliola* 6-8 poll. longa, 2½ poll. lata v. inferiora minora, basi interdum subcordata, apice abrupte acuminata, acumine brevi obtuso, supra nitida, subtus sub lente interdum pilis adpressis subtilissime sericea, nervis primariis subtus prominentibus et reticulatim venulosa, petiolulo 2 lin. longo. *Paniculæ* terminales v. laterales, sæpe pedales. *Flores* conferti albo-virides, pedicellis 1 lin. longis. *Calyx* campanulatus, curvatus, 2 lin. longus, dentibus obtusis. *Vexillum* late obovatum, reflexum, basi biauriculatum, ungue gracili; alæ vexillo æquilongæ, lineari-oblongæ, obtusæ; carina paullo major, obtusa. *Filamentum* vexillare ima basi liberum. *Stylus* fere ad apicem sericeus. *Ovarium* dense sericeum, 4-ovulatum.

A very handsome plant, the racemes and calyxes nearly black, with a velvety pubescence. The pod is unknown, and it is, therefore, doubtful whether it belongs to *Lonchocarpus* or *Milletia*.

1. Milletia *macrophylla,* Hook. fil.; (Tab. XXXII. XXXIII), foliolis 11-15 oblongis subtus ferrugineo-pubescentibus, stipellis subnullis, racemo elongato thyrsoideo ferrugineo-tomentoso, calycis ore truncato vix dentato, vexillo alisque extus glabris, carina apice villosa, filamento vexillari hinc ad medium adnato, legumine tomento brevissimo rufo-sericeo.—Fernando Po, (cultivated), *Vogel.*

Arbor parva. *Folia* 1-2-pedalia; foliola opposita, 3-6 poll. longa, 1½-2½ poll. lata, acuminata, basi angustata v. cuneata, petiolulo 3-4 lin. longo, nervis primariis subtus parallelis prominentibus rufis. *Pedunculi* jam infra medium floriferi, ramulis nodiformibus v. vix evolutis plurifloris inferioribus remotis, summis approximatis. *Flores* quam in cæteris speciebus majores, purpurei, pedicellis 1-2 lin. longis. *Calyx* ferrugineo-velutinus, basi bracteolis 2 parvis appressis stipatus. *Vexillum* pollicare, fere orbiculatum, basi late truncatum, crassum, glabrum; alæ lineari-oblongæ, longe unguiculatæ, vexillo paullo breviores; *carinæ* petala alis subsimilia, paullo latiora. *Filamentum* vexillare versus medium cum cæteris connatum, basi apiceque liberum. *Legumen* lineare v. lineari-lanceolatum, basi angustatum et breviter stipitatum,

plano-compressum, coriaceo-lignosum, ad suturam utramque
præsertim seminiferam incrassatum, 3-6-spermum. *Semina*
orbiculata, funiculo basi carunculato.

Notwithstanding the coherence of the tenth stamen, we have
no hesitation in referring this plant to *Milletia*, an Asiatic and
African genus, numerous in species, including the two which
Hochstetter has endeavoured to distinguish under the name of
Berrebera. The pod, of all the species where it is known, is
intermediate between that of the shrubby *Tephrosiæ* of the
section *Mundulia*, and that of *Sphinctolobium virgilioides* : the
valves adhere closely round the seeds till perfect maturity, when
the pod, in drying up, opens sometimes, if not always, in two
valves.

PLATE XXXII. XXXIII. *Fig.* 1. wing of the corolla ; *f.* 2.
keel ; *f.* 3. stamens and pistil ; *f.* 4. pistil : *all magnified*.

β. *Aboensis*, pube copiosiore, foliolis angustioribus, bracteolis
majoribus ramulis floriferis magis evolutis.—Abòh, *Ansell*.
—Very near the Fernando Po form, and not distinguishable
as a species. The specimens are, however, incomplete and
without fruit.

Besides the above, the Senegambian collection contains
several species of this and allied genera. Among them two, or
perhaps three, have every appearance of *Milletiæ*, though
without fruit : a fourth, agreeing with Schumacher's and Thonning's *Robinia cyanescens*, has a somewhat different habit : a
fifth, the *Lonchocarpus laxiflorus*, Guill. et Perr., appears to be
congener with the Abyssinian and Nubian *Philenoptera ;* and a
sixth, in flower only, is evidently the type of a new genus. To
some of the above may possibly be referable the *Robinia multiflora*, Schum. et Thonn., *R. Thonningii*, Schum., and *R. Guineensis*, Willd., all three Guinea plants.

There is likewise one W. Tropical African species described
of the American genus *Andira*, viz. *A. Africana*, Guill. et Perr.
from Senegambia.

BAPHIA, Afzl. Char. Gen. ref.—*Calyx* spathaceus, antice fissus,
postice integer vel 3-5-denticulatus. *Corollæ* æstivatio papi-
lionacea ; vexillum orbiculatum ; alæ ovato-oblongæ ; carina

falcata. *Stamina* 10, libera, omnia fertilia, antheris oblongis. *Ovarium* subsessile, pluriovulatum, stylo incurvo
brevi apice stigmatifero. *Legumen* oblongo-lineare, plano-
compressum, rectum (v. falcatum?), valvulis coriaceis maturitate dehiscentibus, marginibus leviter incrassatis.—Frutices
v. arbores Africanæ, *Dalhousiæ* affines, foliolis ad apicem
petioli solitariis, pedicellis axillaribus fasciculatis brevibus
unifloris apice bibracteolatis.

1. *B. spathacea,* Hook. fil. ; foliolis oblongis acuminatis cauleque glabris, calyce subcoriaceo ferrugineo-puberulo, bracteolis ovatis.—Bassa Cove, *Ansell.*

Rami ramulique virgati, superne interdum leviter puberuli.
Foliorum petiolus semipollicaris, basi et apice leviter incrassatus et articulatus, foliolum 3-4 poll. longum, 1½ latum,
acuminatum, basi rotundatum, coriaceum, supra nitidulum
et obscure reticulatum, subtus pallidius, venis prominulis
reticulatum. *Folia floralia* superiora minora et caduca, et
flores in racemum bracteatum dispositos apparent. *Pedicelli*
per 2-4 fasciculati, 2 lin. longi. *Bracteolæ* lineam longæ,
calyci adpressæ. *Calyx*s emipollicaris, recurvo-adscendens,
acuminatus, apice obtusiusculus et integer, antice usque ad
basin fissus, extus pube ferruginea subsericeus. *Petala*
tenuia, glabra; vexillum calycem vix æquans, orbiculatum,
emarginato-bifidum, brevissime unguiculatum; alæ vexillo
paullo breviores; carinæ petala alis æquilonga sed latiora.
Ovarium villosum, apice attenuatum.

2. *B. pubescens,* Hook. fil. ; ramulis pedicellisque ferrugineo-
pubescentibus, foliolo obovato-oblongo acuminato supra
glabro nitido subtus pubescente v. glabrato, calyce subcoriaceo puberulo, bracteolis orbiculatis.—In *Vogel's* collection,
without the precise station.

Rami teretes, superne præsertim parce pubescentes. *Foliolum*
2-3 poll. longum, 1-1½ poll. latum, apice in acumen lineare
obtusum productum, basi subacutum; in petiolo 2-3 lin.
longo articulatum. *Inflorescentia B. spathaceæ,* pedicellis
plurimis 4-5 lin. longis sessilibus v. rarius pedunculo communi brevissimo fultis. *Calyx* oblique obovato-oblongus

apice obtusus, tenuissime puberulus. *Vexillum* sessile, orbiculatum, emarginato-bifidum, calycem æquans; alæ lineari-
oblongæ, vexillo sublongiores; carinæ petala alis consimilia
nisi paullo latiora. *Legumen* (vix maturum) 3 poll. longum,
9 lin. latum, rectum, stylo acuminatum, valvulis planis
coriaceis glaberrimis. *Semina* circa 3.
Differs from the preceding species in its smaller flowers,
with a less coriaceous calyx shorter in proportion to.the petals,
besides the general pubescence of the branches.
3. Baphia *hæmatoxylon*, Hook. fil.; glabra, foliolis ovali-
oblongis acuminatis, calyce membranaceo glabro tridentato.—
Podalyria hœmatoxylon, *Schum. et Thonn. Beskr. p.* 202.—
Carpolobia versicolor, *G. Don, Gard. Dict.* 1. *p.* 370?—
Cape Coast, *Don, Vogel;* also Guinea, *Thonning.*
Although we have no hesitation, at the suggestion of Dr.
Planchon, in referring the above three plants to the genus
Baphia of Afzelius, so imperfectly figured and described by
Loddiges (Bot. Cab. t. 367); yet we do not feel justified in
identifying the *B. hæmatoxylon* as a species with the original
B. nitida, which is said to have pinnate leaves. Brunner (Flora
1840, v. 2. Beibl. p. 22), describes a plant which he supposes
may be *B. nitida*, which is evidently the *Dialium nitidum*, and
can hardly be the one Afzelius had in view. Desvaux has
described, under the name of *Delaria pyrifolia*, a Guinea plant
which must belong either to the present or to one of the two
following genera. The calyx and inflorescence described by him
are those of *Baphia*, but the broad leaves and stipitate ovary
refer rather to *Leucomphalus*. The same author's *Delaria
ovalifolia* is probably very different. We know of no Brazilian
plant at all like it.
BRACTEOLARIA, Hochst. Char. Gen. *Calyx* spathaceus antice
fissus, demum bipartitus, segmentis reflexis, æstivatione valvatis, integris v. postico bidentato. *Petala* subæquilonga,
calycem superantia; vexillum orbiculatum, alæ et carinæ
petala ovata. *Stamina* 10, libera, omnia fertilia, antheris
oblongis. *Ovarium* subsessile, villosissimum, biovulatum,

322 FLORA NIGRITIANA.

stylo incurvo glabro apice stigmatifero.—Frutices v. arbores
Africanæ, *Baphiæ* similes, foliolis ad apicem petioli solitariis,
floribus pedicellatis ad axillam bracteæ solitariis in racemos
axillares v. ad apicem caulis paniculatos dispositis.

Though closely allied to *Baphia*, and like that genus very
near to *Dalhousia*, the differences in the calyx and inflorescence
induce us to maintain these genera as distinct, at any rate until
the fruit shall be knowu.

1. Bracteolaria *polygalacea*, Hook. fil. ; puberula, foliolis ovali-
ellipticis acuminatis basi rotundatis cuneatisve supra glabratis, calycis segmentis integris, bracteolis orbiculatis.—Carpolobia dubia, *G. Don, Gard. Dict.* 1. *p.* 370.—Sierra Leone,
Don ; Grand Bassa, *Vogel.*

Frutex scandens, ramulis puberulis, striatis. *Petioli* 1-2-pol-
licares, pilosiusculi. *Foliolum* 3-5 poll. longum, 1½-2 poll.
latum, acuminatum, basi rotundatum v. subcordatum, pallide
virens. *Paniculæ* axillares et terminales, compositæ e racemis gracilibus 2-3 poll. longis, puberulis. *Bracteæ* minutæ.
Pedicelli sparsi, 1-2 lin. longi. *Calyx* 2 lin. longus, ante
anthesin ovoideus, obtusus, per anthesin fissus in segmenta 2
reflexa, membranacea, ovata, concava, extus puberula. *Bracteolæ* calyci appressæ, parvæ, ciliatæ. *Petala* subæquilonga,
calyce dimidio longiora ; vexillum sessile, late obcordatum,
reflexum, maculis luteis notatum ; alæ patentes, late oblongæ
obtusæ, brevissime unguiculatæ. *Ovarium* villosum, stylo
glabro.

There is another unpublished species of *Bracteolaria* amongst
Heudelot's Senegambian plants.

LEUCOMPHALUS, *Benth.* (nov. gen.)

Calyx demum bipartitus, segmentis reflexis, æstivatione valvatis integris. *Petala* subæquilonga, calycem superantia ;
vexillum late obovatum integrum ; alæ lineari-oblongæ ;
carinæ petala alis similia nisi latiora. *Stamina* 10, libera,
omnia fertilia, antheris linearibus. *Ovarium* longe stipitatum, pluriovulatum. *Legumen* longe stipitatum oblique

semi-orbiculatum, subfalcatum, valvulis coriaceis convexis, marginibus vix incrassatis. *Semina* pauca vel solitaria, funiculo in carunculum incrassato.—Frutex Africæ tropicæ, *Baphiæ* et *Bracteolariæ* habitu similis.

1. Leucomphalus *capparideus,* Benth. — *Planch. in Hook. Ic. t.* 784. (Tab. XXXI.) Fernando Po, in woods, *Vogel. Frutex* e basi plerumque ramosus, ramulis gracilibus glabris lævibus. *Folia* glabra, unifoliolata. *Petiolus* ½-1-pollicaris. *Foliolum* ovatum v. ovato-oblongum, acuminatum, basi acutum v. rotundatum, 3-5 poll. longum, 1½-2 poll. latum, rigide chartaceum, nitidulum, reticulatum. *Racemi* v. paniculæ axillares et terminales, ramis brevibus. *Pedicelli* breves, secus ramos solitarii, sparsi. *Bracteolæ* sub calyce parvæ orbiculatæ. *Flores* magnitudine *Bracteolariæ polygalaceæ* et iis subsimiles. *Legumen* ¾ poll. longum, 6 lin. latum, stipite pollicari fultum, valvulis glabris flavo-rufescentibus crassiusculis. *Semen* sæpius solitarium, transverse oblongum, funiculo brevi crasso fungoso.

This again is very near to the two preceding genera, and especially to *Bracteolaria,* of which it has the calyx and inflorescence, but the pod is very remarkable, nearer to that of *Swartzia* than of any other leguminous genus. The æstivation of the corolla, however, leaves no doubt as to its place among *Sophoreæ,* near *Baphia* and the allied genera.

Plate XXXI. *Fig.* 1. calyx unopened; *f.* 2. flower (an inaccurate representation); *f.* 3. anther; *f.* 4. ovary.

1. Sophora *tomentosa,* Linn. *DC. Prod.* 2. *p.* 95.—S. crassifolia, *Duham.,* et S. littoralis, *Schrad. DC. l. c.*—S. nitens, *Schum. et Thonn. Beskr. p.* 201.—Cape Palmas, *Vogel.*—A common sea-coast plant, both in the New and the Old World.

1. Parkinsonia *aculeata,* Linn.—Frequent along the west coast of Africa, as well as in East India, and in Tropical America, from whence it is usually said to have been introduced to the Old World. The occurrence, however, of a second species amongst Zeyher's South African plants, would rather tend to

show that Africa may be also the native country of *P. aculeata*.

1. Guilandina *Bonduc*, Linn.—Cape Palmas, *Vogel.*—Abundant along the west coast, and like the *Parkinsonia,* found also in Asia and America.

1. Cæsalpinia *pulcherrima*, Sw.—Sierra Leone, *Vogel, Don.* —Cultivated here, as in other Tropical countries, for the beauty of its flowers.

An undescribed species of the Asiatic genus *Mezoneurum* occurs in Heudelot's Senegambian collection.

1. Cassia *Sieberiana*, DC. *Prod. 2. p.* 489.—Cathartocarpus conspicua, *G. Don, Gard. Dict.* 2. p. 453.—Sierra Leone, *Don ;* Senegal.

2. Cassia *lævigata,* Willd.—*Vog. Syn. Cass. p.* 19.—Sierra Leone and Fernando Po (cultivated), *Vogel.*—An American species found from Mexico to Brazil, but perhaps in some instances cultivated.

3. Cassia *occidentalis,* Linn.—*Vog. Syn. Cass. p.* 21.—Abundant along the coast, *Vogel, Don,* &c.; common also in Tropical America and East India, but often in cultivation only. Among the numerous varieties observed of this species, two occur most frequently in W. Tropical Africa, the one with smaller leaflets, and shorter and straighter pods, the other with larger leaflets and longer curved pods. To the latter form belongs the *Chamæfistula contorta,* G. Don, Gard. Dict. 2. p. 452.

4. Cassia *obtusifolia,* Linn.—*Vog. Syn. Cass. p.* 24.—Fernando Po, Cape Palmas, &c.; abundant near habitations, *Vogel ;* Senegambia; frequent in America. It may not be specifically distinct from the E. Indian *C. Tora,* Linn., which is also found in Senegambia, and other parts of West Africa.

5. Cassia *Absus,* Linn.—*Vog. Syn. p.* 50.—C. viscosa, *Schum. et Thonn. Beskr. p.* 205.—On the Quorra at Attah, *Vogel ;* Senegambia, also Abyssinia and Egypt, and frequent in East India.

6. Cassia *mimosoides,* Linn.—*Vog. Syn. p.* 68.—Cape Coast,
and Accra, *Vogel;* Cape Palmas, *Ansell.* Common along
the west coast of Africa, as well as in East India. Besides
the *C. microphylla,* Willd., and the other synonyms adduced
by Vogel, this species includes also the *C. geminata,* Vahl,
described by Schumacher and Thonning. The other W. African *Cassiæ,* are: *C. Afzeliana,* Vog. from
Sierra Leone, *C. podocarpa,* Guill. et Perr. from Senegambia,
and *C. obovata,* Linn., and *C. micrantha,* Guill. et Perr., both
of which extend from Senegambia to Egypt and Arabia.

The *Cordyla calycandra* from Senegambia, and the *Swartzia
marginata,* Benth., from Angola, are the only known West
African species of the sub-tribe *Swartzieæ.* The authors of
the Flora Senegambiæ had already recognized, in their addenda
to the first volume, that their *Calycandra* was congener with
Loureiro's *Cordyla,* but it does not appear that Walpers was
justified in identifying the Senegambian plant with Loureiro's
East African species.

The Tamarind (*Tamarindus Indicus,* Linn.) is not uncommon on the W. African coast, but most probably in cultivation
only.

1. Afzelia *bracteata,* Vog. MS. (PLATE XXXIV. XXXV.) ramis
foliisque glabris, foliolis 4-5-jugis oblongo-ellipticis obovatisve obtusis supra nitidis subtus subglaucis, paniculis terminalibus subsimplicibus cano-pubescentibus, bracteis ovatis
reflexis, petali summi ungue lobis calycinis triplo longiore.—
Sierra Leone, *Vogel;* Senegambia, *Heudelot.*

Arbor mediocris, ramis teretibus, ramulis pendulis striatis
verruculatis. *Petiolus* communis 3-4-pollicaris, supra obscure
sulcatus. *Foliola* opposita, brevissime petiolulata, 2-3 poll.
longa, 1-1½ poll. lata, coriacea. *Panicula* erecta, folio æquilonga, basi tantum divisa in racemos dense floridos, rhachide
bracteis calycibusque ut in *A. Africana,* pube brevi canescentibus. *Bracteæ* 3-4 lin. longæ, ovatæ, obtusæ, diu persistentes et per anthesin reflexæ. *Flores* (teste Heudelot) pulchre coccinei et odorati, pedicellis suberectis quam bractea
subtendens paullo longioribus. *Bracteolæ* ad apicem pedi-

celli bracteis similes sed paullo minores. *Calycis* tubus
4-5 lin. longus, cylindricus, lobi 4 lin. longi, ovati, obtusi,
extus pubescentes, intus colorati. *Petalum* summum seu
vexillum erectum, ungue demum ultra pollicem longo puberulo, lamina late rotundata bifida; petala lateralia minima.
Stamina 6-8, vexillo breviora. *Ovarium* stipitatum, oblique
lanceolatum, pubescens, in stylum filiformem incurvum
desinens, 8-10-ovulatum.

PLATE XXXIV. XXXV. *Fig.* 1. flower; *f.* 2. pistil; both magnified.

The original *A. Africana*, Sm., from Sierra Leone, gathered
also by Heudelot in Senegambia, has much larger leaves and
leaflets, and the flowers considerably smaller; a third undescribed species was found by Heudelot in Senegambia, and if,
as is supposed, the *Pancovia* of Willdenow, from Guinea, belongs
to this genus, it constitutes a fourth species.

1. Anthonota, sp. n.?—Sierra Leone, *Don.* The specimen is
not in a state to describe.

Besides the above and the original *A. macrophylla*, Pal. de
Beauv., from Oware, there are two other undescribed species of
Anthonota, both in Heudelot's Senegambian collection.

BERLINIA, *Soland.* (nov. gen.)

Char. Gen.—*Alabastrum* bracteolis 2 tandem bivalvatim apertis
persistentibus inclusum. *Calycis* tubus cylindricus, limbus
5-partitus. *Petala* 5, summum longissime unguiculatum,
cætera lineari-spathulata, sessilia, calyci subæqualia. *Stamina* 10, longe exserta, omnia fertilia. *Ovarium* stipitatum,
pluriovulatum, in stylum filiformem attenuatum.—Arbor
Africæ tropicæ *Afzeliæ* et *Anthonotæ* affinis, ramulis pendulis, foliis abrupte pinnatis, racemis v. paniculis sessilibus
terminalibus.

1. Berlinia *acuminata*, Soland. *in Herb. Banks MS.* (fide
Planchon in Herb. Hook.)—Bassa Cove, *Ansell,* also from
Senegambia, *Heudelot.*

Arbor (fide Heudelot) 20-30-pedalis, ramulis pendulis glabris
novellis ferrugineo-puberulis. *Foliorum* petiolus communis v.

semipedalis, subteres, glaber; foliola 3-4-juga brevissime
petiolulata, pleraque opposita v. subalterna, oblonga, acuminata, basi angustiora at obtusa, sæpius inæquilatera et sub-
falcata 4-7 poll. longa, circa 2 poll. lata, coriacea, nitidula,
glabra, subtus reticulato-venosa. *Racemi* terminales, nunc
6-8-pollicares subsimplices nunc a basi ramosi et breviores,
corymbum foliis breviorem formantes, pube brevissima canescentes. *Bracteæ* brevissimæ, orbiculatæ, ante anthesin caducæ.
Pedicelli fere pollicares, rigiduli. *Bracteolæ* pollicem excedentes, ante anthesin clausæ, alabastrum simulantes oblique
clavatum obtusum basi longe attenuatum, per anthesin apertæ,
obovato-spathulatæ, obtusissimæ, crassæ, intus extusque pubescentes. *Flores* albi, usque ad anthesin intra bracteolas omnino reconditi. *Calycis* tubus semipollicaris, glaber; laciniæ
paullo breviores, oblongæ, obtusæ, fere glabræ, petaloideæ.
Petalum summum 2-2¼-pollicare, extus pilosum, basi in unguem longum canaliculatum et subulatum angustatum, lamina
late orbiculata, profunde bifida, margine crispula; lateralia
et inferiora inter se conformia, lineari-lanceolata, calycem
æquantia. *Stamina* basi pilosula, summum a cæteris dis-
cretum; filamenta apice attenuata; antheræ ovato-oblongæ,
versatiles. *Ovarium* breviter stipitatum, villosum.

Ansell's specimen has larger leaves and flowers, and a longer
simple raceme than Heudelot's, where the raceme is branched
and forms a panicle, and possibly the two may belong to distinct
species, although in every other respect they appear identical.
The fruit is unknown, and therefore its exact relation to the
two preceding African genera, *Afzelia* and *Anthonota*, to the
Asiatic *Intsia Amboinensis*, (*Outea bijuga*, Wall.?), and to the
American *Eperua*, *Parivoa* and *Outea*, cannot at present be
determined, although there is no one of these genera to which it
could be united more than to the other. The distinctions drawn
from the flower between the three African genera may thus be
shortly stated :

Afzelia. Flores jam ante anthesin e bracteolis exserti. Calycis
segmenta 4. Petala 1-5. Stamina fertilia 6-10.

Anthonota. Flores ante anthesin intra bracteolas inclusi. Calycis segmenta 4. Petala 1-3. Stamina fertilia 3-8.

Berlinia. Flores ante anthesin intra bracteolas inclusi. Calycis segmenta 5. Petala 5. Stamina fertilia 10.

Schumacher and Thonning describe a species of *Schotia,* from Guinea, under the name of *S. simplicifolia,* which is entirely unknown to us.

Three species of *Bauhinia* are natives of West Tropical Africa, *B. rufescens,* Lam., from Senegambia, found also in South Africa and the Mauritius; *B. Adansoniana,* Guill. et Perr., and *B. reticulata,* DC., both from Senegal. The *B. Thonningii,* Schum., appears to be a variety of the latter, distinguished by the leaves slightly downy underneath, and the pods densely covered with rusty down, it occurs both in Senegal and Guinea. The Nubian *B. tamarindacea,* Delile, supposed by the authors of the Flora Senegambiæ to be the same species, has a very different pod.

1. Cynometra *Vogelii,* Hook. fil.; glaberrima, foliolis unijugis elliptico-oblongis utrinque obtusis apice emarginatis aveniis, floribus in racemos axillares fasciculatis breve pedicellatis decandris, ovario villoso, legumine oblongo falcato rugosissimo.—On the Niger, at the confluence, *Vogel, Ansell;* also in Senegambia.

Arbor teste Vogelio habitu *Mali,* nunc frutex arborescens (15-20-pedalis ex Heudelot). *Ramuli* cortice griseo tecti verruculosi. *Petiolus* 2 lin. longus. *Foliola* 2-2¾ poll. longa, 9-12 lin. lata, plana, coriacea, eglandulosa : *Gemmæ* floriferæ axillares, obtectæ bracteis imbricatis ovato-cymbiformibus obtusis scariosis ciliatis. *Pedicelli* graciles, 2 lin. longi, puberuli. *Flores* 3 lin. diametro, rosei et odori ex Heudelot, albi teste Vogel. *Sepala* 4, petaloidea, lineari-oblonga, subciliata, extus puberula. *Petala* 4, patula, sepalis paullo longiora, oblonga, obtusa. *Filamenta* petalis dimidio longiora, antheris parvis. *Ovarium* breviter stipitatum, uniovulatum, stylo læviusculo curvato. *Legumen* 1-1½ poll. longum, ¾ poll. latum, utrinque obtusum, valde convexum, dorso canaliculatum, suberosum. *Seminum* testa tenuis, cotyledones siccæ atræ.

2. Cynometra? *tetraphylla*, Hook. fil.; glaberrima, foliolis bijugis valde inæquilateris obtusis coriaceis subtus reticulatis jugi inferioris ovato-rotundatis, jugi superioribus majoribus ovatooblongis.—Sierra Leone, on the ascent to the Sugar-loaf Mountain, *Don.*

A single specimen, in a very imperfect state, without flowers or fruit, but most probably a *Cynometra.* The lower pair of leaflets are from half an inch to an inch long, the upper ones twice as long. In one of the axillæ are several of the scariose bracts, which usually surround the inflorescence in the genus.

Heudelot's collection contains a third undescribed species of *Cynometra,* from Senegambia.

1. Dialium *Guineense*, Willd.—D. nitidum, *Guill. et Perr. Fl. Seneg.* 1. *p.* 267. *t.* 58.—St. Thomas and Sierra Leone, *Don ;* from Senegal to Guinea. All the flowers we have opened have one petal, as observed by Bennett in his valuable notes on the genus, Pl. Jav. Rar. p. 136 et seq.

2. Dialium *discolor, Hook. fil.*—Codarium discolor, *DC. Prod.* 2. *p.* 520.—On the Quorra, *Vogel, Ansell.*

Besides the differences in foliage mentioned by De Candolle, the panicle is looser, with fewer flowers than in *D. Guineense,* the ultimate cymes more compact, the flowers rather smaller, containing two petals instead of one, and the anthers are nearly sessile.

The Senegambian collections contain also two species of *Detarium :* D. *Senegalense,* Gmel., and *D. microcarpon,* Guill. et Perr., and an undescribed species of the American genus, *Crudya.*

1. Parkia *Africana,* Br. *App. Oudn. p.* 234.—Sierra Leone, *Don,* and apparently frequent along the coast.

1. Erythrophleum *Guineense,* G. Don, *Gard. Dict.* 2. *p.* 424.— Fillæa suaveolens, *Guill. et Perr. Fl. Seneg.* 1. *p.* 242. *t.* 55. —Sierra Leone, *Don ;* Senegal and Guinea.

1. Pentaclethra *macrophylla,* Benth. *in Hook. Journ. Bot.* 4. *p.* 330.—Fernando Po, *Vogel ;* Senegal.

The *Entada scandens* occurs among Heudelot's Senegambian plants, and a second species from the same country is described in the Flora Senegambiæ, under the name of *E. Africana.*

1. Piptadenia? *Africana,* Hook. fil.; inermis, ramulis petiolis inflorescentiaque puberulis demum glabratis, glandulis nullis, pinnis 10-12 jugis, foliolis multijugis linearibus supra nitidis, floribus glabris, ovario subsessili glabro.—On the Niger, at Abòh, *Vogel, Ansell.*

Arbor excelsa. *Folia* iis *P. communis* simillima; pinnæ 2-3-pollicares; foliola 4 lin. longa, leviter falcata, obtusa v. acutiuscula, basi valde obliqua, glabra v. minute ciliata, supra nitidula, subtus subrufescentia. *Spicæ* bipollicares, in paniculas axillares v. terminales dispositæ. *Flores* polygami secus rhachin sessiles, albidi, cum staminibus 2 lin. longi. *Calyx* brevis, cupuliformis, ore truncato leviter undulato. *Petala* lineari-lanceolata, acuta. *Stamina* 10, antheris apice glanduliferis. *Ovarium* subsessile, glabrum, pluriovulatum.

These specimens, in flower only, can only be distinguished from the Brazilian *P. communis* by the absence of all glands on the petiole, and by the nearly sessile ovary; characters which may possibly prove not to be absolutely constant in the genus.

Four other *Mimoseæ*, with glanduliferous anthers, are natives of West Tropical Africa, viz.: *Tetrapleura Thonningii*, Benth.,* from Guinea; *Prosopis? oblonga*, Benth., from Senegambia, to which may perhaps be referred the *Coulteria Africana*, Guill. et Perr.; *Dichrostachys nutans*, Benth., from Senegambia, extending through a considerable portion of Tropical Africa; and *Neptunia oleracea*, Lour., a common aquatic plant, both in the New and the Old World.

1. Mimosa *asperata*, Linn.—On the Niger, *Vogel.*—Abundant

---

* In the Spicilegia Gorgonea, above, p. 131, Parlatore has given the name of *Tetrapleura* to an Umbelliferous genus, not being then aware of my having previously published the Mimoseous plant referred to in the text under the same name. The volume of Hooker's Journal of Botany, in which it is described, though published in 1842, did not reach Florence till 1847.—(G. B.)

on the west coast, as well as in East Tropical Africa and America.

1. Schranckia *leptocarpa,* DC. *Prod.* 2. *p.* 443.—Accra, *Don;* Cape Coast, *Vogel.*—An American plant, not unfrequent in Guinea and Brazil, the only species of the genus hitherto found in the Old World.

1. Leucæna *glauca,* Benth. *in Hook. Journ. Bot.* 4. *p.* 416.— Fernando Po, *Vogel.*—Cultivated here, as in many parts of Africa and Asia. Its native country is probably either Tropical America, or some of the islands of the Pacific.

1. Acacia *Farnesiana,* Willd.—Sierra Leone, *Vogel,* probably cultivated here, as in other parts of Africa, Asia, America, and the south of Europe. It is indigenous to America, and said to be also a true native of East India.

2. Acacia *Adansonii,* Guill. et Perr. *Fl. Seneg.* 1. *p.* 251.— Sierra Leone, *Don,* judging from a very imperfect specimen ; Senegal and Guinea.

3. Acacia *Arabica,* Willd., var. *tomentosa,* Benth. *in Lond. Journ. Bot.* 1. *p.* 500.—On the Niger, *Vogel;* Senegal to Egypt and Arabia, and thence to the East Indies.

4. Acacia *ataxacantha,* DC. *Prod.* 2. *p.* 459.—On the Quorra, *Vogel ;* Senegal.

5. Acacia *pentagona,* Hook. fil.—Mimosa pentagona, *Schum. et Thonn. Beskr. p.* 324 ; fruticosa, glabra, aculeis plurimis sparsis recurvis, stipulis lineari-lanceolatis striatis persistentibus, glandula petiolari elevata, pinnis 6-10-jugis, foliolis ultra 20-jugis linearibus, capitulis globosis subpaniculatis, calyce corolla parum breviore, ovario glabrato.—Cape Coast, *Vogel ;* Guinea.

Very near to the Asiatic *A. pennata,* but easily recognized by the rigid stipules, from 2 to 4 lines long.

The other West Tropical African species of *Acacia* are *A. Sieberiana,* DC., *A. Sing,* Guill. et Perr., and *Prosopis dubia,* Guill. et Perr., all three from Senegal, and most probably closely allied to each other, if not forming but one species ; *A. Seyal,* Delile, extending from Senegal to Nubia and Upper Egypt ; *A. fasciculata,* Guill. et Perr., from Senegal ; *A. albida,* Delile,

332 FLORA NIGRITIANA.

from Senegal to Upper Egypt; and *A. saccharata*, Benth.,
A. Verek, Guill. et Perr., and *A. macrostachya*, Reichb., all three
from Senegal.

1. Albizzia *altissima*, Hook. fil.; ramulis petiolisque ferrugineo-
puberulis, stipulis subulatis deciduis, pinnis 5-6-jugis, foliolis
8-multijugis lineari-oblongis inæquilateris glabris, glandula
in petiolo et inter pinnas supremas, pedunculis solitariis axillaribus, capitulis amplis multifloris, floribus sessilibus glabris,
calyce tubuloso 5-fido corolla dimidio breviore.—Cape Coast
and Abòh, *Vogel*.

Arbor ex Vog. altissima. *Folia* 4-6 poll. longa; foliolis 4-5 lin.
longis, 1½-2 lin. latis, obtusis v. acutiusculis. *Pedunculi* graciles, in specimine solitarii (an tamen et fasciculati more
affinium occurrunt ?) pollicares. *Flores* albi. Affinis *A. amaræ*
et *A. myriophyllæ*.

Besides the above, the following *Albizziæ* have been found
in West Tropical Africa : *A. Lebbek*, Benth., from Senegambia,
a common Asiatic and Egyptian plant, perhaps cultivated in
West Africa; *A. rhombifolia*, Benth., from Senegambia, which
may not be distinct from *A. glaberrima*, Benth., from Guinea;
and *A. ferruginea*, Benth., extending from Senegambia to
Abyssinia.

The *Calliandra portoricensis*, Benth., is described by Schumacher and Thonning from Guinea, under the name of *Mimosa
Guineensis*. It is a Tropical American tree, frequently cultivated for ornament in Egypt and other countries bordering on
the Mediterranean, and probably introduced into Guinea.

Two species of *Zygia* complete the list of West Tropical
African *Leguminosæ* : 1. *Inga Zygia*, figured by De Candolle,
and named by Walpers *Zygia Brownei*, although it may be
doubtful whether Browne's Jamaica plant be the same, and
2. *Mimosa adiantifolia*, Schum. et Thonn., which appears to
be identical with E. Meyer's South African *Zygia fastigiata*.

XLVII. CHRYSOBALANEÆ.*

1. Parinarium *excelsum*, Sab.—*DC. Prod. 2. p.* 527.—St. Thomas and Sierra Leone, *Don ;* Senegal.

The structure of the flower, both of this species and of the *P. Senegalense*, has been accurately described in the Flora Senegambiæ, but in the plates of both species the artist has misrepresented several analytical details, especially as to the insertion and arrangement of the stamens.

2. Parinarium *curatellæfolium*, Planch. *in Herb. Hook. ;* foliis oblongo-ellipticis obtusis basi inæqualibus, novellis subtus niveis, adultis concoloribus coriaceis margine undulatis supra scabrellis subtus tomentellis, calycis vix inæqualis laciniis lanceolatis acutis, staminibus fertilibus circa 8 unilateralibus.

—On the Quorra, at Patteh, *Vogel;* Senegal, *Heudelot.*

Arbor mediocris. *Ramuli* teretes tomentosi. *Stipulæ* lineari-lanceolatæ, caducissimæ. *Folia* 2-3 poll. longa, 1-1½ poll. lata, margine undulata et obscure crenata, basi obtusa et inæqualia, petiolo 2-4 lin. longo eglanduloso, novella supra puberula subtus niveo-tomentosa, adulta supra subglabra subtus vix pallidiora. *Paniculæ* 3-4-pollicares, tomentosæ, cymis laxe paucifloris secus ramos racemosim dispositæ. *Bracteæ* ovato-lanceolatæ, acuminatæ, concavæ, caducissimæ. *Flores* magnitudine *P. excelsæ.* *Calyx* molliter tomentosus, tubo basi hinc paullo amplior, laciniis acuminatis tubo æquilongis. *Petala* ovata, calyce breviora. *Stamina* fertilia vix calyce longiora, sterilia plura, brevia, valde inæqualia. *Ovarium* et stylus *P. excelsæ.*

I describe this chiefly from Heudelot's specimen, which is in flower, with young leaves. Vogel's has only old leaves and some peduncles, from which the fruit has fallen off, but I have little doubt of both belonging to one species.

3. Parinarium *polyandrum*, Benth.; foliolis oblongis ovatisve

---

* This and the succeeding Orders have been entirely worked up by Mr. Bentham, although he has generally been much assisted by the previous determinations of most of the species in the Hookerian Herbarium.

coriaceis glabris nitidis basi biglandulosis, calycis subæqualis
crassi laciniis petalisque orbiculatis obtusis, staminibus numerosis (circa 40) in orbem fere completam dispositis.—On the
Quorra, at Attah, *Vogel.*

Frutex arborescens, præter inflorescentiam glaber, ramis verrucosis, ramulis brevibus. *Stipulæ* ovatæ, caducissimæ. *Folia*
breviter petiolata, 3-4 poll. longa, 1-2 poll. lata, apice obtusa,
basi cuneata v. rotundata, margine recurvo, utrinque glaberrima ; glandulis ad apicem petioli sitis parvis scutellæformi-
bus. *Panicula* terminalis, brevis, cymoso-ramosissima, floribunda, ramis crassis, brevibus, pubescentibus. *Bracteæ* orbiculatæ, caducissimæ. *Calyx* fere 4 lin. longus, tubo crasso-
carnoso turbinato-campanulato subæquali, extus tomentoso
intus glabro ; laciniæ 2½ lin. longæ, ovato-orbiculatæ, concavæ, obtusissimæ, intus glabræ, extus pubescentes, æstivatione valde imbricatæ. *Petala* orbiculata, calycem subæquantia,
caducissima. *Stamina* circa 40, fertilia, petalis plus duplo
longiora, basi in orbem completum disposita et brevissime et
oblique connata, ad latus pistilliferum floris dejecta, ad latus
oppositum rariora, sterilia tamen nulla detexi. *Ovaria* in
floribus 2 examinavi, in altero duo vidi in altero tria, distincta,
omnia lateri calycis adnata et villosissima, singula more generis
biovulata, ovulis dissepimento spurio sejunctis. *Stylus* inter
ovaria a basi eorum natus, glaber, adscendens, more generis
flexuosus.

The multiplication of ovaries is singular among *Chrysobalaneæ,* but may not be constant even in the species, as there is
but a single specimen and I could only dissect two unopened
flowers. The glandular leaves, fleshy calyx, and numerous
stamens, might at first suggest the establishmont of a distinct
genus, but in a Malacca plant gathered by the late Dr. Griffith,*
which has also coriaceous biglandular leaves and a similar calyx,
there are but about twenty-five fertile stamens, with somes hort

* *P. Griffithianum,* foliis oblongis acuminatis coriaceis glabris nitidis
basi biglandulosis, calycis subæquali crassi laciniis petalisque orbiculatis
obtusis, staminibus fertilibus numerosis (circa 25) unilateralibus, sterilibus paucis minimis.

abortive ones on the opposite side, forming a gradual passage
through Jack's and other Eastern species to the more common
forms of *Parinarium*. On the other hand, those species which
De Candolle had placed in his section *Neocarya*, characterized
chiefly by the stamens being all fertile, prove to have in fact
fertile ones on one side, and sterile on the other, as in *Petrocarya;* and the genus must, if at all, be divided on other prin-
ciples.

Taking as the essential character of *Parinarium*, among *Chrysobalaneæ*, the spurious dissepiment which separates the ovules,
the species we are acquainted with may be distributed into three
sections, viz. :

§. 1. *Petrocarya* (*Balantium*, Desv.) Calyce æquali v. vix
gibbo, laciniis acutis, staminibus fertilibus sæpius paucis
(7-8 v. rarius 10-15.)—To this section belong the African
P. excelsum and *P. curatellæfolium*, the whole of the
known American species, (four from Guiana or the West
Indies, and three or four from Brazil), and most probably
among Asiatic ones the *P. Sumatranum* (*Petrocarya*, Jack.),
P. glaberrimum, Hassk., and *P. scabrum*, Hassk., which
three last I have not seen.

§. 2. *Sarcostegia*. Calyce æquali v. vix gibbo carnoso laciniis
obtusis, staminibus numerosis (25-40, rarius 11 ?), foliis
basi biglandulosis.—To this belong the African *P. polyandrum*, the Asiatic *P. Griffithianum*, and possibly also
P. Jackianum, (*Petrocarya excelsa*, Jack.) which is unknown to me.

§. 3. *Neocarya*. Calyce basi hinc gibboso-saccato, laciniis
obtusis, staminibus plurimis fertilibus (circa 15).—Confined to the two following African species.

4. Parinarium *macrophyllum*, Brown, *G. Don, Trans. Hort.
Soc. 5. p.* 452.—St. Thomas and Sierra Leone, *Don.*

The leaves are as much as nine inches long and five broad,
closely sessile and broadly cordate at the base ; in other respects
they much resemble those of *P. Senegalense*. The down of
the stem, leaves and inflorescence, the inflorescence itself and
the flowers are so much alike in the two species as to suggest

the idea that, notwithstanding the difference in the form of
the leaves, they may be mere varieties of one.

The *P. Senegalense,* Perr., has hitherto been found in Senegambia only.

1. Chrysobalanus *Icaco,* Linn.—Grand Bassa and Cape Palmas,
very common, *Vogel;* Senegal and Guinea, also in the West
Indies and in Tropical America, but possibly introduced there
from Africa. What is frequently mistaken for it is the *C.
pellocarpus,* which is truly indigenous and frequent in Tropical
America, and which, as well as the *C. oblongifolia,* from the
United States, has a much less fleshy fruit than the African
species.

2. Chrysobalanus *ellipticus,* Soland.—*DC. Prod.* 2. *p.* 526.—
Sierra Leone, *Don.*

The *C. luteus,* from Sierra Leone, mentioned by Sabine, is
not amongst Don's plants.

Besides the above *Chrysobalaneæ,* no species of the extensive
Order of *Rosaceæ* appears to have been founded in West Tropical Africa, although there is scarcely another district of the
globe where it is not more or less numerously represented.

XLVIII. COMBRETACEÆ.

1. Terminalia (Catappa) *glaucescens,* Planch. *in Herb. Hook.;*
foliis sparsis longiuscule petiolatis ovali-ellipticis breviter
supra acuminatis basi obtusis v. inæqualiter acutis eglandulosis nitidis subtus glaucescentibus, drupa samaroidea glau-
cescente ala longa apice retusa.—On the Quorra, at Attah,
Vogel.

Arbor mediocris, ramulis crassiusculis apice pubescentibus mox
glabratis. *Folia* secus ramulos sparsa, adulta semipedalia, 2½-3
poll. lata, in acumen breve obtusum producta, basi sæpius
rotundata, interdum valde inæqualiter angustata, coriacea,
supra glabra et demum lucida, pube brevissima marginata,
subtus glaucescentia ad costam venasque primarias puberula,
cæterum glabra v. pilis brevibus conspersa; petioli 1-1½-pol-
licares. *Flores* desunt. *Pedunculi* fructiferi axillares, pollicares, pubescentes, a medio ad apicem cicatrices ostendunt

crebras florum delapsorum. *Drupæ* samaroideæ, stipite 2-3 lin. longo fultæ, cum alis 1½-2 poll. longæ, 7-9 lin. latæ, apice obtusissimæ et sæpe emarginatæ, basi rotundatæ v. subcuneatæ ; drupa ipsa in media ala oblonga, sarcocarpio crassiusculo carnoso, endocarpio osseo. *Semina* perfecta non vidi.

Two other species are mentioned as West African, *T. macroptera* and *T. avicennoides,* Guill. et Perr., both from Senegambia.

1. Conocarpus *erecta,* Jacq., *β. procumbens,* DC. *Prod.* 3. *p.* 16.—Sierra Leone, *Vogel;* on the Gambia, *Don.*

According to Vogel, the woody stems of this species spread over the ground for a considerable space, sending up a number of erect branches.

The *Anogeissus leiocarpus,* Guill. et Perr., is a native of Senegambia.

1. Laguncularia *racemosa,* Gærtn. fil.—*DC. Prod.* 3. *p.* 17.— Grand Bassa and Fernando Po, *Vogel.* Also in Senegal, and a common maritime plant in Tropical America.

The *Guiera Senegalensis,* Lam., extends from Senegal to Cordofan.

1. Poivrea *grandiflora,* Walp. *Rep.* 2. *p.* 64.—Combretum grandiflorum, *Don, DC. Prod.* 3. *p.* 21.—C. Afzelii, *Don, Linn. Trans.* 15. *p.* 437.—Sierra Leone, *Don, Whitfield, Miss Turner ;* Munrovia, *Vogel.*—Fructus late 5-alatus. Embryo 5-angulatus, cotyledonibus crassiusculis convolutis, minus tamen quam in *P. comosa.*

It appears to have been by mistake that Don placed his *C. grandiflorum* in his octandrous division of *Combretum,* for although a tetramerous flower may have been found by chance, by far the greater part, all indeed that I have seen, are pentamerous, and the seed has the cotyledons certainly convolute. The thorns in this and other *Poivreæ* are formed of the lower portion of the petiole, which remains attached to the stem, becoming hard and prickly after the upper portion has fallen off with the leaf.

2. Poivrea *constricta,* Benth.; glabra in inflorescentiam pu-

338 FLORA NIGRITIANA.

berula, spinifera, foliis obovali-ellipticis vix acuminatis basi
obtusis, spicis brevibus simplicibus confertifloris, bracteis subnullis, calycis tubo adnato elongato, limbo infundibuliformi
supra basin in collum contracto, petalis cuneato-oblongis
extus villosis.—On the Quorra, at Abòh, often growing in the
water, *Vogel.*
Frutex erectus (ramis nunc sarmentosis? ex Vog.) cortice albido.
Petiolorum bases persistentes indurato-spinescentes. *Folia*
alterna v. rarius opposita, adulta 3-4 poll. longa, 2 poll. lata,
obtusa cum acumine brevissimo, basi rotundata v. rarius obsolete cordata, rigidule chartacea, utrinque viridia, reticulato-
venosa et creberrime minuteque punctata. *Spicæ* terminales,
rhachide vix pollice longiore puberula. *Bracteæ* minutissimæ
v. omnino desunt. *Flores* pentameri, semel tamen etiam te-
tramerum vidi. *Ovarium* (seu calycis tubus adnatus) 3 lin.
longum, 5-striatum, tenue, pubescens. *Calycis* pars libera
(limbus seu fauces) infundibuliformis, 6 lin. longa, supra
ovarium inflata, dein in collum longum constricta, apice campanulata et 5-loba, extus parce pubescens, lobis triangularibus
acutis æstivatione valvatis. *Petala* 1½ lin. longa, coccinea?
Stamina longiuscule exserta. *Ovula* 2, a funiculis longis
suspensa. *Fructus* non visus.
3. Poivrea *conferta*, Benth.; scandens, glabra v. vix puberula,
   spinifera, foliis obovali-ellipticis acuminatis basi obtusis, racemis brevissimis multifloris in paniculam capituliformem
   confertis, floribus pedicellatis tubo adnato elongato limbo
   tubuloso-campanulato, petalis late ovatis glabris.—In woods,
   Fernando Po, *Vogel.*
Frutex super arbores volubilis. *Ramuli* et petioli juniores
minute puberuli, mox glabrati. *Petiolorum* bases persistentes, indurato-spinescentes. *Folia* opposita, 3-5 poll.
longa, 2-2¾ poll. lata, chartacea, reticulata, utrinque viridia et creberrime puncticulata, petiolo semipollicari. *Paniculæ* fere corymbiformes, subsessiles v. ramulos breves terminantes, 1½ poll. latæ, e racemulis pluribus compositæ
quorum rhachides vix 5 lin. longi. *Bracteæ* parvæ setaceæ.
Flores pentameri, breviter pedicellati. *Ovarium* (seu calycis

tubus adnatus) 2½ lin. longum, puberulum. *Calycis* limbus (seu pars libera) glaber, 2½ lin. longus, dentibus breviter et late triangularibus. *Petala* coccinea, linea breviora, fere orbiculata. *Stamina* longe exserta. *Ovula* 2.

4. Poivrea *comosa*, Walp. *Rep.* 2. *p.* 64.—Combretum comosum, *G. Don, DC. Prod.* 3. *p.* 20.—Ejusdem var. bracteis latioribus quæ *Combretum intermedium*, G. Don, et DC. l. c.— Sierra Leone, *Don.*

In both varieties the flowers are pentamerous, and in the seeds of the first variety at least, the cotyledons are thin and convolute.

The only other W. Tropical African species is the *P. aculeata*, DC., from Senegal, which is also found in Cordofan.

1. Combretum *spinosum*, Don.—*DC. Prod.* 3. *p.* 20.—Sierra Leone, *Don.*

The bases of the petioles remain on the branches, and harden into thorns, as in most *Poivreæ;* but the flowers are tetramerous, and the cotyledons fleshy and not convolute. As in all the *Combreta* and *Poivreæ* I have examined, the seed is marked with as many angles as the fruit has wings, and even the embryo is moulded as it were into the same form, for the angles retain their position with regard to the seed, whatever be the texture or arrangement of the cotyledons.

2. Combretum *racemosum*, Pal. de Beauv.—*DC. Prod.* 3. *p.* 20.—To the synonyms adduced by Guillemin and Perrottet (Fl. Seneg. p. 285) must be added that of *C. leucophyllum*, Don, DC. l. c.—Sierra Leone, *Don.*

3. Combretum *fuscum*, Planch. in Herb. Hook.; scandens, inermis, ramulis inflorescentiaque ferrugineo-puberulis, foliis oppositis oblongis ellipticisve breviter acuminatis basi rotundatis demum glabratis, spicis brevibus densis paniculatis, bracteis minimis, calycis limbo tuboloso-campanulato, petalis spathulatis limbo calycino pluries brevioribus.—Sierra Leone, *Vogel;* Grand Bassa, *Ansell.*

Frutex scandens, ramis teretibus v. ad nodos compressiusculis. *Folia* majora 6-8 poll. longa, 2½-3 poll. lata, apice in acumen breve angustata, utrinque siccitate fusca et

z 2

glabra, superiora minora, floralia numerosa parva, in vivo ex
Vog. alba et e longinquo nitentia, siccitate tamen pariter ac
caulina fusca et opaca. *Petioli* omnino decidui, 2-4 lin.
longi. *Paniculæ* breves, floribundæ; spicæ ultimæ pedunculatæ, 6-8 lin. longæ, floribus confertis. *Bracteolæ* lineares,
crassæ, 1-2 lin. longæ, caducissimæ. *Flores* tetrameri. *Calycis* tubus adnatus brevissimus, limbus 1½ lin. longus, disco
staminigero tomentoso usque ad medium limbi attingente,
dentibus triangularibus acutis. *Petala* minima, glabra.
Ovula 2.

This agrees in many respects with Don's description of *C.
micranthum* from the same country; but he describes the flower
spikes as simple and axillary; and if the Senegal plant referred
to Don's species in the Flora Senegambiæ be really his, it is
very different from ours in the petals and other details of the
flower.

4. Combretum *cuspidatum*, Planch. in Herb. Hook.; scandens,
inermis, ramulis inflorescentiaque ferrugineo-puberulis, foliis
oppositis ovali-ellipticis acuminatis basi rotundatis demum
glabris, spicis paniculatis, bracteis minimis, calycis limbo cyathiformi minute dentato, petalis orbiculatis concavis integerrimis limbo calycis multo brevioribus.—Sierra Leone, *Vogel.*
Frutex scandens, ramuli ad nodos compressiusculi, novelli uti
petioli et pedunculi ferrugineo-pubescentes, demum glabrati.
Petioli 3-4 lin. longi, cum foliis ex toto decidui. *Folia* majora semipedalia, 3 poll. lata, summa minora, omnia siccitate
supra fusca et glabra, subtus pallidiora, glabra sed creberrime
glanduloso-punctata, apice in acumen semipollicare obtusum
abrupte producta, basi rotundata. *Paniculæ* axillares v.
terminales, basi foliatæ, apice nudæ, semipedales v. longiores,
pluries ramosæ, spicis ultimis 1-2-pollicaribus laxis. *Bracteolæ* perpaucæ adsunt lineares, parvæ, crassæ. *Flores* 4-
meri, parvi. *Calycis* tubus adnatus semilineam longus, hirsutus, limbus (seu fauces) cyathiformi-campanulatus, ¾ lin.
longus, dentibus brevissimis. *Petala* semilinea breviora,
reflexa, glabra, concava, integerrima, vix unguiculata. *Stamina*
fere 2 lin. longa. *Ovula* 2.

5. Combretum *sericeum*, Don.—*DC. Prod.* 3. *p.* 21.—C. herbaceum, *Don, l. c.*—Sierra Leone, *Don.*

Stems herbaceous, simple, a foot, or a foot and a half high, proceeding from a woody trunk, approaching in that respect to the habit of the East Indian *C. nanum.* The flowers are generally tetramerous, but probably sometimes pentamerous, as Don places the species in the decandrous division of the genus, and I have occasionally seen one of the petals much broader than the other, or even divided into two, with a very small calycine tooth between the two; but I have never observed more than eight stamens.

Besides the above species, nine *Combreta* have been described from W. Tropical Africa, viz. : *C. mucronatum*, Thonn., from Senegal and Guinea, from which however must be excluded *C. intermedium*, Don, which is a mere variety of *Poivrea comosa ; C. micranthum*, Don, from Senegal and Sierra Leone ; *C. paniculatum*, Vent., from Senegal, which is certainly distinct from *Poivrea comosa*, referred to it by Guillemin and Perrottet; *C. altum*, Perr.; *C. glutinosum*, Perr.; *C. chrysophyllum*, Guill. et Perr., and *C. nigricans*, Lepr., all from Senegal ; *C. tomentosum*, Don, from Sierra Leone ; and *C. macrocarpon*, Beauv., from Oware.

1. Quisqualis *ebracteata*, Pal. de Beauv.—*DC. Prod.* 3. *p.* 23. —Q. obovata, *Schum. et Thonn. Beskr. p.* 218.—On the Quorra, *Vogel ;* Senegal and Guinea.

XLIX. RHIZOPHOREÆ.

1. Rhizophora *Mangle*, Linn.—*DC. Prod.* 3. *p.* 32.—Senegambian coast, *Don*, and others.
2. Rhizophora *racemosa*, Mey.—*DC. Prod.* 3. *p.* 32.—Sierra Leone and Grand Bassa, *Vogel ;* Guinea Coast, *Herb. Hook.*

Both these are American species, at least as far as the specimens show, I can detect no differences. The Asiatic species, some of which are found on the eastern coasts of Africa do not appear any of them to have spread to the western coast.

1. Cassipourea *Africana*, Benth. ; glabra, foliis obovali-ellipticis oblongisve apice rotundatis vel brevissime obtuseque acumi-

natis integerrimis v. pauciserratis coriaceis.—On the Quorra, opposite Stirling, *Vogel.*

Folia 2 v. raro 3 poll. longa. *Inflorescentia C. ellipticæ. Flores* desunt. *Pedicelli* fructiferi lineam longi. *Calyx* persistens, 2½ lin. longus, coriaceus, glaber, ultra medium in lacinias 5 oblongas valvatim fissus. *Capsula* glabra, calyce brevior, obtuse trigona, apice depressa, trivalvis, trilocularis, dissepimentis membranaceis. *Semina* in loculis solitaria, (adjecto tamen ovulo abortivo) pendula, ovoidea: testa chartacea; albumen carnosum; embryo albumini fere æquilongus, rectus ; cotyledones foliaceæ, late ovatæ ; radicula cotyledonibus æquilonga, ad hilum spectans. *Stylus* in fructu persistens, flexuosus, apice incrassatus.

This may possibly be the species alluded to by Brown in the Appendix to Tuckey's Congo, and named by De Candolle *C. Congensis,* but not described. Without the flowers, (of which the buds in the specimen are still in the earliest stage), this plant can only be distinguished from the West Indian *C. elliptica* by the smaller and more coriaceous leaves.

ANISOPHYLLUM, *Don,* (gen. nov.)

Calyx liber v. basi breviter adnatus, cyathiformis, 4-fidus, lobis æstivatione valvatis. *Discus* carnosus fundum calycis occupans, inter stamina et ovarium glandulosus. *Petala* 4, lobis calycinis alterna, disco inserta, biloba, lobis laciniato-fimbriatis, laciniis subulatis apice inflexis. *Stamina* numero petalorum dupla, disco inserta ; antheræ versatiles, biloculares, loculis longitudinaliter dehiscentibus. *Ovarium* trilobum, triloculare, in speciminibus a me visis semiabortivum. *Frutices* v. arbores, foliis stipulatis alternis integerrimis coriaceis, 5-nervibus, floribus secus pedunculos supra-axillares sessilibus parvis.

1. Anisophyllum *laurinum,* Don MS. ; foliis quintuplinervibus, spicis ebracteatis.—Sierra Leone, *Don ;* Senegambia, *Leprieur,* ex Herb. Hook.

Frutex v. arbor, partibus novellis pube rara minuta appressa conspersis, demum glabratus. *Stipulæ* lanceolatæ, aliæ minimæ, aliæ 2-3 lin. longæ, caducissimæ. *Folia* adulta 3-4 poll. longa, 1-1½ poll. lata, ovata v. oblonga, acuta, basi

cuneata, coriacea, siccitate flavescentia, opaca et impunctata,
costis 5 subtus prominentibus, quarum 2, ad basin attin-
gentes, sæpe fere marginales sunt v. cum margine confluunt
et vix conspicuæ, 2 cum costa media paullo altius et inæqualiter confluunt, venulis transversis numerosis. *Pedunculi*
prope basin innovationum supra axillas inferiores solitarii,
2-3-pollicares, tenues, jam infra medium usque ad apicem
interrupte floriferi. *Bracteæ* omnino deesse videntur. *Flores*
arcte sessiles, 1½ lin. diametro. *Calyx* glaber v. minute puberulus, ultra medium fissus in lobos 4 late ovato-triangulares,
æstivatione valvatos. *Petala* calyce paullo longiora, glabra,
ad medium biloba, lobis irregulariter fissis in lacinias subulatas quarum pleræque apice inflexæ. *Stamina* petala subæquantia, glabra, filamentis apice inflexis, antheris ovatis.
Ovarium disco impositum, villosum, depresso-trilobum et
obsolete triloculare; nec ovula nec stylum detexi.

A remarkable plant, evidently allied to *Cassipourea*, notwithstanding its alternate leaves and apparently polygamous
flowers. Its exact affinity cannot however be determined, until
the perfect ovary and fruit shall have been seen. A second
species,* with the same foliage and structure of flowers, but
unfortunately with the like imperfection in the ovaries in the
only flower I could examine, is among Mrs. General Walker's
Ceylon plants. It differs specifically from the African one in
the presence of small bracts, in the form of the petals and some
other slight points.

L. ONAGRARIÆ.

1. Jussiæa *villosa*, Lam.—*W. et Arn. Fl Penins.* 1. *p.* 336.—
St. Thomas, *Don.*—A common East Indian plant.
2. Jussiæa *acuminata*, Sw.—*DC. Prod.* 3. *p.* 54.—On the
Quorra, at Attah, *Vogel;* St. Thomas, *Don.*—A West Indian
species.
3. Jussiæa *linearis*, Willd.—*DC. Prod.* 3.*p.* 55.—Grand Bassa,
Vogel; Senegal and Guinea.

* Anisophyllum *Zeylanicum;* foliis fere a basi 5-nervibus, spicis tenuiter bracteolatis.

The remaining *Onagrarieæ* described as West Tropical African are *Jussiæa stolonifera*, Lepr. et Perr., and *J. altissima*, Lepr., *Prieurea Senegalensis*, DC., considered by Guillemin and Perrottet as another species of *Jussiæa*, and *Isnardia multiflora*, Guill. et Perr., all from Senegal.

LI. LYTHRARIEÆ.

There are no specimens belonging to this Order either in the Niger collection or in those of Don from West Tropical Africa, although no less than fifteen species are enumerated in the Senegambian Flora, viz. : *Ameletia tenella* and *elatinoides*, (both described under *Ammannia*), *Ammannia filiformis*, DC., *A. Senegalensis*, Lam., *A. auriculata*, Willd., *A. gracilis*, Guill. et Perr., *A. salsuginosa*, Guill. et Perr., *A. floribunda*, Guill. et Perr., *A. pruinosa*, Guill. et Perr., *A. crassicaulis*, Guill. et Perr., *A. aspera*, Guill. et Perr., *Nesea erecta, radicans* and *Candollei*, Guill. et Perr., and *Lawsonia alba*, Lam.

LII. TAMARISEINÆ.

This small Order or genus is also, in West Tropical Africa, as far as hitherto known, confined to Senegambia, from whence a species has been described under the name of *Tamarix Senegalensis*, but which is probably, as suggested by Webb, a mere variety of the *T. Gallica*, so widely diffused over South Europe, North Africa, and the temperate regions of Asia.

LIII. MELASTOMACEÆ.

Nearly the whole of the West African plants of this Order belong to the tribe *Osbeckieæ*, and though hitherto chiefly referred to the two Asiatic genera, *Osbeckia* and *Melastoma*, a closer examination shows them to belong to groups, generic or sectional, perfectly distinct from both the Asiatic and American ones, although perhaps nearer allied to the former. The chief characters which separate them from *Osbeckia* and *Melastoma*, as now limited, will be best seen from the following synopsis :

*Osbeckia.** Calycis laciniæ deciduæ, appendicibus squamæformibus setosis in tubo sparsis. Antheræ uniformes. Capsula loculicide dehiscens. — (Sect. 1. *Osbeckia.* Species Asiaticæ, antherarum loculo in filamentum subsessili connectivo ad insertionem biauriculata.—Sect. 2. *Podocælia.* Species Africanæ, antherarum loculo in filamentum stipitato mediante connectivo basi breviter producto.)

Dissotis. Calycis laciniæ deciduæ, appendicibus squamæformibus setosis in tubo sparsis. Antheræ biformes. Capsula loculicide dehiscens. Species Africanæ.

Heterotis. Calycis laciniæ persistentes, appendicibus squamæformibus vel in tubo sparsis vel sub limbo numero definito in annulum dispositis v. omnino nullis. Antheræ biformes. Capsula loculicide dehiscens. Species Africanæ.

Tristemma. Calycis laciniæ persistentes, tubo ciliarum annulis 1-5 cincto v. nudo. Antheræ uniformes. Capsula irregulariter disrupta. Species Africanæ.

Melastoma. Calycis laciniæ deciduæ, tubo squamis paleaceis setisve imbricatis obtecto. Antheræ biformes. Capsula irregulariter disrupta. Species Asiaticæ.

In all the African species of the above genera, excepting the Senegambian plants of doubtful affinity mentioned below, I have always found the flowers pentamerous; in several of the Asiatic ones, especially among the *Osbeckiæ*, they are tetramerous or variable.

1. Osbeckia *tubulosa*, Sm. *in Rees' Cycl.—DC. Prod.* 3. *p.* 143. Sierra Leone, *Vogel, Don.*

Caulis adscendens, 1-2-pedalis, annuus (nec fruticosus). *Flores* pauci, subsessiles, sæpe in spicam interruptam unilateralem aphyllam dispositi. *Bracteæ* ovatæ, acutæ, membranaceoscariosæ v. coloratæ, calyce multo breviores, deciduæ. *Squamæ* calycis breves, apice palmatim setosæ, duas tertias calycis ob-

---

* It will be observed that I do not concur entirely with M. Ch. Naudin in his generic character of *Osbeckia*, given above, p. 130, which in several points is not applicable to the majority of the Asiatic species. These, whether as genera or sections, must surely be distinguished from the majority at least of the American ones.

tegentes, calycis collo subnudo. *Laciniæ* calycinæ, petala, et
stamina caducissima. *Antheræ* erostres, uniporosæ, uniformes, connectivo infra loculum breviter producto, in filamentum articulato, ad insertionem subintegro.

Although, as already stated, this plant may be separated
from the Asiatic *Osbeckiæ* by the form of the connectivum of
the anther, yet this character is so slight, and the anthers of
some of the Asiatic species are as yet so little known, that I
have preferred considering it as forming with the following
species a section of *Osbeckia* to establishing it as a distinct
genus. The form of the calyx brings it nearest to those
species which Korthals proposes to separate under the name of
Ceramiocalyx, on account of characters which do not appear to
me to be either definite or constant enough to found a genus.

2. Osbeckia *multiflora*, Sm. *in Rees' Cycl.—DC. Prod. 3. p.*
143.—Melastoma Afzelianum, *Don, in Trans. Wern. Soc.—*
DC. l. c. p. 147.—Sierra Leone, *Don.*

The stem of this species, although hard, is probably herbaceous, not frutescent. I have been able to examine the anthers
but very imperfectly, on account of the insufficiency of the
specimens.

The *Osbeckia Senegambiensis,* Guill. et Perr., from Senegambia, if the anthers are really all equal and similar, is probably a third species of this group, the two small appendages,
mentioned as being on the filaments near the summit, being
probably the extremity of the connectivum where it is inserted
on the filament.

DISSOTIS, (gen. nov.)

Calyx ovoideo-tubulosus, ovario mediantibus costis adnatus v.
demum liber, limbi laciniæ 5, deciduæ, apice pluri-setosæ;
squamæ palmatim setosæ in tubum sparsæ v. subseriatim dispositæ. *Petala* 5, ampla. *Stamina* 10, antheris lineari-falcatis rostratis uniporosis, 5, petalis opposita, connectivo longissimo filiformi postice in appendices 2 tenues producto, 5,
laciniis calycinis opposita, antheris dimidio minoribus connectivo brevi sed pariter filiformi et bicalcarato. *Ovarium* disco

setoso coronatum, 5-loculare. *Stylus* æqualis v. superne leviter incrassatus, apice truncato-dilatatus et stigmatosus. *Capsula* calyce inclusa, fere libera, 5-locularis, valvulis 5 loculicide dehiscens. *Semina* numerosa, cochleata.—Herba Africana, erecta, habitu *Chœtogastris Americanis* approximans.
1. Dissotis *grandiflora*, Benth.—Osbeckia grandiflora, *Sm. in
Rees' Cycl.—DC. Prod. 3. p.* 143.—Melastoma elongatum,
Don, in Mem. Wern. Soc.—DC. l. c. p. 147.—Sierra Leone,
Don; and apparently the same species in Heudelot's Senegambian collection.
Radix ex Don tuberosa. *Caules* stricti, parum ramosi, 1-1½-
pedales. *Bracteæ* scarioso-membranaceæ, calyce multo breviores. *Calycis* tubus 4 lin. longus, squamis vix seriatis;
laciniæ 3 lin. longæ, anguste oblongæ, rigidule subscariosæ,
margine ciliolatæ, apice stellato-setosæ. *Flores* ampli. *Antherarum* majorum loculus fere 5 lin. longus, connectivo
semipollicari sustensus.

The very long connectivum and dissimilar anthers are the
chief points which distinguish this plant from *Osbeckia.*

HETEROTIS, (gen. nov.)

Calycis tubus ovatus v. oblongus, ovario mediantibus costis
adnatus v. demum liber; limbi laciniæ 5, persistentes, membranaceæ, reflexæ, apice uni- pluri-setosæ; squamæ seti-
feræ in tubo sparsæ, vel sub limbo numero definito (calycis
laciniarum æquali v. duplo) in annulum dispositæ v. omnino
deficientes. *Petala* 5, ampla. *Stamina* 10, antheris lineari-
falcatis subrostratis uniporosis, 5, petalis opposita, connectivo
elongato filiformi postice in appendices 2 obtusas v. in unam
bifidam producto, 5, laciniis calycinis opposita, antheris minoribus, connectivo brevi postice leviter emarginato. *Ovarium*
disco setoso coronatum, 5-loculare. *Stylus* æqualis v. superne leviter emarginatus, apice truncatus stigmatosus. *Capsula* calyce inclusa, fere libera, 5-locularis, valvulis 5 loculicide
dehiscens. *Semina* numerosa, cochleata.—*Herbæ* suffruticesve Africanæ, procumbentes v. ascendentes, rarius erectæ.
Flores terminales, solitarii v. capitati.

§. 1. *Floribus solitariis v. distinctis, intra bracteas parvas sæpe deciduas breviter pedicellatis, squamis in tubo calycis sparsis.*—Heterotis.

1. Heterotis *lævis,* Benth. ; glabra, procumbens, foliis ovatis orbiculatisve, floribus solitariis, calycis squamis sparsis paucis parvis 1-3-setosis, summis cum laciniis calycinis alternantibus.—On the Nun, *Vogel.*

Caules 1-2-pedales, basi humifusi et radicantes, ramis breviter adscendentibus, præter cilias interpetiolares glaberrimis. *Folia* semipollicaria vel raro fere pollicaria, acutiuscula, basi acuta rotundata v. subtruncata, integerrima, membranacea, glabra, trinervia v. rarius sub-5-nervia, petiolo gracili 3-6 lin. longo glabro nudo v. parce setoso, linea setarum utrinque cum petiolo opposito juncto more plerumque *Osbeckiarum. Flores* ad apices ramulorum solitarii, majusculi, intra folia summa approximata et bractearum par parvum pedicello lineam longo fulti. *Calycis* tubus ovoideus, 4 lin. longus, basi attenuatus, sub limbo leviter contractus, membranaceus ; squamæ in tubo sparsæ paucæ, pleræque ad setam simplicem v. tripartitam reductæ, 5 summæ laciniis calycinis alternantes paullo majores palmatim trisetiferæ ; laciniæ limbi anguste lanceolatæ, tubo fere æquilongæ, reflexæ, margine minute ciliolatæ, apice unisetæ. *Petala* late obovata, purpurea, (8-9 lin. longa). *Staminum* filamenta laciniis calycinis subæquilonga ; connectiva majorum loculo æquilonga, appendicibus posticis dilatatis $\frac{3}{4}$ lin. longis, minorum vix latitudine loculi longiora ; loculus ipse 4 lin. longus. *Capsula* calyce paullo aucto inclusa, supra medium sparse pilosa et annulo pilorum seu setarum coronata, basi plus minus calyci adhærens, superne loculicide dehiscens.

2. Heterotis *plumosa,* Benth. ; procumbens, foliis ovatis suborbiculatisve cauleque pilosis, floribus solitariis, calycis squamis sparsis numerosis plumoso-setosis summis cum laciniis calycinis alternantibus.—Melastoma plumosum, *Don, in Mem. Wern. Soc.*—*DC. Prod.* 3. *p.* 147.—Osbeckia rotundifolia, *Sm. in Rees' Cycl.*—*DC. Prod.* 3. *p.* 143.—Sierra Leone, *Vogel;* Accra, *Don.*

Habitus et foliorum forma *H. lævis*, sed tota planta piloso-setosa, calyces breviores, squamæ numerosæ setis numerosis plumosæ quarum summæ stellatim dispositæ, laciniæ limbi longius ciliatæ. *Flores* rubri.

3. Heterotis *prostrata*, Benth.; caule procumbente strigoso-pubescente, foliis ovatis oblongisve glabriusculis, floribus solitariis, calycis squamis numerosis sparsis ciliatis apice stellato-setosis, summis cum laciniis calycinis alternantibus.—Melastoma prostrata, *Schum. et Thonn. Beskr. p.* 220.—On the Quorra, at Patteh, *Vogel;* Guinea.

Rami elongati, pube in parte inferiore fere pulveracea, superne strigosa. *Folia* 1-1½-pollicaria, utrinque acutiuscula, membranacea, trinervia v. rarius 5-nervia, nervis basi petioloque tenui strigoso-setosis, cæterum glabra. *Calyces* magnitudine eorum *H. lævis*, sed obtecti squamis numerosis linearibus plumoso-ciliatis. *Petala* (ex Vog.) carnea.

The *Melastoma decumbens*, Pal. de Beauv. *Fl. Ow. et Ben. v.* 1. *p.* 69. *t.* 21, or *Osbeckia decumbens*, DC., is evidently another species of *Heterotis*, belonging probably to the present section, but the details of his figure and description are too vague and too little to be relied on to establish the characters satisfactorily without seeing his specimens.

§. 2. *Floribus solitariis v. distinctis intra bracteas parvas brevissime pedicellatis, calycis squamis* 10 *sub limbo in annulum dispositis.*—Cyclostemma.

4. Heterotis *antennina*, Benth.; caule decumbente setoso-ciliato, foliis lanceolatis ovatisve, petiolo dilatato longe ciliato, floribus paucis distinctis, calycis squamis 10 sub limbo in annulum dispositis longe pectinato-ciliatis.—Osbeckia antennina, *Sm. in Rees' Cycl.*—DC. *Prod.* 3. *p.* 143.—Sierra Leone, *Don.*

Caulis ut videtur humilis, basi decumbens, superne divaricato-ramosus, setis longis præsertim ad nodos ciliatus. *Folia* 1-3-pollicaria, supra setis longis hirsuta, subtus glabriora. *Flores* sæpius gemini v. terni, intra bracteas parvas et folia summa approximata brevissime pedicellati. *Calycis* tubus præter squamarum annulum omnino glaber et lævis, membranaceus, fere omnino ab ovario liber; squamæ lineares, basi dilatatæ, lacinias calycinas sæpius æquantes v. superantes; laciniæ parce ciliatæ, apice paucisetæ. *Stamina* et fructus omnino præcedentium.

§. 3. *Floribus solitariis v. distinctis singulis bracteis scariosis involutis, calyce lævi nudo.*—Leiocalyx, *Planch. in Herb. Hook.*

5. Heterotis *segregata,* Benth.; suffruticosa, appresse strigillosa, foliis oblongis subovatisve 5-nervibus, floribus ad apices ramulorum 1-3-nis distinctis, singulis bracteis scariosis involutis, calyce lævi nudo.—On the Nun and at Abòh, *Vogel;* at the confluence, *Ansell.*

Caulis basi frutescens subbipedalis; rami (erecti?) tetragoni, uti folia strigis arcte appressis adnatisque* parum conspicuis obtecti; ciliæ intrapetiolares breves v. obsoletæ. *Folia* 2-4 poll. longa, 8-18 lin. lata, acutiuscula, v. breviter acuminata, basi rotundata v. acutiuscula, rigidule membranacea, strigis paginæ superioris longioribus arcuatis, inferioris ramealibus similibus; costæ subtus valde prominentes. *Flores* ad apices ramulorum 1-3, singuli brevissime pedicellati v. sessiles, at non capitati. *Bracteæ* 4, scariosæ, brunneæ, per paria oppositæ et imbricantes, calycis tubum æquantes v. paullo breviores et eum arcte includentes, obovatæ, concavæ, truncatæ, glabræ, læves; adsunt etiam interdum 2 exteriores angustiores longiores et laxiores subfoliaceæ et dorso strigillosæ. Intra bracteas *ciliæ* nonnullæ observantur ut in plerisque affinibus, calycis basin cingentes sed e receptaculo seu pedicelli summitate nec e calyce ipso ortæ. *Calyx* cæterum omnino nudus, tubus oblongo-ovoideus fere semipollicaris, sub limbo contractus; limbi laciniæ lanceolatæ, acutæ, margine ciliolatæ, reflexæ, vix 3 lin. longæ. *Petala* perfecta mihi desunt, sed ex alabastro ampla videntur, et sec. Vog. pur-

---

* In a large number of *Melastomaceæ,* the hairs of the upper side of the leaves and of other parts, where they appear to be appressed, are in fact adnate in the greater part of their length.

purea sunt. *Capsula* calyce parum aucto inclusa, loculicide dehiscens, calycis tubo membranaceo irregulariter disrupto obtecta. *Semina* numerosissima, minuta.

§. 4. *Floribus capitatis, cum bracteis scariosis calyces includentibus intermixtis, calyce tubo nudo v. squamis paucis stellato-setosis in tubo sparsis.* — Wedeliopsis, *Planch. in Herb. Hook.*

6. Heterotis *theæfolia,* Benth.; caule erecto glabro v. ad angulos scabro, ciliis intrapetiolaribus longis rigidis, foliis oblongis, novellis pilosis, floribus capitatis bracteis scariosis involutis, calyce lævi nudo.—*Melastoma theæfolia,* G. Don, Gard. Dict. 2. p. 764.—Sierra Leone, *Don;* a single very imperfect specimen.

Caulis validus, cortice albo lævissimo, angulis solis exasperatis. *Folia* novella jam pollicaria, ciliis interpetiolaribus usque ad 4 lin. longis, adulta desunt. *Flores* in capitulo ultra 6, arcte imbricati, bracteis calyces æquantibus intermixti, nil tamen superest nisi capsulæ 6-7 lin. longæ, calycium bractearumque reliquiis obtectæ.

7. Heterotis *cornifolia,* Benth.; caule subglabro, ciliis interpetiolaribus tenuibus, foliis brevissime petiolatis oblongis 5-nervibus parce strigosis ad costas ciliatis, floribus capitatis bracteis scariosis involutis, calycis tubo versus medium squamis paucis stellato-setosis onusto, laciniis glabris.—Grand Bassa, *Vogel.*

Suffrutex videtur, ramulis plerisque glabris etiam ad angulos lævibus, ciliis interpetiolaribus paucis 1-2 lin. longis. *Folia* bipollicaria, acuta, basi rotundata et in petiolum vix lineam longum contracta. *Capitula* pauciflora, intra folia suprema sessilia. *Bracteæ* latæ, calyce longiores, apice subfoliaceæ, dorso costatæ et secus costam setosæ. *Calycis* tubus ovoideo-oblongus, 7 lin. longus, sub limbo contractus; squamæ setosæ interdum ad setam unicam reductæ paucæ, in zona calycis medium circumdante sparsæ; limbi laciniæ lanceolatæ, acutæ, glabræ et vix ciliolatæ, fere 3 lin. longæ. *Corolla* ampla, rubro-purpurea. *Stamina* quam in affinibus majora, antheris longiuscule rostratis. *Capsula* calycis tubo inclusa et eo brevior, annulo pilorum more affinium coronata.

8. Heterotis *Vogelii,* Benth.; caule minute strigilloso ciliis interpetiolaribus parvis, foliis petiolatis ovatis 5-7-nervibus supra sparse strigosis subtus pallidis subnudis, floribus capitatis bracteis subfoliaceis involucratis, calycis tubo nudo laciniis ad costam strigillosis.—Sierra Leone, *Vogel.*

Caulis basi lignosus, laxe flexuoso-ramosus. *Folia* pauca, distantia vel ad apices ramorum approximata, 2-3-pollicaria, subtus siccitate pallida v. albido-flavicantia. *Flores* in capitulo pauci, bracteis vel omnino foliaceis vel basi plus minus scarioso-dilatatis involucrati, majusculi sed minores quam in speciebus præcedentibus. *Calycis* tubus ovoideo-globosus, vix 5 lin. longus, basi ciliis numerosis circumdatus, ipse tamen nudus ; laciniæ 4 lin. longæ, acute acuminatæ, margine minute ciliolatæ, dorso percursæ costa strigoso-ciliata.

There is an imperfect specimen in the Senegambian collection which Dr. Planchon considers as belonging to the same species, although it appears to have rather smaller flowers.

9. Heterotis *capitata,* Benth. ; caule pubescente, foliis breviter petiolatis late ovatis 5-nervibus supra sparse strigosis subtus glabriusculis pallidis, floribus capitatis bracteis scariosis v. apice foliaceis involucratis, calycis tubo nudo laciniisque glabris.—Melastoma capitatum, *G. Don, Gard. Dict.* 2. *p.* 764. —Sierra Leone, *Don.*

This species is evidently allied to *H. Vogelii,* but the leaves are broader, the stem covered on the sides, as well as the angles, with numerous short hairs, and the calyx without hairs or appendages, except the long ciliæ, with which it is surrounded at its insertion on the stalk. The flowers appear to be about the size of those of *H. Vogelii,* but are fewer in the head, as far one can judge from the specimens.

Beside the above, the *Melastoma cymosum,* DC., or *M. corymbosum,* Bot. Mag. t. 904, from Sierra Leone, to judge from the figure and description, may be referred to *Heterotis,* although it does not agree in habit or inflorescence with any of the foregoing groups.

TRISTEMMA, *Juss.*

Char. Gen.—*Calycis* tubus ovatus v. oblongus, ovario basi omnino adnatus et altius cohærens mediantibus costis, cinctus annulis ciliarum parallelis 1-5 vel rarius nudus; limbi laciniæ 5, persistentes, reflexæ. *Petala* 5, mediocria. *Stamina* 10, conformia; antheris lineari-falcatis subrostratis uniporosis, connectivo brevi postice bidentato. *Ovarium* disco setoso coronatum, 5-loculare. *Stylus* æqualis v. superne leviter incrassatus, apice truncato-stigmatosus. *Capsula* calyce inclusa et ei plus minus adnata, maturitate irregulariter disrupta, rarius subvalvatim dehiscens. *Semina* numerosa.— *Herbæ* suffruticesve Africanæ, decumbentes ascendentes v. rarius erectæ. *Flores* terminales, capitati, v. rarius solitarii, singuli bractearum paribus 1-3 involuti.

As far as can be judged from dried specimens, the placentæ in this genus appear to be more fleshy than in the preceding genera, or almost pulpy, and the general affinity is much rather with *Melastoma* than with *Osbeckia* or *Heterotis*. The rings of ciliæ in most of the species are very remarkable in their arrangement.

1. Tristemma *littorale*, Benth.; caule adscendente ramoso obtuse tetragono parce strigoso v. lævi, foliis ovatis oblongisve 5-nervibus parce strigosis, floribus capitatis, bracteis imbricatis exterioribus foliaceis calycibusque glabris lævibus nudis. —Fernando Po, on the sea-shore, *Vogel.*

T. albifloro simile sed glabrius, petioli longiores, florum capitula breviter pedicellata, bracteæ angustiores omnino glabræ v. extimæ rarius dorso strigis paucissimis onustæ. *Herba* est ex Vog. annua, procumbens, ramis adscendentibus 1-2-pedalibus. *Petala* rosea. *Folia* in speciminibus aliis semipedalia, in aliis 2-3-pollicaria.

2. Tristemma *albiflorum*, Benth.; caule adscendente ramoso acute tetragono parce strigoso, foliis ovatis oblongisve 5-nervibus supra strigosis, floribus capitatis, bracteis imbricatis latis dorso strigosis, calycibus lævibus nudis v. corona pilorum

A A

incompleta cinctis.—Melastoma albiflorum, *G. Don, Gard. Dict. 2. p.* 764.—Sierra Leone, *Vogel, Don.*

Folia 2-3-pollicaria, ad apices ramulorum approximata. *Capitula* sessilia, 1-6-flora. *Bracteæ* calyces æquantes et ei appressæ v. rarius apice breviter foliaceo-productæ. *Calycis* tubus ovario longior, superne contractus, laciniæ lato-lanceolatæ, tubo duplo breviores, minute ciliolatæ. *Antherarum* connectiva breviter biauriculata.

3. Tristemma *Schumacheri*, Guill. et Perr. *Fl. Seneg. p.* 311.— Benguelen, *Curror, in Herb. Hook.* ; Gaboon coast, *Middleton, in Herb. Benth.* ; Senegambia and Guinea.

This species very much resembles in habit the two preceding and the following one, but is readily known by the single ring of hairs surrounding the calyx. The *T. incompletum*, Br., from Congo, may possibly, as suggested by the authors of the *Flora Senegambiæ*, be a variety of the same species with slight traces of a second ring of hairs, which, however, I have not observed on the more usual form.

4. Tristemma *coronatum*, Benth. ; caule ascendente minute strigoso, foliis ovatis supra strigosis, floribus subcapitatis, bracteis oblongis lanceolatisque laxiusculis apice foliaceis, calycis tubo globoso ciliarum annulis 4 distinctis cincto.— Sierra Leone, *Don.*

This again, judging from a single specimen, is very near the three preceding ones in appearance, but remarkable from the four completely distinct parallel rings of hairs which surround the calyx at the distance of above half a line from each other.

5. Tristemma *hirtum*, Vent. *ex DC. Prod. 3. p.* 144.—*Pal. de Beauv. Fl. Ow. et Ben.* 1. *p.* 94. *t.* 57 ; caule suffruticoso, ramis acute tetragonis vel quadrialatis longe barbato-pilosis, foliis ovatis hirsutis, floribus capitatis, bracteis imbricatis latis strigosis apice subfoliaceis, calycis tubo ovato ciliarum annulis 3 distinctis cincto.—On the Nun, *Vogel.*

6. Tristemma *involucratum*, Benth. ; caule procumbente, ramis longe barbato-pilosis, foliis ovatis acuminatis hirtis, floribus subcapitatis, bracteis scariosis appressis pilosis, calycis tubo

urceolato cincto pilis numerosis in zonam unicam dense
imbricatis.—Melastoma involucratum, *Don, in Mem. Wern.
Soc.—DC. Prod.* 3. *p.* 147.—Sierra Leone. *Don.*
I only find a single flower on the specimen, which is remarkable by the hairs or ciliæ, not arranged in any distinct
rings, but thickly collected in one broad belt surrounding the
middle of the calyx, forming in this respect a transition to the
hairs or scales which cover the calyx of the true *Melastomas:*
the stamens, however, as far as I could judge, appear to be
those of a *Tristemma.*

In the Senegambian Flora another species is described, under
the name of *T. erectum;* and Heudelot's collection contains an
undescribed plant belonging to the tribe of *Osbeckieæ,* but
scarcely referrible to any of the foregoing genera. It is a small
erect annual, with tetramerous flowers, resembling rather the
American *Arthrostemmas* than the Asiatic *Osbeckiæ.*

DINOPHORA, (gen. nov.) e tribu *Rhexiearum.*

Calyx urceolato-turbinatus basi ovario adhærens, apice liber
membranaceus brevissime sinuato-dentatus. *Petala* 5, late
ovata, acutiuscula. *Stamina* 10, conformia, antheris subrostratis uniporosis, connectivo basi breviter producto postice
leviter dilatato emarginato. *Ovarium* 5-loculare, apice
nudum. *Capsula* membranacea, cum calyce membranaceo
maturitate irregulariter disrupta? *Semina* numerosissima,
placentis carnosulis affixa, cochleata, vix rugosula.—*Suffrutex*
Africanus, habitu *Spenneris* simillimus, floribus majoribus.

1. Dinophora *spenneroides,* Benth.—Fernando Po, *Vogel.*

Caulis 3-5-pedalis, ramosissimus, basi frutescens; rami divari-
cati, herbacei, tetragoni, ad angulos pilis interdum glandulosis rariter ciliati, cæterum glabri. *Folia* cujusve paris paullo
inæqualia, majora 2-4-pollicaria, ovata v. oblonga, acuta v.
acuminata, margine ciliato-serrata, basi cordata, membranacea,
5-7-nervia, supra strigis paucis conspersa, cæterum glabra;
petiolo altero sæpe pollicari, altero dimidio breviore. Inflorescentia *Spennerœ. Panicula* laxa, ramis gracilibus oppositis laxe dichotomis. floribus in dichotomiis v. ad

apices ramulorum pedicello 3-4 lin. longo fultis. *Bracteæ*
minutæ, subulatæ. *Calyx* fere 3 lin. longus, glaberrimus,
lævis, basi acutus, supra ovarium paullulum constrictus;
dentes lati obtusissimi brevissimi v. fere obsoleti. *Petala* 3
lin. longa. *Ovarium* glaberrimum, per anthesin mediantibus
costis fere ad apicem calyci adnatum, vertice libero nudo
calyce tamen breviore. *Stylus* basi attenuatus, apice truncatus. *Capsula* membranacea, vix valvatim dehiscere videtur
sed pericarpium membranaceum cum calyce tenuissimo saltem
in speciminibus siccis disrumpitur.

This plant is so very near in most respects to the American
genus *Spennera*, that I should have considered it as a species
of it, but that in the present state of our knowledge of
Melastomaceæ it seems necessary to keep separate those genera
where the cells of the ovary are reduced below the number of
parts of the floral envelope, from those which are strictly iso-
merous.

1. Spathandra *cœrulea*, Guill. et Perr. *Fl. Seneg. p.* 113. *t.* 71.

—Sierra Leone, *Vogel, Don;* Senegambia.

It is to the tribe of *Memecyleæ*, not to that of *Chariantheæ*,
that this genus should be referred; indeed it is only to be
distinguished from *Memecylon* itself by the three-ribbed leaves
and the cotyledons, which are more fleshy and less plaited,
although in *Spathandra cœrulea* they are to a considerable
degree irregularly wrinkled and folded. The unilocular ovary
is not alluded to in the ordinal character either in Endlicher's
Genera or in Lindley's Vegetable Kingdom, although in both
works *Spathandra* and *Memecylon* are included among *Melastomaceæ :* the little importance, however, of this anomaly in
certain calyciflorous Orders is shown, as well by its occasional
occurrence among *Myrtaceæ* and *Rubiaceæ*, as by the very
natural genus *Mouriria*, (the American representative of *Memecylon*), which includes both unilocular and plurilocular species.
This circumstance is much in favour of the supposition, that
in these Orders the placentæ often proceed on Schleiden's
theory from a prolongation of the axis, and not from the
margins of the carpellary leaves.

2. Spathandra *memecyloides*, Benth. ; foliis breviter petiolatis ovatis vix coriaceis, corymbis breviter pedunculatis paucifloris, baccis oblongis.—Fernando Po, *Vogel.*

Tota glabra. *Folia* majora 6-8 poll. longa, 3-4 poll. lata, apice in acumen latum angustata, basi rotundata, rigide membranacea v. subcoriacea, multo tamen tenuiora quam in *S. cærulea,* trinervia v. interdum sub-5-nervia, petiolo 2-3 lin. longo. *Flores* desunt. *Fructus* in corymbis parvis pauci ad axillas foliorum annotinorum. *Pedunculus* communis semipollicaris, pedicellis unifloris oppositis 2-3 lin. longis, bracteis nullis v. jam delapsis. *Baccæ* nondum maturæ ovoideo-oblongæ, 3-4 lin. longæ, apice constrictæ, calycis limbo 4-sinuato coronatæ, intus uniloculares, ovulis paucis quorum unum auctum, cætera abortientia.

3. Spathandra *fascicularis*, Planch. *in Herb. Hook.*; foliis breviter petiolatis ovatis coriaceis, floribus in umbella brevissime pedicellata fasciculatis, bracteis persistentibus, baccis globosis.—Sierra Leone, *Don.*

Tota glabra. *Folia* fere *S. cæruleæ,* nisi petiolo longiore, perfecta tamen non vidi. *Umbellæ* axillares, fere sessiles v. pedunculo communi 1-2 lin. longo fultæ, floribus ultra 10, pedicellis 1 lin. longis receptaculo incrassato insertis et bracteis intermixtis brevibus latis membranaceis fuscis persistentibus. *Baccæ* globosæ, lineam diametro, calycis limbo amplo 4-dentato coronatæ, intus uniloculares. *Ovula* plurima abortiva. *Semen* unum grandefactum, cavitatem implens, ei *S. cærulei* simile, sed in fructu a me aperto nondum maturum erat.

1. Memecylon *Donianum*, Planch. *in Herb. Hort. Soc.*; ramulis teretiusculis, foliis subsessilibus oblongis acuminatis basi acutis coriaceis uninervibus, corymbis brevibus laxe paucifloris, baccis globosis.—Pavetta lateriflora, *G. Don, Gard. Dict. 3. p.* 575.—Sierra Leone, *Don.*

Folia pleraque 5 poll. longa, 2 poll. lata, rigide coriacea, acumine brevi abrupto acutiusculo, marginibus recurvis, costa subtus prominente, venis transversis inconspicuis v. rarius in pagina superiore obscuris in venulam marginalem tenuem confluentibus. *Pedunculi* axillares, gemini, 2-3-chotomi, crassi, breves. *Flores* desunt. *Pedicelli* fructiferi 2-3 lin. longi. *Baccæ* 6-8 lin. diametro, 1-2-spermæ. *Seminum* testa crustacea, cotyledones quam maxime irregulariter plicatæ ; radicula crassa, carnosa, incurva, ad hilum spectans.

LIV. MYRTACEÆ.

1. Psidium *pomiferum*, Linn. (the Guava).—Sierra Leone, *Don ;* Cape Coast, *Vogel,* who states it to be a shrub common in thickets ; although it is probably introduced. The peduncles in both specimens are one-flowered, but the fruit, at least in Don's specimen, is globose.

The *Psidium Guineense,* Sw., cultivated in the West Indies, is said to have been introduced there from Guinea, but it does not appear that any African specimens have been found.

1. Eugenia *Michelii,* Lam.—*DC. Prod.* 3. *p.* 263.—Cape Coast, *Vogel,* cultivated under the name of *Barbadoes Cherry.*

2. Eugenia *coronata,* Vahl,—*DC. Prod.* 3. *p.* 271.—*Schum. et Thonn. Beskr. p.* 230.—Accra, *Don.*
Pedicelli 3 lin. longi, intra bracteas parvas ovatas acutas concavas ciliatas orti, apice sub calyce bibracteolati, bracteolis ovatis rigidulis ½ lin. longis cum pedicello subcontinuis. *Ovarium* biloculare, ovulis in quoque loculo circa 4.

3. Eugenia *calycina,* Benth. ; ramis glabris, foliis ovatis ellipticis oblongisve basi acutis breviter petiolatis glabris nitidis, pedunculis in axillis superioribus unifloris folio brevioribus apice bibracteolatis, calycis tubo hemisphærico tomentoso, lobis amplis orbiculatis glabratis.—Grand Bassa, *Vogel.*
Folia 2-3-pollicaria, obtusa v. brevissime acuminata, marginibus recurvis, costa subtus prominula, venis primariis paucis conspicuis juxta marginém confluentibus. *Pedunculi* in axillis summis solitarii, vel ad apices ramulorum foliis floralibus abortientibus fasciculati, basi bractea oblonga lineam longa stipati, 8-9 lin. longi, rigiduli, minute puberuli. *Bracteolæ* sub flore orbiculatæ, concavæ, 1-1½ lin. longæ. *Alabastrum*

globosum, 4 lin. diametro, album. *Calycis* tubus hemisphæricus, linea paullo longior, pilis brevibus appressis rufescentibus dense obtectus, lobi orbiculati, concavi, fere 5 lin.
lati, 2 exteriores ciliolati, 2 interiores nudi. *Petala* calyce majora,
ciliolata, in specimine nondum expansa. *Ovarium* biloculare,
ovulis in quoque loculo 6-8.

4. Eugenia *memecyloides*, Benth.; glabra, ramulis compressis,
foliis amplis oblongis acuminatis basi acutis breviter petiolatis,
pedunculis petiolo brevioribus axillaribus congestis unifloris
ebracteolatis, bacca depresso-globosa.—Grand Bassa, *Vogel*.

Frutex ex Vog. 4-pedalis, emittens e summitatibus ramulos numerosos virgatos plantam quamdam parasiticam simulantes.
Specimen adest unicum ramuli foliati et fructiferi, ad nodos
compressi. *Folia* 5-6-pollicaria, 2-2½ poll. lata, coriacea at
non nitida, marginibus revolutis, subtus pennivenia et crebre
punctulata. *Pedicelli* fructiferi 1-1½ lin. longi. *Baccæ* 4-5
lin. diametro, calycis limbo brevi 4-lobo coronatæ. *Semen*
abortu solitarium, cotyledonibus crassis conferruminatis.

The *Eugenia caryophylloides*, DC., from Sierra Leone, must
be very near the last species, especially in the form and size of the
leaves and in inflorescence, but the leaves are said to be im-
punctate, and the branches, nerves of the leaves, peduncles and
tube of the calyx to be downy.

1. Jambosa *vulgaris*, DC. *Prod. 3. p.* 286.—Sierra Leone and
St. Thomas, *Don;* probably cultivated.

1. Syzygium *Owariense*, Benth.—Eugenia Owariensis, *Pal.
Beauv. Fl. Ow. et Ben.* 2. *p.* 20. *t.* 70.—Jambosa Owariensis,
DC. Prod. 3. p. 287.—Grand Bassa, *Vogel;* Oware.

The only other West Tropical African species of true *Myrteæ*
hitherto described is the *Syzygium Guineense*, DC., from Senegal
and Guinea, which is very nearly allied to the preceding species.
Both belong to the genus (or section) *Syzygium*, admirably
defined by Wight in his review of Indian *Myrtaceæ*.

(SUBORD. NAPOLEONEÆ.)

1. Napoleona *Vogelii,* Hook. et Planch. (TAB. XLIX, L.) ;
glaberrima, foliis lanceolato-ellipticis breviter et obtusiuscule
cuspidatis basi acutis leviter repandis, glandulis 2 impressis
supra basim folii margini laminæ adnatis v. contiguis puncti-
formibus, floribus axillaribus subsessilibus, corolla flava intus
medio rubra, coronæ externæ laciniis (70-75) linearibus liberis,
internæ 40-fidæ laciniis æqualiter late linearibus integris,
bacca depresso-globosa, seminibus intra loculos solitariis
longe reniformibus v. geminis superpositis brevioribus.—Cape
Palmas, *Vogel.*

Species a *N. imperiali* et *N. Heudelotii* foliis minoribus brevi-
ellipticis nec oblongis et colore florum distinctissima. *Folia*
1½-3 poll. longa, 15-20 lin. lata, petiolis haud crassis 1-2 lin.
longis. *Corolla* generis, obsolete 5-loba, lobulis flabellatis,
crispis, circiter 30. *Coronæ* exterioris laciniæ corolla plus
quam duplo breviores, flavescentes ; internæ flavescenti-albidæ
laciniæ sat profundæ, uninerves, obtusiusculæ. *Antheræ* 10,
lineari-oblongæ, uniloculares. *Discus* urceolatus, crassus,
margine leviter lobatus. *Ovarii* structuram internam in flore
unico imperfecto rite observare non licuit. *Bacca* corticosa,
magnitudine et facie fructus *Punicæ,* cortice extus rubescente
punctulis albis crebre consperso, septis pulposis in speciminibus nostris exsiccatis et semi-collapsis et cum integumento
seminum conglutinatis. *Semina* majora a basi ad apicem
loculi extensa.—(*Planchon.*)

I have here inserted the above brief account of this curious
plant in the words of Dr. Planchon in Hooker's *Icones.* It
had been that botanist's intention to have entered into a detailed
exposition of his views of its affinities in the present work ; but
the completion of this Flora having devolved upon me, we
are deprived of the benefit of his observations. I myself have
had no opportunity of examining the living specimens which have
flowered in this country, and therefore refrain from adding any

speculations of my own to all that has been said on the subject.
I would merely observe that, from the various dissertations
published, especially by Ad. de Jussieu, (Ann. Sc. Nat. Par.
Ser. 3. v. 2. p. 22), Lindley, (Veg. Kingd. p. 728), and Hooker,
(Bot. Mag. t. 387), it would appear that, notwithstanding the
striking discrepancies of detail in the drawings of Beauvois,
Jussieu, Lindley and Hooker, the *N. Heudelotii* and *N. Vogelii*
are not really specifically distinct from the original *N. imperialis*,
and that of all the different affinities suggested, no hypothesis is
more plausible than that suggested by Planchon, and partly con-
curred in by Lindley, that the place of *Napoleona* is near *Barring-
toniea*, among anomalous *Myrtaceæ*. If the *corona*, or rings
of filaments, called by some the intermediate and inner corollæ,
be regarded as sterile stamens, they will be found to be as analogous to the stamens of *Myrtaceæ* in their insertion on the
fleshy disk crowning the ovary as in their folded arrangement
in the bud. In neither respect can they be compared with the
rings of filaments in *Passifloræ*.

PLATE XLIX, L. *Fig* 1. flower, cut open; *f.* 2. upper and
stigmatic portions of the style; *f.* 3. fruit; *f.* 4. the same,
vertical section; *f.* 5. transverse section of the same; *f.* 6.
seed.—*Fig.* 1 and 2 *magnified : the remainder natural size.*

The *Asteranthos* of Desfontaines, which A. de Jussieu has
shown to be probably a native of Angola, does not appear to
have been again found by any collector.

LV. HOMALINEÆ.

1. Blackwellia *Africana*, Hook. fil.; foliis ovato-oblongis breviter calloso-dentatis glabris coriaceis, racemis paniculatis,
floribus secus ramos elongatos subsessilibus parvis pentandris,
petalis calyce subduplo longioribus.—Sierra Leone, *Don.*
Ramuli et inflorescentia pubescentes. *Folia* breviter petiolata,
4-5 poll. longa, 2-2½ poll. lata, obtuse acuminata, marginis
dentibus glanduloso-callosis sæpius distantibus et haud pro-
fundis, basi obtusa, venis primariis validis subtus prominulis,

362   FLORA NIGRITIANA.

utrinque opaca et pallescentia. *Stipulæ* hinc inde ad folia floralia supersunt lineari-cuneatæ, 2-3 lin. longæ; foliorum ramealium omnes delapsæ. *Flores* secus ramos paniculæ folio longioris, ad axillas bracteolarum solitarii v. fasciculati, singuli vix 1½ lin. diametro. *Calycis* tubus breviter et late turbinatus, ovario adhærens, limbi laciniæ crassiusculæ, triangulares, acutæ, ¼ lin. longæ, æstivatione in primo juventute verosimiliter valvatæ, sed omnes quas vidi jam alabastro increscente apertæ erant. *Petala* (perigonii laciniæ interiores) ovata, obtusa, concava, crassa, pubescentia, calycis laciniis duplo longiora, æstivatione leviter imbricata. *Stamina* his opposita et paullo breviora, glabra. *Glandulæ* breves, latæ, tomentosæ. *Styli* brevissimi 4 v. 5. *Ovarii* placentæ totidem, 3- (v. 4- ?) ovulatæ.

DISSOMERIA, (gen. nov.)

Calyx profunde 4-fidus, tubo brevi basi vix ovario adnato, laciniis æstivatione imbricatis. *Petala* 8, persistentia, calycis laciniis majora, cum glandulis totidem alternantia, æstivatione duplici serie imbricata, per anthesin conniventia. *Stamina* plurima in phalanges 8 petalis oppositos (singulis 6-andris) disposita; antheræ subglobosæ, connectivo crasso carnoso, loculis introrsis longitudinaliter dehiscentibus. *Ovarium* fere omnino liberum, uniloculare. *Styli* 4, rarius 3, filiformes, apice minute stigmatosi. *Ovula* ex apice cavitatis ovarii pendula, placentis tot quot styli parvis vix distinctis, singulis 3-4-ovulatis. *Capsula* indehiscens, crasso-coriacea, seminibus abortu solitariis v. paucis.—Frutex? foliis alternis stipulatis, spicis axillaribus interruptis.

Dissomeria *crenata*, Hook. fil.—At the Confluence, on the banks of the Niger, *Ansell.*

Frutex? ramulis verrucosis, novellis inflorescentiaque puberulis. *Folia* alterna, breviter petiolata, ovata, irregulariter crenata et hinc inde sinuato-lobata, crenaturis sæpe minute glandulosis, basi acuta v. obtusa, subcoriacea, penninervia et reticu-

lato-venosa, axillis venarum subtus fasciculum pilorum foventibus, cæterum glabra v. juniora minute puberula. *Stipulæ*
lineares, foliaceæ, 2 lin. longæ, valde caducæ. *Spicæ* in
axillis superioribus cum pedunculo 3-5-pollicares, jam infra
medium interrupte floriferæ, floribus arcte sessilibus, inferioribus dissitis, superioribus approximatis, singulis subglobosis fere 5 lin. diametro. *Bracteæ* 3, fusco-membranaceæ,
una exteriore (bractea subtendens), duæ (bracteolæ) interiores
oppositæ. *Sepala* 3 lin. longa, lato-ovata, obtusissima, basi
breviter connata, margine ciliolata, membranacea, venulosa,
dorso puberula. *Petala* consistentia et pube sepalis similia
sed duplo majora, basi angustata et distincta, circa fructum
persistentia et globoso-conniventia. *Glandulæ* breves, latæ,
cum petalis discum hypogynum et subperigynum margi-
nantes. *Stamina* intra petala fasciculatim inserta, sæpissime
ad quoddam petalum sex, rarius 5 ; filamenta filiformia, pilis
longis patentibus barbata, petalis æquilonga v. vix longiora.
Ovarium dense hirsutum. *Styli* exserti, ultra medium pilis
longis barbati, apicem versus fere glabri.

The almost entirely free ovary, and the number of petals,
as well as that of the glands and of the bundles of stamens,
being double that of the calycine segments, sufficiently distinguish this genus from *Blackwellia* and *Homalium*, independently of the remarkable size and form of the flowers.
R. Brown mentions a Madagascar genus with a free ovary, but
without any further indications to show how far it differs from
ours.

Two other *Homalineæ* have been described from W. Tropical
Africa ; *Homalium angustifolium*, Sm., from Sierra Leone, (of
which the true genus is doubtful), and *Byrsanthus Brownei*,
Guillem., from Senegambia. The generic name of the latter
has been changed by Endlicher to *Anetia*, on account of the
pre-existence of Presl's *Byrsanthus* among *Lobeliaceæ* ; but if,
as is suspected, with every probability of correctness, by A. De
Candolle, the latter should prove to be really not different
from *Siphocampylos*, the name of *Byrsanthus* should be retained for Guillemin's plant.

The close affinity, so well pointed out by Brown between
this Order and *Passifloreæ*, has been objected to by Lindley,
(Veg. Kingd. p. 742), on account of their " inferior ovary, to say
nothing of their want of stipules and glands on the leaves, of
the presence of glands at the base of the floral envelopes, and
of their erect and very different habit ;" but besides *Dissomeria*
and the Madagascar plant alluded to by Brown, in which the
ovary is truly free, the other genera offer, as observed by
Brown, very different degrees of adherence, and in almost all
cases the summit, which is the only placentiferous portion, is
free ; stipules, though very deciduous, may be found on the
young branches of most, if not all the species ; glands exist in
the crenatures of the leaves of several species, and are not universal
in *Passifloreæ* ; and the habit is surely much nearer to that of
Smeathmannia and the allied genera than to any *Cacteæ* or
Loaseæ. The glands of the flower, combined with the insertion
of the stamens, remain the only essential characters which prevent the actual union of *Homalineæ* with *Passifloreæ*.

LVI. PASSIFLOREÆ.

1. Smeathmannia *pubescens*, Sol.—*DC. Prod. 3. p.* 322.—Bu-
   lowia insignis, *Schum. et Thonn. Beskr. p.* 247 ?—Sierra
   Leone, *Don;* Abòh, *Vogel.*
2. Smeathmannia *lævigata*, Sol.—*DC. l. c.*—Sierra Leone,
   Vogel, Don ; Senegambia.

Brown has described a third species, a variety of *S. lævigata*,
from Sierra Leone, under the name of *S. media.*

CROSSOSTEMMA, (gen. nov.)

Calycis tubus brevissimus patelliformis, laciniæ 5, ovatæ, obscure uninerves, æstivatione valde imbricata. *Petala* 5, tubo
calycino inserta, sepalis subconformia sed majora et magis
colorata, et distincte 3-5-nervia. *Corona* e filamentorum
serie unica ad basin petalorum composita. *Gynophorum* brevissimum, expansum in discum depressum, dentibus seu appendicibus marginalibus 5 brevibus erectis acutis cum sta-

minibus alternantibus. *Stamina* 5, ad marginem disci inter
dentes inserta, filamento in alabastro brevissimo, anthera
magna oblongo-lineari. *Ovarium* in medio disco sessile,
oblongo-triquetrum, superne in stylum brevem attenuatum,
stigmate magno dilatato integro coronatum, intus uniloculare,
placentis 3 parietalibus, ovulis in quaque placenta plurimis
biseriatis.

1. Crossostemma *laurifolium*, Planch. *in Herb. Soc. Hort.
Lond.*—Sierra Leone, *Don.*

Frutex scandens, glaberrimus. *Folia* alterna, petiolata, oblonga, breviter acuminata, integerrima, basi acuta, 3-4 poll.
longa, subcoriacea, nitidula, utrinque venosissima, eglandulosa
v. ad apicem petioli obscure biglandulosa. *Cirrhi* ex axillis
sterilibus orti. *Inflorescentiæ* axillares, solitariæ, pedunculo
communi semipollicare cymam ferente laxe paucifloram
flexuosam, ramis ultimis 3 lin. longis, medio articulatis (pedicellis unifloris terminatis). *Bracteæ* minutæ setaceæ v. nullæ.
Flores ex scheda Doniana lutei, (semipollicem diametro?)
in specimine nondum aperti. *Sepala* et *petala* in alabastro
subsimilia late ovata, obtusa, concava, et crebre lineato-punc-
tata, hæc vero evidentius colorata et venis 3-5 percursa, illa
viridiora, extima minora, et vena unica brevi additis rarius 2
brevissimis lateralibus notata. *Fructum* haud vidi.

Though closely allied to some of the numerous forms of
Passiflora, this genus is at once distinguished by the very
short gynophorum, and the entire style. The collection contains but a single specimen, with few buds in a perfect
state. In the only one I dissected I could not ascertain very
precisely whether the delicate ring of filaments forming the
corona at the base of the petals was continuous or interrupted
opposite each petal.

1. Modecca *cissampeloides*, Planch. *in Herb. Hook.*; foliis cordato-orbiculatis 5-nervibus obsolete 5-lobis integerrimisque
membranaceis glabris subtus albidis, petiolo apice glandulifero, fl. fœm. petalis linearibus ad basin loborum calycinorum insertis et iis brevioribus.—Fernando Po, *Vogel.*

Tota planta glaberrima. *Ramuli* teretes, cinerascentes v. albidi.

366   FLORA NIGRITIANA.

Cirrhi axillares, simplices v. ramosi, interdum floriferi. *Folia*
2-3 poll. longa et lata, petiolo subpollicari, glandula majuscula ad summum apicem petioli in facie superiori.   *Paniculæ*
vel terminales vel axillares supra cirrhos enatæ, laxæ, e cymis
paucis secus rhachin alternis breviter pedunculatis compositæ,
rhachi interdum in cirrhum abeunte.   *Flores* fœminei tantum
adsunt, 5-6 lin. diametro, tubo late hemisphærico, laciniis
calycinis ovato-lanceolatis acutis integerrimis, æstivatione subvalvatis, lineis coloratis crebris percursis.   *Petala* fere dimidio breviora, ad basin laciniarum inserta et cum iis alternantia, anguste linearia, persistentia, consistentia calyci similia
v. paullo tenuiora.   *Corona* pilorum brevissimorum sub
petalorum insertione vix conspicua.   *Squamæ* (seu stamina
abortiva?) 5, breves, acutæ, disco hypogyno tenui in fundo
calycis insertæ.   *Ovarium* fere sessile, ovato-globosum, sub-
triquetrum, glaberrimum.   *Stigmata* 3, subsessilia, flabellata,
crenata.   *Placentæ* 3, parietales, pluriovulatæ.

2. Modecca *cynanchifolia*, Benth. ; foliis cordato-ovatis oblongisve acuminatis vix sinuatis membranaceis concoloribus,
petiolo apice glandulifero, fl. masc. petalis oblongo-linearibus
margine fimbricatis ad basin loborum calycinorum insertis et
eos subæquantibus.—Fernando Po, *Vogel.*

Tota planta glaberrima.   *Rami* tenues.   *Cirrhi* et inflorescentia
M. cissampeloidis.   *Folia* lætevirentia, 2-4 poll. longa, 1-2
poll. lata, basi late cordata, subtrinervia, margine integerrima
v. obsolete sinuata.   *Glandula* nunc ad apicem petioli, nunc
in limbo ipso ad insertionem petioli sita.   *Flores* quam in
M. cissampeloidi minora videntur, sed in specimine unico
mascula et vix aperta.   *Calycis* laciniæ anguste ovato-oblongæ,
lineato-punctatæ, margine membranaceo-hyalinæ, integræ,
æstivatione valvatæ v. vix angustissime imbricatæ.   *Petala*
membranaceo-alba, tenuia et vix punctata, obtusa, marginibus
breviter lacero-fimbriatis.   *Corona* e filis paucis brevissimis
compositæ.   *Antheræ* 5, majusculæ, oblongæ, filamentis brevissimis.   *Ovarii* rudimentum minimum.

3. Modecca? *tenuifolia*, Planch. *in Herb. Hook;* foliis cordato-
ovatis acuminatis integris v. sinuato-lobatis membranaceis

concoloribus, petiolo sub apice biglanduloso, fl. masc. petalis
ovatis fimbriatis prope basin tubi calycini insertis et calyce
plus duplo brevioribus.—Sierra Leone, *Vogel.*
Planta tota glaberrima. *Folia* ampla, majora semipedalia, 3-5-
nervia, glandulis petioli majusculis, a limbo distinctis. *Cirrhi*
floriferi, cymis in medio cirrho 2-3 alternis breviter pedunculatis laxifloris. *Flores* quam in præcedentibus majores; calyx
8 lin. longus, tubuloso-campanulatus, ad tertiam partem 5-
lobus, laciniis ovatis obtusis æstivatione leviter imbricatis; 2
exteriores margine integerrimæ, tertia hinc integerrima hinc
inflexa membranacea et leviter fimbriata, 2 interiores ad marginem utrumque inflexa et leviter fimbriata. *Petala* 5, prope
basin calycis inserta, tubo calycino subbreviora, breviter unguiculata, margine pectinato-lacera. *Corona* e filamentis
brevibus composita ad basin petalorum. *Antheræ* 5, oblongæ,
filamentis brevissimis. *Ovarii* rudimentum parvum.

A second specimen from Grand Bassa appears to belong to
the same species. It is a female, of which the fruit is fallen off,
which, according to Vogel, was a berry with three parietal
placentæ.

The only other West Tropical African *Passifloreæ* known are
Modecca diversifolia, Schum. et Thonn., from Guinea, and
M. lobata, Jacq., from Sierra Leone, and the very imperfectly
described *Kolbia elegans*, Pal. de Beauv., from Benin.

LVII. CUCURBITACEÆ.

1. Melothria *triangularis*, Benth.; foliis late deltoideis acutis
denticulatis membranaceis scabris, floribus longe pedicellatis
in eadem axilla solitariis v. geminis monoicis, corolla glabra
lobis integris, fructibus globosis glabris.—Abòh and Fernando Po, *Vogel.*
Caules filiformes, glabri v. scabrelli. *Cirrhi* apice sæpius bifidi.
Folia exacte triangularia v. basi subangulata, lateribus 1½-2
poll. longis, 5-nervia, utrinque glabra sed tuberculis crebris
scabra, denticulis marginalibus irregularibus remotis. *Pedicelli*
filiformes, ex eadem axilla sæpius gemini, alter masculus 8-10

lin. longus, alter fœmineus duplo longior. *Flores masculi* lineam longi, campanulati, tubo calycis corollæque in unum arcte connato, dentes calycini minuti, cum lobis corollinis ovatis integris albis tubo suo paullo longioribus alternantes. *Stamina* 3, medio tubo inserta; filamenta brevia; antheræ biloculares, loculi duarum magis discreti connectivo apiculato, tertiæ magis approximati connectivo crassiore mutico, omnes dorso minute ciliolatæ. *Glandula* in fundo corollæ majuscula, globosa. *Flores fœminei :* ovarium subglobosum, pedicello apice incrassato-calloso insidens, apice in collum brevem acuminatum. *Calyx* et *corolla* quoad pars superior fere maris. *Stamina* nulla. *Discus* ut in mare globosus, stylum fert corollam fere æquantem apice trifidum, lobis crasse dilatato-stigmatosis. *Baccæ* globosæ, 3-4 lin. diametro, seminibus 1½ lin. longis compressis, et forma iis *Cucumeris* subsimilibus.

2. Melothria *deltoidea*, Benth. ; foliis late deltoideis subacutis sinuato-dentatis membranaceis scabriusculis, pedunculis petiolo brevioribus ex eadem axilla geminis monoicis v. solitariis, corolla glabra lobis integris, fructibus (ovoideis ?) subglabris.
—Bryonia deltoidea, *Schum. et Thonn. Beskr. p.* 429? *non Arn.*—Abòh, *Vogel,* and Guinea, if I am correct in referring Vogel's specimen to Thonning's species.
Specimen cum illis *M. triangularis* intermixtum, sed certe diversum. *Caulis* multo firmior, costis elevatis angulatus. *Folia* usque ad 3 poll. lata, evidentius dentata. *Pedicelli* multo breviores. *Flores masculi* paullo majores, dentibus calycinis setiformibus dimidium limbi corollini æquantibus. *Antheræ* latiores, breviores. *Ovarium* pilosiusculum. *Fructus* nondum maturus jam longior et apice attenuatus nec globosus, demum glabratus.

1. Bryoniæ sp. ?—Fernando Po, *Vogel.*
The male flowers, so far as I can judge from very young buds, are those of a *Bryonia ;* but the inflorescence is different, and there are no female flowers : I therefore refrain from describing the plant.

2. Bryoniæ sp. ?—Fernando Po, *Vogel.* A small fragment, with the remains of fruitstalks, but no flowers.

There are three supposed species of *Bryonia* described from
West Tropical Africa, viz. : *B. Perrottetiana*, Ser., from Senegal,
and *B. fœtidissima*, and *B. capillaris*, Schum. et Thonn., from
Guinea. None of them agree with either of the above.

1. Æchmandræ? v. Rhynchocarpæ? sp.—A very imperfect
specimen in fruit, gathered south of the Line by Curror, and
not in a state to be described.

The *Rhynchocarpa fœtida*, Schrad., is a Guinea plant, not in
the collections before us.

1. Momordica *Vogelii*, Planch. *in Herb. Hook.*; foliis ample
cordato-ovatis integris sinuato-dentatis, floribus monoicis
ex diversis axillis, masculis umbellatis in involucro longiuscule pedicellatis, corolla calyce triplo longiore, tubo echinato,
fœmineis solitariis geminisve, fructibus ovato-acuminatis echi-
natis.—Fernando Po, *Vogel.*

Caulis glaber et lævis. *Folia* pleraque 4 poll. longa et lata,
remote denticulata et subsinuata, basi late cordata et in
petiolum angustissime decurrentia, lætevirentia, membranacea,
pedatinervia, ad venas marginesque minute puberula, cæterum
glabra. *Cirrhi* simplices v. bifidi. *Pedunculi* petiolum subæquantes. *Involucrum* orbiculatum, apice sæpius dentatum,
3-4 lin. longum, coloratum. *Pedicelli* masculi sæpius 3-5,
semipollicares. *Calyx* profunde 5-fidus, tubo brevi costato
parce echinato, lobis 2 lin. longis ovatis obtusis medio crassis
et parce puberulis margine membranaceis. *Corolla* lutea,
petalis obovatis, 8-9 lin. longis. *Antheræ* (exsiccatione valde
compressæ) vel distinctæ vel facile separabiles. *Flores fœminei* brevius pedicellati; ovarium oblongum, 3 lin. longum,
echinulatum, in collum attenuatum, intus uniloculare, placentis parietalibus membranaceis ovula involventibus. *Calyx*
et *corolla* quam in mare minores. *Fructus* in specimine im-
maturus pollicaris, utrinque acutus, squamis longis muricatus,
ex Vog. maturus ruber et trilocularis.

The description of *Momordica fœtida*, Schum. et Thonn.,
from Guinea, agrees so well in most respects with the above
plant that I should have considered it the same, but that
Thonning expressly says that the calycine segments are acute.

B B

370 FLORA NIGRITIANA.

2. Momordica *Senegalensis*, Lam.—*DC. Prod.* 3. *p.* 311.—On
the Gambia, *Boteler;* a very bad specimen. The species
altogether may be a mere variety of *M. Charantium.*
3. Momordica *cissoides*, Planch. *in Herb. Hook.;* foliis trisectis,
segmentis petiolulatis denticulatis, intermedio ovato, lateralibus oblique ovato-rhombeis breviter subbilobis, floribus masculis umbellulatis, pedicellis involucro brevioribus, calycis laciniis oblongis uncinato-mucronatis corolla duplo brevioribus.
—In hedges, at Abôh, *Vogel.*
Planta dioica videtur, tota glabra excepta pube rara minima
ad venas foliorum vel inflorescentiam. *Petioli* subpollicares,
apice trifidi, petiolulis 2-4 lin. longis, segmentum intermedium
bipollicare, lateralia breviora et valde obliqua, omnia mucronulata, margine remote denticulata, basi acuta v. truncata,
membranacea, glabra et punctis minutis scabrella. *Pedunculi*
petiolo longiores, floribus in speciminibus nostris omnibus
masculis *Involucrum* reniforme, 9 lin. latum, breviter cre-
nulato-dentatum v. integerrimum, scabro-puberulum, glandulis paucis scutellatis per paginam interiorem dispersis,
præcipue versus marginem. *Pedicelli* 5-6, puberuli. *Calyx*
fere 5-partitus, laciniis ovali-oblongis pubescentibus. *Petala.*
5 lin. longa, papulosa, alba, basi intus purpureo-maculata.
Antheræ flexuosæ, vel distinctæ vel facillime separabiles
4. Momordica *maculata*, Planch. *in Herb. Hook.;* foliis trisectis, segmentis petiolulatis obovali-oblongis sinuato-dentatis
lateralibus bipartitis, floribus fœmineis in involucro solitariis,
calycis tubo glabro laciniis subulato-acuminatis.—On the
Quorra, opposite Adda, *Vogel.*
Planta dioica? caule angulato glabro v. ad nodos villosulo. *Petioli*
crassi, striati, pollicares, apice trifidi uti petioluli (3-6 lin.
longi) pube brevi scabridi. *Segmentum* intermedium 3-3½
poll. longum, 1½ poll. latum, apice obtusum irregulariter
sinuatum et dentibus paucis calloso-mucronatis notatum, basi
acutum, membranaceum, læviusculum; lateralia minora,
usque ad basin in duo divisa, at lobi non v. vix petiolulati.
Pedunculi petiolum subæquantes. *Involucrum* orbiculatum.
Flores examinare nequivi, alabastrum enim unicum tantum

vidi fœmineum, breviter pedicellatum, ovario ovoideo, laciniis calycinis longius acuminatis quam in affinibus.

5. Momordica *guttata*, Planch. *in Herb. Hook.;* foliis trisectis, segmentis petiolulatis ovatis acutis v. acuminatis sinuato-dentatis lateralibus bipartitis, floribus masculis umbellatis, pedicellis involucro brevioribus, calycis laciniis lanceolatis subulato-acutis.—Fernando Po, *Vogel;* Sierra Leone, *Don.*

M. maculatæ valde affinis, sed diversa videtur foliis acutioribus, involucris majoribus reniformibus. *Flores* masculi in umbella 5-6, pedicellis calyce brevioribus. *Calycis* laciniæ 2½ lin. longæ. *Petala* multo majora, basi maculata, in specimine tamen nostro nondum aperta. *Stamina* cohærere videntur in phalanges duo.

Dr. Planchon considers Don's plant to belong to a different species; but I can find no essential difference in the very unsatisfactory specimens in the collections. I have indeed considerable doubts whether *M. maculata* may not be merely the male individual of *M. guttata*. Both these species, as well the *M. cissoides*, differ in some respects from the more usual forms of *Momordica*. They are to all appearance diœcious, the leaves decompound, the flowers very shortly pedicelled in the involucre, the petals yellowish-white with purple spots at the base, and the anthers, if not quite free, yet more distinct than in most species. I have not seen the fruit of either, nor yet been able to examine any female flower. The scutelliform glands on the upper surface of the involucres are very irregular in number and in arrangement, and are often wholly wanting.

The only other known West African species, the *M. anthelmintica*, Schum. et Thonn., from Guinea, is described as having still more divided leaves than either of the foregoing.

1. Luffa *scabra*, Schum. et Thonn. *Beskr. p.* 405. —Abòh, *Vogel;* Accra, *Don;* Fernando Po, very abundantly in hedges, *Vogel.*

Dr. Planchon considers the Fernando Po plant to be specifically distinct, but it agrees as well as the other with Thonning's description; and although at first sight there appear to be some differences, it is probably owing to the Fernando Po specimens being all females, and the Continental ones in the collections

before us, all males. Possibly, indeed, the original *L. scabra*
itself may be but a variety of the common East Indian *L. acu-*
tangula. The rudimentary stamens in the female flowers vary
in number, according to Vogel. In the flowers I opened there
were five.

ADENOPUS, (gen. nov.)

Flores (dioici ?) *masculi :* *Calyx* tubulosus 5-dentatus. *Petala*
5, ad apicem tubi calycis inserta, margine integra v. leviter
crispata. *Stamina* 5, medio tubo calycis inserta, filamentis
brevissimis, antheris longe linearibus flexuosis diadelphis.
Flores fœminei . . . . . Herbæ Africanæ hinc *Luffæ* hinc *Tri-*
chosanthi affines. *Folia* palmatim lobata, petiolo apice biglanduloso. *Flores* masculi racemosi.

1. Adenopus *longiflorus,* Benth. ; foliis 5-lobis vix scabriusculis,
calycis tubo petalis longiore infra medium longe attenuato
basi dilatato.—Sierra Leone, *Vogel, Don.*
Caulis tenuis, striatus, glaber. *Cirrhi* stipulares unilaterales ;
stipula altera parva glanduliformis. *Foliorum* petiolus 6-10
lin. longus, sub lamina auctus glandulis 2 oppositis linearibus
divaricatis ; lamina 3 poll. longa et lata, angulato-5-loba,
lobis 1-3 intermediis productioribus, omnibus acutis et margine sinuato-dentatis, dentibus mucronatis, utrinque subglabra et punctato-scabriuscula, trinervis, nervis lateralibus
bifurcatis. *Flor. masc. ;* pedunculi axillares, folio multo
breviores, breviter 3-5-flori. *Bracteæ* parvæ, dentatæ. *Pe-*
dicelli 2-3 lin. longi. *Calycis* tubus 1¾ poll. longus, incurvus, junior puberulus, demum glaber, apice 2 lin. latus,
sub staminum insertione gradatim contractus, prope basin
abrupte dilatatus ; limbi laciniæ lineari-lanceolatæ, glandu-
loso-dentatæ, vix 2 lin. longæ. *Petala* pollicaria (v. majora ?)
obovata, et, tanquam e specimine male siccato apparet,
margine crispa sed integra. *Antheræ* 8-9 lin. longæ, extrorsæ, mediantibus connectivis in corpuscula 2 connexæ,
loculis a basi ad apicem flexuosis.
2. Adenopus *breviflorus,* Benth. ; foliis 5-lobis scabris, calycis

tubo petalis multo breviore basi vix attenuato.—From the Niger Expedition, without the precise locality.

Habitus *A. longiflorœ*, sed folia profundius lobata, majora et scabra, glandulis petiolaribus crassis conicis. *Stipula* cirrho opposita foliacea, subreniformis, denticulata, 3-4 lin. longa. *Pedunculus* florum masculorum crassus, folia subæquans, apice racemoso-10-12-florus. *Bracteœ* foliaceæ, cuneatæ v. ovatæ, acutæ, denticulatæ, 2-5 lin. longæ. *Pedicelli* semi-pollicares. *Calycis* tubus 9 lin. longus, a basi ad apicem gradatim latior. *Antherœ* infra medium tubi insertæ, ut videtur diadelphæ. *Petala* pollice longiora, ovata, basi in unguem brevem contracta, margine crispa.

It is with great reluctance that I have established this genus upon a knowledge of the male flowers only; but the two species were so evidently congeners, and so remarkable in the peculiar glands of the leaves, that I was unwilling to pass them over, and they could not be united with *Luffa*, *Trichosanthes* or *Gymnopetalum*, with all of which they have more or less of affinity.

There remain two *Cucurbitaceœ* in the collection which I am compelled to leave undetermined; the one, from Fernando Po, appears to be a *Cucumis* with the hairiness of *Benincasa :* the flowers, according to Vogel, are white, but in the specimens are not in a state to be examined. The other, also from Fernando Po, with deeply palmately lobed leaves and long male racemes, has only a few young buds, which in their structure resemble those of *Luffa ;* yet I can scarcely believe the plant to belong to that genus.

The *Cucumis arenarius* and *C. chrysocoma*, Schum. et Thonn., both from Guinea, complete the list of known *Cucurbitaceœ* from West Tropical Africa.

LVIII. Portulaceæ.

1. Portulaca *oleracea*, Linn.—Accra and Fernando Po, *Vogel*. A common sea-coast plant, appearing very early on newly formed islands.

2. Portulaca *foliosa*, Ker.—*DC. Prod. v. 3. p.* 353.—P. prolifera, *Schum. et Thonn. Beskr. p.* 239.—On the Quorra, at Attah, and sandy shores of the Nun, *Vogel;* Guinea.
1. Talinum *crassifolium*, Willd.?—*DC. Prod. v. 3. p.* 357?— Accra, *Vogel.*—A very bad specimen, which appears to be either this or the *T. triangulare*, both of them American maritime plants.
1. Trianthema *obcordata*, Roxb. *Fl. Ind. v.* 2. *p.* 445.—Cape Coast, *Vogel.*—A common East Indian plant.

The *Sesuvium Portulacastrum*, Linn., a common Tropical maritime plant, is also a native of Senegal; and *S. brevifolium*, Schum. et Thonn., from Guinea, may not be specifically distinct from it.

LIX. PARONYCHIEÆ.

1. Polycarpæa *stellata*, DC. *Prod.* 3. *p.* 374.—Accra, *Vogel, Ansell;* on the Quorra, *Vogel.*
This species varies much in the calyx, which is more or less acuminated, and often more or less covered with longish hair.
2. Polycarpæa *glabrifolia*, DC. *Prod.* 3. *p.* 374.—Sierra Leone, *Don;* Grand Bassa, *Vogel;* Senegal.
3. Polycarpæa *linearifolia*, DC. *Prod.* 3. *p.* 374.—On the Quorra, *Vogel, Ansell;* Whydah, *Don;* Senegal.
De Candolle describes a fourth species from Senegal, *P. tenuifolia*.

LX. CRASSULACEÆ.

There are no species belonging to this Order in the collections before us, and the only one published from West Tropical Africa is the *Kalanchoë crenata*, Haw., from Guinea.

LXI. SAXIFRAGEÆ.

1. Vahlia *silenoides*, A. DC. *Prod.* 4. *p.* 54.—On the Gambia, *Don.*
Two other species of *Vahlia*, *V. ramosissima*, A. DC., and *V. tomentosa*, A. DC., are also natives of Senegambia.

LXII. Umbelliferæ.

1. Hydrocotyle *petiolaris,* DC. *Prod.* 4. *p.* 60.—-Cape Palmas, Grand Bassa, and on the River Nun, *Vogel, Ansell;* also East Tropical Africa and Madagascar. Both this plant and the South African *H. Caffra,* Meissn., are very variable in the length of the petioles and in the crenatures of the leaves, as well as in inflorescence; and there is considerable doubt whether either is really distinct from the South American *H. Bonariensis,* Lam.

2. Hydrocotyle *Asiatica,* Linn.—*DC. Prod.* 4. *p.* 61.—St. Thomas, *Don.*—A common plant within and south of the Tropics in both hemispheres.

LXIII. Loranthaceæ.

1. Loranthus *Pentagonia,* DC. *Prod.* 4. *p.* 303. *et Coll. Mem.* 6. *t.* 8.—On the Gambia, *Don.*

2. Loranthus *Belvisii,* DC. *l. c.*—L. lanceolatus, *Pal. Beauv. Fl. Ow. et Ben. t.* 69.—Abòh, Cape Coast and Sierra Leone, *Vogel,* who gathered it at the latter station on *Psidium pyriferum,* and observes that the flowers are red, striped with white and tipped with brown, and the fruit red, assuming a greenish tinge when quite ripe; Grand Bassa, *Ansell;* Sierra Leone, *Don.*

3. Loranthus (Scurrula) *leptolobus,* Benth.; glaber, ramis vix compressis, foliis petiolatis ovatis obtusis basi rotundatis, pedunculis brevissimis fasciculatis umbellatim paucifloris, bracteis parvis appressis, corolla tenui scabra basi æquali hinc fissa limbi lobis 5 angustissime linearibus.— On the Quorra, at Attah, *Vogel.*

Rami crassiusculi, teretes v. ad nodos leviter compressi, verruculoso-punctati. *Folia* opposita, magnitudine et forma varia, pleraque lato-ovata, 3 poll. longa, 2 poll. lata, apice obtusissima, basi rotundata, crassa, penninervia, glabra vel siccitate scabra, superiora sæpe angustiora; petioli semipollicares vel paullo longiores. *Flores* numerosi, ad axillas congesti, fere sessiles; pedicelli nempe rarissime lineam excedunt.

Bracteola ovata, concava, crassa, calyce brevior et ei adnata. *Calyx* vix linea longior, margine breviter libero undulato. *Corolla* sesquipollicaris, tenuis, extus leviter papulosa, basi haud dilatata, apice paullulum clavata, fere usque ad medium fissa in laminam profunde 5-fidam lobis angustissimis; color (ex Vogel) luteus apice ruber. *Ovarium* per anthesin basi cum pedunculo continuum et infra apicem in discum cum calyce connatum expansum, cæterum a calyce liberum. *Bacca* (ex Vog.) cupreo-viridis.

4. Loranthus *Nigritanus,* Hook. fil.; ramis ferrugineo-villosis, foliis ovali-ellipticis oblongis sublanceolatisve supra glabris subtus ferrugineo-tomentosis, pedunculis brevissimis umbellatim 3-5-floris, bracteis orbiculatis patentibus ovario longioribus, corollæ ferrugineo-hirsutæ basi gibbæ apice hinc fissæ laciniis 5 lineari-cuneatis.—On the Quorra, at Patteh, *Vogel;* on the Niger, *Mac William.*

Ramuli teretes v. ad nodos leviter compressi, tomento subfloccoso rubro-ferrugineo vestiti, novelli pilis longioribus villosi; rami vetustiores denudati. *Folia* 1½-2-pollicaria, crassiuscule membranacea, novella utrinque tomentosa, adulta supra glabrata, subtus tomento subfloccoso ferruginea. *Flores* fere sessiles, in axillis congesti, pilis longis rubro-ferrugineis dense vestiti. *Bracteæ* exteriores 2 lin. latæ, crassiusculæ, undulatæ, patentes, orbiculatæ, basi angustatæ, interiores minores angustiores apice breviter patentes. *Calyx* 1¼ lin. longus, basi ovario adnatus, limbo libero truncato integro quam ovarium longiore. *Corolla* 15-16 lin. longa, supra calycem inflata, dein abrupte contracta, fere ad medium fissa, lobis linearibus acutis basi angustatis.

The ovary of this species is similar in general structure to those described by Griffith, in his paper on *Loranthus* and *Viscum,* in the 18th vol. of the Linnæan Transactions. In that of *L. leptolobus,* however, the ovary at the time of flowering, enclosed within the calyx, appears to be free from it, excepting in a transverse line a little below the top, corresponding with the commencement of the epigynous disc, common to so many calyciflorous Orders.

The remaining West Tropical African *Loranthi* are *L. cupulatus*, DC., and *L. dodoneæfolius*, DC., from Senegal, *L. Thonningii*, DC., from Guinea, and *L. sessilifolius*, Beauv., from Oware. It is perhaps the only extensive geographical region, especially in warm climates, from which no species of *Viscum* appears to have as yet been brought.

LXIV. RUBIACEÆ.

This large Order, so natural and well defined, and yet so diversified in detail, is very abundant in Tropical countries, and more especially so in West Africa. Its tribes, as successively proposed by De Candolle, Jussieu and A. Richard, and finally adopted in the Prodromus, are generally easy of determination, but in some of the details perhaps too artificial, and a few slight changes and transpositions might render them more conformable to nature, without interfering with their practical utility. Too much reliance has perhaps been placed on the number of carpellary parts, and not enough on placentation (insertion of ovules), which, with the æstivation of the corolla, might in many instances better serve, both for the definition and for the grouping of genera. The *Naucleæ*, well marked by their inflorescence and seeds, form an excellent tribe, if made to include *Sarcocephalus*, *Anthocephalus* and *Cephalanthus*.* In the *Cinchoneæ* there is little to alter; although the line of demarcation between them and the *Rondeletieæ* is at present very ill defined. The remainder of the many-ovuled tribes require some re-arrangement. The two-celled genera often present a third cell, and the many-celled are not unfrequently reduced to two. *Gardenia* itself has not the characters assigned to the

* The seeds of *Cephalanthus*, although solitary in each cell, are inserted on a pendulous placenta, as in *Stephegyne*, and have, like the rest of *Naucleæ*, a wing-like expansion at their apex, although it be so small as to have escaped the notice of most botanists. It appears to me also, that in the young ovary there are a few additional minute abortive ovules, but of that I do not feel certain.

tribe to which it gives its name, and the hardening of the endocarp, which distinguishes the polypyrenous from the multi-locular berry, is but a vague character. It might be better to suppress the two last tribes (*Isertieæ* and *Hamelieæ*), and re-distribute the whole into three : *Gardenieæ*, with fleshy indehiscent fruits; *Rondeletiæ*, with dehiscent or pluricoccous fruits and interpetiolar stipules; and *Hedyoteæ*, with dehiscent or pluricoccous fruits and conpetiolar stipules. For the subdivision of *Gardenieæ*, the placentation appears to afford good characters, although I have not as yet examined with this view the whole of the genera. Probably three distinct forms will be found : *Eugardenieæ*, with parietal placentæ not reaching the axis; *Randieæ*, with the ovules more or less immersed in thick fleshy placentæ; and *Bertiereæ*, (or say *Hamelieæ ?*) with thinner placentæ, superficial ovules, and little or no pulp to the berry. Some genera of the last group come into close connection with some *Rondeletieæ*, and from the latter the passage is very gradual into *Hedyoteæ ;* yet I am unable to suggest any better distinction between them than those generally adopted. Of the tribe *Isertieæ*, DC., *Metabolos* would take its place among *Hedyoteæ*, next to *Hedyotis* (*Euhedyotis*, Arn.), from which it differs but slightly as a genus; *Gonzalea* among *Rondeletieæ*, close to *Lerchea ; Isertia* and *Bruinsmania* among *Gardenieæ* (*Bertiereæ*). The greater part of the *Hamelieæ* would also range in the last-mentioned subtribe ; for which, on that account, the name *Hamelieæ* might be retained. *Morelia,* however, as well as *Alibertia, Schradera,* and perhaps one or two more, would be classed in the subtribe *Randieæ.* To the same sub-tribe I should refer *Cordiera,* and a few imperfectly known genera allied to it, in which the ovules are said to be large, fleshy and peltate, but which have most probably large fleshy peltate placentæ, with one or more ovules immersed therein, but not easily distinguishable in dried specimens. Among the solitary-ovulated tribes, the only alteration of importance which suggests itself is the consolidation of the two tribes of *Guettardieæ* and *Coffeeæ* into one, as neither the two-

or many-celled ovary, nor the drupaceous or baccate fruit
appear to be sufficiently absolute distinctions to separate them
as tribes. The whole might take the name of *Coffeæ*, and be
divided into four or five subtribes, chiefly according to the
æstivation of the corolla* and insertion of the ovules, viz. :
Vanguerieæ, (including *Morindeæ* and *Canthium*), with a valvate
æstivation and pendulous ovules ; *Guettardeæ*, with an imbricate
æstivation and pendulous ovules ; *Ixoreæ*, with an imbricate æstivation and laterally attached ovules ; *Psychotrieæ*, (including
Cephaëlideæ), with a valvate æstivation and erect ovules. Possibly
a fifth might be inserted between *Guettardeæ* and *Ixoreæ*, with
an imbricate æstivation, like in those two tribes, but differing
from *Guettardeæ* in the baccate, not drupaceous fruit, and from
Ixoreæ in the ovules suspended from the apex, or nearly so, but
I am not acquainted with the fruits of a sufficient number of
species of *Chomelia, Chiococca, Kraussia,* &c., to ascertain
whether they can be really so separated from *Guettardeæ* even as
a subtribe. I am doubtful also whether the few genera with
an imbricate æstivation, and ovules erect or ascending, should
be reckoned among *Ixoreæ*, or form an intermediate subtribe
between them and *Psychotrieæ*. They are chiefly South American, and require further examination.

1. Sarcocephalus *esculentus*, Sab.—*DC. Prod.* 4. *p.* 368.—
Sierra Leone, *Vogel, Don ;* Senegal and Guinea.

The fruit of this plant appears to me to be improperly
described as an agglomeration of berries. The syncarpium,
especially the upper portion, consisting chiefly of the epigynous
discs and summits of the calyxes, is indeed fleshy; but the
divisions, formed of the combined parietes of adjoining fruits,
are scarcely so ; the two cells of each fruit have no pulp, and
the dissepiment which separates them is membranous. The
placentæ are adnate : the seeds (as yet unripe in the fruit I

---

* The imbricately æstivated corollæ of *Rubiaceæ* are usually convolute
as in *Apocyneæ ;* but since it is frequently difficult, if not impossible to ascertain whether it is so constantly, I have preferred using the more general
term *imbricated* (of which the *convolute* is a modification) in opposition to
valvate.

examined) are short, orbicular, flattened, with a narrow
wing at their apex, and are suspended by filiform funi-
culi, sometimes very short, sometimes twice as long as the
seed.

A second species of this genus, with broader leaves and
larger stipules, was found by Kotschy in Nubia.

1. Stephegyne *Africana,* Korth.—Nauclea Africana, *Willd.*—
DC. Prod. 3. p. 345.—Nauclea platanocarpa, *Planch. in
Hook. Ic. Pl. t.* 787.—Platanocarpum Africanum, *Hook. fil.
MS.* (Tab. XXXVII.)—On the Quorra, at Attah, where it
forms a handsome tree of 30 feet high, and at Accra, where
it is a bush of little more than a man's height, with
white flowers, turning yellowish, *Vogel;* Sierra Leone, *Don;*
Senegal and Guinea.

The confusion which has crept into our synonymy of this
plant is owing to Vogel's specimens having been mixed with
those of *Sarcocephalus,* and Dr. Planchon having unfortunately
examined a fruit of the latter plant as belonging to the *Stephegyne.* The heads of the true *Stephegyne* are not far advanced towards maturity ; but the young capsules are precisely
as described by Korthals in his generic character, perfectly
distinct, though closely connected in a head, with linear pendulous placentæ, each bearing several ascending imbricated ovules.
I cannot either agree with Dr. Planchon in considering this
plant and Sieber's Senegal specimens as specifically distinct from
Willdenow's. The receptacle in all our specimens is certainly
pilose, and Willdenow corrected his first character of " capitulis
sessilibus" to that of " subsessilibus," the peduncle being usually
very short beyond the last pair of leaves.

Tab. XXXVII. *Fig.* 1. flower with two bracteolæ ; *f.* 2. the same
in a more advanced stage ; *f.* 3. young fruit.

There is among Vogel's Fernando Po specimens, one without
flowers or fruit, but which has the appearance of a species of
Sarcocephalus or *Stephegyne.* It is a tree, with large leaves
and very large foliaceous stipules, thus far agreeing with
Nauclea stipulosa, DC., a Senegambian plant, which may
very possibly be referrible to *Stephegyne.*

1. Uncaria *Africana*, G. Don, *Gard. Dict.* 3. *p.* 471. (TAB. XLII.)—Sierra Leone, *Vogel, Don.*

TAB. XLII. *Fig.* 1. ripe capsule ; *f.* 2. the same, with one valve (one half of the calycine tube ?) removed, showing the endocarp (or real capsule ?) and columella ; *f.* 3. unripe capsule ; *f.* 4. seeds : *all more or less magnified.*

1. Crossopteryx *febrifuga*, Benth. ; foliis ovatis glabratis, corollæ tubo limbo suo triplo longiore, staminibus semiexsertis. —Rondeletia febrifuga, *Afz. ex G. Don, Gard. Dict.* 3. *p.* 516. —Sierra Leone, *Don.*

The generic character of *Crossopteryx* may be completed as follows :

Calycis limbus 5-partitus, lobis dentiformibus erectis. *Corolla* hypocrateriformis, extus villosa, tubo tenui, limbo 5-lobo, laciniis oblongis obtusis æstivatione imbricatis, fauce intus nuda. *Stamina* ad apicem tubi inserta, antheris oblongis apiculatis plus minus exsertis. *Stylus* filiformis apice clavatobilobus. *Ovarium* biloculare, placentis dissepimento adnatis, ovulis imbricato-appressis.

Heudelot's Senegambian collection contains specimens, both in flower and fruit, of *C. Kotschyana*, a species which extends into Nubia, and specimens, in flower only, of a new species.

In Don's Sierra Leone collection is an imperfect specimen, in fruit only, of a plant belonging to the tribe *Cinchoneæ*, which it is impossible to determine for want of flowers, more especially as from the capsules it would appear to form a new genus allied to *Cascarilla* or *Remijia*.

1. Gardenia (Macrotheca) *Vogelii*, Hook. fil. (TAB. XXXVIII. XXXIX.) ; inermis, glabra, foliis membranaceis acuminatis subrepandis, floribus sessilibus, calycis dentibus limbo tubuloso multo brevioribus strictis acutis, corollæ tubo longissimo, limbi lobis oblongo-linearibus, fructu cylindraceo-fusiformi.— Near Abòh, *Vogel.*

Folia oblonga, 6-10 poll. longa, alia basi acuta petiolo 5-8 lin. longo sustenta, alia, præcipue prope basin ramulorum, petiolo brevissimo basi subcordata, omnia in acumen breve obtusum v. acutiusculum apice producta, rigide membranacea, costa venisque primariis utrinque prominulis, rete venularum subter præcipue conspicua. *Stipulæ* in vaginam brevem irregulariter fissam coalitæ. *Flores* ad apices ramulorum solitarii. *Calycis* limbus fere pollicaris. *Corolla* alba, tubo 5-6 poll. longo, limbi laciniis sesquipollicaribus leviter obliquis patentibus. *Fructus* cylindraceo-fusiformis, tripollicaris, crassitie digitis minoris, sed maturus verosimiliter major, calycis limbo persistente coronatus, intus unilocularis, placentis 3 parietalibus stipitatis sed haud ad medium loculi attingentibus. *Semina* compressa, pulpa mixta, in specimine nondum matura.

The *Gardenia lutea,* Fresen., an Abyssinian and Nubian plant, is also contained in Heudelot's collection ; and there is a *G. nitida,* Hook., from Sierra Leone, which I have not seen, and of which the ovary is undescribed, but which is nevertheless most probably a genuine species of the same genus. The three ternate-leaved species enumerated in the Prodromus, *G. triacantha,* DC., from Senegambia, and *G. ternifolia,* Thonn., and *G. medicinalis,* Vahl, from Guinea, are entirely unknown to me.

This genus, *Gardenia,* well characterised by De Candolle, and subsequently by Wight and Arnott, belongs exclusively to the Old World. It is readily distinguished by its unilocular ovarium and parietal placentæ from all others, except perhaps some American *Amaiouæ* and *Genipæ,* which require further examination. De Candolle, however, in his Prodromus, was obliged to retain under *Gardenia* many anomalous species which he found there, and of which he had no opportunity of examining the ovary, removing to *Randia* only such species as he had ascertained to be bilocular. Thence arose the confusion into which these genera have been again thrown by subsequent botanists.

Amongst the bilocular species collected by authors under *Randia,* may be observed at least five groups, which may be either considered as distinct genera, or as sections of *Randia,* viz. :

1. *Rothmannia,* remarkable for its long funnel-shaped corolla,

contains several African and one East Indian species. Amongst
them, *R. Bowieana* differs from the others by the calycine
limb divided to the base.
2. *Randia (Ceriscus)*. Lobes of the calyx not divided to the
base of the limb, and often foliaceous. Corolla of a thick
consistence, with a short thick tube, and broad, blunt, often
undulate lobes. These species are all African or Asiatic, and
include, amongst others, *R. dumetorum* and its allies, as well
as *Lachnosiphonium*, Hochst., in my specimen of which at
least the ovary has certainly two cells only.
3. *Randia (Genipantha)*. Calycine limb tubular or campanu-
late, with short teeth. Tube of the corolla cylindrical, not
twice as long as the calyx, lobes pointed.—African species,
connecting *Randia* with *Oxyanthus*.
4. *Randia (Oxyceras)*. Calycine lobes long and narrow, usually
divided to the base. Tube of the corolla slender and cylindrical, not much longer than the calycine lobes.—All
American species.
5. *Randia (Euclinia)*. Calycine lobes usually short and not
divided to the base. Tube of the corolla considerably longer
than the calycine lobes, and slightly dilated under the limb.—
The genuine species are all American, excepting perhaps one
undescribed one from Senegambia, which comes very near
to them, and the *R. longistyla*, differing rather more in
the remarkable style, which may probably hereafter be considered as the type of a sixth group.

Oxyanthus, with its very slender tubed corolla and small
calyx, and *Griffithia*, with a very deciduous limb of the calyx are,
again, in close connection with the above groups.

1. Rothmannia *Stanleyana*, Hook.; foliis subcoriaceis nitidis
glaberrimis, calycis laciniis subulatis strictis tubo suo multo
brevioribus, corollæ subglabræ v. vix puberulæ tubo longissimo.—Gardenia Stanleyana, *Hook. Bot. Mag. t.* 4185.—
Sierra Leone, *Whitfield*.
A specimen in fruit in the Hookerian Herbarium, gathered
by Captain Boteler on the Gambia, appears to belong to the

same species. This fruit, the size of *R. Bowieana*, is distinctly ribbed.

The *Rothmannia longiflora*, Salisb., may possibly be the same as *R. Stanleyana*; yet it were better to retain Hooker's specific name, in order to avoid confusion with the *R. Bowieana*, which was called *Randia longiflora* by Salisbury.

2. Rothmannia *malleifera*, Hook.; foliis coriaceis subnitidis glaberrimis, calycis laciniis subulatis strictis tubo suo longioribus, corolla extus dense tomentosa.—Gardenia malleifera, *Hook. Bot. Mag. t.* 4307.—Gardenia Whitfieldi, *Lindl. Bot. Reg.* 1845 *sub t.* 47.—Sierra Leone, *Don, Whitfield, Miss Turner.*

Fruit globose, larger than in *R. Bowieana*, and not ribbed. Lindley's synonym, not being taken up in the indexes to the Register, had been nearly overlooked by myself, as it has been by other botanists.

3. Rothmannia *Bowieana*; foliis membranaceis, novellis hirtellis demum glabratis, calycis limbi 5-partiti laciniis linearibus subdilatatis patentibus, corolla glabra.—Gardenia longiflora, *Ait.*—Randia longiflora, *Salisb.*—Gardenia macrantha, *Rœm. Schult. Syst.* 5. *p.* 237.—Randia macrantha, *DC. Prod.* 4. *p.* 388.—Randia Bowieana, *Hook. Bot. Mag. t.* 3409.— Gardenia longifolia, *G. Don, Gard. Dict.* 3. *p.* 499.—Sierra Leone, *Don, Whitfield*, and others.

Fruit globose, or somewhat pear-shaped, not ribbed, and nearly an inch and a half in diameter.

For the history of the synonymy of this plant see Hook. Bot. Mag. sub t. 4307; to which I would only add that the *Gardenia Devoniana*, Lindl. Bot. Reg. 1846, t. 63, appears to be (from the figure) a well-marked, broad-leaved variety at least, if not a good species; and that although both *R. malleifera* and *R. Bowieana* are in Don's Herbarium, it seems to be the latter species only, with membranous leaves, to which he gave the name of *longifolia*.

The completely bilocular ovary readily distinguishes the genus from *Gardenia*, and the long funnel-shaped throat of the corolla from *Randia.* Among the preceding species, *R. Bowieana* and *Devoniana* differ from the others in their calyx ; all agree in their ovules more or less immersed in a fleshy placenta, which is either adnate to the dissepiment, or sessile and peltate. In the East Indian *R. macrophylla,* Br., the placentæ of the ovary are stipitate and bifid, nearly as in *Mussaenda.* I have not had an opportunity of examining the South African species.

1. Randia (Ceriscus) *Sherborniæ,* Hook.—Gardenia Sherborniæ, *Hook. Bot. Mag. t.* 4044 ; volubilis, foliis glabris coriaceis, pedicellis solitariis axillaribus petiolo brevioribus bracteatis unifloris, calycis lobis lato-cuneatis foliaceis obtusissimis, corolla infundibuliformi-campanulata calyce duplo longiore, stylo staminibusque inclusis.—Sierra Leone, *Herb. Hooker.*

2. Randia (Ceriscus) *Doniana,* Benth.—Gardenia calycina, *G. Don. Gard. Dict.* 3. *p.* 497, (non *Randia calycina,* Cham. Schl.) subscandens ? foliis glabris coriaceis, pedicellis axillaribus solitariis petiolo brevioribus bracteatis unifloris, calycis lobis lato-oblongis cuneatisve foliaceis acutiusculis, corolla infundibuliformi-campanulata calycem breviter superante, stylo staminibusque inclusis.—Sierra Leone, *Don.*

I had some doubts whether this might not be the same as *R. Sherborniæ,* but the leaves are narrower and more pointed, and the corolla considerably smaller.

3. Randia (Genipantha) *acuminata,* Benth.; glabra, foliis subsessilibus longe cuneato-oblongis acuminatis, corymbis axillaribus breviter pedunculatis, calycis limbo tubuloso minute dentato, corollæ tubo cylindrico brevi, lobis acutis.—Gardenia acuminata, *G. Don, Gard. Dict.* 3. *p.* 499.—Pomatium dubium, *G. Don, ibid. p.* 505.—Sierra Leone, *Don ;* on the Nun River, *Vogel.*

Arbor parva. *Folia* 8-9-pollicaria v. fere pedalia, superne 3-4 poll. lata, a medio ad basin angustata, ibidem oblique subcordata, apice abrupte acuminata, rigide membranacea v. chartacea. *Stipulæ* latæ, breviter cuspidatæ. *Pedunculi* solitarii, axillares, ½-1 poll. longi, apice divisi in ramos breves cymosoplurifloros. *Bracteæ* parvæ. *Pedicelli* brevissimi. *Ovarium*

c c

386 FLORA NIGRITIANA.

lineam longum ; calycis limbus tubulosus 3 lin. *Corollam*
apertam non vidi, sed alabastrum mox florendum tubum
ostendit calyce paullo longiorem, et lacinias iis *R. genipæfloræ*
simillimas. *Ovarium* biloculare ; placentæ stipitatæ loculos
iterum fere dividunt. *Ovula* numerosa, placentis semiimmersa.
Bacca ovoidea bipollicaris, utrinque subacutata, longitudinaliter costis numerosis obtusis percursa. *Semina* compressa,
horizontalia, placentis pulposo-carnosis immersa.

This species, if I have made no error in matching the flowering and fruiting specimens, differs remarkably from the three
following in the size of the fruit, which resembles in some
particulars that of *Oxyanthus*, of which, however, it has not
the flowers. The habit is, in some respects, intermediate
between the two genera.

4. Randia (Genipantha) *genipæflora*, DC. *Prod.* 4. *p.* 389.—
Niger Expedition, *Vogel*, without the precise locality ; Senegambia, *Heudelot*.

The fruit is about 4 lines diameter, very much like that of
Morelia in appearance, but that it is two-celled only. It is
crowned with the tubular limb of the calyx till very nearly ripe,
when that part usually falls off.

5. Randia (Genipantha) *reticulata*, Benth. ; foliis ovali-oblongis
acuminatis inter venas transversim reticulatis, floribus axillaribus breviter pedicellatis fasciculatis, bracteolis in cupulam
connatis, calycis limbo tubuloso breviter dentato.—Sierra
Leone, *Vogel, Don;* Senegambia, *Heudelot*.

Ramuli et folia novella minutissime puberula, planta cæterum
glabra. *Folia* breviter petiolata, 3-4-pollicaria, basi acuta, coriacea, iis *R. genipæfloræ* primo intuitu simillima, sed venulæ
transversæ crebræ utrinque prominent dum in *R. genipæfloræ*
evanescunt. *Stipulæ* brevissimæ, latæ, breviter setaceo-cuspidatæ. *Flores* in cymas contractas dispositi, ad axillas congesti, vix petiolum superant. *Bracteæ* lineam longæ, in
cupulam brevem connatæ. *Calyx* 2 lin. longus, truncatus,
dentibus 5-6 brevissimis, in speciminibus floriferis vix conspicuis, in fructiferis evidentioribus. *Corollæ* tubus breviter
exsertus, faux leviter dilatata, laciniæ 5-7, oblongo-lineares,

acutæ, 2 lin. longæ, æstivatione imbricata. *Antheræ* lineares,
acutæ, corollam subæquantes. *Ovarium* biloculare (rarius
3-loculare?) placentis peltatis, ovulis in quoque loculo circa
sex immersis. *Fructus* pisi magnitudine, calycis limbo coronatus.
6. Randia (Genipantha) *coriacea*, Benth.; glabra, foliis ovatis
oblongisve brevissime acuminatis crasso-coriaceis vix venosis,
floribus axillaribus confertis sessilibus, bracteolis in cupulam
connatis, calycis limbo cupulato vix dentato.—Grand Bassa,
Vogel.
Tota planta glaberrima videtur. *Folia* breviter petiolata, 3-4-
pollicaria, basi sæpius acuta, penninervia, rete venularum vix
conspicua. *Stipulæ* breves, latæ, longe setaceo-cuspidatæ.
Flores fere *R. reticulatæ* et pariter 5-7-meræ, sed sessiliores,
calycis limbo latiore. *Ovarium* biloculare. *Ovula* placentis
immersa, in quoque loculo sæpius 4. *Fructus* ovato-globosus,
lævis, 3 lin. longus. *Semina* in quoque loculo 1-3, omnino
Randiæ.
This and the preceding species have much the habit and
flowers of *Coffea Arabica* and its allies, although the structure
of the ovary is so different.
7. Randia *longistyla*, DC. *Prod.* 4. *p.* 388.—Gardenia longistyla, *Hook. Bot. Mag. t.* 4322.—Oxyanthus villosus, *G.
Don. Gard. Dict.* 3. *p.* 494.—Sierra Leone, *Don, Whitfield;*
Senegambia.
In the single ovary I examined, the placentæ appeared
scarcely to cohere in the axis, and I have not seen the fruit.
Possibly this plant may form the type of a distinct genus.
There is in Heudelot's Senegambian collection an unpublished
species of *Randia (Genipantha)* allied to *R. reticulata.*
1. Oxyanthus *speciosus*, DC. *Prod.* 4. *p.* 376.—Sierra Leone,
Don.—Folia coriacea, basi acuta et subæqualia.
2. Oxyanthus *formosus*, Hook. fil. (Tab. XL. XLI.); glaberrimus, foliis amplis oblongis basi obliquis, corymbis multifloris, calycis dentibus brevissimis acutis, corollæ tubo foliis
vix breviore.—Cape Palmas, *Vogel.*

c c 2

Caulis subherbaceus, erectus, subsimplex, ramis tetragonis ad nodos superiores compressiusculis. *Folia* breviter petiolata, majora 8-10 poll. longa, 2½-3 poll. lata, apice breviter acuminata v. acutiuscula, basi oblique obtusa v. cordata, coriacea, penninervia. *Stipulæ* lato-lanceolatæ, interdum fere pollicares. *Corymbus* axillaris, breviter pedunculatus, divisus in ramos plures 2-4-floros et fere in paniculam elongatus. *Pedicelli* ebracteati, circiter semipollicares. *Calycis* tubi pars adnata (seu ovarium) lineam longa, limbus ovario fere æquilongus. *Corollæ* tubus tenuis, 4-5 poll. longus, limbi lobis angustis pollice brevioribus, æstivatione imbricata. *Antheræ* paullo infra apicem tubi subsessiles, lineares, apiculo brevi terminatæ. *Ovarium* biloculare; placentis peltatis, ovulis numerosis haud immersis. *Stylus* tubum corollæ subæquans, apice fusiformis et acute brevissimeque bifidus.

3. Oxyanthus *Thonningii*, Benth.; pubescens, foliis oblongis basi valde obliquis truncatis, racemis brevibus plurifloris, calycis dentibus acuminatissimis.—Ucriana racemosa, *Schum. et Thonn. Beskr. p.* 107.—Accra, *Vogel.*

Although the specimen is only in young fruit, without flowers, I have no hesitation in referring it to Thonning's plant, both from its description, and from the identity of the station. The calyx and the form of the leaves at the base prevent the uniting it with *O. hirsutus,* and the pubescence distinguishes it from *O. speciosus.*

3. Oxyanthus *breviflorus*, Benth.; glaberrimus, ramis ad nodos compressis, foliis amplis oblongis basi obliquis, corymbis multifloris, calycis dentibus brevissimis obsoletisve, corollæ tubo brevi.—Fernando Po, *Vogel.*

The leaves and inflorescence are exactly those of *O. formosus,* the flowers are too young to judge of their exact size, but they are evidently so very much shorter than those of *O. formosus,* and the teeth of the calyx so small or so nearly obsolete, that the specimens cannot belong to that species, nor yet to either of those described in the Prodromus. The branches are also much more compressed under the leaves, and the stipules smaller. One specimen is in fruit, which is a hard, almost dry, pear-shaped berry, and about an inch long, with two cells separated by a thin dissepiment almost obliterated in the pulpy placentæ. Two other described species, *O. tubiflorus*, DC., and *O. hirsutus*, DC., are both from Sierra Leone, and there is an unpublished one in Heudelot's Senegambian collection. The *Megacarpha pyriformis*, Hochst. Flora, 1844, p. 551, from Port Natal, is evidently an *Oxyanthus*, closely allied to *O. breviflorus*.

1. Morelia *Senegalensis*, A. Rich.—*DC. Prod. 4. p. 617.*— Abòh, *Vogel;* Sierra Leone, *Don ;* Senegal.

This genus has considerable affinity with the small-fruited *Randiæ* of the section *Genipantha*, from which it chiefly differs in the ovary being completely divided into four perfect cells. The ovules, of the shape of those of *Randia*, are two or three in number in each cell, and more or less immersed in a fleshy placenta, peltately attached to the internal angle of the cell. The seeds are precisely those of *Randia.*

1. Stylocoryne *conferta*, Benth. ; foliis obovato-oblongis acuminatis basi acutis subtus ad venas ramulisque minute puberulis cæterum glabris, corymbis terminalibus subsessilibus densis, calycis pubescentis limbo acute 5-dentato, corollæ lobis linearibus tubo suo sublongioribus. — On the Nun River, *Vogel.*

Frutex humanæ altitudinis. *Ramuli* juniores compresso-tetragoni, pilis minutis appresse puberuli, demum glabrati subteretes. *Stipulæ* persistentes, latæ, rigide membranaceæ, cuspidatæ integræ v. summæ bicuspidatæ. *Folia* 3-4-pollicaria, pleraque obovata cum acumine semipollicari, rarius anguste oblonga, omnia basi acuta, petiolo 5-6-lineari, siccitate nigricantia, consistentia rigidule chartacea, supra nitidula. *Corymbus* intra folia summa tripartitus, ramis brevibus apice dense cymiferis, cymis singulis 10-20-floris, pedunculis compressis bracteisque minutis pubescentibus. *Calycis* tubus (seu ovarium) semilineam longus, limbus lineam longus, ad medium divisus in dentes 5 acutos subpatentes, in fructu juniore persistens, in maturo obliteratus. *Corolla* extus

glabra, alboviridis ex Vogel; tubus tenuis, 2½ lin. longus,
intus superne leviter pubescens; limbi laciniæ 3 lin. longæ,
acutæ, æstivatione valde imbricatæ. *Antheræ* lineares, ad
faucem mediante filamento brevissimo insertæ, corollæ laciniis
paullo breviores. *Stylus* corolla paullo longior, apice longe
subclavatus, integer, medio minute papuloso-pubescens. *Bacca*
junior calyce coronata, matura nuda, globosa, 4 lin. diametro,
pericarpio tenuiter carnoso. *Semina* in quoque loculo plurima, irregulariter hemisphærica, hilo lato ovato, placentæ
peltatim affixa.

2. Stylocoryne *nitidula*, Benth.; ramis glaberrimis, foliis oblongis longe acuminatis basi acutis ad axillas venarum subtus
barbulatis cæterum glaberrimis, corymbis laxis terminalibus
v. demum axillaribus, calycis glabri v. apice minute puberuli
limbo acute 5-dentato, corollæ lobis oblongo-linearibus tubo
suo sublongioribus.—Sierra Leone, *Vogel, Whitfield.*
Primo intuitu *S. confertæ* simillima, sed folia angustiora, longius
acuminata, axillæ venarum pleræque minute foveolatæ v. bar-
batæ; pubes venarum et ramulorum omnino deest, et vix
in inflorescentia apparet, et inflorescentia laxior. *Calyces*
puberuli, lineam longi, dentibus brevissimis. *Corolla* fere
S. confertæ. Baccæ juniores calyce coronatæ, demum sub-
nudæ.

3. Stylocoryne *grandiflora*, Benth.; glaberrima, foliis oblongis
ellipticisve acuminatis basi acutis, corymbis laxis terminalibus
sessilibus, calycis glabri limbo truncato obsolete dentato.—
Fernando Po, *Vogel.*
Tota planta vel omnino pilis destituta, v. paucissimis munita in
ramulos inflorescentiæ. *Ramuli* læves, teretes v. vix compressi. *Stipulæ* brevissimæ, truncatæ, margine subciliatæ v.
brevissime cuspidatæ. *Folia* 4-6-pollicaria v. etiam longiora,
basi longiuscule angustata, apice breviter acuminata, membranaceo-chartacea, utrinque viridia, lævia. *Corymbi* intra
folia summa v. in axillis supremis sessiles, laxe trichotomo-
ramosi, foliis multo breviores. *Calyx* 1¼ lin. longus, turbi-
nato-globosus, vix obsoletissime dentatus. *Corollæ* tubus
virens, pollice paullo longior, consistentia quam in affinibus

crassiore, vix superne latior; limbi laciniæ 4 lin. longæ, oblongæ, obtusæ, albidæ, extus virescentes, æstivatione valde
imbricata. *Antheræ* ad faucem subsessiles, lineares, lacinias
subæquantes. *Stylus* breviter exsertus, superne leviter incrassatus. *Ovarium* 2-loculare, ovulis in quoque loculo 6,
placenta membranacea v. vix carnosa immersis. *Bacca* depresso-globosa.

The three preceding species, as indeed the whole genus
Stylocoryne, are closely allied to *Pavetta* in habit as well as in
flowers, although so far removed by the many-ovulated cells
of their ovary.

1. Heinsia *jasminiflora*, DC. *Prod.* 4. *p.* 390.—Sierra Leone,
Don, Vogel, and others; Fernando Po, *Vogel.*—A plant very
nearly allied to the *Mussaendæ* of the section *Landia,* especially to *M. elegans.* The flowers, according to Vogel, are
white with a yellow eye.

1. Mussaenda *Afzelii,* G. Don, *Gard. Dict.* 3. *p.* 490; abortu
dioica, ferrugineo-pubescens v. villosa, foliis petiolatis elliptico-oblongis breviter acuminatis basi subobtusis, cymis densis
masculis multifloris corymbosis, fœminies solitariis paucifloris,
sepalo bracteæformi amplo albido villoso, calycis limbi lobis
ovatis foliaceis ovario fœmineo multo brevioribus, corolla
dense villosa.—Sierra Leone, *Don;* Abòh, *Vogel;* South of
the Line, *Curror.*

Frutex videtur subscandens. *Ramuli* teretes, pilis brevibus
ferrugineis plus minus vestiti. *Stipulæ* latæ, bifidæ, caducæ.
Folia 3-5-pollicaria v. raro longiora, latitudine varia, ad venas
utrinque ferrugineo-pubescentia et inter venas pilis parvis
conspersa, supra siccitate fusca, subtus pallida, venis parallelis prominentibus; petioli 3-5 lin. longi. *Cymæ* ad apices
ramorum densifloræ, pilis aureis v. ferrugineis dense vestitæ,
masculæ sæpius 3-4 in corymbum brevem dispositæ, fœmineæ
fere semper solitariæ videntur. *Bracteæ* parvæ. *Calycis*
tubus per anthesin in fl. masculis 1½ lin. longus, turbinatus,
ovarium 2-3-loculare semiabortivum includens; limbus stellato-patens, 4-5 lin. diametro, lobis ovatis acuminatis utrinque
villosis; calycis extimi lobus bracteæformis sæpe 3-pollicaris,

in vivo (ex Vogel) albus, in sicco flavicans ; floris fœminei
calycis tubus 5-6 lin. longus, ovarium perfectum 2-3-loculare
multiovulatum fovens. *Corollæ* tubus pollicaris, basi tenuis,
supra medium amplior, extus pilis reflexis dense vestitus,
intus superne pilis longis aureis reflexis clausus. *Antheræ*
inclusæ, lineares, in flore fœmineo minores. *Discus* parvus
glaber. *Stylus* apice 2-3-fidus, in flore masculo abbreviatus,
nunc brevissimus. *Fructus* oblongo-fusiformis, $1\frac{1}{4}$ poll. longus, calycis vestigiis coronatus, durus, siccus, indehiscens,
extus dense villosus.

2. Mussaenda *grandiflora*, Benth. ; (abortu dioica ?) ferrugineo-
villosa, foliis brevissime petiolatis ovatis oblongisve acuminatis basi obtusis, cymis (masculis ?) confertis corymbosis,
calycis laciniis maximis ovato-oblongis acuminatis foliaceis
tubo (masculo ?) pluries longioribus, extima bracteæformi
maxima villosa, corolla villosissima.—Sierra Leone, *Don.*

Ramuli teretes, superne præsertim pilis longis ferrugineis reflexo-patentibus dense vestiti. *Folia* 3-4-pollicaria, utrinque
præcipue ad venas ferrugineo-pilosa. *Cymæ* in specimine
plures, dense 5-6-floræ. *Bracteæ* lineares 3-5 lin. longæ.
Calycis tubus $1\frac{1}{2}$ lin. longus, limbi lobi 9 lin. longi, 3 lin.
lati, basi angustiores, apice acute acuminati membranaceo-
foliacei (colorati ?) utrinque villosi. *Corollam* apertam non
vidi, alabastra jam calyces excedunt, villosissima sunt, tubo
medio dilatato, pili interiores, stamina et stylus omnino
florum masculorum *M. Afzelii.*

3. Mussaenda *tenuiflora*, Benth. ; abortu dioica, scabro-hirtella,
foliis petiolatis oblongo-ellipticis basi acutis, cymis multifloris
masculis pluribus pedunculatis fœmineis paucioribus, calycis
laciniis subulatis tubo sublongioribus, corolla calyce pluries
longiore tenui extus villosa, fructu oblongo villoso ecostato
calycis laciniis coronato.—Fernando Po, on the sea coast,
Vogel.

Rami et folia pilis brevibus rigidis plus minus consperso v.
vestita. *Folia* 3-5-pollicaria, breviter et acute acuminata,
concoloria v. subtus pallida, petiolo brevi rarius pollicari.
Stipulæ acutæ, bifidæ. *Flores masculi* in cymas densas con-

gesti, sessiles. *Bracteæ* parvæ. *Calycis* tubus turbinatus, 1½ lin. longus, villosus, laciniis subulatis basi subdilatatis 2 lin. longis demum stellato-patentibus ; lacinia bracteæformis longe petiolata 2-3-pollicaris, pubescens, albovirens. *Corolla* pollicaris, tubus tenuis, medio ampliatus, extus viridis, intus a medio ad apicem pilis flavis clausus; limbi laciniæ 1½ lin. longæ, cuspidatæ, reflexo-patentes, aurantiacæ. *Antheræ* lineares, inclusæ. *Stylus* brevis, nunc brevissimus. *Ovarium* biloculare, multiovulatum, sed post anthesin non grandescit et ovula abortiunt. *Flores fœminei* in corymbum laxiusculum simpliciorem dispositi, singuli breviter pedicellati, ovario seu calycis tubo per anthesin ovato-oblongo jam 2-3 lin. longo. *Calycis* laciniæ quam in mare tenuiores. *Corolla* maris sed in medio minus ampliata, antheris minoribus. *Stylus* tubo corollæ æquilongus, apice breviter bifidus. *Fructus* coriaceus, exsuccus, bilocularis, semipollicaris.

4. Mussaenda *Isertiana*, DC. *Prod.* 4. *p.* 371.—Sierra Leone, *Don ;* Fernando Po, *Vogel.*

Though generally smooth, or nearly so, this plant appears to be occasionally hairy. The flowers are like those of *M. tenuiflora,* from which however it is readily distinguished by the very small, broadly-lanceolate, pointed divisions or teeth of the calyx.

β? *laxiflora,* pilis longis hispida, inflorescentia laxa.—Sierra Leone, *Vogel.*—A single specimen, with a few leaves and two fruits remaining on it. These are about 9 lines long, hairy, without ribs, and crowned by the very short teeth of the calyx. It may possibly be a distinct species, or, on the other hand, the mere female individual of *M. Isertiana.*

5. Mussaenda (Landia) *elegans,* Schum.—*DC. Prod.* 4. *p.* 372. —Bertiera coccinea, *G. Don, Gard. Dict.* 3. *p.* 506.—Sierra Leone, Aguapim, and Accra, *Vogel, Don,* and others ; Senegal and Guinea.

A beautiful plant, with fiery red flowers, near 2 inches diameter when expanded, apparently variable in the hairiness and form of the leaves. The calycine lobes are usually slightly dilated above the middle, and occasionally one of them shows a tendency to expand into a green leaf.

394   FLORA NIGRITIANA.

The *M. erythrophylla*, Schum. et Thonn., from Guinea, is the only remaining West African species of *Mussaenda* published.

1. Bertiera *Pomatium*, Benth.—Pomatium spicatum, *Gærtn.*— *DC. Prod.* 4. *p.* 391.—Wendlandia pilosa, *G. Don, Gard. Dict.* 3. *p.* 519.—Sierra Leone, *Vogel, Don;* Senegal.

I cannot find the slightest character to distinguish *Pomatium* as a genus from *Bertiera*. The teeth of the calyx are not really deciduous, but, originally very short, they become worn away as the fruit ripens.

2. Bertiera *laxa*, Benth.; ramulis inflorescentiaque puberulis, foliis oblongis acuminatis basi angustatis supra glabris subtus minute puberulis, thyrsis laxis flexuosis, floribus secus ramos paucis sessilibus, dentibus calycinis minimis, fructibus puberulis obscure striatis.—Fernando Po, *Vogel.*

Ramuli novelli pube minuta subferruginei, rami annotini glabrati. *Stipulæ* foliaceæ, utrinque solitariæ, et inter se basi brevissime connatæ, vaginantes, semipollicares, oblongo-lanceolatæ, acutæ, ferrugineo-puberulæ. *Folia* 4-7 poll. longa, 1½-2 poll. lata, membranacea v. demum subcoriacea, petiolo 3-6 lin. longo. *Paniculæ* thyrsoideæ, terminales, solitariæ v. geminæ, foliis summis sæpius breviores, rhachide flexuosa compressa ferrugineo-pubescente, bracteis parvis subulatis acutis, ramis paucis simplicibus v. bifidis 3-8-floris. *Flores* solitarii, secus ramos sessiles v. ramos terminantes. *Calyx* pubescens, lineam longus, dentibus minutis obtusiusculis vix discum floris superantibus. *Corolla* (quam ipse non vidi) ex icone a J. D. Hookero ad florem unicum depicto, calyce 5-ies longior, infundibuliformis, extus villosa, tubo tenui, limbo 5-lobo, æstivatione imbricata. *Antheræ* lineares, basi sagittatæ. *Fructus* eo *B. Guianensis* paullo major, costis minus prominentibus, cæterum omnino conformis, bilocularis, dissepimento tenui, seminibus angulatis placentæ parvis affixis.

3. Bertiera? *macrocarpa*, Benth.—Wendlandia racemosa, *G. Don, Gard. Dict.* 3. *p.* 519.—Sierra Leone, *Don.*

The fruit, on Don's specimen, is not a capsule, but fleshy and indehiscent, ovate, about 4 lines long, crowned with the remains

of a small cup-shaped calycine limb; it is two-celled, with
stipitate bifid placentæ, and innumerable small flat angular
seeds, not imbedded in pulp. The plant is therefore not a
Wendlandia any more than the other West African species
referred to that genus by Don. The specimen is without
flowers, the habit and inflorescence are those of *Bertiera,* the
leaves are 8 or 10 inches long, the branches of the thyrsus very
short. The fruit, however, is much larger, and differently
shaped from that of any *Bertiera* I am acquainted with; the
genus must therefore remain in some measure doubtful.

1. Pouchetia *Africana,* DC. *Prod.* 4. *p.* 393.—Wendlandia
virgata, *G. Don, Gard. Dict.* 3. *p.* 519.—Sierra Leone,
Vogel, Don ; Senegal.

Pedicelli calyce duplo longiores. *Ovula* in quoque loculo plurima, ab apice loculi pendula.

2. Pouchetia *parviflora,* Benth.; foliis ovatis acuminatis, stipulis
petiolo sublongioribus, floribus sessilibus in paniculis axillaribus fasciculatis.—Fernando Po, *Vogel.*

Arbor parva, ramis horizontalibus foliisque distichis ex Vog.
folia pinnata simulantibus, tota glaberrima excepta pube in
inflorescentia parca, ramulis teretibus ad nodos compressis.
Folia breviter petiolata, 3-4-pollicaria, acumine sæpe semipollicari v. longiore, basi acuta, chartacea, siccitate fusca,
subtus pallida. *Stipulæ* latæ, longiuscule acuminatæ, carinatæ, integræ, 2-4 lin. longæ. *Paniculæ* oppositæ, folia
subæquantes, parum ramosæ, floribus interrupte fasciculatis
sessilibus quam in *P. Africana* multo minoribus. *Bracteæ*
minutæ. *Calyces* semilineam longi, dentibus 5 acutis.
Corolla alba, infundibuliformis, tubo ¾ lin. longo, limbo tubo
æquilongo 5-fido, laciniis oblongis patentibus, æstivatione imbricatis. *Stamina* ad faucem inserta, filamentis brevissimis,
antheris oblongo-linearibus exsertis. *Stylus* supra medium
bifidus. *Ovarium* biloculare, ovulis in quoque loculo geminis, ab apice loculi pendulis. *Fructus* non vidi.

Wendlandia sulcata, of Don, from Sierra Leone, may be
another *Bertiera* or *Pouchetia,* but there is no specimen in his
collection which I can identify with his description. The labels

" Hedyotis sulcata" and " Hedyotis pilosa" were both with the
specimens of *Bertiera Pomatium,* the former, however, had
evidently been misplaced.

1. Urophyllum *rubens,* Benth.; foliis petiolatis elliptico-oblongis
vix acuminatis basi acutis coriaceis nitidis ramisque glaberrimis, cymis multifloris petiolos vix æquantibus, calycis
limbo integerrimo.—Fernando Po, *Vogel.*

Frutex orgyalis, ramis teretibus v. vix compressis. *Stipulæ*
foliaceæ, lineari-oblongæ v. obovatæ, 4-6 lin. longæ. *Folia*
4-5 poll. longa, 2-2½ poll. lata, margine leviter revoluta,
coriacea, nitidula, penninervia, siccitate rubentia, in vivo
utrinque viridia; petiolo semipollicari v. paullo longiore.
Pedunculi axillares, 3-4 lin. longi, minute puberuli, apice
cymam ferentes 10-20-floram. *Flores* fere *U. glabri. Calyx*
brevis, limbo cyathiformi truncato edentulo. *Corolla* 1½ lin.
longa, extus glabra, tubo brevi, intus ad faucem villosissimo,
laciniis 5 æstivatione valvatis. *Ovarium* 4-loculare, disco
crasso umbilicato radiatim sulcato coronatum. *Stylus* 5-sul-
catus, apice fusiformis acutiusculus subbilobus. *Ovula* in
loculis numerosa, placentis axilibus vix carnosis affixa.

There is no doubt that this plant is congener with *Urophyllum,* Wall., (1824), which includes *Axanthes,* Blum.,
(1825), and in many respects allied to *U. glabrum,* although the
ovary is only four celled in the flowers I have examined, and the
style entire, but apparently divisible into two lobes. The
number of cells varies in some of the Eastern species from four
to five, in others I find, as in this species, 4 cells with a two-
lobed style, and sometimes two of the dissepiments have appeared to me to be not quite complete. The external furrows,
in this, as in other thickened styles, depend, not on the number
of its divisions, but upon the pressure of the external organs—
the stamens or the edges of the petals, with which they
generally agree in number. The genus extends over the whole
of the East, from Madagascar to the Philippine Islands, and
is apparently numerous in species; it corresponds among mul-
tiovulate *Rubiaceæ* to *Lasianthus* and *Vangueria* among the
uniovulate genera.

2. Urophyllum *hirtellum*, Benth.; ramis puberulis, foliis amplis obovali-ellipticis subacuminatis basi acutis chartaceis subtus hirtis, cymis multifloris fructiferis laxis petiolos subæquantibus, calycis limbo integro.—Abòh, *Vogel.*

Ramuli obscure quadrilaterales. *Folia* 8-10 poll. longa, 3-4 poll. lata, acumine brevi acutissimo, basi longe angustata in petiolum 1-2-pollicarem, consistentia chartacea, supra minute et sparse puberula, subtus pilis sparsis hirta, venis parallelis more fere *Dilleniacearum* subtus valde prominentibus. Nec *stipulas* nec *flores* vidi. *Baccæ* magnitudine *Ribesios rubræ*, juniores limbo calycino tandem fere obliterato coronatæ, 4-5-loculares. *Semina* numerosa, ut in cæteris speciebus foveolato-exculpta.

1. Sabicea *ferruginea*, Benth.; ramulis villosis, foliis ovatis oblongisve subtus ferrugineo-tomentosis, stipulis subcordato-ovatis magnis, floribus in capitulo longe pedunculato involucrato numerosis sessilibus, calycis laciniis lineari-subulatis tubo subduplo longioribus.—Cephaëlis ferruginea, *G. Don, Gard. Dict. 3. p.* 605.—Sierra Leone, *Don.*

Rami scandentes?, ramuli teretes, uti petioli et nervi foliorum pilis ferrugineis appressis dense vestiti. *Folia* 3-5-pollicaria, petiolo semipollicari, latius v. angustius ovata, acuminata, basi obtusa, supra siccitate nigricantia, ad venas hirtella et pilis paucis parvis conspersa, subtus dense obtecta tomento ferrugineo molli. *Stipulæ* petiolo æquilongæ v. longiores, foliaceæ, intus glabræ, extus pilosæ, 3-5-costatæ, integræ et acutiusculæ v. superiores bicuspidatæ. *Pedunculi* in axillis solitarii, 3-4-pollicares, ad apicem capitulum ferunt depresso-globosum, 1½-2 poll. diametro, 30-50-florum. *Bracteæ* involucrantes 6-10, foliaceæ, ovato-lanceolatæ, acute acuminatæ, flores subæquantes, uninerves, pilosæ, extimæ latiores intimæ angustæ. *Flores* subsessiles, omnino *Sabiceæ*, 4-5-meri. *Calycis* pilosi tubus adhærens 1 lin. longus, pars limbi tubulosa 1 lin., laciniæ subulatæ molles pilosæ 4 lin. longæ. *Corolla* semipollicaris, tubulosa, extus basi glabra superne appresse pilosa, intus basi glabra supra medium pilosa, laciniis extus villosis intus basi glaberrimis, æstivatione valvatis. *Stamina* infra medium tubi inserta, inclusa, filamentis brevissimis, antheris oblongo-linearibus. *Stylus* corollam æquans, apice 4-5-fidus, lobis spathulatis intus stigmatosis. *Ovarium* 4-5-loculare, loculis multiovulatis.

2. Sabicea *capitellata*, Benth.; ramulis villosis, foliis ovatis oblongisve subtus dense albido-tomentosis, stipulis brevibus late rotundatis, floribus in capitulis oppositis breviter pedunculatis pluribus sessilibus, calycis laciniis linearibus tubo vix longioribus.—Fernando Po, on the sea shore, *Vogel*.

Frutex sarmentosus *S. cinereæ*, Aubl., simillima. *Folia* 3-4-pollicaria, late v. anguste ovata, plus minus acuminata, basi rotundata v. superiora angustata, supra pilis brevibus raris conspersa, subtus tomento albido v. leviter ferrugineo vestita, petiolo semipollicari v. longiore. *Stipulæ* (saltem superiores) latiores quam longæ, obtusissimæ, undulatæ. *Capitula* 6-10-flora, pedunculo 1-3 lin. longo fulta, globosa, semipollicem diametro. *Bracteæ* involucrantes 4-6, ovatæ, acutæ, villosissimæ, exteriores latæ concavæ. *Flores* 5-meri. *Calycis* villosissimi laciniæ demum stellato-patentes. *Corolla* 5 lin. longa, extus superne villosa, laciniis patentibus crassiusculis intus glabris, æstivatione valvata. *Fructus* pisi magnitudinis, globosus, villosissimus, calycis limbo coronatus, 5-locularis, polyspermus.

3. Sabicea *Vogelii*, Benth.; foliis ovatis oblongisve utrinque viridibus ramulisque hirtis, stipulis ovatis, umbellis breviter pedunculatis multifloris, calycis laciniis linearibus submembranaceis tubo suo pluries longioribus. — Sierra Leone, *Vogel*.

Frutex sarmentosus, habitu et foliis *S. hirtæ*, Sw., simillimus. *Ramuli* tenues, nunc dense hirsuti, nunc fere glabri. *Folia* 2-3-pollicaria, rarius longiora, plus minus acuminata, basi rotundata v. acuta, utrinque pilis longiusculis hirta, subtus pallidiora, at minime albida nec tomentosa. *Stipulæ* quam in *S. hirta* minus dilatatæ. *Pedunculi* 2-6 lin. longi, apice 10-20-flori. *Bracteæ* involucrantes 2 v. 4, oblongæ. *Flores* omnes pedicellati, pedicellis valde inæqualibus (1-4 lin. longis) hirtis. *Calycis* tubus hirsutissimus,

vix lineam longus, laciniæ 3-4 lin. longæ, subfoliaceæ, supra
glabræ, subtus hirsutæ, ½-¾ lin. latæ. *Corolla* calycis lacinias
æquans (v. breviter superans ?), laciniæ 5, extus villosæ, intus
glabræ, æstivatione valvata. *Ovarium* 5-loculare, multiovulatum. *Fructus* globosus, 1½ lin. diametro, calycis limbo
coronatus.

β. *villosior,* floribus in capitulo subsessilibus.—Sierra Leone,
Don.

4. Sabicea *venosa,* Benth.; foliis ovatis utrinque viridibus ramulisque hirsutis, stipulis ovatis, corymbis breviter pedunculatis oppositis laxe multifloris, calycis laciniis lineari-lanceolatis tubo suo duplo longioribus corollæ tubo multo brevioribus.—Virecta lutea, *G. Don, Gard. Dict. 3. p.* 521 ?—Sierra
Leone, *Don ;* Senegambia, *Heudelot.*

Præcedenti et *S. hirtæ* quoad habitus et folia similis; hæc vero
subtus magis villosa, venis parallelis numerosis in pagina
inferiore prominentibus. *Pedunculi* 2-4 lin. longi. *Flores*
nec in capitulum nec in umbellam conferti, sed pedunculi
ramuli plus minus evolvuntur et cymam formant petiolo
sublongiorem. *Bracteæ* paucæ, parvæ, lanceolatæ. *Pedicelli* ultimi calycis tubo breviores. *Calycis* tubus villosus,
¾ lin. longus; laciniæ 4 v. 5, inæquales, majores 1½ v. raro
2 lin. longæ. *Corollæ* tubus villosus, 4 lin. longus, laciniæ
4-5, intus glabræ. *Fructus* 1½ lin. diametro, 4-5-locularis.

5. Sabicea *calycina,* Benth.; foliis ovali-oblongis cordatis utrinque viridibus ramulisque hirtellis, stipulis ovatis, umbellis
longe pedunculatis multifloris, calycis laciniis majoribus ovatis
coloratis tubo multoties longioribus.—Fernando Po, *Vogel.*

Rami volubiles, pilis appressis plus minus hirti. *Folia* 3-4-
pollicaria, acuminata, basi auriculis rotundatis cordata, membranacea, utrinque viridia, ad venas hirsuta et inter venas
pilis paucis conspersa. *Petioli* sæpe pollicares. *Stipulæ* foliaceæ, 3-4 lin. longæ. *Pedunculi* fere glabri, 2-3-pollicares,
apice 10-20-flori. *Bracteæ* involucrantes semipollicares, obtusæ, 4 exteriores latissime subcordatæ, interiores ovatæ,
membranaceæ, virentes v. subcoloratæ. *Pedicelli* valde inæquales, uti calyces fere glabri. *Calycis* tubus lineam longus,

laciniæ 3-5, valde inæquales, membranaceæ, foliaceæ, rubi-
cundæ, majores 6-7 lin. longæ, 3 lin. latæ. *Corolla* 8-9 lin.
longa, alba, sæpius pentamera, glabra, laciniis brevibus ovatis,
æstivatione ut videtur valvata. *Fructus* 1½ lin. longus, car-
nosulus, 4-5-locularis.

PELTOSPERMUM, (gen. nov.) e tribu *Rondeletiarum.*

Calycis dentes 5, breves, persistentes. *Corollæ* infundibuliformis faux intus villosa, limbi lobi 5, subpatentes, æstivatione valvata. *Stamina* versus apicem tubi inserta, filamentis brevibus. *Ovarium* biloculare loculis pluriovulatis,
disco crasso coronatum, stylo apice oblongo-clavato subintegro. *Capsula* dicocca, coccis apice loculicide dehiscentibus.
Semina plurima, placentæ carnosæ peltatim affixa, orbiculata,
margine hyalina at non alata.—*Frutex* Africanus. *Stipulæ*
utrinque solitariæ latæ cuspidatæ. *Panicula* terminalis,
ramis oppositis bifidis, floribus secus ramulos sessilibus.

1. Peltospermum *paniculatum*, Benth.—Fernando Po, on the
sea coast, *Vogel;* Senegambia, *Heudelot.*

Frutex ex Vog., arborescens, ramis teretibus v. compressiusculis, novellis minute bifariam puberulis, adultis glabratis.
Stipulæ breves, latæ, herbaceæ, acumine subulato brevi
cuspidatæ, glabræ, deciduæ v. potius ætate obliteratæ.
Folia ovato-lanceolata v. oblonga, acuta, basi cuneata, 3-4
poll. longa, membranacea, penninervia, glabra v. ad venas
subtus puberula, petiolo 3-6 lin. longo. *Panicula* terminalis, laxe thyrsoidea, semipedalis ad pedalem, glabra v.
minute puberula, ramis oppositis inter se distantibus rigidis
2-4-pollicaribus infra medium bifidis v. inferioribus iterum
paniculatis. *Folia floralia* infima interdum caulinis subsimilia, cætera ad bracteas reducta. *Bracteæ* ramulorum
minutæ. *Flores* secus ramulos 1-3-natim sessiles, uno sæpe
in bifurcatione alari. *Calyx* ¾ lin. longus, tubo omnino
adnato, limbo in dentes 5 parvos acutos æquales diviso.
Corolla 4 lin. longa, extus glabra, intus supra medium pilosa,
tubus tenuis supra medium in faucem ampliatus; limbi

laciniæ breves, acutæ, vix patentes, intus extusque glabræ, apice brevissime inflexo-mucronatæ. *Antheræ* parvæ, versus apicem tubi (ad basin faucis) subsessiles, omnino inclusæ. *Stylus* apice dilatatus in massam oblongam vix brevissime emarginatam. *Capsula* subglobosa, dura, 1½-2 lin. longa, dentibus calycinis coronata, maturitate septicide bipartita, coccis apice breviter transversim dehiscentibus. *Placentæ* dissepimento peltatim affixæ, crassæ, carnosæ. *Semina* in quaque placenta circa 15, semiimmersa, irregulariter orbiculato-depressa, testa membranacea in marginem periphericam hyalinam sed crassiusculam nec vere alæformem dilatata. *Embryo* brevis.

The affinity of this genus with *Lerchea* is evident, but independently of the great difference in habit and inflorescence, the fruit of *Lerchea* consists rather of the indehiscent cocci of *Metabolos* and *Gonzalea* than of the dehiscent capsules of the majority of *Rondeletieæ*, and the included stamens, the seeds and placentation will supply sufficient distinctive characters between the two genera; whilst those indicated by Bennett as separating *Lerchea* from *Wendlandia* and other *Rondeletieæ*, will also serve to separate *Peltospermum* from them. The margin of the seed shows (at least in the dry state) an approach to the wings of some *Cinchoneous* seeds, but less decided than in several plants retained among *Rondeletieæ*. The inflorescence of our genus is that of *Bertiera*, but the carpological, as well as floral characters, are very different.

1. Virecta *procumbens*, Sm.—Sierra Leone, *Don*.

Of the *Virecta multiflora*, Sm., I have a Sierra Leone specimen, but not from either of the collections now described. The *V. paniculata*, Don, from Sierra Leone, is not in his herbarium under that name, and probably does not belong to the genus. *V. elatior*, DC., from Angola, is unknown to me, and may possibly be a *Pentas*.

The *Argostemma pumilum*, Benn., from Sierra Leone, is not in our collections.

1. Pentas *parviflora*, Benth. *in Bot. Mag. sub t.* 4086.—Accra, *Vogel*.

D D

Rami volubiles, novelli pilis paucis brevibus subglandulosis hirtelli, demum glabrati. *Stipulæ* breviter vaginantes, setis
utrinque 2-3 glandulosis. *Folia* 2-3 poll. longa, circa pollicem lata, acutiuscule acuminata, basi in petiolum brevem
angustata, membranacea, utrinque viridia, subtus pallidiora,
supra pilis brevibus raris conspersa, subtus ad costas pinnatas
elevatas pubescentia, inter costas subavenia, glabra. *Cymæ*
terminales, pedunculatæ v. rarius foliis floralibus 2 stipatæ
quam folia caulina multo angustioribus. *Flores* secus cymæ
ramos breves subfasciculatos subsessiles. *Bracteæ* parvæ,
lanceolatæ. *Calycis* extus glanduloso-hirtelli tubus semilineam longus, limbi laciniæ 6, lineares, subfoliaceæ, inæquales, majores lineam longæ. *Corolla*, ex Vog. cœrulescens,
extus glabra; tubus 3 lin. longus, intus supra medium barbatus, cæterum glaber; laciniæ 5, vix ¾ lin. longæ, acutiusculæ, utrinque glabræ, æstivatione valvata v. vix leviter imbricata. *Stamina* tubo inclusa, in parte superiore intra pilos
inserta, filamentis brevibus, antheris linearibus medifixis.
Ovarium biloculare, placentis peltatis, ovulis numerosis non
immersis. *Stylus* apice clavatus, bifidus. *Capsula* septicide
breviter divisa, coccis intus apice dehiscentibus. *Semina*
angulata.

This genus was founded on the *Pentas carnea* of our gardens,
the origin of which is unknown, but is most probably Tropical
African. Two other W. African species, *Virecta elatior*, DC.,
from Angola, and *Hedyotis pentandra*, Schum. et Thonn., from
Guinea, have been referred to the same genus, though with
some doubt.

1. Kohautia *rigida*, Benth.; ramis trichotomis fasciculatis divaricatis, foliis anguste linearibus obtusis, stipulis vaginantibus breviter setiferis v. nudis, floribus trichotome corymbosis,
corollæ lobis oblongo-linearibus, capsula dentes calycinos superante.—West Africa, south of the Line, *Curror*.
Caules rigidi, ephedroidei, ramuli sæpe fasciculati, foliis abortivis
intermixtis. *Folia* pauca, distantia, 1-1½-pollicaria, crassiuscula, avenia. *Inflorescentia K. grandifloræ*, sed flores minores
K. strictæ. *Calyx* turbinatus, semilineam longus, dentibus

latis obtusiusculis. *Corolla* intus extusque glabra; tubus fere 5 lin. longus, tenuis, superne abrupte ampliatus et sub limbo constrictus; limbi laciniæ 1½ lin. longæ, æstivatione levissime imbricatæ. *Capsula* compresso-globosa, calycis tubo omnino adnato fere duplo longior, apice loculicide dehiscens, ad medium dentibus calycinis parvis circumdata. *Semina* plurima (in quoque loculo 8-10), placentis crassiusculis semiimmersa, ovoidea, lateraliter affixa.

2. Kohautia *parviflora,* Benth.; foliis lineari-subulatis margine subrevolutis, stipulis vaginantibus utrinque bisetis, floribus secus ramos paniculæ 2-3-chotomos breviter pedicellatis subfasciculatis, corollæ parvæ laciniis oblongis, calycis dentibus acutissimis capsulam superantibus.—Attah and Accra, *Vogel, Ansell, Don.*

Caules ramosissimi, erecti, virgati, graciles, glabri. *Folia* tenuia, rigidula, majora pollicaria, pleraque multo minora. *Inflorescentia* irregulariter fasciculato-corymbosa, pedicellis ultimis sæpius calyce brevioribus v. paullo longioribus. *Dentes* calycini tubo *calycis* florentis turbinato longiores, acutissimi. *Corolla* vix 2 lin. longa, laciniis limbi æstivatione subvalvatis, cæterum uti antheræ et stylus omnino *Kohautiæ. Capsula* compressa-globosa, subdidyma, vix apice brevissime libera, dentibus calycinis acutissimis distantibus coronata, loculicide bivalvis. *Placentæ* crassiusculæ, seminibus numerosis angulatis semiimmersis.

This may possibly be the *Hedyotis stricta,* Sm., from Sierra Leone, but it certainly is not the *Kohautia stricta,* DC., from Senegal, which that author describes as having the tube of the corolla 5 lines long.

Besides the *K. stricta,* there are three other W. African species known, all from Senegal, viz.: *K. grandiflora,* DC., *K. Senegalensis,* Cham. Schl., and an undescribed small-flowered one.

1. Oldenlandia *virgata,* DC. *Prod. 4. p.* 425.—Accra, *Don;* Sierra Leone.

2. Oldenlandia *herbacea,* DC. *Prod. 4. p.* 455 ? var. cæspitosa,

procumbens.—Cape Palmas, *Vogel.*—A small plant, with the habit of *O. pumila,* but with the capsules of *O. herbacea* and *corymbosa.* I have not been able to find any flowers. The species to which I have referred it is apparently a common one in Tropical countries.

3. Oldenlandia *corymbosa,* Lam.—*DC. Prod.* 4. *p.* 426.—In various parts along the coast and on the Niger.—It is another of the common Tropical weeds. The lower part of the plant in the African specimen is sometimes hairy, and the stipules vary much in length.

4. Oldenlandia *laxiflora,* Benth.; glaberrima, caule elongato, foliis subsessilibus oblongis lanceolatisve uninervibus flaccidis, pedunculis axillaribus elongatis apice laxe dichotomis, calycis fructiferi truncati dentibus 4 parvis remotis.—On the Nun River, *Vogel.*

Caules flaccidi, pluripedales. *Folia* 2-3 poll. longa, ½-1 poll. lata, tenuia, lætevirentia. *Stipulæ* vaginantes, truncatæ, nudæ v. utrinque 1-2-setæ. *Pedunculi* axillares, ½-2-pollicares, apice dichotomi, ramis divaricatis, pedicellis ultimis 3 lin. longis unifloris. *Corollas* non vidi. *Calyces* fructiferi 1½ lin. longi, paullo latiores, lateraliter compressi, apice truncati, dentibus parvis acutis. *Capsula* calyce paullo brevior, apice rima loculicida dehiscens.

This appears to have some affinity with *O. macrophylla* and *O. pentandra,* but the former is described as having the peduncles racemiferous, with opposite pedicels, and the latter as being pentamerous, on which account it may possibly be a species of *Pentas,* although the number of parts is not the character to be relied on for the distinction of the latter genus.

Besides the four species above enumerated, seven species of *Oldenlandia* are described as West African, viz.: *O. sabulosa,* DC., *O. riparia,* DC., *O. linearis,* DC., and *O. macrophylla,* Lepr. Perr., from Senegal; *O. lancifolia,* DC., *O. longifolia,* DC., and *O. pentandra,* DC., from Guinea; but these annual weeds vary so much in appearance, according to age, season, or station, that it is very probable a more careful investigation would much reduce the supposed number of species.

There is also one true *Hedyotis,* from Senegambia; *H. Goreensis,* DC.

OTOMERIA, (gen. nov.) e tribu *Hedyotidearum.*

Char. Gen.—*Calycis* tubus oblongus, limbus 4-5-partitus, lacinia altera foliacea cæteris multo longiore. *Corollæ* tubus filiformis, apice paullo dilatatus, ad faucem intus pilosus, limbi laciniis 5 lineari-oblongis, æstivatione imbricata. *Antheræ* lineares, intra villos faucis subsessiles. *Stylus* disco carnoso impositus, filiformis, apice vix brevissime bifidus. *Ovarium* biloculare, loculis multiovulatis. *Capsula* oblonga, calycis laciniis coronata, submembranaceo-dicocca, coccis intus longitudinaliter dehiscentibus placentas lineares denudantibus. *Semina* plurima ovoidea v. angulata.—Habitus *Otiophoræ.*

1. Otomeria *Guineensis,* Benth.—Grand Bassa, *Vogel, Ansell.*

Suffrutex ramosus, bipedalis; ramis herbaceis teretibus v. obsolete tetragonis, præsertim ad nodos pilosis, demum glabratis. *Vaginæ stipulares* brevissimæ, divisæ in lacinias ciliæformes pilosas utrinque circa 6. *Folia* petiolata, 1-1½-pollicaria, ovata, acutiuscula, basi cuneata, membranacea, penninervia, ad costas petiolosque parce pilosa, cæterum glabra. *Spicæ* terminales, solitariæ v. ternæ, semipedales, glabræ v. minute puberulæ, interrupte multifloræ. *Flores* secus rhachin geminatim sessiles, fasciculis alternis, in spica juniore approximatis, demum remotis, bracteis minutis fultis. *Calycis* tubus glaber, per anthesin lineam longus, lacinia maxima 2 lin. longa, oblonga v. lanceolata, viridis, glabra, cæteræ multo minores, dentiformes, inter se inæquales, una sæpe deficiente. *Corollæ* tubus tenuissimus, 3 lin. longus, extus subglaber; limbi laciniæ vix lineam longæ, extus hirtellæ, intus glabræ. *Capsula* 2 lin. longa, lateraliter sulcata. *Semina* in quoque loculo 10-20, mutua pressione difformia, testa tenui foveolato-exculpta, albumine carnoso, embryone recto.

This curious genus has so exactly the peculiar inflorescence and flowers of *Otiophora*, Zucc., (from Madagascar), that I could scarcely convince myself that it was really distinct without the examination of a considerable number of ovaria and capsules of both plants, the one being constantly polyspermous, and therefore a *Hedyotidea*, whilst the other has the monospermous cocci of *Spermacoceæ*.

1. Morinda *quadrangularis*, G. Don, *Gard. Dict. 3. p.* 545.— Sierra Leone, *Don, Vogel.*

This is a true *Morinda*, with the leaves nearly of *M. citrifolia*, and is allied to that species, but is remarkable by the quadrangular branches and compressed peduncles. The heads of flowers are also much smaller than in *M. citrifolia*.

2. Morinda *lucida*, Benth.; arborea, glaberrima, ramis subteretibus, foliis brevissime petiolatis ovalibus nitidulis, stipulis abbreviatis, pedunculis oppositifoliis subternis compressiusculis, capitulis parvis globosis.—On the Quorra and Fernando Po, *Vogel.*

Folia fere *M. citrifoliæ*, 5-6-pollicaria, sed petiolus vix 2-2½ lin. longus et consistentia forte subcarnosa. *Stipulæ* in ramulis novellis brevissimæ, latæ, subbifidæ, mox fere obliteratæ. *Pedunculi* tenues, sesquipollicares. *Capitula* quam in *M. citrifolia* multo minora. *Baccæ* in capitulo paucæ, omnino connatæ in syncarpium subglobosum 8-9 lin. diametro, singulæ 4-pyrenæ, pyrenis monospermis.

3. Morinda *longiflora*, G. Don, *Gard. Dict. 3. p.* 555.—Sierra Leone, *Don.*

Frutex scandens, ramulis floriferis supra-axillaribus oppositis. *Pedunculi* in axillis oppositi, et ad apices ramulorum terni, breves, 3-5-flori. *Ovaria* connata, 4-locularia, loculis uniovulatis. *Calycis* limbus cupularis, truncatus, ovario æquilongus. *Corolla* fere bipollicaris.

Another specimen, gathered by Vogel in Fernando Po, looks very much like Don's plant, but it is in leaf only, with the remains of fruit-bearing peduncles. One of these peduncles has at its extremity a singular peltate expansion, the nature of which does not clearly appear.

Three other published *Morindæ* are from W. Tropical Africa:

M. geminata, DC., from Senegal, which appears to differ from
M. lucida chiefly by its pubescence ; *M. palmetorum,* DC.,
from Senegal, and *M. chrysorhiza,* DC., from Guinea, which
latter is very near to *M. quadrangularis.*

1. Cuviera *acutiflora,* DC. *Prod.* 4. *p.* 468.—Sierra Leone,
Don, Whitfield; Grand Bassa, *Vogel.*

Folia alia 4-6 poll. lata, subsessilia ; alia longiora, angustiora,
distincte petiolata, omnia coriacea, basi subæqualia et acuta.
Cymæ dichotomæ, paniculæformes, multifloræ, in axillis superioribus v. ad apices ramulorum oppositæ, breviter pedunculatæ. *Bracteæ* lineares, basi angustatæ, 3-6 lin. longæ.
Flores sessiles. *Calycis* laciniæ 3-4 lin. longæ, bracteis
similes. *Corollæ* tubus 2 lin. latus, late turbinatus, intus
extusque glaber nisi intus medio pilis paleaceis reflexis
densis annulatus ; laciniæ 4 lin. longæ, lanceolatæ, subulato-
acutissimæ, rigidulæ, æstivatione valvata, per anthesin reflexæ.
Antheræ parvæ, oblongæ, mediantibus filamentis brevibus ad
sinus laciniarum corollæ affixæ, per anthesin extus reflexæ.
Ovarium 5-loculare, ovulis solitariis, infra apicem affixis,
pendulis. *Drupa* ovoidea, 9-10 lin. longa, obtuse pentagona,
calycis laciniis coronata, intus pentapyrena.

2. Cuviera *subuliflora,* Benth. ; foliis oblongis basi obliquis
subcordatis, cymis axillaribus subsessilibus floribundis, corollæ
laciniis longissime subulato-acuminatis.—Fernando Po, on
the sea shore, and, apparently the same species, at Abòh,
growing in the water, *Vogel.*

Frutex arborescens, totus glaber, ramulis lævibus subteretibus.
Folia subsessilia, semipedalia ad pedalia, breviter acuminata,
basi valde obliqua lata v. angustata et plus minus semicor-
data, consistentia chartacea v. vix coriacea. *Stipulæ* latæ,
connatæ, in specimine pleræque jam detritæ. *Cymæ* paniculæformes, breviores confertiores et sessiliores quam in *C. acutiflora.* *Bracteæ* et laciniæ calycinæ multo longiores et
acutiores, post anthesin accretæ ; lacinias vidi calycinas usque
ad pollicem longas et bracteas sesquipollicares, per anthesin
tamen breviores et angustiores sunt. *Corollæ,* in speciminibus male desiccatæ, iis *C. acutifloræ* paullo minores sunt,

laciniis tamen acumine subulato 3-4 lin. longo terminatis. *Annulus* internus tubi e paleis brevibus reflexis constat. *Stamina* versus apicem tubi inserta, antheris ovato-oblongis connectivo acuminatis. *Ovarium* 5-loculare, ovulis solitariis pendulis. *Stigma C. acutifloræ.*

The *Pachystigma venosum,* Hochst., from South-east Africa, appears to be a third species of *Cuviera,* a genus which comes very near to *Vangueria* and *Canthium,* as well in inflorescence, æstivation, style and ovules, as in the curious ring of reflexed, often scaly hairs, in the tube of the corolla.

The *Ancylanthus rubiginosus,* Desf., from Angola, is not among our collections.

1. Vangueria? *canthioides,* Benth.; inermis, pubescenti-hirta, foliis ovatis oblongisve acuminatis basi rotundatis, cymis brevissime pedunculatis subtrifloris, corollæ pentameræ tubo supra medium annulato.—Sierra Leone, *Don.*

Habitus fere *Canthii discoloris,* sed ovarii loculi 4 v. rarius 3. *Ramuli* pilis brevibus rufis hirti. *Folia* vix bipollicaria, obtuse acuminata, subtus pallida, utrinque pilis brevibus hirtella. *Flores* breviter pedicellati. *Calyx* hirtellus, ¾ lin. longus, limbo brevissimo 5-dentato. *Corolla* extus glaber; tubus calyce duplo longior, intus supra medium pilis longis reflexis paleaceis annulatus; limbi laciniæ glabræ, reflexæ, tubo vix breviores. *Stylus* apice globoso-capitatus, obsolete 4-dentatus.

As *Vangueria* and *Canthium* are only known from each other, when in flower, by the number of cells of the ovary, this plant must, in the absence of fruit, be referred to the former genus, although in habit it comes so very near to some *Canthia.*

1. Canthium *horizontale,* Benth.—Phallaria horizontalis, *Schum. et Thonn. Beskr. p.* 112.—Cape Coast, *Vogel;* Guinea.

Although I have not seen Thonning's specimens, his description indicates most clearly a *Canthium,* and Vogel's single specimen, although in fruit only, agrees with it as far as it goes. Like the rest of the genus, the fruit is generally very oblique, and reduced to a single cell, although occasionally a perfect didymous two-celled berry is met with.

2. Canthium *discolor*, Benth. ; inerme v. rarius spinosum, foliis
ovatis oblongisve acuminatis basi cuneatis subtus pallidis ad
venas ramulisque novellis rufo-pubescentibus, cymis pedunculatis ramosis multifloris, corollæ pentameræ tubo ad medium annulato.—Sierra Leone, *Don, Vogel.*

Specimina inermia, ramo unico vetustiore excepto, cui spinæ
axillares oppositæ divaricatæ 8-10 lin. longæ (pedunculi abor-
tivi ?) ; partes novelli pube brevi rufescunt, adulti glabrati.
Folia subbipollicaria v. raro longiora, acumine brevi obtuso,
basi acuta v. obtusiuscula at minime cordata, consistentia
membranaceo-chartacea, supra siccitate fusca, subtus pallida
sublutescentia, reticulato-venosa ; petiolo bilineari. *Stipulæ* e
basi latiuscula subulatæ, 1-3 lin. longæ. *Cymæ* breviter pedunculatæ, multifloræ, dimidium folii subæquantes. *Calycis*
limbus 5-fidus, lobis parvis ovatis obtusiusculis patentibus.
Corolla 2 lin. longa, tubuloso-campanulata, intus versus
medium pilis reflexis subpaleaceis annulata, cæterum intus
extusque glabra, lobis 5 reflexis dimidium tubi æquantibus,
æstivatione valvata. *Stylus* exsertus, apice nodoso-incrassatus,
summo apice minute bifidus et stigmatosus, sub stigmate ap-
pendice mitræformi auctus. *Fructus* ut in affinibus nunc
didymus, nunc loculo altero abortiente dimidiatus, obliquus.

I should have taken this for the *Phallaria spinosa,* Schum.
et Thonn., but that the two branches of each cyme are said to
be in that plant *simplicissimi,* and the leaves are described as
larger.

3. Canthium *hispidum*, Benth. ; foliis brevissime petiolatis cordato-ovatis oblongisve ramulisque pilis longis sparsis hispidis,
cymis brevissime pedunculatis dense multifloris, calyce truncato vix dentato, corollæ tubo laciniis longiore.—Sierra Leone,
Don.

Ramuli volubiles, pilis longis fuscis patentibus hirti. *Folia* majora 4-pollicaria, late cordato-ovata, ramealia sæpe vix pollice
longiora, basi leviter cordata v. rotundata, omnia acuminata,
supra siccitate nigricantia, pilis longis consparsa, subtus
pallida pilis rarioribus. *Stipulæ* breves, cuspidatæ, hir-

sutæ, caducæ.   *Inflorescentia C. Cornelia*, v. cymæ brevius pedunculatæ.   *Calyces* parvi, dense villosi.   *Corollæ* 3 lin. longæ, 5-meræ, extus glabræ, intus more affinium pilis longis paleaceis reflexis annulatæ.   *Ovarii* loculi 2.   *Fructus* non vidi.

4. Canthium *anomocarpum*, DC. *Prod.* 4. *p.* 475.—Niger Expedition, without the exact locality, *Vogel;* Senegal.

Vogel's specimens have smaller and more shining leaves than the Senegambian ones, but both appear to belong to one species, both agreeing with De Candolle's characters, and having the young shoots compressed, the stipules narrow, rigid and acute, the cymes reduced to a head or umbel, either all but sessile, or borne on a peduncle about a line long.   The fruit-bearing pedicels are from half an inch to an inch long, but from the remains of flowers, it is evident that they were much shorter at an earlier stage.   The fruit is usually dimidiate, as described by De Candolle, yet occasionally a perfect didymous one may be observed.

The remaining W. Tropical African species of *Canthium* are: *C. Cornelia*, Cham. Schl., *C. subcordatum*, DC., and *C. Senegalense*, A. Rich., (*Plectronia hirsuta*, DC.) from Senegambia; *C. Thonningii*, (*Phallaria spinosa*, Schum. et Thonn.), from Guinea, and two undescribed species in Heudelot's Senegambian collection.

The genus *Canthium* should without doubt be placed next to *Vangueria* and *Cuviera*, in the subtribe for which I have proposed the name of *Vanguieriæ;* as a genus, it should probably include (as partly suggested by Arnott and others) *Plectronia*, Linn., *Psydrax*, Gærtn., *Dondisia*, DC., *Phallaria*, Schum. et Thonn., *Mitrastigma*, Harv., *Psilostema*, Klotzsch., and *Chiococca barbata*, Forst.   On the other hand, *Kraussia*, Harv., (*Carpothalis*, E. Mey.), referred to *Canthium* by Endlicher, is very different, both in the æstivation of the corolla and the form of the fruit.   *Mitriostigma*, Hochst., again, differs still more widely, being a Gardenieous genus, closely allied to *Fernelia*, notwithstanding the character assigned to it by

Hochstetter, which can only be explained by the supposition that he had unfortunately examined flowers and fruit belonging to widely different genera.

CRATERISPERMUM, (nov. gen.) e subtribu *Vanguieriearum.*

Char. Gen.—*Calycis* limbus brevis, subsinuatus. *Corollæ* tubus brevis, limbi laciniæ 5, inflexo-acuminatæ, æstivatione induplicato-valvata. *Stamina* ad faucem inserta ; filamenta brevia, antheræ oblongæ, subexsertæ. *Stylus* apice clavatus subinteger. *Ovarium* biloculare, ovulis solitariis ab apice loculi pendulis. *Bacca* globosa, lævis, abortu (an constanter?) unilocularis, monosperma. *Semen* pendulum, hemisphæricum, extus læve, intus excavatum fere pateriforme, albumine carnoso-subcorneo, embryone parvo.—*Frutex* Africanus, cymis densis multifloris axillaribus pedunculatis.

1. Craterispermum *laurinum,* Benth.—Coffea laurina, *Smeathm., DC. Prod. 4. p. 499.*—Sierra Leone, *Don, Vogel.*

Frutex glaberrimus, siccitate flavicans. *Ramuli* crassi, novelli compressiusculi. *Folia* 5-6-pollicaria, obovali-oblonga, brevissime et obtusa acuminata, basi in petiolum longe angustata, rigida, coriacea, reticulato-venosa. *Stipulæ* latæ, subconnatæ, acutiusculæ, cauli adpressæ. *Pedunculi* axillares v. supra-axillares, ancipites, ½-1 poll. longi, apice cymam densam subcapitatam ferentes, ramis brevissimis crassis compressis. *Flores* arcte sessiles, ad apices ramulorum cymæ dense aggregati. *Calyx* semilineam longus. *Corolla* fere *Psychotriæ,* 1½ lin. longa, (apertam tamen non vidi) ; tubus brevis, intus fere usque ad faucem glaber et nudus, fauce laciniisque intus villosulis. *Bacca* subexsucca, pisiformis, siccitate viridi-flavescens, 2-3 lin. diametro, calycis vestigiis oblique umbilicata. *Semen* loculum fere implens ; albumen sectione transversali hippocrepicum.

With the flower of a *Psychotria,* and the fruit nearly that of *Chasalia,* this plant has the axillary inflorescence and pendulous ovules of the *Vanguierieæ.* From *Coffea,* to which it had been referred from imperfect examination, it differs in several essential characters, both in the flower and fruit.

CREMASPORA, (gen. nov.)

Char. gen.—*Calycis* limbus campanulatus, 5-dentatus. *Corolla* hirta, intus ad faucem leviter pilosa; tubus brevis; limbi patentis laciniæ 5, oblongæ, æstivatione imbricata. *Stamina* ad apicem tubi inserta, filamentis brevibus, antheris oblongo-linearibus. *Stylus* hirtellus apice (integer?) acutiusculus. *Ovarium* biloculare, ovulis solitariis ex apice loculi pendulis. *Bacca* ovoidea, lævis. *Semen* abortu (an constanter?) solitarium, læve, raphe distincto percursum, albumine corneo haud excavato, embryone parvo.—*Frutex?* Africana, floribus parvis, in capitula axillaria subsessilia confertis, bracteis intermixtis.

1. Cremaspora *Africana*, Benth.—Coffea hirsuta, *G. Don, Gard. Dict.* 3. *p.* 581.—Sierra Leone, *Don;* also apparently the same species in the Island of Zanzibar, on the S. E. coast of Africa, *Helsing. and Bojer.*

Ramuli pubescentes v. hirtelli. *Stipulæ* utrinque solitariæ, e basi dilatata subulatæ, deciduæ, 1-3 lin. longæ. *Folia* ovata v. oblongo-elliptica, 3-4 poll. longa, breviter acuminata, basi rotundata v. acutata, supra glabra, nitidula, subtus præsertim ad venas hirtella, petiolo brevi hirtello v. pubescente. *Flores* vix petiolos excedunt, uti bracteæ dense sericeo-villosi. *Calyx* vix 2 lin. longus, dentibus acutis erectis. *Corolla* calyce subduplo longior. *Antheræ* corolla breviores. *Bacca* 4 lin. longa, calycis vestigiis coronata, abortu (in specimine examinato) unilocularis, monosperma. *Semen* læve, ad latus interius raphe tenui percursum.

Although Don's specimens are very imperfect, they are sufficient to show great discrepancies from the character of *Coffea*, to which genus he had referred them. The structure of both flower and fruit, confirmed by very good Zanzibar flowering specimens in the Hookerian Herbarium of what appears to be the same species, shows great affinity to *Kraussia*, also a S. African plant, but the form of the fruit, as well as some peculiarities in the calyx and style, and the general habit, have induced me to establish it as a distinct genus. It is not imposible, however, that the discovery of other species, or a better knowledge of the present one, may hereafter render it necessary to unite *Cremaspora* with *Kraussia*.

1. Baconia *corymbosa*, DC. *Prod.* 4. *p.* 485.—Sierra Leone, *Whitfield*; Senegal.

The Senegambian collection contains also a second unpublished species, with nearly sessile leaves. The genus has the imbricate æstivation and terminal inflorescence of *Ixora* and *Pavetta*, but the ovules are attached much higher up in the cells, so as to be nearly pendulous, besides the slight differences in the flower pointed out by De Candolle.

1. Coffea *Arabica*, Linn.—C. stenophylla, *G. Don, Gard. Dict.* 3. *p.* 581.—Commonly cultivated along the coast, and also perhaps indigenous.

The Coffee plant appears to be very variable in the size and form of its leaves. A specimen of Vogel's, cultivated at Cape Palmas, (with the observation that it is wild in Monrovia), has the leaves 10 inches long and 4 inches broad. In Don's *C. stenophylla*, from Sierra Leone, they are 4 or 5 inches long, by about 1½ broad, with rather long points. Other specimens are intermediate, and some even have them still narrower than Don's. The genus, confined to its proper limits, has the æstivation and placentation of *Ixora* and *Pavetta*, with an axillary inflorescence and a peculiar seed.

The *Coffea microcarpa*, DC., from Senegambia, if a true *Coffea*, is unknown to me.

1. Ixora *brachypoda*, DC. *Prod.* 4. *p.* 488.—Sierra Leone, Grand Bassa, Cape Palmas, &c., *Vogel, Don*, and others; Senegal.

Folia valde variabilia, pleraque descr. Candollei conformia, summa sæpe minora, subsessilia, basi late rotundata, ramealia interdum angusta, basi et apice acuta, longius petiolata. *Corymbus* laxus, terminalis, pedunculo foliis breviore v. longiore. *Calycis* dentes brevissimi. *Corolla* alba, apice rosea; tubus 9-10 lin. longus, limbi laciniæ acutiusculæ, oblongo-lineares, glabræ v. vix pilis paucissimis hirtellæ, more generis æstivatione contorto-imbricatæ. *Fructus* globosus, pisiformis.

414   FLORA NIGRITIANA.

The *Ixora laxiflora*, Sm., which I have from an old Sierra
Leone collection, is very much like the above, but the stipules
are much smaller, and the lobes of the corolla are more or less
hairy on the upper side.

2. Ixora *Guineensis*, Benth.; glabra, foliis breviter petiolatis
elliptico-oblongis vix acuminatis subcoriaceis lævibus, corymbo subsessili foliis multo breviore, calycis limbo truncato
subintegerrimo, corollæ laciniis tubi dimidio subbrevioribus.
—Guinea coast, (*Herb. Hook.*)

Specimen imperfectum, *Pavettis* nonnullis simile, sed stylus
apice bifidus. *Stipulæ* breviter vaginantes, in specimine fere
destructæ. *Folia* 5-6 lin. longa, 2 poll. lata, consistentia
coriacea, et minus venosa quam in plerisque *Ixoris*. *Petioli*
marginati, 3-4 lin. longi. *Corymbus* trichotomus, ramis compressis. *Corollæ* tubus 7 lin. longus, laciniæ glabræ 3 lin.
longæ. *Fructus* non vidi.

1. Pavetta *Owariensis*, Pal. de Beauv.? *Fl. Ow. et Ben.* 1. *p.*
87. *t.* 52.—On the Quorra, *Vogel;* Oware, *Beauvois.*

Specimina fructifera tantum adsunt, tota siccitate luride virentia.
Ramuli juniores compressi. *Stipulæ* latæ, acuminatæ, crassiusculæ; paucissimæ tamen in speciminibus supersunt.
Folia 4-6 poll. longa, 2-3 poll. lata, basi in petiolum sub-
pollicarem angustata, subcoriacea, lævia, penninervia, obscure
venulosa. *Corymbus* foliis summis multo brevior, pedunculo
3-6 lin. longo compresso, ramulis trichotomis pariter compressis, pedicellis ultimis superne incrassatis (fructiferis) 2-3
lin. longis. *Baccæ* ovoideo-oblongæ, 3-4 lin. longæ, enerves,
calyce minuto coronatæ, at nondum maturæ. *Flores* non
vidi.

2. Pavetta *neurocarpa*, Benth.; glabra, stipulis caducis, foliis
petiolatis elliptico-oblongis anguste acuminatis basi acutis,
corymbis terminalibus subsessilibus, baccis depresso-globosis
tenuiter costatis.—Fernando Po, *Vogel.*

Specimina omnia fructifera, primo intuitu iis *P. Owariensis*
simillima, sed fructus forma distincta. *Folia* 6-8 poll. longa,
circa 3 poll. lata, apice abrupte producta in acumen angustum
acutissimum 6-8 lin. longum, basi sæpius inæqualia sub-

coriacea nitidula, nervis pinnatis subtus prominulis, petiolo
$\frac{3}{4}$-1 poll. longo. *Stipulæ* ovatæ, cito deciduæ. *Inflorescentia*
P. Owariensis, at pedunculus brevior, interdum subnullus.
Flores desunt. *Baccæ* exsuccæ, fere 4 lin. diametro, lateraliter subcompressæ, in vivo ex Vog. albidæ, calycis vestigiis
umbilicatæ, costis 8 tenuibus percursæ, pericarpio tenui, intus
biloculares dispermæ. *Semina* hemisphærica, intus plana et
medio dissepimento affixa, extus convexa lævia, testa tenui,
albumine carnoso-corneo intus excavato. *Embryo* parvus.

3. Pavetta *genipæfolia*, Schum. et Thonn. *Beskr. p.* 78.—Sierra
Leone and Cape Coast, *Vogel*; Guinea.

These specimens also are in fruit only, they agree, however,
with Thonning's description. The old leaves become smooth
even underneath in some specimens, in others they are pubescent
underneath. The corymb is dense, but when in fruit cannot be
called *globose*. The seeds are hemisphærical, plane on the
inner surface, with the albumen hollowed out as in *P. neurocarpa*.

4. Pavetta? *lævis*, Benth.; glaberrima, foliis longiuscule petiolatis ellipticis oblongisve breviter acuminatis basi longe angustatis coriaceis lævibus, pedunculis axillaribus petiolo longioribus laxe trichotomis multifloris.—Fernando Po, *Vogel*.

Frutex in omni parte glaberrimus, lævis; ramulis teretibus v.
ad nodos compressis. *Stipulæ* triangulares, acutæ v. bidentatæ, cito destructæ. *Folia* 6-8-pollicaria, paucivenia, crassiuscula, costa media subtus lata prominula, marginibus subtus recurvis. *Pedunculi* sæpius ex una tantum axilla orti,
(an potius terminales ramo axillari excurrente?) 2-3-pollicares,
angulati, apice in ramos tres laxe plurifloros divisi. *Flores*
mihi desunt. *Baccæ* in vivo rubræ, magnitudine *Pisi*, seminibus hemisphæricis, albumine intus non excavato.

5. Pavetta? *tenuifolia*, Benth.; glaberrima, foliis petiolatis
ovalibus v. oblongo-ellipticis breviter et obtuse acuminatis basi
angustatis membranaceis, pedunculis axillaribus petiolo brevioribus trichotomis multifloris.—Fernando Po, *Vogel*.

Præcedenti affinis, sed folia membranacea, petiolis brevioribus, venis magis approximatis; pedunculi multo breviores et flores confertiores. *Baccæ* ex Vog. pariter rubræ.

The two last species differ in some respects from the generality of the genus, for the inflorescence is apparently axillary, and the albumen is not excavated on the inner face. There are no flowers to enable me to ascertain whether there may not be other differences. The other West Tropical African species known are: *P. Smeathmanni*, DC., from Sierra Leone, which appears to be nearly allied to the two last, *P. parviflora*, Afz., and *P. subglabra*, Schum. et Thonn., from Guinea, and *P. canescens*, DC., from Angola. Some of these plants may, however, possibly be referrible to *Stylocoryne*.

1. Rutidea *parviflora*, DC. *Prod. 4. p.* 495.—Sierra Leone, *Don, Whitfield;* Cape Palmas, *Vogel;* Senegal.

Frutex scandens. *Folia* 2-4-pollicaria, acuminata, infra medium angustata, basi obtusa v. subcordata, breve petiolata. *Corollæ* æstivatio imbricata. *Ovarium* uniloculare, ovulis geminis e basi cavitatis ascendentibus. *Fructus* luteus, magnitudine *Pisi.*

With the flowers nearly of a *Pavetta,* but pentamerous and smaller, this plant has the ovary of a *Faramea,* and an albumen much more ruminate than in any *Grumilia.*

1. Grumilia *psychotrioides*, DC. *Prod. 4. p.* 495.—On the Niger, at Abòh, *Vogel;* Senegal.—These specimens, like those described by De Candolle, are in young fruit only, and afford no farther illustration of the real affinity of the plant.

1. Chasalia? *laxiflora*, Benth.; glabra, (scandens?), stipulis brevibus truncatis subconnatis, foliis oblongis acuminatis basi acutis, panicula laxa terminali pauciflora, bracteis subnullis, calycis limbo minute dentato disco oblongo multo breviore.—Sierra Leone, *Don;* Grand Bassa, *Ansell.*

Frutex videtur, ex omni parte glaberrimus, ramis volubilibus teretibus lævibus. *Stipulæ* brevissimæ, membranaceæ, cauli appressæ, cito deciduæ v. destructæ. *Folia* 3-4-pollicaria, nunc anguste oblonga, nunc elliptica, v. fere ovata et obovata, longiuscule acuminata, basi angustata, petiolo subsemipollicari.

Panicula terminalis, pedunculata, foliis paullo longior v. brevior, glabra, ramulis brevibus oppositis subcompressis cymoso-paucifloris. *Bracteæ* ad squamas minimas reductæ, v. rarius par foliorum floralium adest sub ramis infimis paniculæ. *Pedicelli* lineam longi, cum ovario et calyce continui. *Calycis* limbus minutus, dentibus acutis vix conspicuis. *Corollæ* tubus 4-5 lin. longus, angulatus, extus glaber, intus basi glaber, ad faucem pilis paucis villosulus ; limbi laciniæ oblongæ, crassiusculæ, apice inflexæ, æstivatione induplicato-valvatæ. *Stamina* supra medium tubi inserta ; filamenta brevia ; antheræ longe lineares, medifixæ, tubo inclusæ v. vix exsertæ. *Discus* epigynus oblongus, truncatus, fere ¾ lin. longus. *Stylus* apice bifidus, lobis complanatis summo apice triangularibus stigmatosis. *Ovarium* biloculare, ovulis e basi loculorum erectis solitariis.

Although I have not seen the fruit of this species, I have little hesitation in referring it to that group of *Chasaliæ* which includes a number of East Indian plants hitherto placed in *Psychotria*, although differing both in flower and fruit from the genuine *Psychotriæ*, whether East Indian or American. They all agree with *Chasalia* in the lengthened tube of the corolla, and the smooth fruit, with the albumen exceedingly hollowed out on the inner face : they only differ, indeed, from the Mauritius specimens by their laxer habit and more slender flowers.*

2. Chasalia *parviflora*, Benth. ; glabra, stipulis brevibus connatis brevissime acuminatis, foliis oblongis acuminatis basi acutis, corymbo terminali multifloro, floribus aggregatis,

---

* Amongst the East Indian species which should be thus transferred to *Chasalia* may be mentioned *Psychotria ophioxyloides*, Wall., *P. curviflora*, Wall., *P. ambigua*, Wight et Arn., n. 8358, 8361, 8363 and 8390, of Wallich's Catalogue, and n. 2317 of Cuming's Malacca plants. The *P. adenophylla*, Wall., and n. 8345 of his Catalogue are *Grumiliæ*, but *P. elliptica*, Roxb., (*P. Reevesii*, Lindl.), appears to me to be rather a true *Psychotria*. *Pæderia ternata*, Wall., *Psychotria oxyphylla*, Wall., and n. 8342 and 8389 of Wallich's Catalogue, seem to be species of *Gærtnera*.

calycis limbo brevissimo truncato disco late depresso breviore.
—Fernando Po, *Vogel.*

Frutex mediocris, ramis debilibus. *Stipulæ* in cupulam latam brevem et brevissime bifidam connatæ, ramo haud appressæ. *Folia* 2-3-pollicaria, membranacea, acumine vulgo brevi, basi in petiolum 3-6 lin. longum angustata. *Pedunculi* terminales, pollicares, trichotomi, ad ramos inferiores sæpius foliorum floralium pare unico bracteati, ramulis brevibus, floribus ad apices ultimorum subcapitato-congestis, corymbo toto foliis multo breviore. *Corollæ* tubus 1½ lin. longus, limbo 5-partito, lobis late oblongis glabris tubo suo duplo brevioribus, æstivatione ut videtur valvata, flores tamen vix perfectos vidi. *Bacca,* adhuc immatura, globosa, lævis, 2 lin. diametro, calycis vestigiis umbilicata, dipyrena. *Semina* ut in speciebus Asiaticis hemisphærica videntur et albumine excavato concavo-pateriformia, sed matura mihi desunt.

1. Psychotria? *reptans,* Benth.; caule lignoso reptante v. scandente, ramis brevibus erectis hirsutis, foliis oblongis acuminatis subglabris, stipulis utrinque in unam magnam cordatam bifidam subconnatis, paniculis pedunculatis multifloris axillaribus subterminalibusque, corollæ laciniis tubi dimidium vix superantibus apice dorso appendiculatis.—Sierra Leone, *Vogel, Don.*

Caules prostrati radicantes v. subvolubiles, lignosi, aphylli. *Ramuli* erecti semipedales v. vix longiores, uti petioli et pedunculi pilis ferrugineis dense vestiti. *Stipulæ* 4-6 lin. longæ, utrinque solitariæ et bifidæ v. fere ad basin divisæ, lobis stipulisve singulis semicordatis basi breviter adnatis apice acuminatis margine undulatis. *Folia* 2-4-pollicaria, breviter petiolata, basi acuta v. rotundata, membranacea, glabra v. ad venas subtus hirtella. *Paniculæ* ovatæ v. corymbiformes, pedunculatæ, foliis breviores, ramis oppositis apice breviter trichotome cymosis. *Bracteæ* ad ramos primarios oppositæ, angustæ, acutæ, ramis breviores. *Flores* in cymis conferti, parvi. *Calyx* vix semilineam longus, dentibus 4 triangularibus acutis. *Corollæ* tubus glaber, 1½ lin. longus;

faux intus villosa ; laciniæ lanceolatæ, acutæ, extus versus
apicem tuberculo cristæformi appendiculatæ, intus fere glabræ.
Stamina inclusa. *Stylus* exsertus, lobis subteretibus longiusculis apice subcapitato-stigmatosis.

Not having seen the fruit, I am rather doubtful whether this
may not be another species of *Chasalia*, but the general appearance is that of *Psychotria*, and it is evidently allied to the
following :

2. Psychotria *obscura*, Benth. ; glabriuscula, scandens, foliis
breviter petiolatis ovato-lanceolatis oblongisve acuminatis,
stipulis magnis utrinque geminis v. subconnatis singulis ovato-
lanceolatis acuminatis, paniculis pedunculatis laxis axillaribus,
cymis ultimis ebracteatis, corollæ laciniis tubi dimidium sub-
æqantibus apice dorso appendiculatis.—Accra, *Vogel.*

Frutex ramulis elongatis scandentibus glabris v. minute puberulis. *Stipulæ* fere semipollicares, caducæ. *Folia* subtripollicaria, membranacea, glabra v. subtus minute puberula,
petiolis 1-2 lin. longis. *Pedunculi* 1½-2-pollicares, ramis
oppositis apice cymiferis. *Bracteæ* angustæ acutæ ad ramos
primarios. *Flores P. reptantis*, sed paullo majores, tubo fere
2 lin. longo. *Stamina* (in specimine) exserta. *Stylus* inclusus. *Bacca* rubra, putamine 10-costato dipyreno. *Semina*
dorso sulcata, sulcis costisque illis endocarpii oppositis, at
albumen non ruminatum ut in *Grumiliis.*

3. Psychotria *Doniana*, Benth. ; ramis pubescentibus, foliis
amplis oblongo-lanceolatis basi longe angustatis, stipulis latis
apice bidentatis, pedunculis axillaribus brevibus apice trichotomis, cymis ultimis subcapitatis ebracteatis, pyrenis obtuse
costatis.—St. Thomas, *Don*, a single specimen with a single
fruit and remains of abortive flowers.

Rami juniores compressi, demum teretes, pube brevissima scabri.
Folia 7-8 poll. longa, 2 poll. lata, lætevirentia, subtus pallida,
membranacea. *Stipulæ* 2 lin. longæ, membranaceæ v. sub-
cartilagineæ, deciduæ. *Pedunculus* petiolo brevior ; cymis
ultimis capitato-multifloris. *Bacca* 3 lin. diametro.

4. Psychotria *latistipula*, Benth. ; subglabra, foliis amplis

E E 2

ovatis ellipticisve breviter acuminatis, stipulis late obovato-
orbiculatis acutis bifidisve deciduis, paniculis axillaribus breviter pedunculatis cymis ultimis umbellatis, bracteis ovato-
lanceolatis involucrantibus, baccis subglobosis pedicellatis
acute costatis.—Fernando Po, *Vogel.*

Frutex videtur ex omni parte glaber v. tomento ferrugineo parco
in inflorescentia partibusque novellis pubescens. *Folia* 5-6
poll. longa, 2-3 poll. lata, membranacea, glabra, basi in
petiolum semipollicarem angustata. *Stipulæ* semipollicares,
margine fimbriatæ. *Pedunculus* communis vix petiolum
æquans, ramis paniculæ oppositis trichotomis. *Bracteæ* foliaceæ, oblongo- v. ovato-lanceolatæ, inferiores ad ramos paniculæ semipollicares, interiores sub cymis ultimis 2-4 lin.
longæ, acutæ. *Flores* desunt. *Pedicelli* fructiferi 1-1½ lin.
longi, subumbellatim aggregati. *Baccæ* 2 lin. diametro, in
sicco insigniter costatæ. *Semina* dorso lævia, nec sulcata.

5. Psychotria *Vogeliana*, Benth.; ramis pubescentibus, foliis
ovali-ellipticis utrinque angustatis chartaceis supra nitidulis
subtus pubescentibus, stipulis latis utrinque acutis bifidis
lacerisve, paniculis axillaribus longiuscule pedunculatis trichotomis, cymis ultimis subcapitatis, bracteis lato-ovatis,
baccis ovoideis costatis.—On the Quorra, at Abòh, *Vogel.*

Frutex 3-5-pedalis. *Rami* novelli compressi, demum teretes,
breviter pubescentes. *Stipulæ* 3-4-lineares videntur, sed fere
omnes jam delapsæ. *Folia* 4-5-pollicaria, rigidiora quam in
præcedentibus, siccitate ferruginea, costis subtus valde prominentibus pubescentibus, inter costas fere glabra. *Pedunculi*
2-3-pollicares. *Bracteæ* latæ, margine laceræ, 2 lin. longæ,
ultimæ sub cymis ultimis subconnatæ. *Flores* non vidi.
Baccæ 3 lin. longæ, brevissime pedicellatæ, endocarpio 10-
costato, seminibus leviter sulcatis.

This species is nearly intermediate between *Psychotria* and
Cephaëlis, but cannot be separated generically from *P. latifolia.*

Six species of *Psychotria* are enumerated by Schumacher and
Thonning from Guinea, none of which I am able to identify,
and probably some of them do not belong to the genus as now
limited. Amongst them, *P. multiflora* must be near to my *P. ob-*

scura, differing in the entire stipules and short peduncles;
P. obvallata is probably a *Geophila*, *P. chrysorhiza*, a *Morinda*,
and *P. Kolly*, possibly, an *Ixora*; *P. umbellata* and *P. triflora*,
if true *Psychotriæ*, are very different from any species known to
me. *P. angustifolia*, G. Don, from Sierra Leone, is probably
either a *Pavetta* or a *Stylocoryne*.

1. Cephaëlis *coriacea*, G. Don, *Gard. Dict.* 3. *p.* 606; glabra,
foliis oblongo-lanceolatis utrinque angustatis tenuiter coriaceis
nitidulis, stipulis amplis bifidis bipartitisve, pedunculis mo-
nocephalis, bracteis pluribus subconnatis flores æquantibus,
calycibus glabris breviter dentatis.—Sierra Leone, *Don.*
Ramuli juniores et pedunculi compressi, demum teretes. *Folia*
3-4 poll. longa, 1-1½ poll. lata, basi in petiolum brevem angustata, venis pinnatis subtus prominentibus, utrinque nitidula, subtus pallida. *Stipulæ* foliaceæ, semipollicares, deciduæ, summæ angustæ. *Pedunculi* foliis breviores. *Capitulum* hemisphæricum, multiflorum. *Bracteæ* circa 8, majores
fere semipollicares, ovato-lanceolatæ, interiores minores, omnes
foliaceæ. *Calyx* brevis, limbo cupulato irregulariter dentato
non ciliato. *Corollæ* glabræ tubus 3 lin. longus, limbi laciniæ
linea breviores, æstivatione valvatæ; faux pilis paucis annulata.
Antheræ exsertæ.

2. Cephaëlis *bidentata*, Thunb. *in Rœm. et Schult. Syst.* 5.
p. 214? abortu dioica, glabra, foliis ovatis oblongisve chartaceis nitidulis, stipulis amplis bidentatis, pedunculis 1-3-
cephalis, bracteis 2-4 latis connatis floribus brevioribus, calycis
dentibus ciliatis, baccis ovoideis costatis. — Sierra Leone,
Vogel, Don ; Grand Bassa and Cape Palmas, *Vogel ;* Senegambia, *Heudelot.*
Arbor parva v. frutex elatus ramosus, v. interdum reptans v.
subscandens. *Ramuli* novelli compressi, mox teretes. *Stipulæ*
latæ, subsemipollicares, apice acute bidentatæ v. breviter
bifidæ, margine undulatæ, basi cordato-adnatæ. *Folia* nunc
3-4-pollicaria, nunc duplo majora, venis parallelis pinnatis
prominulis, basi angustata, petiolo nunc vix 2-3 lin., nunc
pollicem longo. *Pedunculi* axillares, ancipites, 3-6 poll. longi;
fœminei 1-cephali v. rarius 2-cephali, masculi sæpius 3-

cephali. *Bracteæ* in involucrum bifidum lobis bidentatis connatæ, crassiusculæ, subcarnosæ, in capitulo fœmineo flores fere æquantes, in masculo iis multo breviores. *Capitula* hemisphærica, dense multiflora, floribus subsessilibus, receptaculo carnoso. *Calycis* limbus cupulatus, lineam longus, dentibus 5 longe et irregulariter ciliatis. *Corolla* glabra, alba, 3 lin. longa, fauce ampla, lobis 5 brevibus. *Stamina* in masculis, stylus in fœmineis, exserta. *Pedicelli* fructiferi plus minus evoluti, interdum lineam longi. *Baccæ* 2 lin. longæ, endocarpio seminibusque insigniter costatis sulcatisque.

It is possible that more than one species may be here confounded, although I am unable to distinguish them in the dried specimens.

1. Geophila *reniformis*, Cham. et Schlecht. *DC. Prod.* 4. *p.* 537 ; var. ? foliis obtusissimis retusisve.—On the Nun, *Vogel;* St. Thomas, *Don.*

The leaves are more deserving of the epithet *reniform*, than the generality of the South American specimens, yet as far as the specimens go, they do not show any other distinctive character.

2. Geophila *hirsuta*, Benth. ; foliis cordato-ovatis oblongisve acutiusculis utrinque ad venas petiolisque hirsutis, pedunculis abbreviatis, (bracteis subulatis ?).—On the Nun River, *Vogel.* Affinis *G. violaceæ,* sed folia superiora angustiora, tota planta siccitate nigrescit et multo hirsutior est quam unquam vidi varietates *G. violaceæ. Calycis* laciniæ pilis longis ciliatæ. *Specimina* flores paucissimos ferunt. *Planta,* teste Vogelio, dioica est.

A third species, near *G. reniformis,* with remarkable broad bracteæ, is in Heudelot's Senegambian collection.

1. Octodon *filifolium,* Thonn., *DC. Prod.* 4. *p.* 540.—On the Quorra, *Vogel;* Senegal and Guinea.

1. Borreria *Kohautiana,* Cham. et Schlecht., *DC. Prod.* 4. *p.* 541.—Frequent in cultivated grounds, Senegal and Sierra Leone, *Vogel, Don,* and others; also Cape Verd Isles, (supra p. 133), and apparently S. E. Africa.

2. Borreria *ramisparsa*, DC. *Prod.* 4. *p.* 544.—St. Thomas, *Don*; Senegal.—var. major multiflora.—On the Quorra, at Patteh, *Vogel.*

A common Tropical Brazilian weed, with exactly the appearance of *B. parviflora*, but in the African, as well as in the American specimens, two of the calycine teeth are always very minute.

1. Spermacoce *Ruelliæ*, DC. *Prod.* 4. *p.* 554.—Accra, *Vogel, Don*; on the Quorra, *Vogel, Ansell*; Senegal.

These specimens appear to combine the characters of *S. Ruelliæ* and *S. galeopsidis*, DC. The length of the teeth of the calyx is variable. The species is very different from any other I am acquainted with, and remarkable for the size of the capsules.

2. Spermacoce *palmetorum*, DC. *Prod.* 4. *p.* 553?—Sierra Leone, *Vogel.*

This agrees better with Thonning's description of his *Dioidia scabra* than with De Candolle's character of *S. palmetorum*, to which he refers Thonning's plant with doubt. The leaves are oblong-lanceolate, the calycine teeth very unequal, one or two of them being longer and broader than the rest.

3. Spermacoce *pilosa*, DC. *Prod.* 4. *p.* 553.—Sierra Leone, *Vogel*; Guinea.

4. Spermacoce sp., near *S. phyllocephala*, DC., but with broad leaves.—A single imperfect specimen from Sierra Leone, *Don.*

Three other species have been described from W. Tropical Africa: *S. phyllocephala*, DC., *S. stachydea*, DC., and *S. chætocephala*, DC., all from Senegal.

1. Mitracarpium *Senegalense*, DC. *Prod.* 4. *p.* 553.—Sierra Leone, *Vogel*; Accra, *Don*; Cape Verd Isles, Senegal, Guinea and Nubia.

1. Diodia *arenosa*, DC. *Prod.* 4. *p.* 564?—Sierra Leone, *Don*; a Brazilian species.

The specimen is barely in flower, and insufficient to characterize it if it be really distinct from the Brazilian plant, of which it has all the appearance. It is very near *D. articulata*, but has larger leaves, which are rough on both sides. The flowers are also larger, and the calycine teeth longer. The stipular ciliæ are occasionally expanded into short linear leaves.

2. Diodia *maritima,* Schum.—*DC. Prod.* 4. *p.* 564.—On the Nun River, *Vogel;* Senegal and Guinea.

3. Diodia *breviseta,* Benth.; caule scandente, ramis tetragonis, angulis pubescentibus, foliis ovali-lanceolatis oblongisve utrinque acutatis scabris, stipularum setis fructu brevioribus, verticillis multifloris, calycis laciniis 4 reflexis fructu oblongo lævi ecostato subtriplo brevioribus.—Fernando Po, *Vogel.*

Very near to the West Indian *D. scandens* and *D. sarmentosa,* differing chiefly in its numerous small fruits, with a much thinner pericarp.

4. Diodia (Hexasepalum) *vaginalis,* Benth.; procumbens, glabra, foliis sessilibus oblongo- v. lineari-lanceolatis rigidis, ciliis stipularibus rigidis subdilatatis, fructibus oppositis a dorso compressis acute costatis calycis laciniis subsenis coronatis.—At Grand Bassa and on the Nun, creeping in the sands, *Vogel.*

Folia subbipollicaria, 2-3 lin. lata, basi interdum ad margines brevissime ciliata, costa crassa subtus prominula. *Stipularum* vaginæ 2-3 lin. longæ, nunc internodiis longiores subimbricatæ, nunc dissitæ ramo inter nodos compresso-tetragono; ciliæ rigidæ, acutæ, interdum subfoliaceæ, circa 2 lin. longæ. *Corolla* 6 lin. longa, fauce ampla. *Fructus* dicoccus, coccis a dorso valde compressis, lateribus suberoso-dilatatis, marginibus acutis, dorso tricostato. *Semen* compressum, prope basin affixum. *Embryo* fere albuminis longitudine, radicula recta cotyledonibus sublongiore, ad basin fructus spectante.

This agrees with De Candolle's description of *Hexasepalum angustifolium,* from Mexico, in every respect but in the leaves being much broader. Although in habit it is in some measure intermediate between *Diodia* and *Hydrophylax,* yet there does not appear to be sufficient in the size and more corky consistence of the fruit to separate it generically from *Diodia,* which contains other species with five or six teeth to the calyx.

The list of *Rubiaceæ* known from West Tropical Africa will be completed with two species, one gathered by Vogel in Fernando Po, the other by Don in Sierra Leone, the specimens of which are insufficient even to guess at their genus, and three published plants of doubtful affinity : *Stipularia Africana,* Pal. de Beauv., from the River Galbar, *Hylacium Owariense,* Pal. de Beauv., from Oware, and *Benzonia corymbosa,* Schum. et Thonn., from Guinea.

LXV. COMPOSITÆ.

1. Oiospermum *Nigritanum,* Benth. ; caule puberulo, foliis brevissime petiolatis ovali-oblongis basi angustatis subtus pubescentibus, capitulis nudis, corolla glabra achenio oblongo longiore.—On the Quorra, at Patteh, *Vogel.*

Herba O. Wightianæ habitu subsimilis, sed minus villosa. *Folia* 2-3-pollicaria, rugosula, remote subdentata, supra fere glabra, subtus præsertim ad venas pubescentia. *Inflorescentia* et capitula *O. Wightianæ,* hæc vero bracteis foliaceis destituta, alia terminalia, alia pauca lateralia subsessilia v. pedunculata. *Involucri* squamæ multiseriales, acutissimæ, margine scariosæ, dorso puberulæ, exteriores minores angustæ, interiores 3-4 lin. longæ, flores subæquantes. *Receptaculum* planum nudum. *Flores* numerosi. *Corollæ* tubus tenuis, superne ampliatus. *Styli* rami subulati, acuti, recurvi. *Achenium* oblongum, 10-costatum, inter costas minute glandulosum, apice truncatum, omnino nudum.

1. Sparganophora *Vaillantii,* Gærtn. *DC. Prod.* 5. *p.* 12.— S. Africanus, *Gærtn. DC. l. c.*—St. Thomas, *Don.*—Senegal to Benin, and common in Tropical America.

After a careful comparison of several African with American specimens, I cannot discover the slightest difference. Both vary in the blunter or sharper teeth of the leaves, and both are sometimes smooth, though generally more or less pubescent.

The *Ethulia conyzoides,* Linn., a common East Indian weed, extends through Africa to Senegal.

1. Herderia *stellulifera,* Benth. ; decumbens, ramis laxe co-

rymbosis polycephalis, acheniis hispidulis, pappo simplici coroniformi fimbriato stellato-patente.—Sierra Leone, *Don*; Fernando Po, *Vogel.*

Herba ramosissima, diffusa, habitu et inflorescentia formis decumbentibus *Vernoniæ cinereæ* subsimilis. *Pubes* ferruginea, in ramulis novellis sublanosa, demum parca, in foliorum pagina superiore rara v. nulla. *Folia* petiolata, ovata, 1-1½-pollicaria, integra v. sinuato-dentata, membranacea, ramealia minora interdum obovata, summa parva, oblonga. *Corymbi* laxi, irregulares, foliis paucis parvis bracteati. *Capitula* quam in *Ethulia* minora. *Involucri* squamæ subbiseriales, lineari-oblongæ, acutæ, virides, puberulæ; adjectis nonnullis exterioribus minoribus setaceis. *Receptaculum* planum, nudum. *Flores* in capitulo circa 20, homogami, involucro paullo longiores, violacei. *Corolla* tubuloso-campanulata, superne pilosiuscula, basi attenuata, laciniis 5 tubo æquilongis, apice extus glandulosis. *Antheræ* lineares, ecaudatæ. *Styli* lobi subulati, recurvi, acutiusculi. *Achenia* ¾ lin. longa, obscure angulata, undique hispidula. *Pappus* albus, stellato-patens, constans e paleis brevissimis fimbriato-multifidis; setis exterioribus nullis.

This plant has more the habit of *Ethulia* than of *Herderia truncata*, but differs from the former genus in the involucre, achenia and pappus. The pappus has the scales shorter than in *Herderia truncata*, and there is either no trace at all of outer setæ, or they are so short and slender as to be scarcely perceptible. Perhaps the two genera, *Ethulia* and *Herderia*, ought to be united.

The original *Herderia truncata* appears to be confined to Senegal, whence I have what I believe to be a third species, gathered by Michelin.

1. Vernonia *cinerea*, Less.—DC. Prod. 5. *p.* 24, *Webb, supra,* *p.* 134.—Sierra Leone, *Vogel;* Senegal and Guinea. A most abundant East Indian and Tropical African species.
2. Vernonia *Doniana*, DC. Prod. 5. *p.* 23.—Sierra Leone, *Don, Miss Turner,* who states it to grow about 6 feet high in the Leinster Mountains.

3. Vernonia (Lepidaploa) *conferta*, Benth.; fruticosa? ramis
tomentosis, foliis amplis obovali-oblongis sinuatis basi angustatis petiolatis supra ad venas subtus ubique tomentosis, panicula ampla floribunda aphylla, capitulis subsessilibus 10-
12-floris, involucri squamis brevibus obtusis interioribus
lineari-oblongis acutis, achenio glabro, pappi sordidi setis
exterioribus paucissimis brevibus.—Sierra Leone, *Don.*
Folium unicum adest pedale, 5 poll. latum, subtus sordide tomentosum, petiolo sesquipollicari. *Panicula* pedalis, capitulis
secus ramos numerosos divaricatos subsessilibus, ovoideis, 4 lin.
longis. *Flores* involucrum æquantes, glabri. *Pappus* e setis
sordidis rigidulis vix denticulatis constans, involucro sub-
longior.

4. Vernonia (Lepidaploa) *Guineensis*, Benth.; herbacea? erecta,
ramis subfloccoso-tomentosis, foliis lanceolatis serratis supra
araneosis demum glabris subtus dense tomentosis, corymbis
oligocephalis, capitulis late ovoideis 25-30-floris, involucri
sqamis obtusis dorso tomentosis, pappo exteriore interiori
triplo breviore.—Sierra Leone, *Don.*
Specimen bipedale, virgato-ramosum, apice fastigiatum. *Tomentum* paginæ inferioris foliorum ramorum et pedunculorum
densum, subfloccosum, albo-rubens. *Folia* 2-3-pollicaria,
acutiuscula, grosse serrata, basi angustata, sessilia v. breviter
petiolata. *Capitula* 5-6 lin. longa, 4-5 lin. lata. *Involucri*
squamæ rigidæ, pauciseriatæ, exteriores breves, interiores
oblongæ, pappo vix breviores. *Flores* breviter exserti, glabri.
Styli lobi subulati, acuti. *Receptaculum* vix alveolatum,
planum, nudum. *Achenia* pubescentia, leviter costata,
minute glandulosa. *Pappus* rufescens, nitens, fragilis, setis
subdilatatis, seriei exterioris latioribus ciliatis.

5. Vernonia (Lepidaploa) *Vogeliana*, Benth.; fruticosa, puberula, foliis petiolatis oblongis utrinque longe angustatis subintegerrimis membranaceis, cymis terminalibus paniculæfor-
mibus ramosissimis subaphyllis, capitulis pedicellatis 8-10-
floris, involucri squamis oblongis, exterioribus ovatis obtusis,
achenio pubescenti-glanduloso, pappi serie exteriore interiori
quadruplo breviore.—At Clarence, in Fernando Po, *Vogel.*

428   FLORA NIGRITIANA.

Species quoad habitum inflorescentiam involucrum et pappi
colorem simillima *V. extensæ* et *V. grandi*, DC., et *V. Senegalensi*, Less. A priori differt foliis margine integerrimis v.
vix crispulis nec serratis et capitulis minoribus; involucra
nempe vix 2 lin. nec 3½ lin. longa, squamis minus inæqualibus. A *V. Senegalensi* distinguitur pappi exterioris brevitie
et acheniis pubescentibus. *Frutex* est, teste Vogelio, 6-8
pedalis, floribus albis. *Pappus* rufescens, involucro duplo
longior.

6. Vernonia *Senegalensis*, Less., *Linnæa* 4. *p.* 265.—Decaneurum Senegalense, *DC. Prod.* 5. *p.* 68.—Annabon Island,
south of the Line, *Curror;* Senegal and Guinea.

In this species, the external pappus is much longer than in
any of the preceding, but considerably shorter than the inner
one. In the *V. amygdalina,* Delile, (*Decaneurum,* DC.) from
Senegal and Nubia, which differs from the present species
chiefly in the hispid achenia, the external pappus is also still
longer; but even in that species there is a very perceptible
difference in the length of the two series. In the *Gymnanthemum Abyssinicum,* Schultz Bip. (at least in my specimen, Un.
Itin. 1st. Ser. n. 31), which has the habit of the foregoing, and
which ought, with them, to be referred to *Vernonia,* the external
pappus is, as in most *Vernoniæ,* scarcely more than a fourth of
the length of the inner one, but so very deciduous, that if the
head of flowers is not opened with care, the external setæ will
only be found loose amongst the others. All these species,
moreover, are so closely allied in habit to *V. extensa* and some
other species of the Old World, that they cannot well be
generically separated, especially as they have not the involucral
characters ascribed by De Candolle to *Decaneurum.* In all, the
setæ of the pappus are denticulate, and of a more or less ruddy
colour. To the same set of *Vernoniæ* belongs probably the
Candidia Senegalensis, Ten. Cat. Ort. Nap. p. 79, which is
unknown to me.

The other West African *Vernoniæ* are *V. pandurata,* Link,
from Congo, and *V. pauciflora,* Less., from Senegal; besides
two species of *Webbia: W. serratuloides,* DC., from Senegambia,

and *W. Smithii*, DC., from Congo, which Schultz Bipontinus
reunites with *Vernoniæ.* In this he may be right, as well as in
the restoration of the older name of *Gymnanthemum* to De
Candolle's genus *Decaneurum:* it is only to be regretted that,
as the policy of these changes may not be generally admitted,
he should have created much confusion by transferring the two
names he proposes to suppress, to new genera of *Compositæ.*
1. Gymnanthemum *angustifolium*, Benth.; pubescens, foliis (superioribus) lineari-lanceolatis rugosis subtus albo-tomentosis,
   capitulis laxe subcorymbosis, involucri globosi squamis numerosis, exterioribus setaceo-acuminatis recurvis, interioribus oblongo-lanceolatis scariosis.—Sierra Leone, *Don.*
Pars tantum superior plantæ adest corymboso-paniculata. *Ramuli* pilis laxiusculis glandulis intermixtis pubescentes. *Folia*
nonnisi floralia adsunt 1-2-pollicaria. *Capitula* semipollicem
diametro, ad apices ramulorum solitaria, uno alterove secus
ramulum ad axillam bracteæ linearis breviter pedunculato.
Involucri squamæ seriebus numerosis imbricatæ, exteriores
abeunt in acumen filiforme reflexum, interiores erectæ, acutiusculæ, 3 lin. longæ. *Receptaculum* planum, leviter alveolatum. *Flores* numerosissimi, squamis interioribus breviores.
Corollæ graciles, glabræ, superne vix ampliatæ, laciniis 5
linearibus. *Styli* lobi subulati, acuti. *Achenia* glabra,
costis 10 approximatis, valleculis angustissimis glandulosis.
Pappi setæ subæquales, barbellis rigidis longiusculis plumosæ.
This is a true *Gymnanthemum*, agreeing in every respect with
De Candolle's generic character of *Decaneurum.* It is unfortunate that that distinguished botanist, in uniting two already
published genera, did not follow the usual rule of adopting for
the whole the oldest of the published names, although unmean-
ing and applied to an ill-defined group. Cassini's name having
since been restored by Schultz Bipontinus, we feel compelled to
adopt it.
1. Elephantopus *scaber*, Linn.—St. Thomas, *Don;* Senegambia,
   East India and America.
This species should include the five first of De Candolle

(Prod. 5. p. 86) as proposed by Schultz Bipontinus, (Linnæa 18. p. 594) ; for the characters, by which he afterwards (Linnæa 20. p. 515) thought that the East Indian one might be distintinguished, do not hold good. The African specimens before me come nearest to the form described by De Candolle as *E. Martii*, which is chiefly Brazilian, but is also found in other parts of the world.

1. Ageratum *conyzoides*, Linn.—As common along the coast, from Senegal to Benin, *Vogel, Don*, and others, as it is in other parts of Africa, Asia and America.

1. Adenostemma *Perrottetii*, DC., *Prod. 5. p.* 110?—St. Thomas, *Don*.

There are two specimens, the one nearly smooth, the other more hairy, both belonging apparently to the Senegambian species described by De Candolle, and probably not really distinct from the common West Indian *A. Swartzii*.

1. Mikania *chenopodiifolia*, Willd. *DC. Prod. 5. p.* 201.— Common from Senegal to Fernando Po, *Vogel, Don*, and others.

This plant is sometimes quite smooth, in other specimens the upper surface of the leaves and the angles of the branches are rough, with short hairs, and the inflorescence has more or less of glandular pubescence. It is closely allied, both to the common American *M. scandens* and to the East Indian *M. volubilis*, and all three may be varieties of one species.

1. Erigeron *persicæfolium*, Benth. ; herbaceum, elatum, glabrum v. apice puberulum, foliis sessilibus subamplexicaulibus elongato-lanceolatis acutissimis remote denticulatis, corymbo composito polycephalo, involucri squamis linearibus acutis marginatis discum æquantibus, floribus fœmineis numerosissimis disco brevioribus extimis tenuissime ligulatis, interioribus truncatis subdentatis, centralibus paucis hermaphroditis. — St. Thomas, *Don*.

Caulis bipedalis, in specimine simplex. *Folia* 3-5 poll. longa, versus medium 5-8 lin. lata, denticulis minutis remotis v. raro nullis. *Ramus* apice confertim corymbosus, ramulis corymbi exterioribus gracilibus, pedunculis ultimis brevissimis. *Capitula* quam in *E. Canadensi* paullo majora. *Involucri* squamæ inæquales, pauciseriales, dorso virides puberulæ, margine siccæ subscariosæ. *Flores* fœminei omnes pappo breviores, exteriores in ligulam angustissimam brevem producti, interiores stylo suo dimidio breviores; masculi in medio capitulo vix deni, pappo suo subæquales, basi tenues, superne incrassati, angulati, 5-fidi. *Achenia* compressa, glabra. *Pappus* tenuis rufescens.

The *Erigeron spathulatum,* Schum. et Thonn., and *E. exstipulatum,* Schum. et Thonn., both from Guinea, are unknown to me. The latter is most probably a *Blumea.*

1. Microglossa *petiolaris,* DC. *Prod.* 5. *p.* 321.—Sierra Leone, *Don;* Accra and Fernando Po, *Vogel;* Senegambia.

The largest leaves are above three inches long and two broad. The Sierra Leone variety of *M. volubilis* mentioned by De Candolle is probably rather the present species, which has occasionally a slight tendency to climb, and is certainly in other respects very near to the true *M. volubilis.* The generic name has been altered to *Frivaldia* by Endlicher, on the plea of the pre-existence of a *Microglossa* among Birds: a circumstance not now generally considered as requiring the change. The whole genus is, by Schultz Bipontinus, proposed to be united with *Erigeron,* and there is no doubt that the circumscription of *Erigeron, Blumea, Pluchea* and their allies requires considerable modification; but the wholesale alterations proposed cannot be adopted until the characters of the new groups shall have been given, and shown to have been verified on the very numerous species now known from both hemispheres.

The *Microtrichia Perrottetii,* DC., is not in our collections. A Senegal plant in the Hookerian Herbarium, evidently a congener, does not precisely agree with De Candolle's character, and probably forms a second species, and there is in the same Senegambian collection a specimen of an apparently new genus, allied to *Pteronia* and *Chrysocoma.*

1. Sphæranthus *Senegalensis,* DC. *Prod.* 5. *p.* 370.—Senegal and Sierra Leone, *Don,* and others.

The *Grangea ceruanoides,* Cass., and *G. procumbens,* DC.,

are confined to Senegal; the *Berthelotia lanceolata*, DC., extends from Senegal to the Ganges, the *Conyza fastigiata*, Willd., is common to Senegal and Mauritania. *C. dentata*, Willd., is only known from Senegal, and *C. amæna*, Link, from Congo.

1. Blumea *Perrottetiana*, DC. *Prod.* 5. *p.* 443. var. ? *latifolia.*— Sierra Leone, *Don.*—The same variety, as well as the form originally described, are found in Senegal.

This plant is probably a mere variety of a widely spread species, including *Conyza thyrsoidea*, Pers., from Tropical Africa, *Blumea Drègeana*, DC., from South Africa, and two or three supposed species of *Blumea* from East India.

2. Blumea *Senegalensis*, DC. *Prod.* 5. *p.* 449.—On the Gambia, *Don.*

The remaining W. Tropical African species are *B. solidaginoides*, DC., from Sierra Leone, *B. oloptera*, DC., from Senegal, and *B. Guineensis*, DC., from Senegal and Guinea. Of the allied genus, *Pluchea*, there is one species, the *P. ovalis*, DC., from Senegal.

1. Epaltes *Brasiliensis*, DC. *Prod.* 5. *p.* 461.—St. Thomas, *Don*; African coast, south of the Line, *Curror*; a common Brazilian plant.

A plant, in every respect similar to *Vicoa Indica*, DC., but without rays, which would bring it nearer to *Varthemia*, is in the Hookerian Herbarium, gathered on the Gambia by Captain Boteler; and the *Francœuria crispa*, DC., a common Egyptico-Arabian plant, and *Pulicaria incisa*, DC., are both natives of Senegal.

1. Pegolettia *mucronata*, Benth.; puberula, subviscida, ramis elongatis virgatis, foliis sessilibus lineari-sublanceolatis integerrimis v. 1-2-dentatis mucronatis, pappo interiore plumoso achænio plus duplo longiore, exterioris paleis setaceo-multifidis.—Elephant's Bay, south of the Line, *Curror*.

Rami adscendentes sesquipedales. *Pubes* brevior et rarior quam in *P. Senegalense*. *Folia* secus ramos sparsa, majora semipollicaria, omnia præsertim superiora rigidule mucronulata. *Capitula* ad apices ramulorum solitaria, magnitudine

P. Senegalensis, ramulo sub capitulo incrassato. *Achenia* quam in *P. Senegalensi* breviora, striata, leviter hirtella; pappus interior siccitate cœrulescens; exterioris paleæ hyalinæ, ultra medium in setas tenuissimas fissæ. *Antheræ* longe bicaudatæ. *Stylus* exsertus, lobis apice valde dilatatis fere clavatis. The *Pegolettia Senegalensis*, DC., extends from Senegal through Nubia, to Arabia, if, as it appears to me, the *Kuhnia Arabica*, Hochst. (DC. Prod. 7. p. 267), be really the same species. The *Ceruana Senegalensis*, DC., is only known from Senegal.

1. Eclipta *erecta*, Linn., *DC. Prod.* 5. *p.* 490.—Senegal to Benin. As common a weed in Africa as in Asia and America.

1. Coronocarpus *Prieureanus*, Benth.; foliis subsessilibus oblongo-lanceolatis capitulo ovoideo brevioribus, ligulis circa 8. —*Blainvillea Prieureana*, DC. Prod. 5. p. 492.—On the Quorra, at Attah, *Vogel;* Senegal.

The specimens are in a half-rotten state, but agree with De Candolle's character. They closely resemble also the *Coronocarpus Kotschyi*, (*Dipterotheca Kotschyi*, Schultz Bip.), from Nubia, but are more hairy, the leaves are narrower, the heads of flowers rather smaller, the ligulæ much smaller, and yellow or orange, (not purple), and the achenia much shorter and less hairy.

The genus *Coronocarpus*, Schum. et Thonn., accidentally overlooked by De Candolle and subsequent botanists, differs from *Blainvillea* chiefly in the sterility of the ligulæ or florets of the ray. It is evidently identical with *Dipterotheca*, Schultz Bip., and possibly also with *Harpephora*, of Endlicher. The minute appendages at the base of the achenia, on which Schultz Bipontinus proposes to establish a distinct subtribe, are curious, but probably of less importance than he seems to attach to them. Traces of them may be seen in some of the Asiatic species of *Blainvillea;* and, probably, in other *Compositæ* with very paleaceous receptacles they will be found, if carefully sought for. Taking the flower-head as a contracted spike, the

paleæ of the receptacle represent the subtending bracts, and the
appendages in question a pair of opposite bracteolæ on the
pedicel (callus basilaris) of the flower; and it is well known
how very rarely the presence or absence of such bracteolæ can
be made available even as a good generic character. In the
Coronocarpus Prieureanus, they are generally smaller than in
C. Kotschyi, but occasionally as much as ¾ of a line long.
  Whether the genus should be referred to *Eclipteæ* among
Asteroideæ, or to *Coreopsideæ* among *Senecionideæ,* is a matter
of doubt.   Its close affinity is evident with *Blainvillea, Wedelia*
and *Viguiera,* genera now classed in three different subtribes.
2. Coronocarpus *Gayanus,* Benth.; foliis petiolatis ovatis v.
  ovato-lanceolatis, pedunculis capitulo ovoideo subbrevioribus,
  ligulis circa 8.—Blainvillea Gayana, *Cass.—DC. Prod.* 5.
  p. 492, *Webb, supra, p.* 141.—On the Quorra, *Vogel;* Confluence of the Niger, *Ansell;* Senegal and Cape Verd Isles.
  var. *β? peduncularis,* pedunculo capitulo 2-3-plo longiore.—
  Accra, *Don.*
Herba annua, erecta v. adscendens, 1-2-pedalis, undique scabra
  v. hispida.   *Folia* 2-3-pollicaria, petiolo 2-4 lin. longo. *Involucri* squamæ imbricatæ, siccæ, striatæ, appressæ; exteriores
  plus minus foliaceæ et dorso villosæ, interiores margine
  ciliatæ.   *Paleæ* complicatæ, flores includentes et iis breviores.   *Ligulæ* parvæ, orbiculatæ, ex Vog. albidæ v. pallide
  roseæ.   *Achenia* compressiuscula, villosa, pappo brevi ca-
  lyculato-dentato, aristis (in his speciminibus) vix pappum
  interiorem superantibus.   *Squamæ* basilares brevissimæ, sed
  in acheniis plerisque certe adsunt.   Specimen Donianum
  valde mancum est.
3. Coronocarpus *helianthoides,* Schum. et Thonn. *Beskr.p.* 393;
  foliis breviter petiolatis, capitulis longe pedunculatis hemisphæricis, ligulis circa 20.—Wedelia Africana, *Pers.—DC.
  Prod.* 5. *p.* 539.—Sierra Leone, *Don, Vogel;* Accra, *Vogel;*
  Guinea and Oware.
Habitus et folia fere *C. Gayani.   Petioli* breviores.   *Pedunculi*
  1-6-pollicares, nudi v. sub apice monophylli.   *Capituli* forma

a præcedentibus facile distinguenda. *Achenia* pubescentia, apice calyculata, aristis minimis sæpius vix conspicuis, squamis basilaribus brevissimis.

1. Cryphiospermum *repens*, Beauv.—*DC. Prod.* 5. *p.* 497.— Fernando Po, *Vogel;* Guinea and Benin.

1. Ambrosia *Senegalensis*, DC. *Prod.* 5. *p.* 525.—St. Thomas, *Don;* Senegal.

1. Lipotriche *Brownei*, DC. *Prod.* 5. *p.* 544.—Buphthalmum scandens, *Schum. et Thonn. Beskr. p.* 392.—Cape Coast, Niger River, at Abòh, Fernando Po, *Vogel;* Guinea Coast. *Folia* longiuscule petiolata, ovata, acuminata, irregulariter dentata v. rarius angulato-lobata, basi hastata cordata v. rarius rotundata et sæpius inæqualia, membranacea, plus minus hispidula, 2-4 poll. longa.

It is possible that there may be more than one species, but the specimens do not afford any positive characters to distinguish them. It is a very different species from the South African *Psathurochæta Drègei*, although Schultz Bipontinus appears to be right in reducing the latter as a genus to *Lipotriche*.

1. Sclerocarpus *Africanus*, Jacq.—*DC. Prod.* 5. *p.* 566.—Cape Palmas, *Ansell;* Senegal and Guinea, and thence through Nubia, Abyssinia and Arabia to East India.

1. Bidens *pilosa*, Linn.—*Webb, supra, p.* 142.—B. abortiva, *Schum. et Thonn. Beskr. p.* 383.—Sierra Leone and Accra, *Vogel;* St. Thomas, *Don.*—A common American plant, which has spread over a great part of Africa.

Besides the varieties occasioned by the presence or absence of the ray, several of these specimens vary much in their leaves, which are more decidedly pinnate than usual.

2. Bidens *bipinnata*, Linn.—*DC. Prod. v.* 5. *p.* 603.—Sierra Leone, *Don.*

Another common American plant, now found also in Senegal and in several parts of North Africa and South Europe. Don's specimen is a mere fragment, but appears to belong to this species.

The *Verbesina ciliata,* Schum. et Thonn., from Guinea, is
insufficiently described to determine the genus into which more
recent classifications would place it.

1. Spilanthes *caulirhiza,* DC. *Prod.* 5. *p.* 623.—Eclipta filicaulis, *Schum. et Thonn. Beskr. p.* 390 ?—St. Thomas, *Don;*
Guinea and Nubia.

2. Spilanthes *costata,* Benth.; glabriuscula, foliis petiolatis oblongis ovatisve subintegerrimis, capitulis ovato-conicis dis-
coideis, involucri squamis ovatis obtusis flores subæquantibus,
acheniis puberulis 3-4-costatis, costis crassis, 2 validioribus
apice in acumina brevia productis.—Cape Palmas and Cape
Coast, *Vogel.*

Caules elongati, basi radicantes, uti folia et inflorescentia glabri
v. vix pilis minutis scabrelli. *Folia* 1-2-pollicaria, obtusa,
basi angustata, integerrima v. rariter et minute dentata. *Pedunculi* 2-4-pollicares, superne paullo incrassati. *Capitula*
quam in *S. oleracea* paullo minora. *Involucri* squamæ late
ovatæ, 1½ lin. longæ, exteriores subfoliaceæ, interiores membranaceæ angustiores, in paleas obovali-oblongas leviter concavas abeuntes. *Flores* omnes tubulosi subcampanulati, 5-
dentati, hermaphroditi. *Antheræ* inclusæ. *Stylus* basi bul-
boso-callosus, ramis apice truncatis. *Achenia* a latere compressa, utrinque tamen convexa et latere uno costato sub-
triquetra, v. utrinque costata et tetragona; costis duabus
(antica posticaque) validioribus demum fere suberosis ciliolatis,
in aristam seu mucronem brevem crassam productis.

I am not acquainted with any other *Spilanthes* with the
peculiar achenia of this species, which has entirely the habit of
the genus.

1. Chrysanthellum *Senegalense,* DC. *Prod.* 5. *p.* 631.—On the
Quorra, at Patteh, *Vogel;* Senegal.

It is most probable that the four species of *Chrysanthellum*
described by De Candolle, as well as the *Hinterhubera Kotschyi,*
Schultz Bipont., are mere varieties of one species, in which
the achenia vary very much in the development of their pericarp. If such be the case, this small annual is common to the
West Indies, Brazil, Tropical Africa and East India.

A variety of *Cotula anthemoides*, Linn., is described from Senegal, and *C. sphæranthus*, Link, probably not a true *Cotula*, is from the Congo River. One species of *Helichrysum, H. glumaceum*, DC., has been found on the sea coast of Senegal, and another, apparently new, was gathered at Little Fish Bay, in 15⁰ S. lat., by Mr. Thwaites, but should probably be considered as a stray plant of the South African Flora. Three species of *Gnaphalium* have been found in Senegal; the cosmopolite *G. luteo-album*, Linn.; the common Egyptian *G. Niliacum*, Del.; and the *G. gracillimum*, Perr., peculiar to Senegal.

1. Gynura *cernua*, Benth.—Cremocephalum cernuum, *Cass.*— *DC. Prod.* 6. *p.* 298.— Cacalia uniflora, *Schum. et Thonn. Beskr. p.* 382.—Confluence of the Niger, *Vogel.*—An East Indian and Mauritius species.

Neither in these specimens, nor yet in cultivated ones from the Berlin Botanical Garden, can I find any florets entirely without anthers; although the florets generally are much more slender than in most *Gynuræ*, and those of the circumference more especially so, with their anthers probably sterile. As, moreover, the *G. aurantiaca*, described below, is intermediate in habit, as well as in the thickness of the florets, between *Cremocephalum* and *Gynura*, it seems advisable to unite the two genera as proposed by Lessing. Mœnch's name of *Crassocephalum* has the right of priority, but has been rejected as being compounded of a Latin and a Greek word; and, of the two proposed by Cassini, *Cremocephalum* has only been given to one species, and is not applicable to the majority, whilst *Gynura* is not only very expressive of the principal character of the genus, but its rejection would necessitate the changing nearly five-and-twenty well established names.

2. Gynura *polycephala*, Benth.; erecta, elata, foliis amplis lyrato-pinnatifidis, corymbi compositi ramis elongatis pleiocephalis, capitulis breviter pedunculatis nutantibus, involucri squamis interioribus circa 30, flosculis tenuibus (exteriorum antheris cassis).—Fernando Po, *Vogel.*

Herba 2-3-pedalis, minute puberula, caule striato. *Folia* petiolata, semipedalia, membranacea, lobis inciso-dentatis acutis,

ultimo maximo, superiora minora minus incisa, summa (in
ramulos paniculæ) linearia. *Inflorescentia Ericthitis. Capitula* cernua, iis *G. cernuæ* simillima. *Flores* ex Vogel e
brunneo rubri. *Flosculi* iis *G. cernuæ* similes; antheris
tamen evidentius acuminatis.

3. Gynura *crepidioides*, Benth.; erecta, scabro-pubescens, foliis
ovatis inciso-dentatis basi longe angustatis, capitulis longe
pedunculatis subcorymbosis, involucri squamis interioribus
circa 15, flosculis tenuibus (exteriorum antheris cassis).—
Sierra Leone, *Don;* and, apparently the same species, Senegal, *Heudelot.*

A smaller plant even than *G. cernua,* and usually more
hispid. The inflorescence is less simple than in that plant,
less compound than in *G. polycephala,* and it differs from both
in the narrower heads of the flowers, with about half the
number of scales to the involucre.

4. Gynura *vitellina,* Benth.; glabriuscula, caule diffuso radicante,
ramis adscendentibus parce ramosis, foliis petiolatis ovatis
grosse dentatis, capitulis ad apices ramorum solitariis cernuis,
involucri squamis interioribus circa 20, flosculis omnibus
conformibus hermaphroditis involucrum breviter superantibus.—Fernando Po, *Vogel.*

Rami e caule prostrato 1-2-pedales, angulati, compressi v. subteretes, in ramulos 2-4 monocephalos divisi. *Folia* longiuscule
petiolata, membranacea, 1-3 poll. longa, basi cuneata truncata v. cordata, juniora pubescentia, demum fere glabra;
petiolo ½-1-pollicari, sæpe præsertim ad folia superiora basi
biauriculato. *Ramuli* in pedunculos remote bracteatos abeunt.
Capitula quam in speciebus Indicis latiora. *Involucrum* 5
lin. longum, squamis interioribus acutis margine membranaceis et ultra medium diu connatis, dorso herbaceis striatis
et minute puberulis; bracteæ seu squamæ exteriores parvæ,
setaceæ. *Flores* (ex Vogel), aurantiaci vel vitellini, numerosissimi, omnes hermaphroditi, corolla superne latiore quam in
præcedentibus. *Styli* rami superati cono subulato acuto
pubescente. *Receptaculum* minute fimbrilliferum. *Achenia*
striata, minute puberula.

This species is certainly intermediate between the three preceding and the East Indian *Gynuræ*, the corollas are not so slender as in the former, and always all alike, and not so thick, although more numerous than in most *Gynuræ*. Their colour is different from that of the generality of species in which it is recorded.

1. Emilia *sonchifolia*, Cass.—*DC. Prod.* 6. *p.* 302.—Sierra Leone, *Don.*—A common East Indian and Tropical African plant, which has become naturalized also in some parts of America.

The almost universally prevalent genus, *Senecio*, is represented in West Tropical Africa by two species only, *S. strictus*, DC., and *S. Perrottetii*, DC., both from Senegal; and the two large sub-orders of *Carduaceæ* and *Labiatifloræ* by three species, two of *Centaurea, C. Perrottetii*, DC., and *C. Senegalensis*, DC.; and one of *Dicoma, D. tomentosa*, Cass.; all also from Senegal.

1. Cichorium *Intybus*, Linn.—St. Thomas, *Don ;* to all appearance identical with the common European plant.

The other *Cichoraceæ* from West Tropical Africa are : *Picris humilis*, DC., from Senegal; *Lactuca taraxacifolia*, Schum. et Thonn., from Senegal and Guinea; *Brachyramphus Goreensis*, DC., and *Rhabdotheca Brunneri*, Webb, both from Senegal.

LXVI. CAMPANULACEÆ.

Of this Order there are no specimens in the collections before us, nor does it appear that any have been found within the hot regions of Guinea and the Niger, although Senegambia has furnished five, *Lobelia Senegalensis*, A. DC., *Cephalostigma Perrottetii*, A. DC., *C. Prieurii*, A. DC., *Wahlenbergia riparia*, A. DC., and *W. cervicina*, A. DC., the latter species extending also into Egypt.

LXVII. GOODENOVIEÆ.

1. Scævola *Senegalensis*, Presl ? *A. DC. Prod.* 7. *p.* 507 ?— Grand Bassa, *Vogel;* Elephant Bay, south of the Line, *Curror ;* Senegal.

The specimens are very imperfect. The tube of the corolla is very woolly inside, but the lobes appear to be nearly smooth, as in the South African *S. Thunbergii*, Roth. Both species are probably not distinct from the *S. Plumieri*, a sea-coast plant belonging to both the New and the Old World.

LXVIII. UTRICULARINEÆ.

Of this Order also we have no specimens from the Niger Expedition, but the following eight W. Tropical African species have been published: *Utricularia stellaris*, which, from Senegambia, extends over nearly the whole of Africa and East India; *U. inflexa*, Forsk., confined to Africa, from Guinea to Nubia and Egypt; *U. ambigua*, A. DC., and *U. arenaria*, A. DC., from Senegambia; and *U. spiralis*, Sm., *U. micropetala*, Sm., *U. striatula*, Sm., and *U. pubescens*, Sm., from Sierra Leone.

LXIX. SAPOTACEÆ.

1. Chrysophyllum *albidum*, G. Don, *A. DC. Prod.* 8. *p.* 162. —St. Thomas, *Don.*
Folia semipedalia, subtus tomento minuto vix sub lente distincto argenteo-micantia. *Pedicelli* umbellato-fasciculati, vix 2 lin. longi. *Sepala* orbiculata, vix linea longiora, coriacea, æstivatione valde imbricata. *Corolla* calyce paullo longior, lobis margine tomentosis leviter imbricatis. *Appendices* inter lobos minutæ, inflexæ, vix conspicuæ. *Stamina* medio tubo inserta, corollam subæquantia, glabra; *filamenta* antheris duplo longiora; *antheræ* extrorsæ, connectivo supra loculos longiuscule producto. *Ovarium* villosum, depressum, 5-loculare. *Ovula* solitaria, lateraliter prope basin loculi affixa, adscendentia, fere orbiculata, a latere compressa. *Stylus* glaber, 5-sulcatus, apice obtusus, punctis stigmatosis vix conspicuis.

The presence of appendages (or abortive stamina) alternating with the lobes of the corolla, would probably remove this plant from *Chrysophyllum*: these appendages are however so minute as to be readily overlooked; and the absence of fruit does not admit of determining as yet into what other genus the species should be placed.

2. Chrysophyllum *obovatum*, G. Don, *A. DC. Prod.* 8. *p.* 163. —Sierra Leone, *Don, Vogel* (?) Neither specimen has flower or fruit, and the identity of Vogel's with Don's is therefore doubtful. The species itself remains very uncertain.

3. Chrysophyllum *Africanum*, A. DC. *Prod.* 8. *p.* 163.—Sierra Leone; Fernando Po, *Vogel?* Don's Herbarium contains no specimen of the Sierra Leone plant, and Vogel's specimens have no flowers. This species also must therefore be doubtful.

The other West Tropical African *Sapotaceæ* are *Sapota sericea*, A. DC., from Guinea; *Sideroxylon dulcificum*, A. DC., from Guinea; *Bassia Parkii*, G. Don, from Bambara; and *Omphalocarpum procerum*, Pal. de Beauv., from Oware, most of them very imperfectly known.

LXX. Ebenaceæ.

1. Euclea *angustifolia*, Benth.; ramis pubescentibus, foliis linearibus crassiusculis tomentellis mox glabratis, floribus fœmineis subsolitariis, calyce pubescente, corolla subglobosa hirta brevissime 5-6-loba.—West Africa, south of the Line, *Curror.*

Frutex habitu *E. pseudebeno*, E. Mey., affinis, pube ramulorum densa canescente facile distinctus. *Folia* 1-2 poll. longa, 1-1½ lin. lata, apice rotundata et mucrone brevi terminata, basi angustata, coriacea, costa media tenui vix prominula, cæterum avenia, pleraque alterna, rarius subopposita. *Flores* fœminei tantum adsunt, et sæpissime solitarii, nutantes, 1½ lin. longa, pedicello lineam longo recurvo rarissime bifloro. *Calyx* brevissimus, 5-6-lobus. *Corolla* fere globosa, extus densissime hirsuta, lobis brevissimis latis obtusis, intus sub fauce leviter hirta, cæterum glabra. *Staminum* vestigia nulla. *Ovarium* sessile, globosum, dense villosum, intus 4-loculare. *Stylus* brevis, glaber, fere ad basin bipartitus, ramis crassis cuneato-dilatatis emarginato-bilobis. *Ovula* in loculis solitaria, pendula.

1. Diospyros *Senegalensis*, Perr. *in DC. Prod.* 8. *p.* 234 ?—On the Quorra, at Stirling, and in Fernando Po, *Vogel;* Senegambia, *Heudelot.*

The specimens from these three different localities may belong to distinct species, but, if so, they are insufficient to distinguish them. The Quorra specimen, in leaf, with a portion of a single fruit, agrees best with De Candolle's description; that from Fernando Po has only a few small leaves, with the remains of the calyx after the fruit has fallen off. The Senegal specimen in the Hookerian Herbarium, has smaller and less coriaceous leaves than those mentioned in the Prodromus, but they are younger. It is a branch of a male plant in bud : the flowers are in axillary bunches of three or four, with a short quadrifid calyx, and about twelve stamens.

The *Noltea tricolor*, Schum. et Thonn., from Guinea, is probably a species of *Diospyros.*

1. Maba *vacciniæfolia*, Benth. ; ramulis hirsutis, foliis ellipticis acutis subcoriaceis supra ad costam subtusque sparse hirtellis, calycis fœminei hirti lobis obtusissimis, baccis oblique ellipsoideis.—St. Thomas, about 2,000 feet up the peak of the island, *Don.*

Fruticulus diffuse ramosissimus, ramulis novellis pilis longiusculis rigidis vestitis, ramis annotinis jam glabratis. *Folia* pleraque subpollicaria, nonnulla paullo majora, alia vix semipollicem excedunt, omnia apice obtusa, basi subcuneata, petiolo brevissimo hirto ; pagina superior præter costam hirtellam glabra, inferior pilis rigidulis conspersa, costa prominente hirsutiore. *Flores* desunt. *Ovarium* ex ovulis in fructu persistentibus biloculare videtur, ovulis in quoque loculo geminis pendulis. *Baccæ* solitariæ, subsessiles, calyce persistente lineam longo late et brevissime trilobo fultæ, 4 lin. longæ, hinc contractæ, illinc valde convexæ, pilis appressis conspersæ ; pericarpio tenui ; intus abortu uniloculares. *Semen* abortu unicum, ellipsoideum, ab apice loculi pendulum, hinc longitudinaliter sulcatum ; testa membranacea; albumen nullum; cotyledones crassæ, carnosæ, lateri sulcato parallelæ; radicula brevissima ad hilum spectans.

Besides the above, there are two W. African species of *Maba* published, *M. Guineensis*, A. DC., from Guinea, and *M. Smeathmanni*, A. DC., from Sierra Leone.

I omit the *Styraceæ*, because the *Styrax Guineensis*, G. Don, the only supposed W. African species published, does not belong to the Order.

LXXI. JASMINEÆ.

1. Jasminum *noctiflorum*, Afz., *DC. Prod.* 8. *p.* 309.—Sierra Leone, *Vogel.*
Leaves usually ternately verticillate, the petioles 4 to 5 lines long, articulate near the base.
2. Jasminum *dichotomum*, Vahl, *DC. Prod.* 8. *p.* 307.—J. Guineense, *G. Don, Gard. Dict.* 4. *p.* 60.—Whydah, *Don;* Senegal and Guinea.
Very near *J. noctiflorum;* but the petioles are shorter, and the lobes of the corolla much longer and more pointed. The teeth of the calyx are also rather longer.
3. Jasminum *pauciflorum*, Benth.; ramulis hirsutis, foliis oppositis breviter petiolatis ovatis acuminatis subtus petiolisque pubescentibus supra demum glabris, pedunculis subbifloris, pedicellis elongatis superne incrassatis, calycis lobis circa 6 subulatis.—Cape Coast, *Vogel.*
Frutex volubilis, quoad folia et pubem a *J. Sambac* et *J. pubescente* haud absimilis, sed pedicelli in pedunculo brevi axillari v. terminali sæpius gemini, 6-8 lin. longi, nec calyce subbreviores. *Calycis* tubus 1 lin., laciniæ 2 lin. longæ. *Corolla* deest. *Bacca* subglobosa, calycem subæquans.

LXXII. APOCYNEÆ.

1. Landolphia *Owariensis*, Beauv. *A. DC. Prod.* 8. *p.* 320.— Sierra Leone, *Don;* Oware.
There are two other West African genuine species of *Lan-*

dolphia, the *L. Heudelotii,* A. DC., and an unpublished one in
my herbarium, communicated by Michelin,* both from Senegambia, besides the following, with a rather different aspect, yet
apparently belonging to the genus.

2. Landolphia *florida,* Benth.; ramis foliisque glabris, cymis
pedunculatis multifloris, staminibus infra medium tubi corollæ elongati insertis.—On the Quorra, *Vogel.*

Frutex alte scandens, ramis glabris verruculosis. *Folia* petiolata, elliptica, 3-6 poll. longa, 2-3 poll. lata, utrinque obtusa,
glabra, chartacea v. subcoriacea, venis primariis distantibus,
venulis transversis crebris reticulatis. *Cymæ* corymbosæ,
terminales, dense multifloræ, breviter pedunculatæ. *Flores*
uti tota inflorescentia tomento brevi velutini. *Bracteæ* pedicellos subæquantes, ovatæ, squamæformes. *Pedicelli* circa
lineam longi, crassi. *Calyx* linea paullo longior, fere ad
basin 5-partitus, lobis ovato-oblongis, parum inæqualibus,
intus eglandulosis. *Corollæ* tubus 7-8 lin. longus, tenuis,
circa stamina paullo incrassatus, intus superne pilosus ; laciniæ anguste oblongæ, tubo subæquilongæ, fere glabræ,
albæ basi intus lutescentes, æstivatione dextrorsum† convolutæ. *Stamina* paullo infra medium tubi inserta ; antheræ
oblongæ, filamento paullo longiores, altera sæpius cæteris
majore. *Glandulæ* hypogynæ nullæ. *Ovarium* depresso-
globosum, dense villosum, uniloculare, placentis duobus parietalibus pluriovulatis. *Stylus* filiformis, superne fusiformi-in-
crassatus, summo apice divisus in lobos 2 breves latos præcipue ad margines stigmatosos.

In the absence of fruit, this plant agreeing as well with the
character of *Willughbeia* as with that of *Landolphia,* I refer it
provisionally to the latter as being a West African genus; but
it is most probable that, when better known, the two genera

* *L. Michelini,* Benth.; foliis subtus ramulisque velutino-pubescentibus, cymis subsessilibus densis, antheris medio tubo insertis.—Flores fere
L. Owariensis.

† We use this word in the sense adopted by De Candolle, from left *to*
right, (supposing oneself in the centre of the flower), not *from* the right,
as it is used by some botanists.

will be united. Both appear to be climbers, although the published *Landolphiæ* are not described as such. The tube of the corolla is much longer in our plant than in other *Landolphiæ*, and the stamens inserted lower down, but these cannot be generic distinctions if the fruit coincides.

Heudelot's Senegambian collection comprises what appears to be a fifth species, in some respects resembling the *L. florida*, but with rather smaller flowers and an entirely smooth ovary, which nevertheless is unilocular, as in all the other species.

The *Vahea Senegalensis*, A. DC., another Senegambian plant, is unknown to me.

CLITANDRA, gen. nov. e tribu *Carissearum*.

Calyx parvus, 5-partitus, eglandulosus. *Corollæ* tubus supra basin contractus, dein ventricosus, ad faucem intus pilosam esquamatam contractus, limbi laciniæ angustæ, dextrorsum convolutæ. *Stamina* ad basin partis ventricosi tubi inserta, filamentis tenuibus, antheris nutantibus ovatis obtusis filamentis æquilongis. *Nectarium* nullum. *Ovarium* unicum biloculare, dissepimento tenui (vix completo?) *Ovula* pauca. *Stylus* brevis, conicus, apice dilatatus, supra dilatationem vix productus, vertice integro stigmatosus. — *Frutex* dichotomus, floribus parvis in cymas axillares oppositas dispositis.

1. Clitandra *cymulosa*, Benth.—Sierra Leone, *Don*.

Frutex, inflorescentia excepta, glaber v. punctis minutis (resinosis?) irroratus. *Rami* verrucosi. *Folia* opposita, ellipticooblonga, 3-4-pollicaria, abrupte acuminata, basi in petiolum angustata, coriacea at non nitida, venis primariis a costa valida divergentibus crebris parallelis. *Cymæ* oppositæ, multifloræ, petiolos (corollis neglectis) vix æquantes, minute velutino-puberulæ. *Bracteæ* minutæ. *Calyx* minute velutino-pubescens, semilineam longus, lobis acutis. *Corollæ* extus glabræ tubus vix 2 lin. longus, supra ovarium valde contractus dein abrupte ampliatus; laciniæ lineari-oblongæ, tubo æquilongæ. *Stamina* in parte dilatata tubi nidulantia, antheris a stylo liberis nutantibus et minoribus quam in plerisque *Apocyneis*. *Ovarium* glabrum. *Stylus* vix ovario longior.

This genus is allied in some respects to *Landolphia* and *Couma*, in others to *Carissa*. Its really axillary inflorescence is different from that of most of the allied genera. The placentæ of the ovary, although they meet in the centre, scarcely appear to cohere, and the fruit is unknown : the genus therefore cannot be very exactly defined ; yet I am unable to refer the plant to any of those hitherto published.

1. Carpodinus *dulcis*, G. Don.—*A. DC. Prod.* 8. *p.* 329.— Sierra Leone, *Don*.

The specimens marked in Don's collection by the above name, as also by that of Sweet Pishamin, have neither flower nor fruit; the stems are pubescent, and the leaves are not perfectly smooth. The tendrils proceed from the forks of the branches, and appear to represent transformed peduncles. Of the other species mentioned, *C. acida*, Don, there is no specimen in the herbarium.

1. Carissa *edulis*, Schum. et Thonn.—*A. DC. Prod.* 8. *p.* 332. —Accra, *Vogel;* rather common in Guinea, *Thonning*.

These specimens, in fruit, with very young buds, agree better with Thonning's description than with Schumacher's character ; for the leaves are not cordate. Vogel observes that the berries are black and edible. Thonning says that they have a very agreeable flavour, much like sweet cherries, and make an excellent soup for the sick.

The *C. pubescens*, A. DC., from Senegambia, appears to be very near the preceding.

1. Rauwolfia *Senegambiæ*, A. DC. *Prod.* 8. *p.* 340.—Sierra Leone and Grand Bassa, *Vogel, Don;* Senegambia.

Baccæ (folliculi carnosi) 2, distinctæ, substipitatæ, magnitudine Pisi, obovoideo-globosæ, intus 1-2-spermæ. *Semina* matura desunt.

It is probable that the *R. vomitoria*, very imperfectly described by Sprengel, and stated to be from Guinea, is the same species. Vogel, on one of his labels, describes the plant as a

branching shrub, on another as a tree, which he states to be
cultivated.

The Senegambian collection contains a new species of the
South African genus, *Piptolæna*, so remarkable for the form
of the calyx, with its numerous glands, elegantly arranged in a
a double row withinside.

1. Tabernæmontana *longiflora*, Benth.; glabra, foliis oblongo-
ellipticis abrupte acuminatis basi acutis, petiolis basi dilatatis,
pedunculis laxe subtrifloris, calycis lobis ovali-oblongis, corollæ tubo longissimo paullo infra medium ventricoso con-
torto et staminifero.—Sierra Leone, *Vogel;* Senegambia.

Ramuli crassiusculi, uti tota planta glaberrimi, ad nodos gum-
mam resinosam sæpe scatentes. *Folia* crassiuscula, 4-6-
pollicaria, latitudine varia, venis paucis a costa divergentibus,
venulis inconspicuis. *Petioli* ad caulem in vaginam expansi,
et linea transversali connexi. *Pedunculi* e dichotomiis solitarii, crassiusculi, folio multo breviores, in pedicellos 2-3
vix pollicares unifloros sub flore incrassatos divisi. *Calycis*
lobi inæquales, 3-5 lin. longi, obtusissimi, glandulis ad
quemque lobum ultra 12. *Corollæ* tubus tripollicaris.
Ovarium et genitalia *Tabernæmontanæ.*

I describe this plant from Heudelot's Senegambian specimen.
Vogel's, from Sierra Leone, has every appearance of belonging
to the same species, although the leaves are rather narrower.
There is no corolla; but two persistent though shrivelled calyces,
enclosing imperfect ovaries and the remains of some fruit-
stalks, show that, if not the same species as Heudelot's, it is
closely allied to it. Vogel states it to be a handsome tree, with
the aspect of a *Citrus* and a milky juice.

2. Tabernæmontana *crassa*, Benth.; glabra, foliis ellipticis oblongisve breviter acuminatis basi acutis, petiolis basi dilatatis,
cymis confertim plurifloris, calycis lobis breviter orbiculatis,
corollæ tubo longiusculo paullo infra medium ventricoso con-
torto et staminifero.—Grand Bassa, *Vogel.*

Frutex arborescens, succo lactescente, affinis *T. longifloræ* et
petioli pariter in vaginam expansi et linea transversa connexi.
Rami crassiores. *Folia* majora, crassiora, breviter petiolata.

Cymæ ad apicem pedunculi 1½-2-pollicaris 12-15-floræ, ramis pedicellisque vix 2 lin. longis. *Calycis* lobi 1¼ lin. longi et lati, glandulis 7-9 ad basin cujusve lobi. *Corollam* apertam non vidi, alabastrum omnino *T. longifloræ* nisi brevius et crassius. *Folliculi* ex Vog. oblique obovoideo-globosi, carnosi, seminibus creberrimis in pulpa nidulantibus.

The size of the fruit is not stated on Vogel's label, but this is probably the plant he alludes to in his Journal as "a genus apparently new and near *Tabernæmontana*, remarkable for its double fruit, as large as a child's head, the seeds nestling in the almost woody pulp." Both the above species are nearly allied to *T. ventricosa*, Hochst., from Port Natal, and with it form a very distinct group of *Tabernæmontana*, which from the published description I should suspect to be allied to Du Petit-Thouars' Madagascar genus, *Voacanga*.

3. Tabernæmontana *subsessilis*, Benth.; glabra, ramulis dichotomis, foliis obovali-oblongis acuminatis inferne angustis et
   ima basi obtusis membranaceis, pedunculis subbifloris, lobis
   calycinis amplis oblongis, corollæ tubo calyce subtriplo longiore supra medium ampliato et staminifero, folliculis ovoideo-
   oblongis acuminatis.—Liberia, *Vogel*.

Valde affinis *T. grandifloræ*, Jacq., (ex America Tropica.) *Folia* iis simillima nisi basi sæpius obtusa, cujusve paris uti in specie citata inæqualia. *Pedunculi* communes longiores. *Calycis* laciniæ minores (vix 5 lin. longæ) et angustiores; glandulis in genere normalibus. *Corollæ* tubus paullo longior. *Folliculi* 1½-2-pollicares, carnosi, iis *T. grandifloræ* valde similes.

4. Tabernæmontana? sp.—Sierra Leone, *Don.*

A single specimen, without flowers, with a pair of follicles very much like those of the preceding species, but the leaves are very different in consistence and venation, and the inflorescence appears to have been a compact sessile cyme. As the genus must remain doubtful, I refrain from giving it a name.

I have not seen the *T. Africana*, Hook., from Senegal, Sir W. Hooker not having received any specimen from the traveller for whom he described it.

ROUPELLIA, *Wall. et Hook.* gen. nov. e tribu *Tabernæ-montanearum.*

Calyx 5-partitus, glandulis baseos (circa 12) in annulum dispositis. *Corollæ* tubus infundibuliformis; faux coronata ligulis 10 æquidistantibus basi in annulum connatis; limbi laciniæ 5, latæ, æstivatione sinistrorsum convolutæ. *Stamina* tubo inserta, inclusa, filamentis brevissimis, antheris sagittatis longe acuminatis. *Nectarium* nullum. *Ovaria* 2, adpressa, glabra. *Stylus* filiformis, apice in massam 5-sulcatam antheris cohærentem dilatatus, ultra dilatationem vix productus et brevissime emarginatus, (summo vertice stigmatifer?) *Fructus* ex R. Br. *Voacangæ* v. *Urceolæ.*

1. Roupellia *grata*, Wall. et Hook. *in Bot. Mag. tab.* 4466. *ined.*—Sierra Leone, *Whitfield.*

Frutex glaberrimus, habitu *Tabernæmontanas Africanas* referens, quoad formam corollæ diversissimus et insignis. *Folia* opposita, breviter petiolata, semipedalia v. etiam majora, ovalia v. oblongo-elliptica, breviter acuminata, basi acuta, crassiuscula, venis primariis a costa media divergentibus subtus prominulis, rete venularum parum conspicua; petiolus ad caulem parum dilatatus et intus glandulis 2 parvis acutis quasi stipulatus. *Cymæ* terminales, sessiles, 6-8-floræ, fere umbellæformes. *Bracteæ* ovato-lanceolatæ, acute acuminatæ, dorso carinatæ, 1½-2 lin. longæ. *Pedicelli* bracteis longiores, calyce breviores. *Calycis* lobi obovati, 6-8 lin. longi, membranacei, apice colorati. *Corolla* alba, roseo tincta; tubus sesquipollicaris, superne ampliatus, intus extusque glaber; laciniæ late obovatæ, margine crispæ, pollice paullo longiores; coronæ ligulæ lanceolato-lineares, erectæ, pulchre roseæ, 4-5 lin. longæ. *Stamina* ad originem partis ampliatæ tubi inserta; filamenta brevia, crassa, leviter papulosa; antheræ in acumen tubum corollæ subsuperantem productæ, medio tantum polliniferæ. *Ovarium* insidens disco crasso, haud tamen in nectarium producto.

This handsome plant, now flowering in our stoves, was recognized by Brown as the Cream Fruit of Afzelius, referred

G G

to in the Congo Appendix as a new genus, with a flower
resembling that of *Vahea* and the fruit that of *Voacanga,* or
Urceola. Some other new plants from Congo and Sierra Leone
are also alluded to on the same occasion : they are however unknown to us.

The *Malouetia Heudelotii,* A. DC., appears to be confined to
Senegambia.

1. Vinca *rosea,* Linn.—Cape Coast, *Vogel.*—A common plant
in the warmer regions of both hemispheres, especially near
the sea, but said to have been introduced only into Africa
and Asia from America.   It must be observed, however, that
all the other species of *Vinca* belong to the Old World.

The *Plumiera Africana,* Mill., said to have been raised from
seeds sent by Adanson from Senegal, has not since been found
in that country. The *Adenium Honghel,* A. DC., is a Senegalese
plant, differing but very slightly from the *A. obesum,* R. et S.,
which extends from Nubia to Delagoa Bay.

1. Holarrhena *Africana,* A. DC. *Prod.* 8. *p.* 414.—Rondeletia
floribunda, *G. Don, Gard. Dict.* 3. *p.* 516.—Sierra Leone, *Don.*
The two other W. African species, *H. landolphioides,* A. DC.,
and *H. ovata,* A. DC., are both from Senegal.

1. Isonema *Smeathmanni,* Rœm. et Schult.—*A. DC. Prod.* 8.
p. 415.—Grand Bassa, *Vogel, Ansell* ; Senegambia, *Heudelot.*

Ramuli juniores, inflorescentia et flores pilis brevibus pubescentes, rami adulti glabrati.   *Folia* brevissime petiolata,
obovali-oblonga, vix acuminata, basi obtusa, 3-4 poll. longa,
1-1½ poll. lata, rigidule membranacea, supra glabra, subtus
ad venas hirta et inter venas pilis nonnullis conspersa. *Cymæ*
oppositæ, breviter pedunculatæ, in thyrsum seu paniculam
terminalem 3-4-pollicarem dispositæ, singulæ 6-10-floræ.
Bracteæ parvæ, acutæ.   *Flores* a pedicello brevi recurvo
nutantes. *Calyx* linea paullo longior, lobis acutiusculis ;
glandulæ baseos geminatim v. interdum ternatim approximatæ, cum laciniis calycinis alternantes. *Corollæ* tubus
cylindricus, 4 lin. longus, extus velutinus, intus ad medium
annulo denso pilorum clausus et lineis 5 pilorum a stamini-

bus usque ad annulum decurrentibus notatus, cæterum glaber; lobis ovato-oblongis tubo brevioribus æstivatione sinistrorsum contortis. *Ovaria* apice hispida, stylo glabro.

1. Strophanthus *sarmentosus*, DC.—*A. DC. Prod.* 8. *p.* 418.— —S. Senegambiæ, *A. DC. l. c.*—Sierra Leone, *Don, Miss Turner;* Senegal. Flowers whitish, with deep red stripes. The other W. African species are *S. pendulus,* Kummer, from Senegambia, closely allied to if not the same as *S. sarmentosus : S. hispidus,* DC., a very distinct species from Sierra Leone, and *S. laurifolius,* DC., the exact station of which is not given.

The *Nerium scandens,* Schum. et Thonn., the genus of which, according to recent definitions, is uncertain, and *Motandra Guineensis,* A. DC., both from Guinea, are unknown to me.

ONCINOTIS, gen. nov. e tribu *Echitearum.*

Calyx 5-partitus, eglandulosus. *Corollæ* hypocraterimorphæ tubus subcylindricus, faux coronata ligulis 5 integris cum lobis alternantibus, lobis æstivatione sinistrorsum contortis. *Stamina* prope basin corollæ inserta; filamenta brevissima; antheræ lineari-sagittatæ, apice nudæ, auriculis baseos externe uncinato-recurvis. *Nectarium* e glandulis 5 basi connatis ovatis obtusis. *Ovaria* 2, apice pilosa, nectario sublongiora. *Stylus* brevis, superne fusiformi-dilatatus sulcatus et antheris cohærens, ultra dilatationem productus et in lobos 2 lanceolatos stigmatosos divisus.

1. Oncinotis *nitida,* Benth.—Sierra Leone, *Vogel.*
Frutex scandens, glaber, ramulis compressis demum teretibus. *Folia* opposita, linea tenui connexa, breviter petiolata, obovali-oblonga, pleraque subtripollicaria, abrupte acuminata, margine subrecurva, basi acuta, glaberrima, nitida, eleganter venosa; venæ primariæ a costa valde divergentes prope marginem confluunt in venulam in medio spatio versus costam recurvam et mox ramulosam; axillæ venarum majorum sæpe foveolatæ at non pilosæ. *Cymulæ* oppositæ, in

thyrsos axillares breves v. ad apicem rami paniculatos dispositæ; rhachide thyrsi valde compressa; cymæ ipsæ uti calyces et corollæ sæpius breviter puberulæ. *Bracteæ* minutæ. *Pedicelli* lineam longi. *Calycis* lobi linea paullo longiores, ovati, obtusi, laxi, membranacei, ciliolati, interiores exterioribus paullo minores. *Corollæ* tubus fere 2 lin. longus, superne amplior; tubus intus a fauce ad insertionem staminum pilis reflexis dense villosus, infra stamina glaber; laciniæ oblongæ, tubo subæquilongæ; faucis ligulæ erectæ, ¾ lin. longæ. *Filamenta* intus fasciculo pilorum penicillata; antheræ omnino nudæ, summo apice acutæ et polline destitutæ, auriculis baseos clavellatis. *Ovaria* nectario vix longiora, ovulis numerosissimis amphitropis.

The externally hooked bases of the anthers, the calyx and nectary, bring this plant very near to *Motandra* as characterized by A. De Candolle; but the scales in the throat of the corolla and the absence of those tufts of hairs which suggested the name of *Motandra* prevent the uniting it with that genus.

1. Baissea *Leonensis*, Benth.; glabra, cymis paucifloris folio multo brevioribus.—Sierra Leone, *Vogel, Don.*

Frutex alte scandens. *Folia* petiolata, ovali-oblonga v. elliptica, acuminata, 2-3-pollicaria, more *B. multifloræ* eleganter venulosa, venis ultimis crebris transverse reticulatis, sed basi omnia acuta et consistentia tenuiore quam in specie citata. *Pedunculi* nunc cymam unicam paucifloram ferunt, nunc adduntur etiam 2 oppositæ in medio pedunculo; inflorescentia tota glabra v. vix minutissime puberula. *Bracteæ* minutæ. *Flores B. multifloræ* forma similes sed multo minores. *Calyx* vix semilineam longus, eglandulosus. *Corollæ* albæ v. roseæ tubus 1½ lin. longus, fere campanulatus; laciniæ ligulatæ, æstivatione sinistrorsum contortæ. *Nectarium* brevissimum. *Stylus* supra dilatationem subulato-productus.

The *Baissea multiflora*, A. DC., from Senegambia, has cylindrical coriaceous follicles, above a foot long, scarcely more than a quarter of an inch thick, and clothed with a rusty down: the seeds are numerous, about 9 lines long, truncate at the upper end, with a long and exceedingly dense coma.

There is also, in Vogel's collection, a specimen in fruit of
some plant of the tribe of *Echiteæ*, which agrees in foliage and
in the glands of the petiole with *Strophanthus sarmentosus*,
but the inflorescence appears to be different.

LXXIII. ASCLEPIADEÆ.

There are no specimens in the collection belonging to the
first tribe, *Periploceæ*; but two species have been described from
Angola, the *Zucchellia Angolensis*, Dcne., and *Æchmolepis
myrtifolia*, Dcne.

1. Secamone *myrtifolia*, Benth.; volubilis, glabra, foliis ovatis
acutiusculis v. subacuminatis basi rotundatis cuneatisve
utrinque glabris novellis vix punctatis, cymis ferrugineo-puberulis in paniculas axillares folio longiores dispositis, corolla
glabra, coronæ stamineæ foliolis gynostegii dimidium æquantibus compressis falcinulatis apice incurvo-hamatis, stigmate
brevi obtuso.—Cape Coast, *Vogel.*

Affinis *S. multifloræ*, sed inflorescentia diversa. *Folia* breviter
petiolata, 1½-2½ poll. longa, 1¼ poll. lata, subcoriacea, costa
media subtus prominente, venis obscuris, sub lente minutissime punctulata. *Cymæ* pedunculatæ, secus pedunculum
communem oppositæ, una terminali, v. in axillis supremis
solitariæ et breviter pedunculatæ. *Flores* magnitudine
eorum *S. Thunbergii*, extus rubri, intus flavi.

The *Ichnocarpus Afzelii*, Rœm. et Schult., from Sierra
Leone and Guinea, is probably an Asclepiadeous plant, and
possibly a *Secamone*, judging from the words quoted from
Afzelius that the internal parts of the minute flowers " formant
columnam petalis dimidio breviorem, teretem, superne cras-
siorem, apice rotundatam et inferne cinctam, ut apparet, filis
brevioribus subulatis gracillimis erectis."

There are two W. African species of *Xysmalobium*; *X. Heu-
delotianum*, Dcne., from Senegambia, and *X. sessile*, Dcne.,
from Angola.

1. Cynoctonum *acuminatum*, Benth.; volubile, glabrum, foliis

ovali-oblongis acuminatis sinu lato cordatis auriculis rotundatis subtus glaucis supra ad petiolum glanduliferis, pedunculis folio multo brevioribus multifloris, floribus subumbellatis, corona staminea ore 10-loba, lobis antheris oppositis
crasso-clavatis integris, alternis dimidio minoribus, stigmate
breviter apiculato subintegro.—Sierra Leone, *Don.*

Folia 1½-2½-pollicaria, a petiolo palmatim 5-nervia, acumine
longiusculo acuto. *Pedunculi* semipollicares; inflorescentiæ
axis paullulum elongata in racemum brevissimum umbellæ-
formem. *Pedicelli* 1½ lin. longi. *Flores* vix linea longiores.
Massæ pollinis parvæ, apice attenuato affixæ, pendulæ quidem
sed valde divaricatæ et fere horizontales.

The *Calotropis procera*, Br., common over a great part of
Africa and E. India, is also found in Senegal. *Pentatropis spiralis*, Dcne., extends from Senegal to Nubia.

1. Sarcostemma, sp.—Sierra Leone, *Don ;* Cape Coast, *Vogel.*

A leafless climber, which from the single flower preserved I
am unable to distinguish from the E. Indian *S. brevistigma*, of
which it appears to have the corona and stigmata, but with
rather a larger corolla.

Two species of *Oxystelma* are from W. Africa, and have been
named *O. Senegalense* and *O. Bornuense* by Decaisne, after
their native countries.

1. Dæmia *Angolensis*, Dcne. *in DC. Prod.* 8. *p.* 544.—Common
from Senegal to Angola, *Vogel, Ansell, Don,* and others.

There are two varieties of this plant : one, the *Ascl. con-
volvulacea*, Schum. et Thonn., having the corolla of a deep purple at the base with greenish-white divisions, is the more northern
form found in Senegambia and Guinea, as far as Accra ; the
other, with large leaves and a pure white corolla, extends from
Cape Coast, southwards. This is the *Ascl. scandens* figured by
Palisot de Beauvois, and the *A. muricata* (not *echinata*, as mis-
quoted in the Prodromus) of Schumacher and Thonning. The
Senegambian collection contains also a third form, probably a
distinct species, with a longer tube to the corolla.

The four W. Tropical African species of *Gomphocarpus*, are

all from Angola, viz.: *G. pulchellus*, Dcne., *G. lineolatus*, Dcne., *G. cristatus*, Dcne., and *G. chironioides*, Dcne.

1. Tylophora *sylvatica*, Dcne. *in DC. Prod.* 8. *p.* 610.—Cape Palmas and Fernando Po, *Vogel;* Senegambia.

Dr. Planchon considers the Fernando Po specimens as belonging to a distinct species, with larger and deeper coloured scales to the staminal corona; but on a careful comparison with the Senegalese specimen in the Hookerian Herbarium, the only differences I can perceive appear to me to arise from the Fernando Po specimens being in full flower, whilst on Heudelot's there are only young buds.

1. Marsdenia *Leonensis*, Benth.; volubilis, subglabra, foliis cordatis oblongis v. ovato-lanceolatis acuminatis, cymis petiolo brevioribus laxiusculis, corollæ laciniis tubo intus dense piloso brevioribus, coronæ stamineæ foliolis bilobis, lobo interiore antheræ alte adnato lineari gynostegium subæquante, exteriore brevi ovato obtuso.—Sierra Leone, *Vogel.*

Folia subtripollicaria, longiuscule petiolata, supra petiolum minute glandulifera, subtus ad costas ramuli inflorescentiaque minute puberula, planta cæterum glabra. *Cymæ* bifidæ, breviter pedunculatæ. *Calyx* semilinea brevior. *Corollæ* tubus subgloboso-campanulatus, lineam longus, præsertim ad faucem intus dense barbatus.

In the structure of the flower, this species appears to come near to *M. Calesiana*, Dcne, which is unknown to me.

2. Marsdenia *glabriflora*, Benth.; volubilis? glabra, foliis cordato-oblongis v. ovato-lanceolatis acuminatis, cymis multifloris in paniculam terminalem dispositis, corollæ undique glabræ laciniis ovatis tubo longioribus, coronæ stamineæ foliolis integris lanceolatis ad medium gynostegii adnatis eumque subæquantibus.—Sierra Leone, *Vogel.*

Folia fere *M. Leonensis*, at minora. *Cymæ* pedunculatæ, densifloræ, minus tamen confertæ quam in *M. tenuissima.* *Corollæ* lineam longæ, fere globosæ (tandem rotato-expansæ?) profunde 5-fidæ, intus omnino nudæ.

1. Gymnema *subvolubile*, Dcne. *in DC. Prod.* 8. *p.* 621.—Cape

Coast, and on the Quorra, *Vogel ;* Accra, *Ansell ;* common in Senegal and Guinea.

2. Gymnema *nitidum,* Benth. ; subvolubile ? glaberrimum, foliis breviter petiolatis ovatis oblongisve acuminatis nitidis, cymis subsessilibus paucifloris, corollæ parvæ squamis brevissimis in tubum decurrentibus, stigmate umbonato antherarum membranas longe superante.—Cape Palmas, *Ansell ;* Sierra Leone, *Vogel ?*

Folia 3-4-pollicaria, basi rotundata v. acuta, eglandulosa, subcoriacea, penninervia, superiora non omnia exacte opposita. *Cymæ* minimæ, vix brevissime pedunculatæ, pedicellis 2-6, linea paullo longioribus. *Corolla* ¾ lin. longa, laciniis tubo subæquilongis, squamis usque ad medium tubi lineis pilosis decurrentibus. *Genitalia* omnino generis.

I was only able to examine a single flower on my own specimen from Ansell ; another specimen of Ansell's in the Hookerian Herbarium had lost them all, as well as Vogel's specimen, (gathered and given to him by Mr. Roscher), which makes me uncertain as to the identity of the latter.

1. Gongronema *latifolia,* Benth. ; puberula, foliis longe petiolatis late cordato-ovatis supra petiolum glanduliferis, cymis pedunculatis laxis 2-3-fidis, floribus secus ramos demum elongatos fasciculatis pedicellatis, corolla introrsum pilosula, gynostegio tuberculis 5 carnosis ad basin munito.— St. Thomas, *Don.*

Caulis volubilis, pilis brevibus patentibus haud crebris pubescens. *Folia* 3-4-pollicaria, breviter acuminata, membranacea, utrinque pilis paucis conspersa, petiolo sæpe 2-3-pollicari. *Inflorescentia* pubescenti-hirta, pedunculo communi 1-1½-pollicari, ramis demum pedunculo longioribus. *Corolla* linea paullo longior, subrotata, extus pubescens, intus versus basin laciniarum pilis haud numerosis munita. *Tuberculi* gynostegii (seu foliola coronæ) parvi, patentes. *Massæ pollinis* oblongæ, erectæ, longe stipitatæ.

1. Leptadenia *lancifolia,* Dcne. *in DC. Prod.* 8. *p.* 628.— Tylophora incana, *Sprun. Flora,* 1840. 2. *Beibl. p.* 26.

—Accra, *Don;* extends from Senegal to Nubia and Abyssinia.

Of two other Senegalese species, *L. pyrotechnica,* Dcne., and *L. gracilis,* Dcne., the former is also found in Egypt and Arabia, the latter is confined to Senegambia.

The *Hoya Africana,* Dcne., is common to Senegal, Nubia and Abyssinia.

1. Ceropegia *campanulata,* G. Don, *Gard. Dict.* 4. *p.* 111; caule humili pubescente, foliis linearibus puberulis glabratisve, corollæ tubo (violaceo?) basi leviter ventricoso ad faucem ampliato, limbi lobis lineari-lanceolatis piloso-ciliatis.— Accra, *Don.*

A small bulbous-rooted plant, probably allied to *C. linearis,* E. Mey., which is unknown to me. The tube of the corolla is about an inch long, the lobes rather shorter. I have been unable to dissect the single flower which the specimen bears, so as to give a more accurate character.

Another species, *C. aristolochioides,* Dcne., is found in Senegal.

CURRORIA, Planch., nov. gen. e tribu *Stapeliearum.*

Calyx 5-partitus, sepalis ovato-lanceolatis. *Corollæ* tubus brevis, subglobosus, laciniis lanceolato-ligulatis, æstivatione sinistrorsum contortis, fauce ligulis 5 linearibus squamata. *Gynostegium* inclusum. *Corona staminea* nulla? *Antheræ* apiculo lineari terminatæ. *Massæ pollinis* tenuiter stipitatæ, erectæ, (apice pellucidæ?) *Stigma* breve.

1. Curroria *decidua,* Planch. *in Herb. Hook.*—West Africa, south of the Line, *Curror.*

Rami stricti, lignosi, nodis distantibus. *Ramuli* floriferi ad axillas foliorum delapsorum brevissimi, *folia* ferunt nonnulla novella quasi fasciculata, linearia v. lineari-cuneata, obtusa, basi in petiolum brevem angustata, subpollicaria, membranacea, uninervia, glabra; petiolorum basibus dilatatis imbricatis post folia delapsa persistentibus. *Pedunculus* solitarius uniflorus, e centro fasciculi (ex apice ramuli) foliis subbrevior, sub calyce incrassatus, glaber. *Flores* glabri. *Calycis* laciniæ lineam longæ, membranaceæ, striatæ. *Corollæ* tubus calycem æquans, laciniæ duplo longiores (patentes?); ligulæ faucis laciniis breviores.

The flowers on the specimen are so few and so much crushed in drying, that I am not certain of having very accurately ascertained their structure. They appear to be allied in many respects to those of *Pentasacme* and *Barrowia*, although the very twisted æstivation of the corolla shows more affinity to some of the subdivisions of *Gymnemeæ*.

1. Hoodia *Currori*, Dcne. *in DC. Prod.* 8. *p.* 665.—Scytanthus Currori, *Hook. Ic. t.* 505-506.—W. Africa, 14⁰ south of the Line, *Curror*.

The list of W. Tropical African *Asclepiadeæ* hitherto known is closed by two species of *Boucerosia, B. acutangula*, Dcne., and *B. Decaisneana*, Lem., both from Senegal, and by two doubtful plants, the *Pergularia sanguinolenta*, Lindl., from Sierra Leone, and an incomplete specimen of Don's from Sierra Leone.

LXXIV. LOGANIACEÆ.

1. Strychnos, sp.—On the Quorra, *Vogel*.

There are three specimens, belonging perhaps to three, or at any rate to two different species, but none of them in a state actually to determine. The one, gathered at Attah, is described by Vogel as a tall climber, with an apple-shaped, glaucous fruit; it is in leaf only, with the remains of fruit-stalks. I should have taken it to be the *S. scandens*, Schum. et Thonn., a Guinea plant, but that the racemes or panicles appear to have been very short and few-flowered. Some of the peduncles are converted into hooks. A second specimen, in leaf only, and without any precise station, is very much like the first, but has very blunt leaves. The third specimen, gathered at Patteh, is in leaf only; these leaves are shorter and rounder than in the two others, and here and there are a pair of opposite spines, proceeding from the axillæ of the leaves, at right angles to the stem. The foliage, and a sketch of the fruit given by Vogel, indicate either a *Strychnos*, or more probably a *Brehmia*.

1. Usteria *Guineensis*, Willd.—*A. DC. Prod.* 9. *p.* 22. (TAB. XLV).—Rondeletia loniceroides, *G. Don, Gard. Dict.* 3. *p.* 516.—Sierra Leone, Don; Senegal and Guinea. PLATE XLV. *Fig.* 1. bud ready to open; *f.* 2. flower expanded; *f.* 3. ovary, vertical section; *f.* 4. ovary, outside view; *all magnified.*

1. Gærtnera *paniculata,* Benth.; foliis ellipticis acuminatis basi cuneatis breviter petiolatis, vaginis stipularibus truncatis minute plurisetis, panicula laxa trichotoma, calyce patente brevissime repando-5-dentato.—Grand Bassa, *Vogel.*

Frutex glaber, ramulis lævibus. *Folia* 5-6-pollicaria, venis arcuatis subtus prominentibus costaque media in foliis novellis strigillosis, venularum rete tenue. *Vaginæ stipulares* 3-4 lin. longæ, laxiusculæ, setis inæqualibus sæpius minutis raro lineam longis. *Panicula* pyramidata, 4-6-pollicaris. *Calyx* vix semilineam longus. *Corolla* in vivo ex Vog. flavo-viridis, in sicco extus tomento minutissimo canescens; tubus 1½ lin. longus, superne latior, intus glaber nisi ad faucem densissime hirsutum; limbi laciniæ oblongæ, tubo subbreviores, æstivatione valvata. *Staminum* filamenta brevia, antheræ oblongæ, tubum corollæ vix superantes. *Ovarium* depresso-globosum, hirsutum, loculis sub anthesi minimis, ovulis solitariis erectis. *Stylus* corollam subæquans, apice breviter bifidus, lobis recurvis.

1. Anthocleista *Vogelii*, Planch. *in Hook. Ic. t.* 793, 794. (TAB. XLIII, XLIV), glaberrima, foliis amplis obovato-oblongis obtusiusculis v. subacutiusculis v. subacutis basi longe cuneatis margine leviter revoluto subrepandis utrinque impressopunctatis, petiolis brevibus basi auriculatis, aculeis supraaxillaribus geminatis, corolla calyce triplo longiore, limbo 15-partito tubo subæquali, bacca (immatura) ovoideo-obtusa. (Planch.)—On the Quorra, at Abòh and Attah, *Vogel.*

The above character, drawn up by Dr. Planchon, is followed by a detailed description, in which, however, he does not state

more distinctly the points which induced him to separate this
from Afzelius' original species, gathered at Sierra Leone also by
Don, who has published it under the name of *A. nobilis*. The
foliage is the same in both. The very singular spines do not
appear on Don's specimen, but it is cut off immediately above
the place where they should be; for these spines do not appear
to me to be correctly designated as supra-axillary, but are rather
laterally infra-foliaceous, for if they have any connection with
the leaves, it must be with the pair above them, being placed,
as represented in the Plate, immediately under the petiolar
expansions on each side. The chief absolute distinction relied
on between the two species, is the number of divisions of the
corolla and of stamens, said in Don's plant to be twelve, in
Vogel's fifteen; but I find that number variable in both cases;
one of Don's flowers has only eleven, another has thirteen, and
the remaining five or six have twelve each. Vogel's vary still
more; fifteen is indeed the prevailing number, but I have in
several found either sixteen or fourteen, and in one case only
thirteen. A third supposed species, published by Don under
the name of *A. macrophylla*, is again, most probably, the same
plant described from a cultivated specimen. If further investigation confirms these suppositions, there would be but one
species known, which should retain Don's name of *A. nobilis*.

LXXV. GENTIANEÆ.

1. Canscora *diffusa*, Br.—*Griseb. in DC. Prod. 9. p.* 64.—
Sierra Leone, *Don;* Senegal; a common East Indian plant,
found also in Abyssinia.

The Senegambian collection contains also an unpublished
Gentianeous plant, allied to *Microcala* and *Slevogtia*, but not
agreeing precisely with any one of Grisebach's genera. It is a
slender annual, with solitary, axillary, opposite flowers, an
8-ribbed, 4-toothed calyx, and a regular corolla.

LXXVI. BIGNONIACEÆ.

1. Spathodea *campanulata*, Beauv.—*DC. Prod.* 9. *p.* 208.—
S. tulipifera, *G. Don.*—*DC. l. c. p.* 207.—Bignonia tulipi-
fera, *Schum. et Thonn. Beskr. p.* 273.—Confluence of the
Niger at Stirling, *Ansell;* Guinea and Benin.

Although the descriptions differ in several points, there is
every reason to conclude that Beauvois' and Thonning's plants
belong to one species. Beauvois' characters are generally
drawn up from mere fragments, his drawings made on the spot
of this and other plants having been destroyed by fire at
St. Domingo, and he is very likely to have committed the mistake of describing the leaves as alternate instead of opposite.
The corollas in Ansell's specimens are fully as large as that
figured by Beauvois; those which are well dried, are even
larger; Thonning says they are as large as the largest tulips.
The leaflets in Ansell's plants are rather broader than in
Beauvois'; they are covered on the under side with a minute
tomentum, which is scarcely perceptible in the older leaves, they
are also marked on the same side with innumerable small black
dots, only visible under a lens. Thonning's detailed description
is very accurate.

2. Spathodea *lutea*, Benth.; arborescens, foliis oppositis ramisque glabris v. vix puberulis, foliolis 9-11 oblongis acuminatis
integerrimis v. obsolete denticulatis, racemis terminalibus
tomentosis subpaniculatis, corolla infundibuliformi incurva
glabra calyce duplo longiore, capsula longissima tenuissime
ferrugineo-tomentella.—On the Quorra, at Patteh, and Fernando Po, *Vogel.*

Arbor mediocris. *Folia* pedalia; foliola 3-4-pollicaria, nonnulla versus apicem dentibus paucis minutis instructa, membranacea, supra glabra, subtus ad venas sæpius puberula et
glandulis minutis conspersa, basi obtusa et sessilia v. in petiolum brevissimum angustata, terminale interdum ad apicem
petioli articulatum et ab ultimis lateralibus distans, sæpius
vero addatur unum alterumve e lateralibus pariter ad apicem
petioli sessilibus, et sic folia variare videntur pari- v. impari-

pinnata.   *Racemi* 3-4-pollicares, ad apices ramorum solitarii
v. plures paniculati.   *Pedicelli* breves, fasciculati v. in race-
mulos brevissimos dispositi.   *Calyx* pollicaris, arcuatus,
hinc fissus, inde acuminatus, integer, extus tomentellus.
Corolla bipollicaris, proportione multo angustior quam in
S. campanulata, at forma illi propinquior quam *S. lœvi*,
pallide lutea, intus sulphurea, *stamina* iis *S. campanulatæ*
similia, sed fauce breviora, cum rudimento quinti.   *Ovarium*
disco crasso circumdatum, biloculare, compressum, placenta
columnari.   *Stylus* apice bilamellatus.   *Capsula* bipedalis,
plano-compressa, 5 lin. lata ; loculicide dehiscens in valvulas
2 septam nudantibus angustam, medio dilatatam in dissepi-
mentum spurium plano-suberosum valvulis parallelum, capsulam in loculos spurios 4 dividens, utrinque percursum
nervo longitudinali (septa vera) seminifero.   *Semina* more
plerumque *Bignoniacearum* plana, transverse oblonga, subquadrata, 2½ lin. longa, alis neglectis 5 lin. lata ; testa ad
margines lineamque centralem calloso-incrassata, in disco
tenuior, ad utramque latus expansa in alam membranaceam
5-6 lin. longam ; membrana interna testa multo minor,
tenuis et viridis, embryone conformis, nisi ad basin ubi cuneata
est radiculam includens, apice sphacelata.   *Embryo* planus,
2½ lin. latus ; cotyledones didymæ (forma fere fructus *Biscu-
tellæ*), apice leviter, basi profunde emarginatæ, radicula ex
emarginaturæ recta, ½ lin. longa, ad hilum spectans.

I am not aware whether the membrane which closely envelopes the embryo is universal in *Bignoniaceæ*, as it does not
appear to have been usually noticed.   I find it, however, in
Bignonia venusta, B. tubiflora, and two or three others, of which
I happen to have ripe seeds.   It always includes the whole of
the radicle, so as to make that part appear much shorter than it
really is.

3. Spathodea *tomentosa,* Benth.; foliis oppositis, foliolis 9-11
ovatis oblongisve acuminatis integerrimis supra glabris subtus
ferrugineo-tomentosis, racemis terminalibus tomentosis, corolla
glabra.—From Vogel's collection, without the precise locality,
probably Fernando Po.

In the form and size of the leaves, inflorescence and calyx,
this is very near *S. lutea,* but the leaves are thickly clothed
underneath with a soft, rusty down. Of the corolla there are
only fragments remaining, insufficient to show its form or size.
4. Spathodea *adenantha,* Don, *DC. Prod.* 9. *p.* 207.—Bignonia
glandulosa, *Schum. et Thonn. Beskr. p.* 275.—Sierra Leone,
Don ; Guinea.

Don's specimen is very bad, but the scars show that the
leaves are ternately verticillate, and in other respects it agrees,
as far as it goes, with Thonning's description.

The *Spathodea lævis,* Pal. de Beauv., from Oware, appears to
be a distinct species from any of the above. The *Stereospermum
Kunthianum,* Cham., is confined to Senegal.

1. Kigelia *Africana,* Benth.—Bignonia Africana, *Lam.—DC.
Prod.* 9. *p.* 166.—Cape Coast, *Vogel ;* Senegal.

The specimen has but a single flower, which I was unable to
dissect, but it agrees so well in every particular with Lamarck's
description, except that the leaflets are rather more numerous,
that I have no hesitation in considering it identical. It is also
very near to the *K. Æthiopica* of Decaisne, from Nubia, but
the flowers are not quite so large. The leaves of the original
K. pinnata, from Madagascar, are stated to be alternate ; in our
plant they are certainly opposite, neither the figure nor
Kotschy's specimens of *K. Æthiopica* afford any information as
to their insertion on that species. The W. African plant is
described by Vogel as a tree of considerable height, with spreading branches, and a whitish, rugged bark. The flowers hanging several together from the end of a long peduncle ; the
calyx, 8-9 lines long, not so full as in *K. Æthiopica;* the
corolla about 2½ inches long, of a deep red inside, paler outside,
marked with stripes of a golden yellow. The fruits, hanging
something like large cucumbers, about 2 feet long and 5 inches
broad, somewhat compressed laterally, are filled inside with a hard
kind of fleshy pulp, traversed by almost woody fibres, obscurely
two-celled, and containing numerous seeds nestling in the
pulp.

1. Sesamum *Indicum,* Linn.—*DC. Prod.* 9. *p.* 250.—Antha-

denia sesamoides, *Van Houtte; Walp. Rep.* 6. *p.* 518.—
Common about habitations, from Senegal to Benin and Fernando Po, having spread, probably from cultivation, here as
in other parts of Africa and Asia.

Van Houtte has ascertained that the small globular bodies on
each side of the pedicels, usually described as glands, are, in
fact, abortive flowers, and has corrected, in a few other particulars, the character usually given of this plant, which is certainly
identical, both specifically and generically, with the common
Sesame.

The other W. Tropical African species of the tribe (or, as
some will have it, of the Order) of *Sesameæ* are : *Sesamopteris
radiata,* DC., from Guinea ; *S. alata,* DC., from Senegal and
Guinea ; *Ceratotheca sesamoides,* Endl., from Senegal, Nubia
and Abyssinia, and *Rogeria adenophylla,* Gay, from Senegal and
Nubia.

LXXVII. CONVOLVULACEÆ.

1. Batatas *incurva,* Benth.—Convolvulus incurvus, *Schum. et
Thonn. Beskr. p.* 99.—Ipomæa humilis, *G. Don, Gard. Dict.*
4. *p.* 267.—Sierra Leone, *Don;* on the Nun River, *Vogel.*
Glaberrima. *Caulis* repens, radicans. *Folia* nunc omnia integra, 2-4 poll. longa, 3-6 lin. lata; nunc basi aucta lobis
2 lineari-oblongis angulo recto divaricatis v. sursum incurvis. *Corolla,* ex Vog., alba, basin versus lutescens, ima basi
rubro-fucata. *Ovarium* certe 4-loculare.

This may be a variety of the common American sea-coast
species, *B. acetosæfolia,* Chois., but as well from Thonning's
description as from the few imperfect specimens before me, it
appears to me to be distinct. It can hardly be the same as
Ipomæa Clappertoni, Br., to which Choisy has referred Thonning's plant.

2. Batatas *paniculata,* Chois. *in DC. Prod.* 9. *p.* 339.—Cultivated at Cape Palmas, *Vogel;* a common Tropical plant.—
Ejusdem var. foliis integris v. rarius lobatis ;—Ipomæa erio-
sperma, *Beauv. Fl. Ow. et Ben.* 2. *p.* 73. *t.* 105.—Grand
Bassa, Cape Coast, and on the Nun River, *Vogel.*

3. **Batatas** *pentaphylla*, Chois. *in DC., Prod.* 9. *p.* 339.—
St. Thomas, *Don;* a common Tropical species in both hemispheres.

The common *Batata*, *Batatas edulis*, Chois., is said to be
cultivated in Tropical Africa, as well as in India and America;
but the cultivated specimens, brought by Vogel, certainly
belong to the *B. paniculata.*

1. Pharbitis *Nil*, Chois. *in DC. Prod.* 9. *p.* 343.—On the
Quorra, *Vogel;* a common Tropical plant.

1. Calonyction *speciosum*, Chois. *in DC. Prod.* 9. *p.* 345, *pro
parte.*—Ipomœa bona-nox, *Linn.* var ?—Chonemorpha convolvuloides, *G. Don, Gard. Dict.* 4. *p.* 76.—Abòh, *Vogel ;*
St. Thomas, *Don.*

The confusion of characters and synonyms accumulated
under the name of *Calonyction speciosum* is so great, that it is
difficult to find any specimen agreeing both with the generic
and specific characters given. The original *Ipomœa bona-nox*,
Linn., has the outer sepals (exclusively of their long points)
much shorter than the inner ones, which are either blunt or
have very short points; the corolla is rather hypocrateriform
than infundibuliform, with a slender green tube, 4 or 5 inches
long, and an almost flat, broad-spreading, white limb; the
stamens project considerably beyond the mouth of the tube,
and the capsule is as large as a good-sized nut. In these
respects, the Asiatic and American plants appear to agree;
unless it be that the corolla is rather larger in the American
one. Our African specimens have the calyx and corolla of the
same form and colour, but smaller, and the stamens are
scarcely, if at all, longer than the tube of the corolla. The
Ipomœa muricata, Roxb., of which the calyx could with less
impropriety be said to be " sepalis aristatis æqualibus," and the
corolla " infundibuliformis," has already been shown to be
totally distinct from the *bona-nox*, not only by this form of the
calyx and corolla, but also by the colour of the latter, the tube
of which is of a deep purple, by the stamens always included,
and by the small fruit. It is a true *Ipomœa*, and so is the
Ipomœa acanthocarpa, Hochst. (*Calonyction ?* Chois.) from

H H

Nubia. The genus *Calonyction*, if retained at all, should probably be confined to *C. speciosum* and *grandiflorum*, unless perhaps some of Choisy's *Exogonia* be added to them.

1. Ipomœa *reptans*, Poir., *Chois. in DC. Prod.* 9. *p.* 349.—On the Quorra, among the ruins of Addanda, *Vogel.*

This is a very luxuriant form, with leaves on very long petioles, and often 5 inches long, and 3 broad at the base.

2. Ipomæa *pes-capræ*, Sw.—*Chois. in DC. Prod.* 9. *p.* 349.— Cape Palmas and Fernando Po, *Vogel;* common on this as on other Tropical sea coasts.

The *I. asarifolia*, Rœm. et Schult., a species closely allied to the Asiatic *I. rugosa* and to the American *I. urbica*, is indicated as a Senegal plant. I have not seen any specimens of it, but in the Hookerian Herbarium is one, apparently of *rugosa*, gathered by Macrae, at St. Yago (Cape de Verd Isles),* but remarkable for its peduncles, which are rather longer than usual, and thickly covered near the base with a rusty pubescence. The *I. Clappertoni*, Br., from Central Africa, belonging to the same group, is unknown to me.

3. Ipomœa *filicaulis*, Bl.—*Chois. in DC. Prod.* 9. *p.* 353.— Along the whole Guinea Coast to the Niger and St. Thomas, *Vogel, Don ;* common in Tropical countries.

4. Ipomœa *ovalifolia*, Chois. *in DC. Prod.* 9. *p.* 357.—Accra, *Vogel, Don ;* Guinea, Angola and East India.

To the same group Choisy refers two Angola species, *I. dendroidea*, Chois., and *I. verbascoidea*, Chois.

5. Ipomœa *involucrata*, Beauv.—*Chois. in DC. Prod.* 9. *p.* 365. —Sierra Leone, *Don, Miss Turner ;* Cape Palmas, *Vogel ;* Senegal to Oware, also Madagascar and Java, according to Choisy.—Var. *hirsutior*, Fernando Po, *Vogel.*

This species can scarcely be distinguished from the common East Indian *I. pileata.*

6. Ipomœa *amœna*, Chois. *in DC. Prod.* 9. *p.* 365.—Savannahs

---

* This species has to be added to the Spicilegia Gorgonea (supra, p. 152), where also the reference, under *I. pes-capræ*, to the figure of *I. maritima*, should be to the Bot. Reg. t. 319, not Bot. Mag.; a mistake apparently copied from the Prodromus.

of the Quorra, at Addanda, *Vogel*, who states the flowers to be purplish-white ; Senegal and Guinea.

7. Ipomœa *capitata*, Chois. *in DC. Prod.* 9. *p.* 365.—Cape Palmas, *Vogel;* Accra, *Ansell;* throughout Tropical Africa, and closely allied to the E. Indian *I. capitellata* and the American *I. tamnifolia.*

8. Ipomœa *sessiliflora*, Roth.—*Chois. in DC. Prod.* 9. *p.* 366. —St. Thomas, *Don ;* a common East African and East Indian species.

9. Ipomœa sp., near the last, and with similar small flowers, but very hispid ; the specimens nearly rotten.—St. Thomas, *Don.*

To the *Capitate* group Choisy refers also the *I. dichroa,* Chois. from Senegal.

10. Ipomœa *sagittata*, Desf.—*Chois. in DC. Prod.* 9. *p.* 372.— Accra, *Don;* a N. American, S. European, and N. African plant.

11. Ipomœa *teretistigma, β setifera,* Chois. *in DC. Prod.* 9. *p.* 373 ?—Sierra Leone, *Don ;* Senegambia, *Heudelot.*

This plant, which I am not quite certain of having correctly identified with Choisy's *I. teretistigma,* from Guinea, is remarkable for the very strong prominent nerves of the calyx. The capsules and seeds are both smooth in Don's specimen, which has no flowers. Heudelot's has a single flower, which I was unable to examine.

12. Ipomœa *umbellata*, Mey.—*Chois. in DC. Prod.* 9. *p.* 377.— I. primulæflora, *G. Don, Gard. Dict.* 4. *p.* 270.—Sierra Leone, *Don ;* Fernando Po, *Vogel;* a common American plant, distinguished from the Asiatic *I. cymosa* almost exclusively by its corolla, yellow, not white.

13. Ipomœa *Baclei,* Chois. *in DC. Prod.* 9. *p.* 381.—I. riparia, *G. Don, Gard. Dict.* 4. *p.* 265.—Flowers purple, whitish inside.—Fernando Po, on the sea coast, *Vogel ;* St. Thomas, *Don ;* Senegal.

These specimens agree perfectly with Choisy's character and figure. The *I. Lindleyi,* Chois. from Madagascar, must be very near it.

The *I. ochracea*, Don, from the Gold Coast, *I. Owariensis*, Beauv., from Oware, *I. Afra*, Chois., from Guinea, and *I. Rogeri*, Chois., and *I. zebrina*, Perr., both from Senegal, are unknown to me.

14. Ipomœa *palmata*, Forsk., *Chois. in DC. Prod.* 9. *p.* 386.— Senegal to the Niger and Fernando Po, *Vogel, Don* and others ; also in East Africa.

The *I. vesiculosa*, Beauv., appears to be the same species, with some accidental deformity of the epidermis.

15. Ipomœa *ennealoba*, Beauv.—*Chois. in DC. Prod.* 9. *p.* 388. —Sierra Leone, *Vogel ;* Oware.

The *I. Coptica*, Roth, an East African and East Indian plant, extends also into Senegal.

16. Ipomœa *sinuata*, Ort.—*Chois. in DC. Prod.* 9. *p.* 362.— Fernando Po, *Vogel ;* a common American plant, from the Southern United States to Brazil.

These African specimens belong to some of the larger forms included by Choisy within the limits of the species. The peduncles, as in some of the Brazilian specimens, are often much longer than the petioles, with from three to seven or eight white flowers, the corolla half as long again as the calyx. The peculiar anthers of this and some allied species, would surely justify their separation into a distinct section.

The two remaining *Ipomœæ*, cited as West African, *I. Senegambiæ*, Chois., from Senegal, and *I. Afzelii*, Chois., from Sierra Leone, must both be very near *Breweria secunda*, described below.

There is no genuine species of *Convolvulus* described from W. Tropical Africa ; and the only one known to me is a Senegambian plant, in Heudelot's collection, which is either a long-leaved luxuriant variety of, or a new species allied to, the Egyptian and Nubian *C. microphyllus*, Sieb.

1. Aniseia *uniflora*, Chois. *in DC. Prod.* 9. *p.* 431.—Ipomœa lanceolata, *G. Don, Gard. Dict.* 4. *p.* 282.—Sierra Leone, *Don ;* Madagascar and East India.

1. Hewittia *bicolor*, Wight et Arn.—Shutereia bicolor, *Chois. in DC. Prod.* 9. *p.* 435—Aniseia Afzelii, *G. Don, Gard.*

Dict. 4. *p.* 295.—Sierra Leone, *Don;* Cape Palmas, *Vogel;* common in South-east Africa and East India.

1. Neuropeltis *acuminata,* Benth.—Porana acuminata, *Pal. Beauv. Fl. Ow. et Ben.* 1. *p.* 65, *t.* 39.—*Chois. in DC. Prod.* 9. *p.* 436.—Sierra Leone, *Don;* Oware.

The styles being perfectly distinct to the base, had already led Beauvois to suspect that this might not be a true congener to Burmann's *Porana;* and Don's specimen, though without flowers, and indifferent as to foliage, yet being in full fruit, with enough of leaves to identify it, shows it to be an additional species of Wallich's East Indian genus *Neuropeltis.* It is, indeed, near *N. ovata,* but with a climbing habit and paniculate inflorescence. The enlarged fruit-bearing bracteas are broadly ovate, and often above an inch long; the capsule small and one- or two-seeded.

1. Prevostea *Africana,* Benth.; foliis oblongis longe acuminatis basi cuneatis rigide membranaceis supra glabris, pedicellis unifloris in axillis confertis, corolla sepalum externum duplo excedente.—Codonanthus Africanus, *G. Don, Gard. Dict.* 4. *p.* 166.—C. alternifolius, *Planch. in Hook. Ic. t.* 796. (Tab. XLVI.)—Sierra Leone, *Don.*

Frutex ut videtur subvolubilis. *Folia* semipedalia, breviter petiolata, supra glabra, subtus oculo armato pube minuta conspersa. *Florum* fasciculi axillares, 3-5-flori, pedicellis 3-6 lin. longis, bracteolis minutis angustis ferrugineo-pubescentibus. *Sepala* exteriora late cordato-ovata, 4-4½ lin. longa, minute puberula, interiora multo minora, acuta. *Corollæ* (ex Don albæ) forma valde similis illi *P. sericeæ,* a Kunthio depictæ; tubus tamen paullo amplior et sub limbo evidentius constrictus. *Styli* inæquales, sæpius paullo ultra medium connati. *Ovarium* glabrum, semiseptis ad axin haud attingentibus incomplete biloculare.

I do not see how this plant can be distinguished from *Prevostea,* so well described by Kunth under the name of *Dufourea.* I have not, indeed, been enabled to examine any American species; but Kunth's figure is very satisfactory. The form of

FLORA NIGRITIANA.

the corolla is all but identical, so is the structure of the calyx. The styles in the American plant are more deeply separated than in the African one, but Choisy considers that a variable character. The ovary of *Prevostea* is described as bilocular, but the figure represents the dissepiment as incomplete, precisely as it is in Don's plant. I have restored Don's specific name, for although it be not attached to the specimen, the memoranda on the label leave no doubt but that this is the plant he had in view; and the mistake as to the "opposite leaves," must have arisen from his having noted that it is either *Gentianeous* or *Asclepiadeous*.

PLATE XLVI. *Fig.* 1. unopened corolla; *f.* 2. flower; *f.* 3. corolla opened laterally, showing the stamens; *f.* 4. ovary and styles; *f.* 5. section of the ovary; *all magnified.*

There is another unpublished species of the same genus, in Heudelot's Senegambian collection, with the inflorescence rather more developed than in Don's plant.

1. Breweria *secunda,* Benth.; volubilis, foliis ovato-lanceolatis oblongisve acuminatis retusisve subcoriaceis supra glabris v. parce pilosis subtus ferrugineo-villosis, cymis densis multifloris axillaribus pedunculatis v. ad apices ramorum subsessilibus confertis, sepalis acutis rufo-sericeis corolla pilosa dimidio brevioribus, capsula globosa apice puberula, seminibus glabris.—Ipomœa secunda, *G. Don, Gard. Dict.* 4. *p.* 282.— Sierra Leone, *Don.*

Caulis glaber v. ferrugineo-pubescens. *Folia* breviter petiolata, basi obtusa, nunc 2-3-pollicaria elliptico-oblonga et obtusissima, nunc 1½-pollicaria ovato-lanceolata et apice longiuscule acutata, summo tamen apice semper obtusa cum mucrone. *Cymæ* aliæ axillares distantes, aliæ ad apices ramulorum in cymam v. fasciculum subsecundum densissime confertæ. *Sepala* 3-4 lin. longa, ovato-lanceolata, parum inæqualia. *Corolla* campanulata (ex Don alba). *Stylus* ad tertiam partem fissus, stigmatibus majusculis capitatis. *Ovarium* apice pilosum. *Capsula* calyce brevior.

1. Evolvulus *alsinoides,* Linn.—*Chois. in DC. Prod.* 9. *p.* 447.

—Accra, *Ansell;* on the Quorra, *Vogel;* a common Tropical species, of which the *E. linifolius* appears to be only a narrow-leaved form.

West Tropical Africa seems the only known region from whence no species of *Cuscuta* has as yet been brought.

LXXVIII. BORAGINEÆ.

There are two West African species of *Cordia,* named after their respective stations, *C. Senegalensis,* Juss., and *C. Guineensis,* Schum. et Thonn.

1. Ehretia *cymosa,* Schum. et Thonn.—*DC. Prod.* 9. *p.* 508.— Cape Palmas, Accra and Aguapim, *Vogel;* Sierra Leone, *Don.*

A common shrub, attaining the height of a man, with white flowers, and agreeing well with Thonning's description, except that the leaves are usually much larger, being often 4 inches long and 3 wide.

Senegal has one species of *Tournefortia,* the *T. subulata,* Hochst., extending to Nubia and Abyssinia.

1. Heliotropium *strigosum,* Willd., *DC. Prod.* 9. *p.* 546.— Cape Coast, *Don;* Accra, *Vogel.*

The other West African species are *H. undulatum,* Vahl, common to Senegal, the deserts of North Africa, and Nubia; *H. Kunzei,* Lehm., from Senegal, Nubia and Abyssinia; *H. Coromandelianum,* Lehm., extending from Senegal over the greater part of Africa and East India, to the Philippine Islands and Tropical Australia; *H. Baclei,* DC., from Senegal; and *H. Africanum,* Schum. et Thonn., from Guinea.

1. Heliophytum *Indicum,* DC., *Prod,* 9. *p.* 557.—Common in West Tropical Africa, and in nearly all Tropical countries in both hemispheres.

The common African *Trichodesma Africanum,* Br., for which Webb proposes (above, p. 153) to restore Medik's original name of *Pollichia,* extends into Senegal, and is the only species of the numerous tribe of *Borageæ* proper hitherto found in W. Africa within the Tropics.

LXXIX. Solanaceæ.

1. Physalis *somnifera*, Linn.—Cape Palmas, *Vogel;* a common N. and E. African, S. European, and E. Indian plant.
2. Physalis *angulata*, Linn.—*N. ab E., Linnæa*, 6. *p.* 474?—Common in cultivated places, from Sierra Leone to the Niger.

I am not quite certain whether this should be referred to *P. angulata* or to *P. æquata*, both of them widely spread in Tropical regions. The specimens are perfectly smooth, as is usual with *P. angulata;* but the fruit-calyx, so far as I can judge in their badly-dried state, is not so decidedly angled. On most of the labels, Vogel states the flowers to be yellow; on one, however, attached to a specimen not otherwise distinguishable from the remainder, he has noted: " Flores lutei, basi brunnei."

3. Physalis *minima*, Linn.—*N. ab E., Linnæa*, 6. *p.* 479.—On the Quorra, at Attah, *Vogel;* an East Indian species.
1. Capsicum *annuum*, Linn.—Common in cultivated places at Sierra Leone, *Vogel;* an East Indian and American plant.
1. Lycopersicum *esculentum*, Mill.—*Dun. Syn. Sol. p.* 4.—Common in cultivated grounds at Fernando Po, *Vogel.*—This, the *Tomato*, or *Love-apple*, of American origin, appears to be frequently found growing naturally in the Old World, escaped from cultivation.
1. Solanum *nodiflorum*, Jacq.—*Dun. Monogr. Sol. p.* 151.—Sierra Leone, *Don, Vogel.*

This common East Indian species can hardly be distinguished, except by its straggling habit and perennial stem, from the ubiquitous *S. nigrum*, Linn., of which the *S. Guineense*, Lam., appears to be a large-fruited variety.

2. Solanum sp.—Accra, and ruins of Addanda, on the Quorra, *Vogel.*

An unarmed, suffrutescent, stellately-tomentose and small-flowered species, which I cannot identify with any published

one, nor venture to describe as new in the present state of confusion which prevails among the five or six hundred species of
the genus.

3. Solanum *distichum*, Schum. et Thonn. *Beskr. p.* 122?—
Accra, *Ansell;* a single small specimen, allied to the last-
mentioned species of Vogel's, but shrubby, and bears a single
small prickle.

4. Solanum *Melongena*, Linn.—*N. ab E. Linn. Trans.* 17.
p. 48.—Common in cultivated grounds.

The native country of this, the *Bringall* or *Aubergine*, is
doubtful, it being in universal cultivation and frequently naturalized all over the Tropics, as well as in Southern and some
parts of Central Europe. The *S. edule* and *S. Atropa* of
Schum. et Thonn., are probably very near it, if not mere
varieties.

5. Solanum *anomalum*, Schum. et Thonn. *Beskr. p.* 126.—
Grand Bassa, Cape Palmas, and Fernando Po, *Vogel;* Sierra
Leone, *Don.*

Vogel says that the flowers are nodding, and usually pentamerous, although sometimes tetramerous, as described by Thonning. The berries scarlet and erect.

Two other species of *Solanum*, from Guinea, are published by
Schum. et Thonn., under the names of *S. dasyphyllum*, and
S. geminifolium.

LXXX. Scrophularineæ.

1. Schwenckia *Americana*, Linn.—*Benth. in DC. Prod.* 10.
p. 194.—Sierra Leone, Cape Coast, and on the Quorra,
Vogel, Don; Senegambia, and common in East Tropical
America.

One species of *Linaria, L. spartioides*, Brouss., is found at
Cape Verd, within the extreme northern limits of Senegambia.

1. Alectra *Vogelii*, Benth., *in DC. Prod.* 10. *p.* 339.—On the
Quorra, at Patteh, *Vogel.* Flowers yellow.

The *Alectra Senegalensis*, Benth., is confined to Senegal.

Doratanthera linearis, Benth., and *Stemodia serrata*, Benth.,
extend from Senegal to Nubia and Egypt; and the common
East Indian *Limnophila gratioloides*, Br., is also found in
Senegal.

1. Herpestis *calycina*, Benth.; *in DC. Prod.* 10. *p.* 399. var. ?
Accra, *Don*.

Don's specimen is very bad, but appears to belong to the
narrow-leaved Sierra Leone variety, with rather longer pedicels
than in other specimens I have seen. The broad-leaved variety
is from Senegal, as well as three other species, *H. Hamilto-
niana*, Benth., *H. floribunda*, Br., and *H. Monniera*, H. B. K.
All three are also East Indian, the *H. floribunda* extending to
Tropical Australia, and *H. Monniera* being common to all the
warmer regions of the globe. Senegal supplies also one species
of *Dopatrium*, *D. Senegalense*, Benth.

1. Vandellia *diffusa*, Linn.—*Benth. in DC. Prod.* 10. *p.* 416.
—Sierra Leone, *Vogel*, both the sessile-flowered and pedun-
culate varieties; a common East Tropical American plant.

2. Vandellia *Senegalensis*, Benth. *in DC. Prod.* 10. *p.* 416.
—Sierra Leone, *Don;* on the Nun River, *Vogel;* Senegal.

The little East Indian *Glossostigma spathulatum*, Arn., is
also found in Senegal.

1. Capraria *biflora*, Linn.—*Benth. in DC. Prod.* 10. *p.* 429.—
Cape Coast, *Vogel, Don;* frequent in America.

1. Scoparia *dulcis*, Linn.—*Benth. in DC. Prod.* 10. *p.* 431.—
Sierra Leone to the Niger, *Vogel, Don, &c.;* as common here
as in all the warmer parts of Africa, Asia and America.

There are three Senegambian species of *Buchnera;* *B. dura*,
Benth., common to S. Africa and Madagascar; *B. hispida*,
Hamilt., extending into Abyssinia, Madagascar and East India;
and *B. leptostachya*, Benth., found also in Madagascar. The
Guinea plant described as *B. linearifolia*, Schum. et Thonn.,
belongs probably to some other genus.

1. Striga *orobanchoides*, Benth. *in DC. Prod.* 10. *p.* 501.—On
the Quorra, *Vogel;* Senegal, East Africa and East India.

2. Striga *aspera*, β. *filiformis*, Benth. *in DC. Prod.* 10. *p.* 501.

—On the Quorra, *Vogel.* The type of the species is from Senegal and Guinea.

3. Striga *Senegalensis,* Benth. *in DC. Prod.* 10. *p.* 502.—On the Quorra, at Stirling, *Vogel, Ansell;* Senegal and East Africa. Possibly a mere small-flowered variety of *S. hermonthica.*

Two other *Strigæ* belong to W. Tropical Africa, the *S. Forbesii,* Benth., from Senegal and Madagascar, and *S. macrantha,* Benth., from Senegal and Sierra Leone. The *Rhamphicarpa fistulosa,* Benth., an Abyssinian and Nubian plant, is found also in Senegal ; and *Sopubia filiformis,* G. Don, from Guinea, completes the list of W. Tropical African *Scrophularineæ.*

LXXXI. Acanthaceæ.

1. Thunbergia *geraniifolia,* Benth.; scandens, pilis longis hirta, foliis latis cordatis palmato-5-lobis acuminatis, pedunculis axillaribus unifloris, calyce truncato integro.—Sierra Leone, *Don.*

Pili longi rigiduli appressi v. subpatentes, ad ramulos infra nodos et in petiolos reflexi, supra nodos et in pedunculos arrecti, in foliorum pagina superiore bracteisque numerosi sparsi, in pagina inferiore rariores præcipue ad venas dispositi. *Folia* multo minora, tenuiora et magis lobata quam in *T. grandiflora,* et tomentum breve scabrum illius speciei omnino deest. *Bracteæ* 1¼-pollicares, acuminatæ, membranaceæ nec coriaceæ. *Corollam* non vidi.

2. Thunbergia *chrysops,* Hook.—*N., ab E. in DC. Prod.* 11. *p.* 55.—Sierra Leone, *Whitfield.*

3. Thunbergia *cynanchifolia,* Benth. ; volubilis, pilosula, gracilis, foliis petiolatis cordato-sagittatis membranaceis, calyce 6-7-fido, corollæ limbo tubo suo breviore.— On the Quorra, *Vogel.*

Caules tenues, parce pilosuli v. glabrati. *Folia* 1-3-pollicaria, forma fere *Cynanchi acuti,* apice mucronulata, margine ciliolata, basi profunde cordata, auriculis rotundatis v. sæpius sub angulatis, utrinque viridia et sparse pilosula v. demum glabrata. *Flores* axillares, pedunculati, solitarii v. ad apices ramulorum pauci, pro genere parvi. *Bracteæ* membranaceæ, acutæ, 5 lin. longæ. *Calyx* brevissimus, late et obtuse usque ad medium divisus. *Corolla* alba, pollice brevior. *Antheræ* basi pilosæ, inter se subæquales, loculo altero calcarato, altero (an in omnibus?) mutico. *Stylus* apice infundibuliformis. *Capsula* globosa, vix minutissime tomentella, rostro capsula ipsa longiore.

There is but a single corolla with the specimen, and in this one I saw *five* spurred cells of anthers, but I was unable to ascertain whether the fifth arose from a fifth stamen accidentally developped, or from the second cell of one of the other anthers, the whole being much crushed in drying.

4. Thunbergia? *Vogeliana*, Benth.; scandens?, glabra, foliis ovatis oblongisve integerrimis basi angustatis rotundatisve utrinque papuloso-scabris, pedicellis axillaribus unifloris, calyce sub-12-fido.—Fernando Po, *Vogel.*

Rami lignosi, angulati, flexuosi et scandentes videntur etsi vix volubiles, glabri præter pubem ad nodos juniores fasciculatam demum evanidam. *Folia* 3-4-pollicaria, breviter petiolata, acuminata v. obtusa, venosa, rigidule chartacea, utrinque tactu scabra. *Pedicelli* pollicares. *Bracteæ* pollicares, ovatæ, obtusæ, primum rubræ, demum rubro-albæ v. albæ. *Calyx* brevis, lobis circa 12 inæqualibus subulatis. *Corolla* (quam ipse non vidi) sec. Vogel infundibuliformis est, tubo extus albo intus flavo, limbo fusco-cœruleo. *Stamina* non vidi et ideo genus incertum. *Habitus* potius *Meyeniæ*, sed *stylus* superest apice more *Thunbergiæ* ad partem stigmatosam infundibuliformis.

1. Meyenia *erecta*, Benth.; glabra, foliis petiolatis ovatis oblongisve acuminatis basi angustatis, calyce brevissimo sub-12-fido, corollæ tubo bracteis quadruplo longiore.—Cape Coast, *Vogel.*

Frutex 6-8-pedalis, ramulis tenuibus tetragonis. *Folia* 1-2-pollicaria, integerrima v. obsolete angulata, membranacea. *Pedunculi* axillares, uniflori, pollicares. *Bracteæ* membranaceæ, semipollicares. *Calyx* cum lobis raro lineam longus.

Corollæ tubus (tubus cum faucibus, N. ab E.) fere bipollicaris, supra ovarium contractus, dein ventricosus, ad faucem
ampliatus ; limbus subæqualis. *Antheræ* muticæ, omnes subsimiles, loculis ciliatis inæqualibus, altero altius inserto breviore et magis divergente. *Stylus* apice divisus in lobos
stigmatiferos 2 cuneato-dilatatos emarginatos. *Flores* ex
Vogel erecti, corollis basi luteo-albidis, apice purpureis.

1. Elytraria *marginata*, Beauv. ; *N. ab E., in DC. Prod.* 11. *p.*
   63.—Grand Bassa and Fernando Po, *Vogel ;* St. Thomas,
   Don ; Senegal to Oware, and scarcely to be distinguished
   from the common East Indian *E. crenata*, or the American
   E. virgata.

   The *Nelsonia canescens*, N. ab E., a native of Senegal and of
   a great portion of Africa, appears to grow also in Tropical
   America and Australia. Senegal has likewise supplied the five
   following *Acanthaceæ*, hitherto confined to that country : *Physichilus Senegalensis* and *P. barbatus*, N. ab E., *Polyechma
   micranthum* and *P. odorum*, N. ab E., and *Nomaphila lævis*,
   N. ab E.

1. Brillaintaisia *Lamium*, Benth.—Leucoraphis Lamium, *N.
   ab E., in DC. Prod.* 11. *p.* 97.—Fernando Po, *Vogel, Ansell ;*
   Sierra Leone, *Don.*

2. Brillaintaisia *Vogeliana*, Benth.—Leucoraphis Vogeliana, *N.
   ab E., in DC. Prod.* 11. *p.* 97.—Fernando Po, *Vogel ;* St.
   Thomas, *Don.*

   There seems little doubt that these two plants belong to
   Beauvois' genus *Brillaintaisia*, as characterized in the Fl. Ow.
   et Ben., and accidentally overlooked by Nees. It is also very
   probable that the species figured by Beauvois (*B. Owariensis*,
   Beauv. *Fl. Ow. et Ben.* 2. *p.* 68. *t.* 100. *f.* 2.) is the *Belanthera
   Belvisiana*, N. ab E., notwithstanding the discrepancy of the
   presence of the sterile stamens in Beauvois' *Brillaintaisia*, and
   their absence in Nees' *Belanthera*. Beauvois' specimens are
   in general made up of mere fragments ; and several cases are
   known where the details of the flowers in his f.gures are compiled
   from different plants. The fragments seen by Nees, left him
   also in some uncertainty as to their all belonging to one plant.

At any rate, the two genera of Nees, *Leucoraphis* and *Belanthera*, are so very closely allied, that it seems better to reunite them under Beauvois' name.

The two Senegalese species of *Calophanes*, *C. Perrottetii*, N. ab E., and *C. Heudelotianus*, N. ab E., are confined to that country.

1. Dipteracanthus *elongatus*, N. ab E., *in DC. Prod.* 11. *p.* 140. —Fernando Po, *Vogel;* Oware.

Vogel's specimen is in very young bud only, but appears to belong to this species.

1. Asystasia *Coromandeliana*, N. ab E., *in DC. Prod.* 11. *p.* 165, var. hirsuta, parvifolia, parviflora, calycis laciniis subulatis, corolla roseo-carnea.—Cape Coast, *Vogel;* a common East Indian species, extending over Eastern and Central Africa to Senegal.

2. Asystasia *quaterna*, N. ab E., *in DC. Prod.* 11. *p.* 166, var. calycis laciniis subulatis, corolla alba.—Cape Palmas and Sierra Leone, *Vogel;* Sierra Leone, *Don;* var. corolla rosea. —Sierra Leone, *Don;* the species common to Senegal and Guinea.

3. Asystasia *calycina*, Benth; diffusa, parce pubescens, foliis ovatis, racemis axillaribus elongatis secundis strictis, calycis laciniis lineari-lanceolatis membranaceis. — Grand Bassa, *Vogel.*

The above three species closely resemble each other, and may possibly be all varieties of the common *A. Coromandeliana;* if the size and form of the leaves, the hairiness, the size of the flowers, &c., be really as variable as the usually indifferent specimens appear to indicate. The *A. calycina* is the most distinct, by the divisions of the calyx, which are about ¾ of a line broad. Its flowers are said to be greenish-white.

4. Asystasia *scandens*, Hook. *Bot. Mag. t.* 4449.—Henfreya scandens, *Lindl. Bot. Reg.* 1847, *t.* 31.—Sierra Leone, *Don, Whitfield.*

It is merely from inadvertence, in the hurry of drawing up *Addenda* without seeing the specimens, that Nees referred this plant to *A. quaterna*, after having in the text of the Prodromus well alluded to the evidently close affinity of Thonning's plant to the *A. Coromandeliana*. In the *A. scandens*, not only is the general size of the plant and of the leaves very different from those described by Thonning, but the inflorescence and flowers at once preclude the possibility of uniting them. Schumacher moreover expressly states that the *R. quaterna* only differs from *R. intrusa*, Vahl., (another species scarcely distinguishable from *A. Coromandeliana*), by the number of bracts. With regard to the genus, our plant cannot be separated from *Asystasia ;* to which Lindley himself would probably have referred it, had the volume of the Prodromus been then published.

5. Asystasia *Vogeliana*, Benth.; glabra, foliis ovatis ellipticisve acuminatis basi angustatis breviter petiolatis, racemis terminalibus subramosis glabriusculis, floribus unilateralibus solitariis pedicellatis.—Fernando Po, *Vogel.*

Caulis herbaceus, erectus, subsimplex v. parce ramosus, pluripedalis, ima basi frutescens. *Folia* semipedalia, supra nitidula, longe acuminata. *Inflorescentia* fere *A. Coromandelianæ*, pedicellis tamen sublongioribus. *Calyces* paullo minores, laciniis angustis. *Corolla* tenuis, omnino apertam non vidi, sed alabastrum adest jam 1½ poll. longum. *Antheræ* oblongæ, loculis parallelis basi vix callosis, uno paullo altius inserto.

1. Paulo-Wilhelmia *polysperma*, Benth.; caule glabriusculo, foliis ovatis acuminatis eroso-dentatis basi cuneatis villosis, capsulis 8-10-spermis.—On the Sugarloaf Mountain, Sierra Leone, *Don.*

Rami lignosi. *Folia* 2-4-pollicaria, longe petiolata, rugosa, venis primariis a costa media angula valde acuta divergentibus, secundariis transversis crebris. *Cymæ* subsessiles, bifidæ v. dichotomæ, in thyrsum terminalem dispositæ. *Calycis* laciniæ 5, æquales, lineares, siccæ, glanduloso-puberulæ, per anthesin 5 lin., in fructu 7 lin. longæ. *Corollæ* tubus tenuis 9-10 lin. longus; limbus tubo brevior late expansus 5-

partitus, laciniis obovatis omnibus ad unum latus dejectis. *Stamina* subdidynama, exserta; filamenta basi per paria lateralia connexa; antherarum loculi angusti, contigui, æquales. *Capsula* 9 lin. longa, subtetragona, a basi 2-locularis.

1. Whitfieldia *lateritia*, Hook.—*N. ab E., in DC. Prod.* 11. *p.* 211.—Sierra Leone, *Whitfield.*

1. Barleria *opaca*, N. ab E., *in DC. Prod.* 11. *p.* 230.—Accra, *Don;* Guinea.

2. Barleria *halimioides*, N. ab E., *in DC. Prod.* 11. *p.* 231.— —West Africa, south of the Line, *Curror.*

A third species of *Barleria, B. Senegalensis,* N. ab E., is confined to Senegal.

1. Asteracantha *auriculata*, N. ab E., *in DC. Prod.* 11. *p.* 248.

—Senegal and Guinea, *Don, Vogel* and others; frequent also in East Africa, and closely allied to the common East Indian *A. longifolia.*

Two species of *Lepidagathis, L. Heudelotiana,* N. ab E., and *L. anobrya,* N. ab E., are confined to Senegal.

1. Ætheilema *reniforme*, Br.—*N. ab E., in DC. Prod.* 11. *p.* 261.—Sierra Leone, *Don;* from Senegal, through East Africa to East India.

2. Ætheilema *imbricatum*, Br.—*N. ab E., in DC. Prod.* 11. *p.* 262.—Sierra Leone, *Don;* East Africa.

3. Ætheilema *micrantha*, Benth.; caule flexuoso ramoso patentim piloso, foliis cujusve paris valde inæqualibus, majoribus petiolatis ovatis oblongisve basi inæqualiter angustatis, floribus axillaribus 1-3 confertis, calycis lacinia superiore ovata v. oblonga acuta membranacea ciliata.—St. Thomas, *Don.*

Specimen pedale adest, ramosum. *Folia* tenuiter membranacea, pilis raris conspersa, majora 1-1½-pollicaria, longiuscule petiolata, alterum cujusve paris sæpius minimum est, inferiora tamen mihi desunt. *Bracteæ* late ovatæ, circa 2 lin. longæ. *Calycis* lacinia summa herbacea, membranacea, 2 lin. longa, inferiores angustæ, laterales angustissimæ. *Corolla* 2½ lin.

longa, labio superiore bifido, inferiore duplo longiore trifido. *Capsula* 2 lin. longa, basi brevissime angustata, supra basin tetrasperma. This species would appear to be very near *Æ. rupestre*, N. ab E., from Madagascar, only known to me by Nees' character.

1. Teliostachya *laguroidea*, N. ab E. *in DC. Prod.* 11. *p.* 264. —Accra, *Vogel.*

2. Teliostachya *hyssopifolia*, Benth.; humilis, pilis raris strigillosa v. glabrata, foliis oblongo-lanceolatis angustis breviter petiolatis, spicis brevibus densis, calycis lacinia superiore lanceolata, capsula calyce breviore.—Sierra Leone, *Don.* *Herba* 3-4-pollicaris v. rarius semipedalis, basi diffuse ramosa v. reptans. *Folia* vix pollicaria, pleraque obtusiuscula, utrinque strigis minimis conspersa. *Spicæ* ovatæ v. oblongo-cylindricæ, tenuiores quam in cæteris speciebus etsi densifloræ. *Flores* iis *T. laguroidis* similes, sed laciniæ calycinæ angustiores.

1. Blepharis *boërhaaviæfolia*, Juss.—*N. ab E. in DC. Prod.* 11. *p.* 266.—Sierra Leone, *Don;* Senegal, Guinea, throughout Africa and East India.

1. Cheilopsis *montana*, N. ab E. *in DC. Prod.* 11. *p.* 272. —Common in the mountains of Fernando Po, *Vogel.* The *Acanthodium hirtum*, Hochst., a Nubian plant, extends into Senegal.

1. Isacanthus *Vogelii*, N. ab E. *in DC. Prod.* 11. *p.* 279.— Cape Palmas, *Vogel, Ansell.* The *Crossandra Guineensis*, N. ab E., from Guinea, is not in our collections. The *Amphiscopia Middletoni*, N. ab E., is probably not African. The label does not refer to the Gaboon Coast, but merely to *Johanna P.*, with the date 1782. I have as yet found no clue to the signification of this abbreviation, which occurs on several of Captain Middleton's specimens; but the Gaboon Coast specimens are usually dated 1787.

1. Rostellaria *parviflora*, Benth.; caule repente ramoso gracili, foliis ovatis petiolatis supra scabrellis, spicis axillaribus pedunculatis secundis folio brevioribus, bracteis obovato-orbi-

culatis membranaceis, calycibus 5-fidis lacinia suprema multo breviore.—St. Thomas, *Don.*

Caules tenues, ramis floriferis suberectis vix semipedalibus. *Folia* pollicaria, membranacea, viridia, pilis minutis conspersa. *Pedunculi* filiformes, solitarii v. gemini, petiolo plerumque breviores. *Spicæ* circa semipollicem longæ. *Bracteæ* 1¼ lin. longæ, late obovatæ, obtusissimæ, integerrimæ v. eroso-denticulatæ, membranaceæ, pallide virentes et venulosæ, basi angustatæ et paullo induratæ, in spica oblique imbricatæ, alternæ florem solitarium foventes et superantes, alternæ paullo minores et steriles. *Bracteolæ* et calycis laciniæ subulatæ, basi parum latiores et hyalinæ, lacinia summa vix tertiam partem longitudinis æquans. *Corolla* et *stamina* omnino *Rostellariæ. Capsula* basi sterilis, a medio ad apicem tetrasperma.

This plant seems to connect the genera *Rostellaria* and *Anisostachya,* having the inflorescence of the latter, with the calyx of the former.

The *Rostellaria tenella,* N. ab E., is common to Senegal and Madagascar; *Leptostachya virens,* N. ab E., was gathered on the Gaboon Coast, by Captain Middleton, in 1787; *Schwabea ciliaris,* N. ab E., extends from Senegal to Nubia and Abyssinia.

1. Adhatoda (Amblyanthus) *paniculata,* Benth.; foliis amplis oblongo-ellipticis glabris v. subtus ad venas puberulis, cymis laxis oppositis in paniculam pyramidatam dispositis, bracteis calycisque laciniis herbaceis oblongo-linearibus, antherarum loculis connectivo lato sejunctis subparallelis, inferiore basi mucronato.—Fernando Po, *Vogel.*

Frutex ramosus, 4-5-pedalis. *Ramuli* uti petioli, costæ foliorum et inflorescentia pube minuta tomentelli; planta cæterum glabra. *Folia* semipedalia, acuminata, versus basin angustata, ima basi sæpius obtusa, v. emarginata, rarius in petiolum decurrentia, supra nitidula, petiolo 1-2-pollicari; floralia sub cymis minora, ovata, subsessilia, usque ad apicem paniculæ decrescentia; summa semipollicaria. *Cymæ* oppositæ, dichotomæ, patentes, floribus ad dichotomias et apices ramulorum sessilibus erectis. *Bracteolæ* inferiores calyce paullo lon-

giores, summæ breviores angustiores. *Calyces* 3 lin. longi,
laciniis apice rotundatis et brevissime mucronulatis inter se
subæqualibus. *Corolla* viridis fundo rubescente; tubus
(tubus cum faucibus N. ab E.) 2½ lin. longus, intus basi
brevissime glaber, dein antice gibbus, intus lineis 5 elevatis
pilosis notatus, labia inter se et tubo subæqualia, palato callis
2 brevibus elevatis glabris ad faucem conniventibus notato.
Stamina galeam subæquantia. *Ovarium* supra medium 4-
ovulatum, basi annulo carnoso cinctum. *Capsula* coriacea,
8-9 lin. longa, a basi ad medium sterilis, supra medium
2-4-sperma.

Of the five elevated, hairy, longitudinal lines in the tube of
the corolla, the upper one branches into the two middle nerves
of the upper lip of the corolla, the two lateral ones terminate in
the filaments of the stamens, the two lower converge to a point
where they meet the two callosities of the palate.

2. Adhatoda *Kotschyi*, N. ab E. *in DC. Prod.* 11. *p.* 397.—A
Senegalese and Nubian plant, of which there appears to be a
fragment in Don's collection, without any indication of the
station.

3. Adhatoda (Tyloglossa) *diffusa*, Benth.; caule elongato herbaceo diffuso ramoso piloso, foliis distantibus ovali-oblongis
pilosulis, floribus verticillato-subsenis sessilibus, bracteis obovatis orbiculatisve ciliatis, bracteolis exiguis, laciniis calycinis lineari-lanceolatis subciliatis, capsulis glabris calyce
duplo longioribus.—Fernando Po, *Vogel.*

Affinis *A. Rostellaria*, imprimis bracteis distinguitur. *Caulis*
pluripedalis, basi radicans, ramis floriferis adscendentibus,
pilis patentibus rigidulis; internodia inferiora valde elongata.
Folia petiolata, 1½-3 poll. longa, utrinque acutata, membranacea, pilis utrinque paucis conspersa, cujusve paris sæpius
inæqualia. *Bracteæ* petiolo multo breviores, membranaceæ,
margine longe ciliatæ, calyces sæpius excedentes. *Bracteolæ*
minutæ. *Calyx* 2 lin. longus, laciniis parum inæqualibus margine subscariosis et plus minus ciliatis. *Corolla* 6-7 lin.
longa (rosea?), extus apice pilosa; forma floris et genitalia
fere *A. Kotschyi.*

4. Adhatoda *plicata,* N. ab E. *in DC. Prod.* 11. *p.* 401.—On the Quorra and Nun Rivers, *Vogel;* Guinea.

5. Adhatoda *Anselliana,* N. ab E. *in DC. Prod.* 11. *p.* 403.— Cape Palmas, *Ansell, Vogel;* at Abòh, *Vogel.*

6. Adhatoda *tristis,* N. ab E. *in DC. Prod.* 11. *p.* 404.—Fernando Po, *Vogel.*

1. Eranthemum *hypocrateriforme,* Br.—*N. ab E. in DC. Prod.* 11. *p.* 454.—Accra, *Don;* Sierra Leone, *Vogel.* Folia 1½-3-pollicaria, in specimine Vogeliano subtus glauca, in Doniano subtus pallida.

2. Eranthemum *hispidum,* N. ab E. *in DC. Prod.* 11. *p.* 456. —Sierra Leone, *Don,* who says it is a small trailing plant, with yellow and white flowers. The specimens are about a foot long.

A third species, *E. elegans,* Br., is from Oware.

Senegal contains two species of *Dicliptera, D. umbellata,* N. ab E., and *D. maculata,* N. ab E., the latter of which is also found in Abyssinia; and one species of *Peristrophe,* the common East Indian and East African *P. bicalyculata,* N. ab E.

1. Hypoestes *rosea,* Pal. de Beauv. *Fl. Ow. et Ben.* 2. *p.* 69. *t.* 100. *f.* 1.—*N. ab E. in DC. Prod.* 11. *p.* 506, where the reference to Beauvois is accidentally omitted. St. Thomas, *Don;* Oware. Don's specimen is not in flower, and somewhat doubtful.

The other W. African species are *H. cancellata,* N. ab E., from Sierra Leone, and *H. latifolia,* Hochst., from Senegal and Nubia.

There remain, as *Acanthaceæ* of doubtful genera, *Justicia tunicata,* Vahl, from Sierra Leone, and a Sierra Leone plant in Don's collection, unlike any I am acquainted with, but so much pressed in drying as to preclude all examination.

LXXXII. VERBENACEÆ.

1. Stachytarpheta *Jamaicensis,* Vahl. — *Schauer, in DC. Prod.* 11. *p.* 564.—Cape Coast, *Vogel.*

I am unable to distinguish the common E. Indian *S. Indica*

from the American *S. Jamaicensis :* in both, the angles of the
stem are blunt and often obliterated with age, and the form of
the leaves is variable. This African plant has more of the aspect
of the generality of the American than of the Asiatic specimens.
Another species, *S. angustifolia,* Vahl, is found in Senegambia.

1. Lippia *nodiflora,* Rich.—*Schau. in DC. Prod.* 11. *p.* 585.—
On the Gambia, *Don ;* south of the Line, *Curror ;* and
throughout the Tropics and warmer regions of both hemispheres.

1. Lantana *antidotalis,* Schum. et Thonn.—*Schau. in DC. Prod.*
11. *p.* 598.—Cape Coast and Accra, *Vogel ;* Senegal and
Guinea.

The leaves are opposite, not ternately verticillate ; but in all
other respects these very bad specimens agree better with the
character of *L. antidotalis* than with that of the common
L. Camara, which they resemble. The bracteæ are linear, not
subulate, but the exterior ones are fully as long as the tube of
the corolla.

1. Premna *quadrifolia,* Schum. et Thonn.—*Schau. in DC.
Prod.*11. *p.* 633.—Variat foliis basi obtuse rotundatis v.
etiam acutiusculis.—Accra, *Don ;* in Vogel's collection without the precise station.

2. Premna (Premnos) *hispida,* Benth. ; ramulis cymis foliisque
piloso-hispidis, foliis obovatis oblongisve acuminatis basi angustatis, panicula cymosa terminali subsessili, calyce 5-dentato,
corollæ tubo calyce longiore.—Sierra Leone, *Don ;* Senegambia, *Heudelot.*

Pili longi ferruginei ad apices ramulorum costas foliorum inflo-
rescentiamque copiosi, in utraque pagina foliorum sparsi.
Folia ad apices ramorum conferta, 3-5-pollicaria, integerrima,
chartacea, glandulis paginæ inferioris crebris. *Cyma* foliis
multo brevior, multiflora. *Flores* extus hirti et glanduliferi.
Calyx vix semilineam longus, dentibus 5 mucronulatis v.
muticis parum inæqualibus. *Corollæ* tubus linea paullo longior, limbus subbilabiatus, labium superius integrum, laciniis

inferioribus majus, et certo æstivatione intimum. *Stamina* e tubo vix exserta. *Fructus* non vidi.

The *Volkameria aculeata*, Linn., is in some Senegalese collections, but only from gardens.

1. Clerodendron *volubile*, Beauv.—*Schauer in DC. Prod.* 11. *p.* 661.—C. multiflorum, *G. Don, in Edinb. Phil. Journ.* 1824. *p.* 350.—Sierra Leone, *Don;* Oware.

Don's specimen is in fruit only, and the panicle is rather larger and fuller than represented in Beauvois' figure, but it appears to belong to the same species.

2. Clerodendron *scandens*, Beauv.—*Schau. in DC. Prod.* 11. *p.* 662.—*Bot. Mag. t.* 4354.—Cape Palmas and Fernando Po, *Vogel;* Sierra Leone and Oware.—Ejusdem var. *pubescens.*—C. hirsutum et C. simplex, *G. Don, in Edinb. Phil. Journ.* 1824. *p.* 349.—Sierra Leone, *Don, Miss Turner;* Senegambia, *Heudelot.*

3. Clerodendron *splendens*, G. Don.—*Schau. in DC. Prod.* 11. *p.* 662.—C. aurantium, *G. Don, l. c.*—Sierra Leone, *Don.*

4. Clerodendron *sinuatum*, Hook.—*Schau. in DC. Prod.* 11. *p.* 665.—Sierra Leone, *Whitfield.*

5. Clerodendron *capitatum*, Schum. et Thonn.—*Hook. Bot. Mag. t.* 4355.—Cape Coast and Aguapim, *Vogel;* Sierra Leone, *Whitfield.*

1. Vitex (Chrysomallum) *chrysocarpa*, Planch. *in Herb. Hook. ;* foliolis 3-5 obovatis oblongisve supra glabris subtus ramulis petiolis inflorescentiaque ferrugineo-tomentosis, cymis pedunculatis axillaribus, calyce fructifero brevissime pedicellato late cyathiformi, drupa dense tomentosa.—On the Quorra, *Vogel.* *Tomentum* breve. *Foliolum* intermedium 2-4-pollicare, lateralia minora, petiolo communi 1½-2-pollicari. *Flores* desunt. *Fructus* pedunculus petiolo paullo brevior, calyx amplus in specimine fere destructus ; drupa magnitudine *Nucis Avellanæ* extus dense aureo-tomentosa, putamine lignoso 4-loculari.

2. Vitex sp.—Either a variety of *V. cuneata*, Schum. et Thonn., or a species closely allied to it, but the specimen insufficient

to determine.—Sierra Leone, *Don;* and apparently the same species from Senegambia, *Heudelot.*

3. Vitex sp.—A single leaf, brought by a native from the River Sann to Captain Trotter as the African Oak or Teak.

The wood, so well known in our Navy under the name of *African Oak* or *African Teak,* is a remarkable instance of a highly valuable and most extensively used timber, of which the tree that supplies it is wholly unknown to science. Botanical collectors have frequently made it the object of their researches and inquiries; but ,on the one hand, no botanist appears to have actually visited the forests which furnish it; and on the other, the natives who have brought leaves as from the trees, either by ignorance or carelessness, or more probably from ill-judged interested motives, have evidently in most cases deceived us. Thus we have heard that among various leaves brought to Mr. Brown, as found amongst the timber, the principal part appeared to be those of a *Laurinea.* The plant, brought home by the Earl of Derby's collector, and now in Kew Gardens, is too young to determine, but looks more like a *Sapotaceous* plant. The greater probability, however, is in favour of the Vitex-looking leaf, given to Captain Trotter: it is perfectly smooth, palmately compound, with six (probably seven, of which one is lost) folioles, of which the longest are above 5 inches long, and much narrowed at both ends. With every appearance of a *Vitex,* it is quite distinct from any described species.

Besides the above, there is a *Vitex ferruginea,* Schum. et Thonn., from Guinea, and a species apparently allied to it, though distinct, in Heudelot's Senegambian collection.

1. Avicennia *Africana,* Beauv.—*Schau. in DC. Prod.* 11. *p.* 669.—Grand Bassa and Cape Palmas, *Vogel;* Sierra Leone, *Don;* Senegal to Benin. Probably not distinct from the American *A. nitida,* Jacq.

LXXXIII. Labiatæ.

1. Ocymum *canum,* Sims.—*Benth. in DC. Prod.* 12. *p.* 32.

— Fernando Po, *Vogel;* Guinea, and all over Tropical Africa and Asia, and even in some parts of South America.

2. Ocymum *Basilicum,* Linn.—*Benth. in DC. Prod.* 12. *p.* 32.— Sierra Leone and St. Thomas, *Don;* frequently sent from Tropical and warm countries, but so generally cultivated that it is difficult to say whether the African specimens are indigenous or not.

3. Ocymum *viride,* Willd.—*Benth. in DC. Prod.* 12. *p.* 34.— Sierra Leone and Fernando Po, *Vogel;* St. Thomas, *Don;* Benin.

4. Ocymum *tereticaule,* Poir.—*Benth. in DC. Prod.* 12. *p.* 41. —Accra, *Don;* Senegal and Guinea.

5. Ocymum sp.—A single specimen, not in a state to be determined; Accra, *Don.*

The other West African species are *O. bracteosum,* Benth., from Senegal; and *O. membranaceum,* Benth., and *O. rigidum,* Benth., from Angola.

1. Platostoma *Africanum,* Beauv.—*Benth. in DC. Prod.* 12. *p.* 47.—Fernando Po, *Vogel;* Benin. Ejusdem var. glabrior. —Confluence of the Niger, *Vogel, Ansell;* Guinea and Congo.

1. Moschosma *polystachyum,* Benth. *in DC. Prod.* 12. *p.* 48.— Cape Coast, *Vogel;* common in Tropical Africa and East India.

The *M. dimidiatum,* which I formerly described as distinct from Thonning's specimen of his *Ocymum dimidiatum,* is probably a mere variety of *M. polystachyum.*

The *Orthosiphon glabratus,* Benth., an East Indian, Arabian and Madagascar plant, is also found in Guinea.

1. Hoslundia *opposita,* Vahl.—*Benth. in DC. Prod.* 12. *p.* 54. —Cape Coast and Aguapim, *Vogel;* Guinea.

The *H. verticillata,* Vahl, is from Senegal, and, apparently the same species was also found in Mozambique by Forbes.

1. Coleus? *Africanus,* Benth. *in DC. Prod.* 12. *p.* 74.—Plectranthus? Palisoti, *Benth. l. c. p.* 69.—Grand Bassa and on the Nun, *Vogel;* Sierra Leone and St. Thomas, *Don;*

Benin; Bahia in Brazil, where however it is probably introduced.

In luxuriant specimens the stems are two to three feet high, the leaves as much as four inches long and three broad, and the flowers numerous. All that I have seen are very badly dried, but as far as they go, I feel now convinced that the two species I had formerly distinguished were but forms of one, which varies in the number of flowers in the cymes, and in the lower lip of the calyx, entire or more or less toothed. I still, however, find no flowers in a state to decide whether it be really a *Coleus* or a *Plectranthus*.

1. Æolanthus *pubescens*, Benth. *in DC. Prod.* 12. *p.* 80.—On the Quorra, at Patteh, *Vogel.*

1. Hyptis *brevipes*, Poit.—*Benth. in DC. Prod.* 12. *p.* 107.— Fernando Po, *Vogel.*

2. Hyptis *atrorubens*, Poit.—*Benth. in DC. Prod.* 12. *p.* 108. —Sierra Leone, *Don.*

Both the above species, as well as *H. pectinata*, Poit., and *H. spicigera*, Lam.,* are American plants which have spread into Tropical Africa and Asia: the only *Hyptis* hitherto indicated as exclusively African is the *H. lanceæfolia*, Schum. et Thonn., from Guinea; but that again may be a mere variety of the common *H. brevipes.*

The *Leonurus Sibiricus*, Linn., and *Leucas Martinicensis*, Br., common Tropical plants, and both probably of Asiatic origin, notwithstanding the specific name of the latter, are also found in Senegal and Guinea.

1. Leonotis *nepetæfolia*, Br.—*Benth. in DC. Prod.* 12. *p.* 335. —Sierra Leone, *Vogel.*—A common Tropical species.

---

* I cannot subscribe to the statement in the Spicilegia Gorgonea, (supra, p. 157), that the floral leaves in this species are at first ovate and entire, and afterwards divided into three or four linear partitions, which I am said to have mistaken for bracts. The ovate and entire floral leaves are inserted on the main axis of the spike, and often fade and disappear as the spike advances in age. The linear bracts are inserted in the cymes themselves, at the basis of their branches, and may be seen at the very earliest stage of inflorescence.

2. Leonotis *pallida,* Benth. *in DC. Prod.* 12. *p.* 535.—Cape Palmas, *Vogel;* Senegal to Benin ; Nubia and Abyssinia.

LXXXIV. PLUMBAGINEÆ.

1. Plumbago *Zeylanica,* Linn.—*Boiss. in DC. Prod.* 12. *p.* 692.—Senegal to Benin and Fernando Po, *Vogel, Don* and others : a common species in Tropical Asia and Africa.

LXXXV. PHYTOLACCEÆ.

I find no specimens in the collections before me belonging to this Order, but the following seven species are recorded as having been found in West Tropical Africa: *Mohlana Guineensis,* Moq., from Guinea ; *Semonvillea pterocarpa,* Gay, Senegal ; *Limeum viscosum,* Fenzl, Senegal, Nubia and South Africa; *L. linifolium,* Fenzl, Senegal and South Africa; *Gisekia linearifolia,* Schum. et Thonn., Guinea ; *G. pharnaceoides,* Linn., Senegal and Guinea, and over nearly all Africa and East India, and *G. congesta,* Moq., from Senegal.

LXXXVI. CHENOPODIEÆ.*

1. Chenopodium *album,* Linn.—*Moq. in DC. Prod.* 2. 13. *p.* 70. —St. Thomas, *Don.*
This is stated by Don to be a shrub 10 feet high ; but it is a very common mistake among collectors to describe hard-stemmed, tall-growing annuals as shrubs; and in every other respect the specimens agree perfectly with the universally diffused *C. album.*
2. Chenopodium *ambrosioides,* Linn.—*Moq. in DC. Prod.* 13. 2. *p.* 70.—St. Thomas, *Don.*
This is another weed diffused over a great portion of the globe, as well as *C. murale,* Linn., which is also found in Senegal.

* We know of no adequate reason for changing the name of this Order to that of *Salsolaceæ,* as now first proposed by Moquin.

The remaining *Chenopodieæ*, quoted as West Tropical African, are, *Arthrocnemum fruticosum*, Moq., from Senegal, a European and African plant, which has found its way to Timor and California; *A. Indicum*, Moq., from Senegal, an Egyptian and East Indian species; and *Suæda fruticosa*, Forsk., also from Senegal, a South European and North African plant, seen occasionally in the American colonies.

LXXXVII. AMARANTACEÆ.

1. Celosia (Lestibudesia) *leptostachya*, Benth.; caule herbaceo (diffuso ?) glabro, foliis petiolatis ovatis acutiusculis glabris, spicis elongatis gracilibus interruptis, floribus per 3-5 glomeratis trigynis, sepalis vix acutiusculis uninervibus, utriculis ovoideo-globosis.—Fernando Po, *Vogel*.
Caules parum ramosi, tenues, 2-3-pedales, internodiis elongatis. *Folia* 1-2-pollicaria, nunc acute acuminata, nunc fere obtusa, basi angustata rotundata v. etiam subcordata, tenuiter membranacea. *Spicæ* demum semipedales et longiores, glomerulis inferioribus valde remotis. *Flores* parvi, siccitate pallide brunnei. *Stylus* apice in ramos stigmatiferos sæpius (an constanter ?) 3-divisus. *Utriculus* viridis, calyce dimidio longior. *Semina* pauca (8-10).

2. Celosia *trigyna*, Linn.—*Moq. in DC. Prod.* 13. 2. *p.* 241. var. *parviflora*, Fernando Po, *Vogel;*—var. *fasciculiflora*, Moq., Accra, *Don, Ansell;* on the Quorra, *Vogel, Ansell.*

3. Celosia *laxa*, Schum. et Thonn.—*Moq. in DC. Prod.* 13. 2. *p.* 241.—Accra, Grand Bassa, and Fernando Po, *Vogel;* Senegambia, *Heudelot.*

These specimens agree well with Thonning's description, and some of them are from the locality where he says the *C. laxa* is common. They differ chiefly from *C. trigyna*, in the size of the flowers, the calyx being nearly two lines long, and about twice the length of the bracts. The Senegambian specimen is, however, marked by Moquin, *C. trigyna*, var. *densiflora*.

4. Celosia *argentea*, Linn.—*Moq. in DC. Prod.* 13. 2. *p.* 242.

—Accra, and on the Quorra, *Vogel;* St. Thomas, *Don;* East
Africa and East India.

1. Amarantus *paniculatus,* Linn.—*Moq. in DC. Prod.* 13. 2.
p. 257.—On the Quorra, *Vogel.*—A common East Indian
species, to which belongs the variety cultivated there for the
grain, originally named by Roxburgh *A. farinaceus,* though
afterwards published by him as *A. frumentaceus.*

2. Amarantus *spinosus,* Linn.—*Moq. in DC. Prod.* 13. 2.
p. 260.—Sierra Leone and Fernando Po, about dwellings
and cultivated places; Senegal and Tropical and warmer
regions of both hemispheres.

Another common Tropical weed, *Amblogyne polygonoides,*
Raf., is also found in Senegal.

1. Euxolus *polygamus,* Moq. *in DC. Prod.* 13. 2. *p.* 272.—
Accra, *Vogel;* Senegal, Eastern Africa and East India.

2. Euxolus *viridis,* Moq. *in DC. Prod.* 13. 2. *p.* 273.—On the
Nun, *Vogel;* common over a great portion of the globe.

The *Euxolus caudatus,* Moq., frequent throughout the
Tropics, is found in Senegal and Sierra Leone; and two other
common African and East Indian plants, *Ærua Javanica,* Juss.,
and *Æ. brachiata,* Mart., are also in Senegal.

1. Achyranthes *involucrata,* Moq. *in DC. Prod.* 13. 2. *p.* 310.—
On the Quorra, at Stirling and Pandiaki, *Vogel, Ansell.*

2. Achyranthes (Pandiaka) *angustifolia,* Benth.; caule herbaceo elongato subramoso appresse pubescente, foliis lineari-
sublanceolatis strigoso-pubescentibus viridibus subtus pallidis,
capitulis ovatis oblongisve obtusis, floribus albidis, sepalis
bracteas laterales subæquantibus enervibus dorso dense pilosis.—On the Quorra, *Vogel.*

Pluribus notis cum descriptione *A. Heudelotii* convenit. *Herba*
videtur annua, ejusdem staturæ. *Folia* longiora et angustiora. *Capitula* ejusdem forma et magnitudine, intra folia
2-2½-pollicaria sessilia. *Bracteæ* albæ, nitidulæ, aristato-
mucronatæ, infima glabra, laterales ad costam dorsalem longe
ciliatæ. *Sepala* in mucronem rigidum producta, quam arista
bractearum breviora (bracteam ipsam arista neglecta supe-

rantia), dorso dense obtecta pilis longis, cartilaginea, albida
v. subvirentia. *Filamenta* haud ciliata, staminodiis quam
filamenta multo brevioribus, plano-depressis, quadratis, ad
angulos in appendiculos breves obtusas productis. *Antheræ*
oblongæ.

The *A. Heudelotii*, Moq., from Senegal, and *A. nodosa*, Vahl,
gathered by Isert at Whydah, in Guinea, belong to the same
section. Of the section *Cadelari* there are two species in
Senegal, the *A. aspera*, Linn., a common African and East
Indian plant, and *A. argentea*, Lam., a North African and
South European species. The *Centrostachys aquatica*, Wall.,
an East Indian plant, is also in Senegambia.

1. Cyathula *prostrata*, Bl.—*Moq. in DC. Prod.* 13. 2. *p.* 326.
—Achyranthes Thonningii, *Schum. Beskr. p.* 139.—Sierra
Leone to Fernando Po, *Vogel, Don* and others. A common
Tropical African and East Indian plant, found also in Tro-
pical America.

I am disposed to agree with Vogel in considering this as the *A.
Thonningii* of Schumacher, although Thonning omits all mention
of the staminodia, on which account Moquin enumerates his plant
among doubtful *Pupaliæ.* In all other respects our specimens,
which undoubtedly belong to the common *C. prostrata,* agree
perfectly with Thonning's description, in which, moreover, the
stamens are not expressly said to be without staminodia.

In the generic characters assigned to this and allied genera,
there appears to be a slight inaccuracy in the expression " Flores
subternati, intermedius fertilis, laterales steriles demum in
aristas uncinatas (glochides) mutati." Whereas, in *Cyathula
prostrata,* the total number of glochides is usually from 14 to
20 on each side of the fertile flower, the number of flowers,
perfect or rudimentary, forming the fascicle, is at least seven ;
the central one is perfect and fertile ; the next in order, one on
each side, are sometimes complete in their parts, although
sterile, sometimes more or less reduced, occasionally with the
sexual organs rudimentary or obsolete, and the sepals reduced
to two or three *glochides ;* the remaining four flowers, one on
each side of each of these lateral flowers, are without sexual

organs, with all their sepals, as well as the bract, reduced to *glochides*. In the luxuriant West African specimens, most of the fascicles consist of one perfect fertile flower, two perfect sterile ones, and four reduced to *glochides*, and this is the state described by Thonning. In the East Indian specimens, all the lateral flowers are usually reduced to glochides, with the exception of two or three sepals, and occasionally a small ovary and andrœcium in the centre. But I have observed all these states in different parts of one spike. The South African *Cyathula cylindrica* affords a very good example of many-flowered fascicles, with a number of sterile flowers in different degrees of abortion.

2. Cyathula *geminata*, Moq. *in DC. Prod.* 13. 2. *p.* 330?— Fernando Po, *Vogel.*

Near *C. prostrata*, but the spikes are shorter and more dense, and the flowers nearly as large as in *Achyranthes aspera*, the sepals being from 1½ to 1¾ lines long. The fascicles usually consist of one perfect and fertile flower, a sterile one (on one side of it), more or less complete in its parts, with a rudimentary one on each side, reduced each to two or three glochides; on the other side of the fertile flower are usually two or three glochides only. The staminodia are rather more conspicuous than in *C. prostrata*. The lateral spikes are occasionally unequal, one being nearly sessile, the other on a longish stalk, as described by Thonning; but more frequently they are equal, and both nearly sessile. In other respects his description agrees.

1. Pupalia *lappacea*, Moq. *in DC. Prod.* 13. 2. *p.* 331.—Cape Coast, *Vogel;* Senegal, Nubia, Abyssinia, and East India.

The *P. atropurpurea*, Moq., extends likewise from Senegal and Guinea over East Africa and East India.

1. Iresine (Philoxerus) *vermicularis*, Moq. *in DC. Prod.* 13. 2. *p.* 339.—Sands of the sea-shore, from Senegal to Benin, *Vogel, Don* and others; also in Tropical America.

The *I. aggregata*, Moq., from the same localities, does not appear to have any character to distinguish it, the breadth and thickness of the leaves being very variable.

1. Alternanthera *nodiflora*, Br.—Moq. *in DC. Prod.* 13. 2.

p. 356.—Cape Coast, *Vogel;* Senegal, Nubia, Abyssinia and East India.

2. Alternanthera *sessilis,* Br. — Moq. *in DC. Prod.* 13. 2. *p.* 357.—Cape Palmas, Abòh and Fernando Po, *Vogel;* Sierra Leone, *Don;* over a great portion of Africa, Asia, America and Australia.

The *A. denticulata,* Br., from the same localities, is probably a mere variety of *A. sessilis.* The *Illecebrum obliquum,* Schum. et Thonn., from Guinea, would appear from Thonning's description to be the *Alternanthera Achyrantha,* an American species, which has spread to the Canary Islands and to some parts of Europe.

1. Telanthera *maritima,* Moq. *in DC. Prod.* 13. 2. *p.* 364.— On the sands of the sea-coast, from Senegal to Benin, *Vogel, Don* and others ; eastern sea-coast of Tropical America.

LXXXVIII. Nyctagineæ.

1. Boërhaavia *ascendens,* Willd.—*Chois. in DC. Prod.* 13. 2. *p.* 451.—On the Quorra, *Vogel* (both the smooth and the hirsute forms) ; St. Thomas, *Don ;* an African species.

2. Boërhaavia *paniculata,* Rich.—*Chois in DC. Prod.* 13. 2. *p.* 450.—On the Quorra, at Stirling, *Vogel, Ansell;* a species chiefly from Tropical America.

This genus, so unattractive to the botanist in a dry state, notwithstanding the elegance of some of the species when fresh, had been in a lamentable state of confusion, till recently worked up with great care by Choisy, in the last part of the Prodromus. His characters derived from the fruit are excellent, although some of the species may not be so strictly geographical as he supposes. Our specimens of *B. paniculata* agree precisely with the generality of what we have seen from America, although some of those included under the name by Choisy (as, for instance, Gardner's n. 2292, from Brazil), have the fruit not rounded at the extremity, but truncate, with the ends of the ribs slightly prominent.

The *Boërhaavia verticillata,* Poir. and *B. dichotoma,* Vahl,

both of them Nubian and Abyssinian species, are also quoted
from Senegal. Vogel has also in his collection a specimen of
Mirabilis Jalapa, Linn., from Sierra Leone, but probably from
some garden there.

LXXXIX. POLYGONACEÆ.*

1. Polygonum *Senegalense,* Meisn. *Monogr. Polyg. p.* 54.—
St. Thomas, *Don;* Senegal.

This species has, at first sight, much the appearance of the
Asiatic *P. glabrum* and the American *P. acuminatum,* but the
cotyledons are certainly incumbent, not accumbent, besides
some slight differences in other points. The leaves are remarkable for their very long points. The ochreæ have occasionally a
few very small ciliæ on their edge.

The only other West Tropical African Polygonaceous plant
known, is the *Polygonum exiguum,* Meisn., from Senegal.

XC. THYMELEÆ.

DICRANOLEPIS, *Planch.* (gen. nov.)

Flores hermaphroditi. *Perianthium* hypocrateriforme, tubo
longo, limbo 5-partito, laciniis oblongis, æstivatione imbricatis. *Squamæ* 5, petaliformes, fauci insertæ, laciniis perianthii oppositæ, profunde bifidæ. *Stamina* 10, fauci inserta,
5 longiora, squamis opposita; filamenta brevia, filiformia;
antheræ lineari-oblongæ, basifixæ, loculis connectivo adnatis,
rima introrsa dehiscentibus. *Discus* cupuliformi-tubulosus
ovarii stipitem includens, apice leviter 5-lobus. *Ovarium*

---

* In this and several of the following Orders, which the Prodromus
has not yet reached, and which have not been the subject of any recent
monograph, the geographical indications are necessarily very imperfect,
and I have been obliged to leave many more species either doubtful or
in genera to which they may not properly belong; neither time nor
space admitting of the monographical labour which has been bestowed
on the preceding Orders by the authors of the corresponding portions of
the Prodromus.

brevi-stipitatum, hinc gibbosum, uniloculare, ovulo unico ex apice loculi pendulo. *Stylus* filiformis, in stigma lineari-clavatum desinens. *Drupa ?* exsucca, brevistipitata, meso-carpio e filamentis nitentibus contexto. *Semen* suspensum, globosum, anatropum, integumento membranaceo; embryonis recti cotyledones hemisphæricæ carnosæ, radicula semi-exserta minutissima. (*Planchon*).

1. Dicranolepis *disticha*, Planch. *in Hook. Ic. t.* 798. (TAB. XLVIII).—Sierra Leone, *Don.*

Frutex ? ramulis tenuibus virgatis foliisque distichis, gemmis ramulis novellis petiolis perianthiisque extus pilis adpressis v. patentibus subsericeis v. hispidulis. *Folia* crebra, alterna, oblique subtrapezoidea-lanceolata, cuspidata, integerrima, brevissime petiolata, 1-1½-pollicaria, rigide membranacea, nervis lateralibus tenuibus sat crebris, glabra, nitida, supra in sicco læte viridia, subtus viridi-flavescentia. *Stipulæ* nullæ. *Flores* axillares, solitarii, subsessiles, folio non multo breviores. *Fructus* mole seminis *Coryli Avellanæ*, præter styli basin persistentem mamilliformem pilosulum glaberrimus. (*Plan-chon*).

PLATE XLVIII. *Fig.* 1. flower; *f.* 2. stamen; *f.* 3. pistillum with the disk sheathing its base; *f.* 4. ovary and sheath, vertical section; *f.* 5. fruit; *f.* 6. seed:—*all, but the fruit, more or less magnified.*

XCI. LAURINEÆ.

1. Cassyta *Guineensis*, Schum. et Thonn. *Beskr. p.* 199.— Sierra Leone, *Don;* Grand Bassa, on the Nun and the Quorra, *Vogel;* Guinea.

This species has been accidentally overlooked by Nees in his elaborate monograph. It appears scarcely distinguishable either from the common Brazilian and Guiana species, well described by Nees as *C. Brasiliensis*, Mart., or from the South African specimens distributed by Drège as *C. pubescens*, Br.; but it does not quite agree with Brown's short character of his South-east Australian *C. pubescens*, nor yet with Schlechten-

K K

dahl's *C. pubescens,* from South-west Australia. Brown himself, however, observes that the Congo species can scarcely be
distinguished either from the West Indian one or from his own
C. pubescens. The stems of the West African plant are sometimes thickly pubescent, sometimes nearly smooth, the flowers
usually distinct and rather distant, occupying the upper half of
the peduncle, and agreeing in structure with those of the Brazilian plant.

No species of true *Laurineæ* has been hitherto recorded from
West Tropical Africa; nor have I seen any specimen from
thence, excepting one in leaf only of the *Cinnamon* (*Cinnamo-
mum Zeylanicum*), in Don's collection, from St. Thomas. But
this is evidently a cultivated plant, as is also a specimen, in
leaf only, in the same collection, of *Myristica sebifera,* Sw.,
no species being known from this region of the Order of
Myristiceæ.

XCII. EUPHORBIACEÆ.

1. Euphorbia *prostrata,* Ait.—*Willd. Spec. 2. p.* 895.—Sierra
Leone, *Don, Vogel;* Grand Bassa and Fernando Po, *Vogel;*
also West Indies and South America.

This may be the Guinea plant referred to *E. Chamæsyce* by
Schumacher and Thonning, and is certainly very near that
species. The leaves are, however, more oblique, the flowers
very much smaller, and usually two or more together in each
axilla, although often, in reality, solitary in the axillæ of very
much reduced floral leaves, on axillary flowering branches
much shorter than the subtending leaf. The capsule is much
smaller than in *E. Chamæsyce,* always ciliate on the dorsal ribs,
and generally without hairs on the sides of the carpels. The
African specimens precisely correspond with the South American
ones.

The *E. scordifolia,* Jacq., to which Planchon refers *E. tomentosa,* Poit., extends from Senegal to Nubia and Arabia.

2. Euphorbia *trinervia,* Schum. et Thonn. *Beskr. p.* 253.—
E. glaucophylla, *Sieb. Pl. Seneg. Exs., non Pers.*—Common

on the sandy shores from Senegal to Benin, *Don, Vogel* and others ; and extending south of the Line, *Curror.*

The leaves are more frequently blunt than pointed : in every other respect, Thonning's description is very accurate.

3. Euphorbia (Anisophyllum) *convolvuloides,* Hochst. *in Kotsch. Pl. Nub. Exs. n.* 242 ; herbacea, stipulata, prostrata, ramis villosis, foliis oppositis sessilibus ovatis oblongisve basi valde obliquis crassiusculis supra viridibus parce pilosis subtus appresse tomentosis, floribus in capitula axillaria sessilia nunc in ramulos breves foliatos abeuntia confertis, involucri dentibus exterioribus orbiculato-reniformibus parvis, capsula dense tomentosa, seminibus subtetragonis transverse rugosis.
—On the Quorra, at Attah, *Vogel ;* at the Confluence, *Ansell.*

Habitu *E. piluliferæ* accedit, foliorum consistentia et indumento facile distincta. *Folia* ½-1 poll. longa, 3-4 lin. lata, margine interdum obscure denticulata, basi valde inæqualia et oblique truncata v. subcordata ; pagina superior oculo nudo glabra videtur, sub lente pili appressi sparsi apparent, pagina inferior tomento arcte appresso canescit v. rubescit. *Flores* nunc in capitulum densum subaphyllum dispositi, nunc ad axillas foliorum plus minus evolutorum sessiles, spicas formant unilaterales foliatas ad axillam folii majoris. *Involucra* semilineam longa, dense tomentosa, dentibus exterioribus nunc fere obsoletis v. quam glandulæ interiores brevioribus, nunc eas duplo triplove superantibus subpetaloideis albis v. rubentibus. *Capsula* quam in *E. Chamæsyce* paullo major, obtusangula, tomento albido v. rubescente dense vestita.

I am not aware that Hochstetter's name has been otherwise published than on Kotschy's labels. In order, however, to avoid confusion, I adopt it ; although the fancied resemblance to a *Convolvulus* does not strike me. In Ansell's specimens the flowers are all capitate ; in Kotschy's they are arranged in leafy, axillary spikes ; in Vogel's, both inflorescences may be seen on the same stem.

4. Euphorbia *pilulifera,* Linn.—E. purpurascens, *Schum. et*

500 FLORA NIGRITIANA.

Thonn. Beskr. p. 253.—Senegal to Benin, and over most Tropical regions.

5. Euphorbia *hypericifolia,* Linn.—E. glaucophylla, *Poir. Dict. Suppl.* 2. *p.* 613?—Cape Palmas, *Ansell ;* St. Thomas, *Don ;* a species nearly as common as the last, in Tropical countries.

Two other species of *Euphorbia* from Guinea, *E. lateriflora* and *E. drupifera,* are published by Schumacher and Thonning; and two are quoted from Sierra Leone, *E. toxicaria,* Afz., and *E. grandifolia,* Haw. One or two of the above may be that alluded to by Brown, as being frequently planted over graves by the natives.

The *Anthostema,* of Jussieu, is only known from Senegal; unless it be the same as the new Congo genus alluded to by Brown, as explaining the structure of *Euphorbia.*

1. Dalechampia *ipomœæfolia,* Benth. ; foliis cordatis integris trilobisve membranaceis subtus ad costas petiolis caulibusque pilosulis, stipulis lineari-lanceolatis, involucri foliolis integris acuminatis subglabris, stylis apice clavatis.—On the Quorra, *Vogel.*

Caules scandentes, striatuli, pilis brevibus mollibus plus minus vestiti. *Stipulæ* angustæ, fere semipollicares. *Petioli* 1-2-pollicares, præsertim apice hirsuti. *Folia* 2½-3½ poll. longa, 2-2½ poll. lata, apice acuminata, margine irregulariter sinuato-dentata, basi auriculis rotundatis cordata, præter venas fere glabra, ad petiolum 5-nervia et in pagina superiore glandulis 2 linearibus munita, pleraque indivisa, nonnulla irregulariter triloba. *Pedunculi* graciles, folio paullo breviores. *Involucri* foliola membranacea, non colorata, late cordato-ovata, venis vix prominulis, basi bistipulata. *Flores* in capitulo unico juniore vix aperti nec rite examinare potui ; stamina tamen et styli omnino *Dalechampiæ* vidi.

In the form of the leaves this plant approaches the *D. heterophylla,* Poir., from Guiana ; but independently of some differences in the flowers, which I was unable very accurately to ascertain, the want of the thick, whitish down on the leaves and

stems, and of the remarkably prominent ribs on the involucre of
D. heterophylla, will at once distinguish our plant.
The *D. Senegalensis*, A. Juss., (supra p. 174) is probably,
from its name, a Senegalese as well as a Cape Verd plant.
1. Stillingia *Guineensis*, Benth.; foliis ovali-oblongis ellipticisve
acuminatis integerrimis paucidentatisque basi rotundatis an-
gustatisve, spicis subsimplicibus, floribus masculis plerisque
solitariis trifidis triandris.—Sierra Leone, *Don.*
Frutex videtur glaberrimus, ramulis novellis angulato-compres-
sis demum teretibus. *Folia* breviter petiolata, forma et magnitudine varia, nunc bipollicaria basi rotundata, nunc semipedalia, basi longe angustata, pleraque dentibus paucis obtusis præsertim apicem versus notata, basi supra petiolum 1-2-
glandulosa. *Stipulæ* minutæ v. obsoletæ. *Gemmæ* axillares
v. supra-axillares constantes e squamellis pluribus, secus
caulem linea verticali suprapositis, acuminatis et spiraliter
tortis. *Spicæ* terminales, solitariæ v. geminæ, simplices
(v. basi ramosæ?), 1-2-pollicares. *Bracteæ* ovatæ, acuminatæ,
denticulatæ, basi utrinque glandula peltata stipulatæ, inferiores paucæ fœmineæ 1-1½ lin. longæ, superiores plurimæ
masculæ ½ lin. longæ. *Flores* intra bracteas solitarii (v.
mares rarissime gemini?), pedicello brevi apice incrassato,
calyce fere tripartito, laciniis ovatis acuminatis, æstivatione
leviter imbricatis. *Glandulæ* intra flores nullæ. *Stamina*
fl. masc. 3 v. rarissime 2, filamentis calyce duplo longioribus
ima basi subconnatis. *Ovarium* floris fœminei mari majoris
sessile, glabrum, calycem æquans. *Styli* 3, longiusculi, revo-
luti, basi incrassati, et facie interiore undique stigmatosi.
Fructus nonnisi reliquios vidi, quorum axis 4 lin. longa est,
stipite 1 lin. longo fulta.
1. Microstachys *chamelæa*, A. Juss.—Tragia chamelæa, *Linn.*
—Accra, and on the Quorra at Attah, *Vogel;* at the Confluence, *Ansell;* a common East Indian plant.
1. Tragia *cordifolia*, Vahl, *Symb. 1. p.* 67 ?—Cape Coast and
Fernando Po, *Vogel;* an Egyptian species.
Caulis volubilis. *Folia* longe petiolata, sinu late cordata, acuminata, serrata, 2-3-pollicaria, membranacea, supra minute

strigillosa, subtus pallida, ad venas uti petioli pedunculi et ramuli juniores pilis (ex Vog. valde urentibus) hispida. *Stipulæ* parvæ. *Racemi* axillares v. terminales, 1-1½-pollicares. *Bracteæ* lineares, pedicellos masculos æquantes v. superantes. *Flores* superiores *masculi* parvi, solitarii v. gemini, pedicello ¾ lin. longo basi bracteolato fulti. *Calyx* tripartitus, æstivatione valvata. *Stamina* 3. *Flores fœminei* in parte inferiore racemi subsessiles. *Calycis* laciniæ 6 (an interdum 5 ?), 2-3 lin. longæ, profunde pinnatifidæ, sub fructu stellato-patentes, uti capsulæ setis urentibus horridæ. *Stylus* apice trifidus.

2. Tragia *tenuifolia*, Benth. ; volubilis, foliis petiolatis cordato-oblongis acuminatis membranaceis, supra sparse subtus densius piloso-pubescentibus, bracteis lato-ovatis, floris fœminei calyce 6-partito laciniis cuneatis hirsutissimis integris subdentatisque.—St. Thomas, *Don.*

Habitus *T. cordifoliæ. Folia* tenuiora, sinu angusto cordata, pilis utriusque paginæ crebrioribus sed multo minus rigidis, pagina inferior pallida interdum fere tomentosa. *Racemi* breves, longiuscule pedunculati. *Floris masculi* bracteæ concavæ, acuminatæ, ½ lin. longæ ; calycis laciniæ crassiusculæ. *Floris fœminei* bractea lata et sæpe 2-3-loba, bracteolæ 2 ovato-oblongæ ; calycis laciniæ sub fructu fere 2 lin. longæ, omnes hispidæ, minus tamen quam in *T. cordifolia*, 3 paullo majores sæpe hinc inde dentatæ, omnes basi in stipitem brevissimum contractæ. *Capsula* hispida.

3. Tragia *sphathulata*, Benth. ; volubilis, foliis petiolatis cordatis acuminatis membranaceis supra pilosis subtus pubescentibus, bracteis oblongis linearibusve, floris fœminei calyce 6-partito laciniis integris stipulatis late spathulatis hirsutis.—Cape Coast, *Vogel.*

Affinis *T. tenuifoliæ. Folia* latiora videntur, bracteæ angustiores longiores, et species facillime distinguitur laciniis calycinis floris fœminei quæ basi abrupte contractæ sunt in stipitem distinctam sub fructu lineam longam, dum lamina ipsa 2 lin. longa est et lata.

4. Tragia *angustifolia*, Benth. ; volubilis, foliis breviter pctio-

latis longe cordato-lanceolatis puberulis v. subtus ad venas hispidis, bracteis ovato-lanceolatis, floris fœminei calyce tripartito, laciniis latis profunde palmato-pinnatifidis.—On the Quorra at Addaenda, *Vogel.*

Indumentum fere *T. cordifoliæ,* sed setæ pauciores. *Petioli* raro pollicem longi, sæpius multo breviores. *Folia* pleraque 3-4 poll. longa, ½-1 poll. lata, nonnulla fere semipedalia, consistentia præcedentibus firmiora. *Racemi* breviter pedunculati, densiflori. *Bracteæ* serrato-ciliolatæ, fœmineæ more affinium ternæ, exteriore latiore. *Calycis* laciniæ constanter 3, sessiles, fere ad basin divisæ in lacinias 11-15 lineares quasi digitatim dispositas, per anthesin 1½ lin. longæ et latæ, sub fructu hispido stellatim patentes, 3-4 lin. longæ et fere 5 lin. latæ, laciniis dense setosis.

The *T. pedunculata,* Pal. Beauv. Fl. Ow. et Ben. t. 54, if correctly represented, is different from all the above in inflorescence. The *T. monadelpha,* Schum. et Thonn., from Guinea, would appear from the description, as suggested by Schumacher, not to be a congener.

MICROCOCCA. (gen. nov.) e tribu *Acalyphearum.*

Flores monoici. *Masculi :* *Calyx* tripartitus, laciniis æstivatione valvatis. *Stamina* 6, filamentis liberis receptaculo insertis cum pilorum fasciculis (seu squamellis plumosis?) intermixtis ; antherarum loculis ovoideo-globosis erectis discretis. *Fl. fœminei :* *Calyx* maris (semel vidi abnorme 4-partitum). *Squamæ* 3, lineares, e receptaculo ortæ, ovario appressæ. *Ovarium* subtrilobum, triloculare, ovulis in loculis solitariis lateraliter affixis. *Stigmata* 3, sessilia, plumoso-ramosa. *Fructus* (sæpius hispidus) tricoccus, coccis bivalvibus.— *Herba* annua, Indica v. Africana, foliis alternis, stipulis inconspicuis. *Flores* parvi in racemos filiformes dispositi ad axillas bractearum fasciculati ; fœminei intra quemquam bracteam solitarii, longiuscule pedicellati, masculi ex eadem bractea pauci pedicellis brevissimis.

1. *M. mercurialis,* Benth.—Tragia mercurialis, *Linn.*—On the

Quorra, at Patteh and Stirling, *Vogel;* common in East India.

The African specimens have not the young shoots and capsules so hispid as the East Indian, but they otherwise do not differ. Adrien de Jussieu, Euph. Tent. p. 46, suggested the removal of this plant from *Tragia,* and its approximation to *Acalypha,* of which it has the styles, but not the stamens. It appears to me necessary to consider it as a distinct genus, allied on the one hand to *Acalypha,* and on the other, and perhaps more closely, to *Mercurialis* and *Adenocline.*

An alternate-leaved, tricoccous species of *Mercurialis,* from Senegal, is referred to by A. de Jussieu, but not having seen the specimen, I cannot tell whether it belongs to that genus as now modified, or to *Adenocline,* of Turczaninow.

1. Acalypha *Indica,* Linn.—On the Quorra, at the Confluence, *Ansell;* Nubia, Abyssinia, and all over East India.

These specimens approach nearly that state of the plant to which Hochstetter has given the name of *A. abortiva,* on Kotschy's Nubian tickets. The male part of the spike is very short, and often bears one or more ebracteate female flowers, which are usually reduced to a single carpel, and produce small, globose, echinate, monospermous fruits.

2. Acalypha ciliata, Forsk., *Fl. Æg. Arab. p.* 162.—*Vahl, Symb.* 1. *p.* 77. *t.* 20.—*Wight in Ann. Nat. Hist.* 2. *t.* 5.— A. fimbriata, *Schum. et Thonn. Beskr. p.* 409.—*Hochst. in Kotsch. Pl. Nub. exs.*—On the Quorra, at Patteh, and on the Nun, *Vogel;* St. Thomas, *Don;* Nubia, Arabia, and East India.

3. Acalypha *Leonensis,* Benth.; fruticosa, foliis oblongis breviter et obtuse acuminatis subdentatis basi longe angustatis glabris v. subtus ad costas puberulis, spicis axillaribus gracilibus androgynis, bracteis parvis integris inferioribus masculis superioribus androgynis v. summis fœmineis.—Sierra Leone, *Vogel, Don;* Senegambia, *Heudelot.*

Ramuli novelli angulati, minute tomentelli, demum subteretes, glabrati. *Folia* in summitatibus conferta, a 5-6 poll. usque ad 9-10 poll. longa, 2-3 poll. lata, acumine lato sæpe emar-

ginato, dentibus paucis obtusis nonnunquam obscure glandulosis v. minute penicilliferis, rigide chartacea, basi supra glandulis paucis vix elevatis instructa, petiolo ½-1-pollicari sustensa. *Spicæ* numerosæ, in axillis superioribus geminæ
ternæ v. interdum plurimæ, foliis paullo breviores, canescenti-
tomentellæ, interrupte multifloræ. *Bracteæ* vix semilineam
longæ, latæ, concavæ, tomentosæ. *Flores* ad bracteas inferiores omnes masculi, 6-10 v. numerosiores ; in media spica
bracteæ florem fœmineum unicum fovent cum masculis paucis,
in summa spica flores fœminei sæpe solitarii sunt. *Calyx*
marium 4-partitus, staminibus circa 20, antheris omnino
generis, loculis ad apices connectivi ramorum disjunctis.
Flos fœmineus ante anthesin bractearum paribus duobus involutus ; calyx profunde trifidus, lobis ovario brevioribus obtusis
v. retusis ciliatis. *Ovarium* superne hispidum, stylis omnino
generis.

This is, if I rightly understand an abbreviation on Vogel's
label, a climbing shrub. It is very different in appearance from
the more common *Acalyphæ;* but, amidst all the variations of
habit and inflorescence, the genus is one of the most distinctly
characterized by the anthers and the styles.

4. Acalypha *micrantha*, Benth.; fruticosa, foliis obovali-oblongis vix obtuse acuminatis integerrimis v. rarissime dentatis
glabris, spicis axillaribus gracillimis androgynis, bracteis
minimis integris, inferioribus masculis superioribus androgynis.—Sierra Leone and Fernando Po, *Vogel.*

Very near to the preceding, *A. Leonensis,* but smoother, the
leaves shorter, broader, whitish and almost shining on the under
side, the spikes more slender, and the bracts and flowers still
smaller.

The *A. dentata,* Schum. et Thonn., a shrubby species from
Guinea, appears, from the description, to be different from either
of the above.

ERYTHROCOCCA (gen. nov.) e tribu *Acalyphearum.*

Flores dioici. *Masculi* in axillis fasciculato-racemosi. *Calyx*
profunde trifidus, æstivatione valvata. *Petala* nulla. *Stamina* 6, filamentis basi in annulum connatis brevibus, antheris erectis, loculis contiguis subdistinctis. *Flores fœminei*
in axillis fasciculati vix racemosi. *Calyx* maris. *Ovarium*
sessile, biloculare, loculis uniovulatis, stylis 2 recurvis a basi
plumoso-ramosis. *Fructus* (drupaceus ?) abortu unilocularis,
monospermus, globosus.

1. Erythrococca *aculeata*, Benth.—Adelia anomala, *Poir. Dict.*
Suppl. 1. *p.* 132.—" *Claoxylo* affinior quam *Adeliæ*," *A.*
Juss. Euph. Tent. p. 32.—Sierra Leone, *Vogel;* Senegambia,
Heudelot.

Frutex parvulus, glaber. *Aculei* stipulares conici, incurvi. *Folia*
alterna, 1-2-pollicaria, breviter petiolata, ovata, crenulata v.
integra. *Racemi* masculi nunc brevissimi, nunc 4-5 lin.
longi, floribus fasciculatis pedicellatis minutis. *Flores fœminei* in axillis pauci, pedicellati. *Fructus* magnitudine grani
piperis, indehiscens videtur pericarpio tenuiter carnoso, endocarpio crustaceo ; in vivo ex Vog. intense coccineus est.

1. Claoxylon *cordifolium*, Benth.; dioicum, foliis suboppositis
petiolatis ovatis marium cordatis novellis inflorescentiaque
lepidoto-tomentellis demum glabratis, racemis axillaribus
simplicibus, floribus masculis 4-partitis, fœmineis tripartitis
inappendiculatis.—Cape Coast, Accra, on the Quorra, and at
Abòh, *Vogel.*

Frutex orgyalis, ramulis elongatis ad nodos compressis, partibus
novellis tomento tenui lepidoto mox evanido canescentibus.
Folia fere omnia opposita, longe petiolata, cujusve paris inæqualia, majora sæpe semipedalia, apice acuminata, margine
integerrima v. sinuato-dentata, basi trinervia, in maribus
auriculis rotundatis cordata, in fœmineis rotundato-truncata
v. cordata, membranacea, supra punctis minutis sæpe scabriuscula, subtus glandulis crebris pellucido-punctata. *Spicæ*
graciles, solitariæ, fœmineæ foliis breviores, masculæ sæpius
longiores. *Flores masculi* ad axillam bracteæ ovatæ acumi-

minatæ ½ lin. longæ glomerati. *Calyx* valvatim 4-partitus. *Stamina* numerosissima, receptaculo globoso inserta; filamenta brevia, antherarum loculi paralleli et contigui. *Flores fœminei* solitarii, minute tribracteati. *Pedicellus* e bracteis exsertus, ½ lin. longus. *Calycis* laciniæ tres, lanceolatæ. *Ovarium* globosum, hispidulum, triloculare (rarissime 4-loculare), loculis uniovulatis. *Styli* tres, (rarissime 4), recurvi, intus longiuscule et dense a basi plumoso-ramosi. *Fructus* lævis, glaber, coccis bivalvibus, singulis magnitudine seminis *Lathyri odorati.*

This differs from the Asiatic *Claoxyla* in the opposite leaves and the want of the three appendages alternating with the divisions of the female calyx mentioned by Jussieu: the male flowers are also smaller than usual; but the general habit and other characters are so much those of *Claoxylon,* of which the Eastern species are probably numerous and variable in aspect, that these differences do not appear to be of generic importance.

2. Claoxylon? sp.—Fernando Po, *Vogel;* leaves opposite and shaped as in *C. cordifolium,* but longer and thinner. The specimens are too imperfect to describe.

1. Alchornea *cordata,* Benth.—Schousbœa cordifolia, *Schum. et Thonn. Beskr. p.* 429; foliis ample ovatis basi sinu clauso cordatis sub-5-nervibus 4-6-glandulosis subtus ad venas ramulis petiolisque hirtellis, spicis masculis axillaribus paniculatis, fœmineis e ramis annotinis pendulis simplicibus ramosisve, floribus octandris, calycibus bipartitis laciniis subbifidis. —On the Nun and at Abòh, *Vogel;* Sierra Leone, *Vogel, Don.—Christmas-bush* of the negroes.

2. Alchornea *hirtella,* Benth.; foliis ellipticis oblongisve obtuse acuminatis remote dentatis rigide membranaceis penninervibus v. basi subtrinervibus subtus petiolis ramulisque hirsutis, glandulis baseos obsoletis, spicis masculis gracilibus ramosis, fœmineis simplicibus, floribus octandris, calycibus bipartitis. —Grand Bassa, *Vogel;* Senegambia, *Heudelot.*

Frutex bipedalis v. paullo altior, ramis paucis junioribus hirsutis. *Folia* breviter petiolata, tripollicaria v. paullo majora, acumine lato obtuso v. retuso, ad margines serraturis obtusis nonnunquam glanduliferis notata, versus basin angustata, ima basi obtusa, utrinque viridia, penninervia et transverse reticulato-venosa; venæ primariæ 2 interdum more generis ad basin folii oppositæ validioresque, sæpius tamen obsoletæ et glandulæ in earum axillis omnino deesse videntur. *Stipulæ* rigidæ, subulatæ, petiolo breviores, deciduæ. *Flores* dioici. *Spicæ masculæ* 4-5-pollicares, in paniculam gracilem dispositæ. *Spicam fœmineam* unicam vidi in altero specimine 2½-pollicarem jam fructiferam, carpellis 2 v. rarius 3.

Jussieu refers to two Senegalese and Guinea species, whether the same or not as either of the above, I do not know. There is also in Vogel's collection from Fernando Po, a fragment, consisting of a bit of stalk and a single leaf, which may be that of another *Alchornea*.

A specimen of a branching shrub, with small oval or obovate leaves and a few remains of fruit, gathered by Vogel, at Stirling, on the Quorra, indicates an affinity to *Alchornea* or *Claoxylon*, but is insufficient to determine the genus.

PYCNOCOMA (gen. nov.) e tribu *Crotonearum*.

Flores monoici. *Masculi* racemosi. *Calyx* 5-partitus laciniis æstivatione subvalvatis, per anthesin arcte reflexis. *Corolla* nulla. *Stamina* numerosissima, disco carnoso pulvinato inserta; filamenta filiformia libera apice incrassata; antheræ parvæ globosæ, loculis parallelis ad apicem filamenti transverse adnatis. *Flos fœmineus* ad apicem racemi solitarius, sessilis. *Calyx* 5-partitus, laciniis patentibus. *Corolla* nulla. *Ovarium* sessile, triloculare, ovulis solitariis. *Stylus* erectus, ad medium trifidus, ramis recurvis summo apice peltato-stigmatiferis. *Capsula* tricocca, coccis (bivalvibus?) dorso obtuse bicostatis.

1. Pycnocoma *macrophylla*, Benth.—Fernando Po, *Vogel.*
Arbor 8-pedalis, trunco tenui nudo coma densa foliorum terminato. *Ramuli* (crassitie digitis minoris) glabri. *Folia* ad

summitates conferta, sessilia, 1½-2-pedalia, obovali-oblonga, acuminata, basi longissime angustata, integerrima v. obsolete sinuata, membranacea, penninervia, glabra, subtus pallida, eglandulosa. *Stipulæ* lineari-lanceolatæ, mox deciduæ v. obliteratæ. *Racemi* in axillis supremis solitarii, vix semipedales, rachide glabra v. pilis paucis conspersa. *Bracteæ* orbiculato-concavæ, 2 lin. latæ, cartilagineæ, extus adpresse tomentosæ, secus rachin dissitæ, inferiores steriles. *Flores masculi* intra bracteas superiores solitarii gemini v. terni, pedicello 6-8 lin. longo fulti. *Calyx* roseus, laciniis late oblongis, 3 lin. longis membranaceis, obtusis v. acutiusculis, in pedicellum arcte reflexis. *Stamina* calyce dimidio longiora. *Floris fœminei* laciniæ calycinæ quam in mare angustiores crassioresque, adpresse puberulæ et apice tomentosæ, patentes sed non reflexæ. *Ovarium* tomentosum. *Stylus* fere glaber, infra divisionem 2. lin. longus, ramis paullo longioribus.

1. Manihot *utilissima*, Pohl.—Jatropha Manihot, *Linn.* var. ? *heterophylla.*—Grand Bassa, *Vogel,* cultivated.

This variety of the *Cassava* is the one described by Thonning as commonly cultivated in Guinea. It is not precisely identical with any of those distinguished by Pohl, but probably with the *M. Aipi;* and perhaps one or two others of that author should be included in one species, which must necessarily have varied much by the effect of long and extensive cultivation in all Tropical regions.

1. Jatropha *multifida*, Linn.—Sierra Leone, *Vogel.*

2. Jatropha *gossypifolia*, Linn.—Cape Coast, *Vogel;* Accra, *Don.*

Both the above are South American plants, either introduced or cultivated in Africa, as is also the *Curcas purgans,* Med. (*Jatropha Curcas,* Linn.), said by Thonning to be found here and there in Guinea.

1. Astræa *lobata*, Klotzsch *in Erichs. Archiv. v. 7. p.* 194.—Croton lobatum, *Linn.*—*Pal. Beauv. Fl. Ow. et Ben. t.* 36.—Fernando Po, *Vogel;* Tropical Africa and America.

Two species of *Crozophora* are mentioned by A. de Jussieu
as Senegalese.

1. Phyllanthus *Niruri,* Linn.—P. amarus, *Schum. et Thonn.
Beskr. p.* 421 ?—As common in this as in other parts of Tropical Africa and Asia.

2. Phyllanthus sp., possibly a large-leaved variety of *P. Niruri.*
—St. Thomas, *Don.*

The extensive genus *Phyllanthus,* as well as *Glochidion* and
some other allied ones, are at present in such a state of confusion, and several of the common almost cosmopolite species
so ill defined, that I do not venture to name as new any of the
specimens before me; although there are some which I am
unable absolutely to identify, as is the case with the present
one.

3. Phyllanthus *piluliferus,* Fenzl. *Flora,* 1844. *p.* 312.—P. li-
noides, *Hochst. Pl. Kotsch. Nub. exs. n.* 303.—On the
Quorra, at Patteh, Attah, and Stirling, *Vogel, Ansell;* Senegal and Nubia.

I am not aware that either of the above names are regularly
published, but I can scarcely believe that a plant, apparently
so common in Tropical Africa, should be yet undescribed. In
many respects, Thonning's description of his *P. pentandrus*
agrees with it; but our plant is certainly an annual.

4. Phyllanthus *dioicus,* Schum. et Thonn. *Beskr. p.* 416 ?—
P. Reichenbachianus, *Sieb. Pl. Seneg. exs.*—Cape Coast and
on the Quorra, *Vogel;* Senegal and Guinea.

Most probably, on a revision of the genus, this plant would
be placed in a distinct section or genus from the three preceding.

There is, besides, a species in the Senegambian collection
with very singular, recurved, stipulary thorns; and Schumacher
and Thonning, besides those mentioned above, have described
four others from Guinea : *P. angulatus, P. capillaris, P. Thonningii* and *P. sublanatus.*

1. Glochidion sp.—Phyllanthus polyspermus, *Schum. et Thonn.
Beskr. p.* 416.—On the Quorra, at Stirling, *Vogel;* Guinea.

2. Glochidion, sp., very near the last, but the young branches clothed with rusty hairs.—On the Quorra, at Attah, *Vogel.*

1. Bridelia *ferruginea,* Benth.; fruticosa, foliis ovali-ellipticis coriaceis supra glabris subtus ramulisque ferrugineo-puberulis, baccis parvis oblongis monospermis.—On the Quorra, at Attah, *Vogel.*

Frutex arborescens, divaricato-ramosus. *Folia* breviter petiolata, 2-3-pollicaria, obtusa v. brevissime acuminata, margine recurva, basi rotundata, supra glabra at non nitida, subtus venis primariis pinnatis elevatis venulisque transversis reticulata, ferruginea, præsertim ad costas pubescentia. *Stipulæ* parvæ. *Flores* desunt, sed e cicatricibus fructuque persistente ad axillas glomerati erant. *Calyx* sub fructu persistens, parvus, 5-fidus, laciniis acutis; squamellæ (petala?) 5, parvæ, acutæ, cum laciniis calycinis alternantes. *Drupa* 3 lin. longa, nigra, intus semibilocularis; dissepimenti disrupti axis persistit semen perforans et margines intra sulcos seminis leviter prominent. *Semen* fenestratum, albumine valde involuto fructum totum implens et axin dissepimenti amplectens.

The leaves of this species, especially in their consistence, venation and pubescence, have some resemblance to those of *B. scandens.* I regret much there being no flowers to examine the ovary, for the appearance of the fruit seems to indicate that the one-celled and one-seeded fruit, produced by a two-celled, four-ovuled ovary, is not formed in the usual way, by the fertile cell increasing so as to occupy the whole interior cavity, the abortive one either wholly disappearing or being rejected to one side; but both cells increase equally, the fruit remaining regular in form, with the original axis in its centre; one ovule alone enlarges, and penetrating through the broken dissepiment on each side of the axis, occupies the whole of the two cavities, enclosing the axis in its centre.

Brown mentions two new species of *Bridelia* from Congo; whether or not either may be the same as the above, I have no means of ascertaining.

CLEISTANTHUS, *Hook. fil.* (gen. nov.)

Flores dioici.   *Masculi : Calyx* 5-partitus, laciniis late linearibus
æstivatione valvatis demum patentibus crassiusculis.   *Squamulæ* 5, lineari-oblongæ, laciniis calycinis alternantes, hypogynæ.   *Stamina* 5, squamulis alterna, filamentis inferne in
columnam crassam connatis, superne liberis, subulatis ; antheris oblongis, supra basin dorso affixis, loculis rima introrsa dehiscentibus.   *Rudimentum* pistilli, intra stamina insidens, ovato-oblongum, apice trifidum villosum.   *Planta
fœminea* desideratur.—*Frutex : Folia* alterna, disticha, breviter petiolata, oblongo-lanceolata, acuminata, basi acuta,
integerrima, coriacea, penninervia, reticulato-venosa, glabra.
Stipulæ foliorum non visæ, florales lineari-subfalcatæ glabræ.
Racemi masculi axillares breves, densiflori, axi pedicellisque
ferrugineo-pubescentibus.   *Flores* brevissime pedicellati, 2-3-
natim fasciculati, fasciculo bracteis 2 (stipulaceis ?) stipato.
Calyx extus ferrugineo-pubescens.

1. Cleistanthus *polystachyus*, Hook. fil.—*Planch. in Hook. Ic.
t.* 779. (TAB. XXXVI.)—Sierra Leone, *Whitfield.*

Dr. Planchon, whose generic character is here given, suggests
an affinity between this plant and *Bridelia;* but whether it
belongs or not even to the Order, must remain doubtful until
the female plant shall have been seen.

PLATE XXXVI. *Fig.* 1. bud ; *f.* 2. expanded flower ; *f.* 3.
scale or petal ; *f.* 4. rudiment of the pistil :—*all more or less
magnified.*

1. Amanoa *bracteosa*, Planch (TAB. XLVII.) ; monoica, glaberrima, foliis anguste oblongis acuminatis basi acutiusculis
integerrimis coriaceis nitidis, fasciculis florum bracteis 3 arcte
cinctis in spicas abbreviatas quasi amentaceas distiche confertis, pedicello floris fœminei e bracteis longe exserto, florum
masculorum inclusis, staminibus 5, capsula subglobosa nuce
juglandis paullo minore, seminibus castaneis nigris.—(*Planch.
in Hook. Ic. t.* 797.)—Sierra Leone, *Don.*

Folia alterna, 3-4 poll. longa, 1-1½ poll. lata ; petiolo 4-5 lin.

longo. *Stipulæ* in unam brevem intra-axillarem obtusam concretæ. *Inflorescentiæ* terminales v. axillares, subsessiles. *Bractea* fasciculi singuli inferior late ovata biloba, dorso sub apice mucronulata (revera stipularis), laterales semiovatæ. *Flores* in fasciculo 3 exteriores intra bracteas exteriores solitarii, cæteri in fasciculos circa 3 subdistiche congesti, omnes bractea membranacea fulva suffulti; fœmineus in fasciculo unicus, masculi 6-8. *Perianthium* floris masculi 5-partitum, laciniis angustis æstivatione leviter imbricatis. *Squamulæ* 5, laciniis perianthii oppositæ. *Stamina* 5, squamulis alterna, disco elevato inserta. *Rudimentum* pistilli minutum, trilobum. *Ovarium* floris fœminei triloculare, loculis biovulatis. *Stigma* sessile, pileiforme, obsolete trilobum. *Semina* in loculis gemina, ecarunculata.

PLATE XLVII. *Fig.* 1. fascicle of flowers seen laterally; *f.* 2. the same seen from the rachis; *f.* 6. one of the lateral bracts; *f.* 4. fascicle of flowers without the outer bracts; *f.* 5. portion of a fascicle without the female flower; *f.* 6. male flower, opened; *f.* 7. disk, with the rudiment of the ovary; *f.* 8. pistil, vertical section; *f.* 9. valve of the capsule; *f.* 10. seed :—*all but the two last magnified.*

There are, in Vogel's Fernando Po collection, specimens of a shrub, which appears to constitute a new genus of *Buxeæ*, but as they are females only, I am unable to characterize it. The leaves are alternate, resembling those of *Gelonium.* The young tricoccous capsular fruits are in axillary racemes, surmounted by three bifid styles, each cell containing, in the youngest, two ovules half-immersed in a thick fleshy placenta, only one of which appears to come to perfection.

The following genus, *Microdesmis*, has been referred by Planchon to anomalous *Flacourtianeæ;* but as he also admits its relationship to *Euphorbiaceæ*, with which it appears to me to have much closer connection, I here insert the West African species with Dr. Planchon's specific character, referring to Hooker's Icones, t. 758, for the detailed generic character, taken from this and a Malacca species, which I have not the opportunity of examining.

1. Microdesmis *puberula,* Hook. fil. (TAB. XXVI.) ; staminibus 5, antheris muticis.—Fernando Po, *Vogel.*

Frutex, ramis virgatis gracilibus pubescentibus. *Folia* alterna, brevi-petiolata, lanceolata, cuspidata, 3-4-pollicaria, obsolete serrulata, penninervia, reticulato-venosa, rigide membranacea, pellucido-punctata, subtus ad venas puberula. *Stipulæ* minutæ, persistentes. *Flores* minuti, in fasciculos axillares aggregati ; masculi in fasciculo plures, pedicello vix lineam longo ; fœminei pauciores, pedicello adhuc breviore. PLATE XXVI. *Fig.* 1. bud of the male flower ; *f.* 2. male flower ; *f.* 3. petal ; *f.* 4. stamens, with the rudiments of the pistil ; *f.* 5. fruit ; *f.* 6. the same, vertical section ; *f.* 7. the same, transverse section ; *f.* 8. seed ; *f.* 9. the same, vertical section :—*all magnified.*

XCIII. PIPERACEÆ.

1. Peperomia *Vogelii,* Miq. *Lond. Journ. Bot.* 4. *p.* 413.—On the Quorra, *Vogel ;* St. Thomas, *Don.*

Don's specimens are much more slender than Vogel's, with longer and slenderer spikes, one of them is branched, and eleven inches high. Some of the leaves come very near in form to those of the common *P. pellucida,* from which this species is scarcely distinct.

1. Pothomorphe *subpeltata,* Miq. *Pip. p.* 213. *et Lond. Journ. Bot.* 4. *p.* 431.—Sierra Leone, *Don ;* Fernando Po, *Vogel ;* Tropical Asia and Africa.

1. Cubeba *Clusii,* Miq. *Pip. p.* 304. *et Lond. Journ. Bot.* 4. *p.* 434.—Sierra Leone, very common on trees, *Don ;* Fernando Po, *Vogel.*

Schumacher and Thonning describe a *Piper Guineense,* known in the country under the name of *Dooje* and *Ashantee Pepper,* which appears to agree with the above in the fruit, but to differ in the long petioles, and in the veins of the leaves pubescent underneath. Thonning had not seen it himself, but loose fragments were brought to him by the natives, and possibly the leaves may not have been from the same plant as the fruit.

XCIV. Antidesmeæ.

This Order is more than any one in need of a thorough revision, both as to the number and limitation of the genera it should contain, and as to the character of the Order itself; and indeed its very existence as a separate Order is doubtful. This labour requires too wide an enquiry to be undertaken on the present occasion, the West African species being but very few. None, indeed, appear to have been as yet indicated from that region, and the few specimens in the collections before us are most of them in too imperfect a state to be satisfactorily described. They are:

1. An *Antidesma*, in fruit only, from Abòh, *Vogel;* closely allied to the East Indian *A. Bunius.*

2. An *Antidesma*, with a few fruits, and inflorescences deformed by a kind of proliferous monstrosity, from Vogel's collection without the precise indication of the locality. It can hardly be distinguished from the common East Indian *A. paniculata.*

3. A specimen in leaf only from Sierra Leone, *Don*, called by him *Antidesma Guineensis,* and another in the same state, evidently belonging to the same species from Fernando Po, *Vogel,* but both very different in appearance from any *Antidesma* known to me.

4. A specimen without flowers or fruit from Cape Palmas, *Vogel,* with the aspect of *Bennettia,* Br., and which Dr. Planchon believes to be a new *Sarcostigma.*

5. Another, in the same state, from Grand Bassa, *Vogel,* which Dr. Planchon considers a new *Pyrenacantha.*

There is also in the Senegambian collection a male specimen of a *Lepidostachya,* or some allied genus.

XCV. Urticeæ.

1. Urera *oblongifolia,* Benth.; dioica, caule glabro scandente, foliis longe petiolatis elliptico- v. obovali-oblongis acuminatis basi obtusis, cymis axillaribus laxis, masculis petiolo longioribus floribus pentameris, fœmineis brevioribus perianthio

ovario vix breviore minute 3-4-dentato demum baccato.—
Sierra Leone, *Vogel.*

Frutex scandens, totus glaber exceptis setis nonnullis in cyma
fœminea præsertim sub fructu conspicuis. *Stipulæ* parvæ,
caducæ. *Folia* 2-3-pollicaria, acumine abrupto obtuso, margine integerrima v. superne leviter serrata, basi obtusa v.
rarius leviter cordata, membranaceo-chartacea, rhaphidibus
sub epidermide tenuissima piliformibus scabriuscula, venis
primariis a costa divergentibus paucis, 2 inferioribus oppositis
cæteris paullo validioribus, petiolo ½-1-pollicari. *Cymæ masculæ* paniculæformes, longiuscule pedunculatæ, folium interdum superantes. *Flores* in ramis ultimis fasciculati, brevissime pedicellati, ¾ lin. diametro. *Perianthium* 5-parti-
tum. *Filamenta* brevissima sub ovarii rudimento pulvinato
inserta, antheris parvis rotundatis. *Cymæ fœmineæ* corymbosæ, masculis densiores, breviter pedunculatæ. *Flores* sessiles, ovati, subcompressi, vix obliqui, semilinea breviores.
Perianthium membranaceum, ovarium fere ad apicem includens. *Stigma* globoso-villosum, subpenicillatum, coc-
cineum, fere rectum. *Perianthium fructiferum* baccans,
linea paullo longius, rubro-aurantiacum (ex. Vog.), achenium
arcte includens.

2. Urera *obovata*, Benth.; dioica, caule scandente subsetoso,
foliis petiolatis obovatis acuminatis glabris v. vix setosis,
cymis fœmineis axillaribus laxis subpaniculatis petiolos superantibus, perianthio ovarii dimidium vix superante sub-4-
dentato.—Sierra Leone, *Vogel.*

Affinis *U. oblongæ.* *Rami* crassiores, novelli setis brevibus
rigidis urentibus horridi. *Stipulæ* 2-3 lin. longæ, caducæ.
Folia 3-5 lin. longa, 2-2½ poll. lata, nonnunquam prope
basin ad venas petiolosque pauciseta, cæterum iis *U. oblongæ* similia. Stirps mascula deest. *Cymæ* fœmineæ potius
paniculatæ quam corymbosæ, 1½-3-pollicares, multifloræ, se-
tulosæ. *Perianthium* inæqualiter et obtuse 4-dentatum, per
anthesin ovarii dimidium vix æquans, post anthesin auctum
et teste Vogelio aurantiacum. *Ovarium* apice in stylum
brevissimum attenuatum, stigmate capitato subpenicillato leviter obliquo.

The diœcious flowers and the form of the leaves would separate the above two species from the generality of *Ureræ*; although as a section only. Whether this be the same section as that suggested by Gaudichaud for some Madagascar and Mauritian species under the name of *Obetia*, I have no means of judging. The female perianth is very different from that described by Endlicher in his sectional character of *Urera*, which indeed does not agree with the generality of *Ureræ*, where the perianth is very variable. It is better described by Gaudichaud, excepting that, even in the original *U. baccifera*, it is rather 3-4-lobed than 3-4-partite, and in our species the divisions are no more than short teeth. The rhaphides having the appearance of hairs, mentioned by Dr. Hooker in the Flora Antarctica, are very conspicuous in both the above species.

1. Fleurya sp.—Urtica mitis, *E. Mey. Pl. Dr. exs. ?*—Cape Palmas, *Ansell;* a small specimen, with a single male cyme.

2. Fleurya sp.—Urtica Caravelhana, *Schranck ex Mart. Herb. Fl. Bras. p. 93. n. 84.*—Urtica hirsuta, *Vahl, Symb. 1. p. 77 ?*— On the Quorra, at Patteh and Addaenda, *Vogel;* St. Thomas, *Don.*

3. Fleurya sp.—Urtica villosa, *Salzm. Pl. Bras. exs.*—U. hirsuta, *Vahl ?*—Fernando Po, *Vogel.*

The genus *Fleurya*, if made to include *Laportea*, is a natural one, and readily distinguished from *Urtica*, whether by its alternate leaves, by the shape of the ovary and the uncinate style, the stigmatic portion being more or less reflexed. Endlicher, indeed, in reducing Gaudichaud's genera to the rank of sections, has suppressed this also, but evidently through inadvertence, because he had himself, in another place, established it as a distinct genus, under the name of *Schychowskya*, adopted as such in his Genera Plantarum. Some species, especially the South American and African ones, appear to be diœcious, or nearly so: the N. American *F. Canadensis* and the Oceanic *F. ruderalis* are, however, certainly monœcious. In all, the male inflorescences are much more compact than the female, the male calyx, as in other genera, varies in the number of its sepals and stamens 4 or 5; the female flower has, like *Urtica*,

two sepals flatly applied to the sides of the compressed ovary,
with either one or two minute outer leaflets opposite to the
edges of the ovary, which may be more correctly called bracts
(as suggested by Henslow, whose *Urera Gaudichaudiana* is
rather a *Fleurya*) than sepals. The fruit is generally exceedingly oblique and bent down upon the rachis (or receptacle), whilst the style is bent back on the upper edge, sometimes long and subulate, with a short stigmatic portion at its
extremity, more frequently short, with a more or less elongated
or subulate stigmatic tuft. In the *F. ruderalis*, the type, as it
were, both of *Fleurya* and of *Schychowskya*, it is much less
hooked and the fruit less oblique than in all other species I am
acquainted with. Still it is very different from the style and
stigmate of either *Urtica* or *Urera*, besides having the habit of
the other *Fleuryæ*. To the three above species I have not
given names or character; because I have not before me, at
present, materials for extricating the synonymy and determining
the limits of older species, to which they bear considerable
relation, if they are not identical. The first is a very poor male
specimen, very likely distinct from the S. African *U. mitis*, but
affords no clue to separate it. The two others, as far as my
specimens go, closely resemble the Brazilian ones, to which I
have compared them, but have most probably already received
more than one name as S. American or as African plants.

1. Pouzolsia *Guineensis*, Benth.; pubescens, caule elongato ramoso, foliis alternis petiolatis ovato-lanceolatis, glomerulis
multifloris androgynis, floribus masculis exterioribus minute
bracteatis, fœmineis intra bracteas orbiculatas subreconditis,
fructibus costatis exalatis.—Parietaria Guineensis, *G. Don,
Herb.*—St. Thomas, *Don*; Fernando Po, *Vogel.*

Herba annua videtur, caulibus pluripedalibus adscendentibus
inter frutices subsarmentosis; ramuli, præsertim novelli,
pilis brevibus pubescentes. *Folia* inferiora ovato-lanceolata,
longiuscule acuminata, bipollicaria; floralia breviora; summa
ovata, semipollicaria; omnia integerrima, trinervia, præsertim ad margines setulis parvis pilosula. *Stipulæ* fusco-membranaceæ, acuminatæ, foliorum floralium pleræque basi latæ,

glomerulum bracteantes. *Flores* in glomerulos petiolo breviores arcte conferti, exteriores 6-10 masculi bracteolis minimis subtensi et breviter pedicellati; interiores nunc 2-3, nunc
12-15, arcte sessiles, singuli bractea subfoliacea orbiculato-
reniformi concava hispida bistipellata subtensi. *Ovarium*
perianthio subclauso minute bidentato extus hispido arcte
inclusum. *Stigma* exsertum, elongatum, hinc longiuscule
plumoso-villosum. *Fructus* perianthio persistenti aucto adnatus, ovato-compressus, hispidus, lineam longus, angulis 3
acutis at non in alas productis, faciebus convexis 5-cos-
tatis.

This plant belongs to Bennett's second section, to which
he proposes that the name of *Pouzolsia* should more especially apply. Turczaninow's genus *Gonostegia* belongs to
Bennett's first section, for which the latter botanist adopts
Hamilton's name of *Memorialis*.

The *Haynea ovalifolia*, Schum, et Thonn., from Guinea, which
is unknown to me, is usually referred to *Pilea;* but the three
styles described, render it a very doubtful plant till it has been
re-examined.

1. Bœhmeria (Procris ?) *rigida*, Benth.; fruticosa, dioica, ramis
   aculeato-setosis, foliis alternis late ovatis obovatisve serratis
   basi obtusis glabris subtus albicantibus, cymis masculis axillaribus laxe paniculatis folia æquantibus.—Urtica rigida,
   G. Don, Herb.—Sierra Leone, *Don.*

Specimen unicum masculum. *Ramus* crassus, glaber, setis
crassis aculeiformibus echinatus. *Folia* breviter petiolata,
3 poll. longa, 2-2½ poll. lata, a basi ad apicem grosse serrata.
Inflorescentia eadem ac in stirpibus masculis *B. niveæ* vel
B. Puyæ, sed laxior. *Flores* pentameri, ovarii rudimento
clavato.

1. Musanga *Smithii*, Br. *in Horsf. Pl. Jav. Rar. p.* 49.—
   Sierra Leone, *Don.*

1. Dicranostachys sp. ?—A specimen is in Vogel's collection,
   without any label, in leaf, with one young fruit; the leaves
   are much broader than those described by Trécul in the
   original *D. serrata,* a Senegalese species.

The *Myrianthus arborescens*, Beauv., from Benin, is conjectured by Trécul to be allied to *Dicranostachys*. The *Treculia Africana*, Decaisne, belonging to the same tribe, is from Senegal only.

1. Urostigma *Vogelii*, Miq.* (*Hook. Lond. Journ. Bot. v.* 7. *t.* 12. *f. A.*); foliis longiuscule petiolatis obovato- v. lanceolato-oblongis breviter plerumque obtuse acuminatis basi obtusis v. leviter emarginatis coriaceis supra glabris subtus puberulis trinervibus costulisque utrinque 4-5 patulo-adscendentibus reticulatis, receptaculis axillaribus geminis subsessilibus globosis glabriusculis, involucri trilobi lobis medio hirtis.—Grand Bassa and Cape Palmas, *Vogel.*

Arbor 20-30-pedalis ex Vog. *Petioli* 1-2 poll. longi, uti ramuli in sicco rimulosi. *Stipulæ* breves, latæ, coriaceæ. *Folia* crassa, coriacea, 6-8 poll. longa, 3-5 poll. lata, in sicco supra nigricantia, subtus fuscescentia. *Receptacula* in ramulis nascentibus apice foliatis plerumque gemina, quandoque sena, brunnea, piso paullo majora, fere sessilia, intus bracteis lanceolatis occlusa et inter flores bracteolata. *Flores* fœminei plerumque pedicellati, perigonio tripartito, lobis concavis ellipticis carinatis fuscis pallide marginatis. *Ovarium* brevi gynophoro sustensum, ellipticum; stylo a medio latere brevi, stigmate paullo obliquo elongato compressiusculo. *Flores masculi* in superiore receptaculi parte fœmineis intermixti iisque subbreviores. *Stamen* 1, filamento brevi crasso; anthera oblonga (inclusa) bilocularis, loculis connectivo utrinque adnatis rimaque apertis.

2. Urostigma *rubicundum*, Miq. (*Hook. Lond. Journ. Bot. v.* 7. *t.* 12. *f. B.*); foliis modice petiolatis ovatis ellipticis v. obovato-ellipticis rotundato-obtusis v. breviter apiculatis basi

* The description of the species belonging to the old genus of *Ficus* is by Dr. Miquel, to whom we are indebted for an elaborate and careful monograph of this most difficult group. It is only to be regretted that he had not left so natural and readily-recognised a genus entire, giving sectional names only to his otherwise excellent divisions. I have here, however, followed his own words, only reducing his admeasurements of *centimetres* and *millimetres*, for conformity sake, into *inches* and *lines*.

æquali leviter cordatis integerrimis coriaceis adultis utrinque glabris junioribus ciliolatis trinervibus costulisque utrinque 5-7, petiolis antice puberulis glabrescentibus, stipulis lanceolatis acuminatis convolutis puberulis antice deorsum hirsutis, receptaculis axillaribus geminis v. solitariis breviter pedunculatis obovato-globosis sericeo-villosis, umbilici rima subhiante, involucro subtrilobo pubescente circumscisse deciduo.—On the Quorra, at Addaenda, *Vogel.*

Arbor mediocris. *Petioli* 6-12 lin. longi. *Folia* 2-4 poll. longa, 1-2 poll. lata, siccatione fuscescentia. *Stipulæ* 4 lin. longæ. *Pedunculi* hirto-tomentelli, lineam longi. *Receptacula* viridi-rubescentia, 3-4 lin. crassa, intus sub ore dense bracteata. *Flores fœminei* e tubo brevi trilobi, coriacei, fusci, lobis ellipticis acutis versus basin carinatis. *Ovarium* brevi gynophoro suffultum, obovatum ; stylo ex apice laterali ; stigmate lineari. *Achenia* purpureo-fusca, variegata. *Flores masculi* desunt, forsan deflorati.

3. Urostigma *elegans,* Miq. (*Hook. Lond. Journ. Bot.* 7. *t.* 13. *f. A.*) ; foliis oblongo-obovatis ovatisve breviter obtuseque apiculatis ima basi angustata obtusis integerrimis coriaceo-membranaceis glabris, costulis utrinque circiter 10, stipulis lanceolatis acuminatis convolutis glabris, receptaculis (ex Vog.) longe pedunculatis pendulis.—Cape Coast, *Vogel.*

Arbor pulchra. *Petioli* ¾-1½ poll. longi. *Folia* 4-5 poll. longa, 2-2½ prope apicem lata, costulis patulis prominulis. *Stipulæ* fere 5 lineas æquantes.

4. Urostigma *ottoniæfolium,* Miq. (*Hook. Lond. Journ. Bot.* 7. *t.* 13. *f. B.*) ; foliis longiuscule petiolatis oblongis v. obovato-oblongis breviter et obtuse acuminatis basi acutis integerrimis coriaceo-membranaceis glabris trinervibus costulisque utrinque 4-7 patulis, stipulis lanceolatis glabris apice puberulis convolutis, receptaculis axillaribus 1-2 pedunculatis globosis tenere puberulis glabrescentibus, involucro trilobo demum circumscisso.—Fernando Po, *Vogel.*

Folia in petiolo 1-1½-pollicari 3-4½ poll. longa, 1-2 poll. lata, fere membranacea, haud rigida, in acumen modicum rectum latiusculum terminata, venis in arcum marginalem confluentibus. *Stipulæ* 4 lin. longæ. *Receptacula* sæpe gemina (v. ex tuberculo axillari plura ?), obovato-globosa, 3½ lin. diametro, pedunculos circiter æquantia, involucro membranaceo extus puberulo sustensa.

5. Urostigma *Thonningii,* Miq. (*Hook. Lond. Journ. Bot.* 7. *t.* 13. *f. C.*); foliis longiuscule petiolatis oblongis v. obovato-oblongis obtusis rotundatis v. emarginatis basi acutis v. obtusiusculis integerrimis coriaceis glabris, venulis distinctioribus utrinque fere 12, receptaculis axillaribus geminis breviter pedunculatis globosis versus basin pilosulis v. glabris, involucri trilobi lobis rotundatis puberulis et ciliatis.—Ficus microcarpa, *Vahl, Enum.* 2. *p.* 188. *haud Linn.*—F. Thonningii, *Blume in Rumphia.*—On the Nun, *Vogel;* Guinea.

Arbor mediocris. *Petioli* ½-1½ poll. longi. *Folia* 2-4 poll. longa, 1-1½ poll. lata, rigidiuscule membranaceo-coriacea, supra lævia, in sicco nigricantia, venis crebris patulis parallelis. *Stipulæ* coriaceo-membranaceæ, ovatæ, convolutæ, 1½ lin. longæ. *Receptacula* magnitudine pisi majoris, viridi-lutescentia, punctis rubris, fere tota glabra, intus sub ore bracteis lanceolatis fuscescentibus occlusa. *Flores* sessiles, fœminei tripartiti, lobis 2 posticis lanceolatis acuminatis, dorsali tertio concavo. *Ovarium* sessile, obovatum, stylo modico, stigmate simplici, sed cujus forma accurate distingui nequit, cum omnia stigmata sibi adglutinata sint. *Achenia* fusca, nitidula. *Flores masculi* verosimiliter deflorati.

6. Urostigma? sp.—Specimens in leaf only, not seen by Dr. Miquel, but apparently allied to *U. politum.*—On the Quorra, at Attah and Abòh, *Vogel.*

The other W. African species of the genus (or section) are *U. ovatum,* Miq., *U. politum,* Miq., *U. calyptratum,* Miq., and *U. luteum,* Miq., all from Guinea.

1. Sycomorus *Vogeliana,* Miq.; ramis glabris, ramulis petiolis foliisque subtus præsertim in nervo medio costisque subbarbato-hirtis, his breviter v. longe petiolatis obovato-oblongis acutis v. breviter apiculatis versus basin attenuatis ipsaque leviter emarginatis v. subtruncatis præsertim versus apicem crenato-serratis v. repandis coriaceo-membranaceis trinervibus

et utrinque 5-6-costatis, junioribus supra pilis raris longiusculis inspersis, dein glabris lævigatis subtus fuscescentibus
et sub lente punctulatis, receptaculis supra ramos aphyllos
racemoso-paniculatis junioribus ovatis v. globosis parce hirtellis, bracteis 3 ovatis pedunculoque hirtis.—On the Quorra,
and Fernando Po, *Vogel*.

Arbor 30-40-pedalis cortice albo, paniculis e trunco ramisque
vetustis erumpentibus, ramis racemosis gracilibus usque ad
2⅓ pedes longis. *Petioli* ½-2 poll. longi, cum ramulis foliisque subtus in sicco fuscescentes. *Folia* 3-7 poll. longa, 2-4
poll. lata, majora dentata v. crenato-serrata, minora quandoque
fere integerrima ; costæ subtus prominentes, erecto-patulæ,
prope margines subconfluentes, anastomosibus transversis
parallelis prominentibus junctæ. *Stipulæ* membranaceæ,
ovato-lanceolatæ, 6 lin. longæ, demum glabræ. *Paniculæ*
rami alterni, juniores hirti, stipulis bracteati, e quorum axillis
receptacula solitaria, altero nunc rudimentario.

2. Sycomorus *Guineensis*, Miq. (*Hook. Lond. Journ. Bot.* 7.
t. 14. *f. B.*) ; ramulis petiolis foliisque subtus in nervis pilis
parcis fugacibus inspersis, his longiuscule petiolatis oblongis
acutis v. obtusiusculis basi acutis præsertim versus apicem
dentato-serratis membranaceo-coriaceis lævibus v. senectute
asperiusculis trinervibus et utrinque 4-5-costatis, stipulis lanceolatis dorso appresse hirtis marginibus ciliatis, receptaculis
supra ramos laterales e trunco v. vetustis ramis protrusis
racemosis, maturis rubris.—Ficus Brassii, *Br. in Trans. Soc.
Hort. Lond.* 5. *p.* 448.—Cape Palmas, *Vogel ;* Sierra Leone,
Don.

A præcedente foliis basi magis attenuatis nec excisis, paniculæ
ramis brevibus statim discernitur. *Rami* vetustiores cylin-
drici, parce verrucosi ; ramuli subangulati, parce puberuli,
dein glabrati quandoque rimulosi. *Petioli* ½-2 poll. longi,
antice canaliculati, cito glaberrimi. *Folia* 3-5 poll. longa,
2-2½ poll. lata, rigidiuscula, subtus pallidiora, versus basin
integerrima, raro uno alterove dente instructa e nervo medio
utrinque una costa ad ⅓ alt. perducta, reliquæ inde ab ¼ alt.
ortæ patulo-adscendentes ; anastomoses tenues, paucæ va-

lidiores. *Stipulæ* vix 4 lin. longæ, exteriores perulaceæ gla-
briores dorso versus basin hirtæ ciliatæque, interiores breviores imbricatæ subsericeæ. *Inflorescentia* stricta, aphylla,
6 poll. longa, ramulis subsericeis racemoso-dispositis serius
glabrescentibus.

Dr. Miquel observes that this is most nearly allied to the
Sycomorus Capensis, but is not only much larger in stature, but
differs also in its pubescence. Don's specimen, without fructification, which Miquel had not seen, resembles still more the
South African ones.

The *Ficus umbellata,* Vahl, from Guinea, is referred by Miquel
to *Sycomorus.*

1. Ficus (Sycidium) *asperifolia,* Miq. (*Hook. Lond. Journ. Bot.*
7. *t.* 15. *f. B.*) ; foliis ellipticis breviter acuminatis v. acutis
subdenticulatis integris v. utrinque sinu diversiformi excisis
coriaceis utrinque scabro-asperrimis supra nitidulis demum
subareolatis, subtus in sicco lutescentibus costulisque venosis
5-6 patulis reticulatis, receptaculis axillaribus plerumque
geminis pedunculatis obovato-turbinatis verrucosis asperulis
basi in stipitem constrictis tribracteatis, ore prominule bracteato.—Abòh, often growing in the water, *Vogel.*

Frutex ramis teretibus glabris lævibus, ramulis angulatis glabriusculis. *Petioli* antice sulcati, 4-6 lin. longi. *Folia* 2-3
poll. longa, 1-2 poll. lata, pleraque integra, præsertim versus
apicem minute et æqualiter denticulata v. fere integerrima,
alia ad dimidium altitudinis utrinque sinu acuto v. obtuso
excisa, hinc subtriloba, omnia æquilatera, basi acuta. *Pedunculi* parce puberuli, 2 lin. longi. *Receptacula* pilis minutis
rigidissimis scaberrima, ore prominulo puberula, basi stipiti-
formi bracteis 3 basi cohærentibus puberulis deciduis instructa,
3-5 lin. diametro, intus sub ore bracteis occlusa et inter flores
bracteolata. *Flores fœminei* sessiles v. pedicellati, perigonii
phyllis 5 lineari-lanceolatis coriaceis ovario arcte applicitis, in
floribus sessilibus parieti receptaculi partim adnatis. *Ovarium* subsessile, stylo laterali sursum muriculatum. *Achenia*
lævia, pallide lutescentia. *Flores masculos* non reperi.

Miquel observes that this species is very near the East

Indian *F. heterophylla*, Linn., but is sufficiently distinguished, amongst other characters, by the constricted base of the receptacles.

There are besides, specimens, in leaf only, of three species, apparently of *Ficus* or some allied genus. 1. A tree from Sierra Leone, in Don's collection, with large obovate leaves, nearly smooth. 2. A tree from Sierra Leone, in both Don's and Vogel's collections, with the young branches thick and densely clothed with long hairs, large obovate-oblong leaves, and the fruits (receptacles?), according to Vogel, closely sessile on the branches. 3. A tall tree, from the Quorra, opposite Stirling, with milky juice, and shining, oval, entire, coriaceous leaves, a fruit (receptacles) twice as big as a man's head, and conjectured by Vogel to be an *Artocarpus*, known by the Kroomen under the name of *Oqua*.

1. Celtis *integrifolia*, Lam.—*Planch. in Ann. Sc. Nat. Ser.* 3. *v.* 10. *p.* 308.—On the Gambia, *Don*; Senegal.

2. Celtis sp. Planch. *l. c. p.* 307.—St. Thomas, *Don.*

1. Sponia *strigosa*, Planch. *in Ann. Sc. Nat. Ser.* 3. *v.* 10. *p.* 320.—On the Quorra, *Vogel.*

2. Sponia *Africana*, Planch. *l. c.*—Sierra Leone, *Vogel, Don*; Senegal.

3. Sponia *affinis*, Planch. *l. c. p.* 329.—Sierra Leone, *Vogel.*

The above three species appear to me far too much alike to be considered specifically distinct from each other, or from the *Celtis Guineensis*, Schum. et Thonn. from Guinea, a synonym apparently overlooked by Planchon.

4. Sponia *nitens*, Planch. *in Ann. Sc. Nat. Ser.* 3. *v.* 10. *p.* 325. —Fernando, Po, *Vogel.*

DICOTYLEDONES *incertæ sedis.*

1. Ceratophyllum *vulgare*, Schleid., *Linnæa* 11. *p.* 540. *t.* 11. —In the Quorra and other waters of West Tropical Africa, as well as of the greater portion of the globe.

The careful observations of Schleiden and his elaborate dissertations above quoted, whilst they make us thoroughly acquainted with the real structure of this plant, have shown that no one satisfactory affinity with it has yet been indicated, and it remains an isolated species which cannot be associated with any known Order. The calling it an Order of itself, does not appear to me to throw any additional light on the matter.

DICOTYLEDONES *dubiæ.*

There remain six or seven fragmentary specimens, which we have successively rejected from all known forms in each natural Order with which we had at first compared them, and far too incomplete for any specific mention. They show, however, as well as the large number of species hereinbefore mentioned as imperfectly known, how much there is still for botanists to do who may in future risk their lives in these regions, so pestiferous for the human race, but so favourable for vegetable development.

XCVI. PALMÆ.

Of this Order, the only fragment in the collection is a leaf of the *Elais Guineensis;* and not more than six species appear to have been observed in West Tropical Africa, viz.: 1. *Calamus secundiflorus,* Beauv., extending from Senegal to Benin. 2. *Borassus Æthiopum,* Mart., Senegal to Benin. 3. *Hyphæne Guineensis,* Schum. et Thonn., referred by Kunth to the Egyptian Doum Palm, (*Hyphæne Thebaica,* Mart.) : Brown suggests, however, that the Palm called by Prof. C. Smith *Hyphæne,* is probably not of that genus, as having a single stem. Thonning, in his description, does not mention the branching stem, and his detailed account of the male flowers does not agree with Martius' generic character of *Hyphæne.* 4. *Raphia vinifera,* Beauv., one of the *Wine-palms,* from Sierra Leone to Benin. 5. *Phœnix spinosa,* Schum. et Thonn., another of the Wine-palms, according to Thonning; Guinea, and said also to be found in Senegal and at the Cape of Good Hope. 6. *Elais Guineensis,* Linn., or Oil-palm, also a Wine-palm, and, according to Thonning, common, both wild and cultivated. Hornemann is said by Nees v. Esenbeck, to observe that Thonning's

description confirms Gærtner's assertion that this Palm is diœcious; although Brown and others had satisfactory evidence of
the male and female spadices being produced on the same stem.
But although the *Elais* is placed in *Diœcia* in Schumacher's
Beskrivelse, yet in Thonning's own printed description, the
point in question is not alluded to.

XCVII. PANDANEÆ.

The *Pandanus Candelabrum,* Beauv., common along the coast,
was observed also by Vogel, but no specimen was gathered.

XCVIII. AROIDEÆ.

1. Pistia *Stratiotes,* Linn.—Abòh, *Vogel.*
    The fine set of specimens, as to leaf and flower, collected by
Vogel, still further confirm the opinion stated, amongst others
by myself, (Bot. Sulph. p. 170), that the nine supposed species
of *Pistia* are really forms of one species, common in most of the
warmer regions of the globe, and very variable, like the generality of aquatic plants.
1. Culcasia *scandens,* Beauv.—*Kunth, Enum.* 3. *p.* 46.—On
    the Quorra, opposite Stirling, *Vogel;* Oware.
1. Philodendron? sp., not in a state to determine.—Grand
    Bassa, *Vogel;* Senegal, *Heudelot.*
    The remaining *Aroideæ* published from West Tropical Africa,
are *Stylochæton hypogæum,* Lepr., from Senegal; *Pythonium?*
Hookeri, Kunth, (*Caladium petiolatum,* Hook.), from Fernando Po;* *Amorphophallus difformis,* Bl., from Oware; *A.*
consimilis, Bl., and *A.? Fontanesii,* Kunth, from Senegambia.
There is also a leaf of a cultivated *Colocasia* from Sierra Leone
in Don's collection, probably the same as the one mentioned by
Thonning under the old name of *Caladium esculentum.*

XCIX. TYPHACEÆ.

The *Typha angustifolia,* L., is found in the waters of the

---

* In Kunth's Enumeratio, this island is, by a slip of the pen, stated to
be off *New Guinea* instead of *Guinea.*

Guinea Coast, as in most other parts of the world. It is not in the collection before us, but is described by Schumacher and Thonning under the name of *T. australis.*

C. NAIADEÆ.

1. Potamogeton *pusillus,* Linn.—*Kunth, Enum.* 3. *p.* 136.— Cape Coast, *Don;* Angola, and in most parts of the globe.

The only other species of this Order I can find any mention of are *Ouvirandra Heudelotii,* Kunth, from Senegal, and *Aponogeton subconjugatus,* Schum. et Thonn., (probably an *Ouvirandra*), from Guinea.

CI. ALISMACEÆ.

Of this Order two species only are given as West Tropical African: *Alisma sagittifolium,* Willd., from Guinea, and *A. humile,* Kunth., from Senegal.

CII. HYDROCHARIDEÆ.

1. Vallisneria *spiralis,* Linn.?—Sierra Leone, deep in the water, *Don.* There is no fructification on the specimens.

The extraordinary paucity of aquatic plants in all collections from West Tropical Africa, as compared with other Tropical regions, may be owing rather to the inattention of collectors, and the difficulty of gathering them without a longer stay in the country than the climate permits, than to any real deficiency of vegetation in these waters.

CIII. BURMANNIACEÆ.

1. Dictyostegia *longistyla,* Benth.; caule erecto simplici, floribus solitariis pluribusve laxe cymosis v. unilateraliter racemosis, stylo perianthium subæquante, ovario ecostato.—On the Nun, growing apparently on dead roots, *Vogel.*

Herba tenera, alba, 3-6-pollicaris, *Dictyostegiis Americanis* simillima. *Folia* squamæformia, pauca, erecto-patentia, semilineam longa. *Inflorescentia* revera cymosa centrifuga, cyma tamen nunc ad florem terminalem reducta, nunc bifida 4-flora flore alari deficiente, nunc floribus alaribus solis evolutis et ramo altero cujusve bifurcationis abortiente in racemum unilateralem mutata, pedicellis bracteæ oppositis. *Flores* magnitudine et forma *D. orobanchoidis,* at erecti videntur, lobis alternis perianthii minimis. *Stamina* eadem. *Stylus* multo longior, apice dilatatus et trifidus, lobis stigmatosis bifidis. *Capsula* apice subvalvatim dehiscere videtur, seminibus plurimis placentis 3 parietalibus affixis.

CIV. ORCHIDEÆ.

The difficulty of drying plants which require any extra care in the operation, during the hasty visits to the hot damp regions of West Tropical Africa, has probably been the cause of the paucity of thick-leaved *Monocotyledoneæ* in general, and of *Orchideæ* in particular, in all collections brought from thence. Vogel says, in his Journal, that *Orchideæ* are frequent in Sierra Leone, and that he saw more than thirty species at one time at Mr. Whitfield's, which that collector meant to take to Europe. I cannot, however, find any record of more than twenty-six already published from the whole region, of which at least half-a-dozen are very doubtful. Of the collections before us, Don's contains the *Polystachya puberula,* Lindl., *Eulophia lurida,* Lindl., and two specimens without flowers, not determinable; Vogel's has the *Ansellia Africana,* Lindl., *Eulophia Guineensis,* Lindl., or a species closely allied to it; a new *Lissochilus,* described below, and three undeterminable specimens. The published species, as far as I can collect from the very irregular manner in which garden plants are described in periodicals of all kinds, are the following:

Megaclinium *falcatum,* Lindl., Sierra Leone; *M. maximum,* Lindl., Sierra Leone; *M. velutinum,* Lindl., Cape Coast.

Bolbophyllum *recurvum,* Lindl., Sierra Leone and Rio Janeiro; *B. tetragonum,* Lindl., *B.? pumilum,* Lindl., and *B.? galeatum,* Lindl., Sierra Leone.

M M

Polystachya *puberula*, Lindl., and *P. affinis*, Lindl., both from Sierra Leone, and both probably varieties of one species, as the specimen in the Horticultural Society's herbarium is branched, as in *P. puberula*, with the broad leaves of *P. affinis*.

Dendrobium? *paniculatum*, Sw., and *D.? roseum*, Sw., Sierra Leone, both very doubtful.

Ansellia *Africana*, Lindl., Fernando Po, *Vogel* and *Ansell*.

Eulophia *lurida*, Lindl., Sierra Leone; *E. Guineensis*, Lindl., Sierra Leone, and apparently the same species, on the Quorra, *Vogel; E. articulata*, Lindl., Guinea.

Limodorum *cristatum*, Afz., and *L. cucullatum*, Afz., Sierra Leone, both to be excluded from *Limodorum*, but doubtful as to what genus they should be referred to.

Galeandra *gracilis*, Lindl., and *G. extinctoria*, Lindl., Sierra Leone.

Lissochilus *macranthus*, Lindl., Boney.

Lissochilus *longifolius*, Benth.; foliis carinatis longe lineari-lanceolatis, scapo radicali 6-9-floro, bracteis pedicello multo brevioribus, sepalis lineari-cuneatis obtusis reflexis æqualibus, petalis ovato-oblongis obtusissimis, labello sublibero basi infundibulari, lobis lateralibus abbreviatis intermedio ovato obtusissimo margine undulato medio basi lineis longitudinalibus cristato.— Grand Bassa, in open meadow-like morasses, *Vogel*.

Scapus pedalis, racemo laxo ovariis longe pedicellatis, bractea ovata v. lanceolata vix 2 lin. longa. *Sepala* 6 lin. longa, vix lineam lata, basi angustata. *Petala* iis æquilonga, sed triplo latiora et consistentia multo tenuiora. Lineæ *labelli* fimbriis numerosis cristatæ.

Zygopetalum *Africanum*, Hook., Sierra Leone.

Gymnadenia *macrantha*, Lindl., Sierra Leone.

Habenaria *ichneumonea*, Lindl., Sierra Leone; *H. membranacea*, Lindl., Sierra Leone; *H. filicornis*, Lindl., Guinea; *H. procera*, Lindl., Sierra Leone.

CV. SCITAMINEÆ.

1. Canna *Orientalis*, Rosc.—*Bouch. Linnæa*, 8. *p*. 152. *et* 18. *p*. 490.—Sierra Leone, *Don, Vogel*, perhaps cultivated.

1. Maranta *arundinacea*, Linn.—Cultivated at Cape Coast, *Vogel*.

2. Maranta ? *brachystachys*, Benth. ; culmo basi vaginato apice monophyllo ad basin petioli elongati lateraliter florifero, spicis brevibus distichis subternis, bacca disperma.—Grand Bassa, Cape Palmas, and on the Nun, *Vogel;* St. Thomas, (a leaf only), *Don.*

Culmi crassi, subcompressi, 1-2-pedales, basi vaginis 1-2 arcte appressis inclusi. *Folium* ample ovatum v. oblongum ; petiolus 6-9-pollicaris, erectus et cum culmo subcontinuus, basi hinc breviter fissus inflorescentiam vaginans, dein clausus subteres intus cavus, superne breviter solidus ; lamina 8-10-pollicaris, brevissime acuminata, basi acuta, consistentia et venatione *Phryniorum*. *Spica* intermedia bipollicaris, laterales breviores ; rhachis sæpe pubescens ; spiculæ sessiles, plurifloræ. *Bracteæ* imbricatæ, interiores exterioribus breviores. *Pedicelli* (1-2-flori ?) bracteas subæquantes. *Flores* desunt. *Bacca* magnitudine *Cerasi*, rubra ex Vog., abortu bilocularis (loculo tertio minimo vacuo). *Semina* subglobosa, albumine albo fere osseo.

I have referred this to *Maranta* on account of the baccate fruit ; possibly when the flower is known it may prove to belong to some new genus.

There is a *Maranta cuspidata*, published by Roscoe, from Sierra Leone, but not in our collections.

1. Phrynium *flexuosum*, Benth. ; culmo foliato, foliis longe petiolatis ovato-lanceolatis, racemis gracilibus ramosis, ovario villosissimo, sepalis lanceolatis, staminodiis 2 unguiculatis latissimis, tertio erecto concavo obtuse subtrilobo, intimis erectis angustis apice cucullatis breviter connatis altero hinc antherifero.—Maranta flexuosa, *Don, in Herb. Soc. Hort. Lond.*—Sierra Leone, *Don.*

Specimina adsunt (culmi summitates ?) pedalia, 1-2-foliata. *Petioli* basi vaginantes, medio teretes cavi, apice breviter solidi. *Folia* ipsa 6-10-pollicaria, basi obtusa. *Racemus* semipedalis v. longior, basi ramo uno alterove instructus. *Bracteæ* dissitæ, distiche pollicares, internodiis paullo longiores, cauli sub-appressæ. *Pedicelli* sæpius biflori. *Ovarium* globosum, pilis longis mollibus dense vestitum. *Sepala* 4-4½ lin. longa, acutissima. *Petala* tubo stamineo breviter adnata, sepalis duplo longiora, superne reflexo-patentia. *Tubus stamineus* sepala æquans; staminodia petaliformia (corolla interna et nectarium seu labellum auct.) 2 exteriora patentia lamina lata una ab altera dissimili, cætera erecta et minus connata quam in plerisque speciebus, forma valde irregulari. *Stylus* columnaris, apice reflexus concavo-dilatatus et lamina brevi auctus quasi bilobus. *Ovula* in loculis ovuli solitaria, erecta. *Capsulam* maturam non vidi, junior sicca est et facillime in valvulas 3 separatur.

2. Phrynium *ramosissimum*, Benth.; culmo ramosissimo foliato, foliis supra vaginas breviter petiolatis ovali-oblongis, racemis terminalibus subsimplicibus, ovario pubescente, sepalis lanceolatis, staminodiis 2 vix unguiculatis ovato-oblongis basi connatis, tertio brevi latissimo, intimis erectis angustis apice cucullatis alteconnatis altero hinc antherifero.—Fernando Po, *Vogel.* *Culmi* brachiato-ramosissimi, 2-3-pedales (ex Vog.), vaginis foliorum plus minus obtecti, ad nodos sæpe puberuli, cæterum planta glabra. *Folia* semipedalia ad pedalia, anguste et abrupte acuminata, basi obtusa, petiolo longe vaginante, dein brevissime cavo, parte solida vix pollicari. *Inflorescentia P. flexuosi*, nisi brevior, simplicior, bracteis longioribus. *Flores* ejusdem magnitudine. *Petala* tubo stamineo alte connata. *Stylus P. flexuosi. Capsula* obovoidea, trivalvis, fere glabra.

3. Phrynium *filipes*, Benth.; culmo ramoso foliato, foliis supra vaginam brevissime petiolatis ovatis v. ovato-lanceolatis cuspidatis, racemis gracilibus paniculato-ramosis paucifloris, pedunculis bracteas superantibus, floribus parvis, ovario pubescente, sepalis lanceolatis, staminodiis omnibus apice concavis cucullatisve, 2 obovatis, tertio valde difformi bilobo, intimis angustis, antheræ filamento sublibero.—Fernando Po, *Vogel.* *Rhizoma* e tuberibus pluribus horizontaliter dispositis compositus, fibris elongatis intermixtis. *Culmi* herbacei, suberecti, brachiato-ramosi, 2-4-pedales, vaginis foliorum fere obtecti. *Folia* 4-5 poll. longa, cuspide sæpe semipollicari; petiolo

supra vaginam longam 1-2 lin. longo. *Panicula* ex apice
vaginæ foliis summi exserta, folio brevior. *Bracteæ* angustæ, pollicares, patentes. *Pedunculi* filiformes, plerique
biflori. *Flores* omnium minimi. *Sepala* 1½ lin. longa,
acuta, striata. *Petala* tubo stamineo breviter adnata,
sepalis subdimidio longiora. *Tubus stamineus* lineam longus. *Staminodia* 2 exteriora obovata, breviter unguiculata, obliqua, apice concava subcucullata, uno altero paullo
majore; tertium carinato-concavum, lobo majore apice concavo, minore plano intus tamen sub apice transverse ap-
pendiculato, 2 interiora valde inæqualia, lineari-cuneata,
altero apice cucullato. *Anthera* unilocularis, filamento ad
apicem tubi staminei inserto. *Stylus* columnaris, apice cu-
cullato-incurvus. *Capsula* obovoideo-triquetra, pallide lutea,
trilocularis, trivalvis, rarius loculo uno alterove abortiente.
Semina in loculis solitaria, globoso-triquetra, badia, lævia,
breviter arillata.

The genus *Phrynium*, to include the above three species,
must also comprehend the greater number of the *Calatheæ* of
modern authors, though perhaps not the original *Calathea* of
G. F. W. Meyer, if correctly described. The petaloid sterile
stamens or *staminodia* of Nees (inner corolla, nectarium and
labellum of other authors) are always very irregular in their
form, size and degree of connection with each other and with
the fertile stamen, very difficult to observe accurately in dried
specimens, and consequently only well known in a very few
species, scarcely ever similar in two different species, and apparently variable in some cases, in different flowers in one species.
These differences cannot therefore be available for generic distinctions, and all the *Canneæ* with 3-celled ovaries, solitary
ovules, columnar style and 3-valved capsule may be referred to
Phrynium. The inflorescence, which will be most probably
found to be connected with other differences, will characterize
the most natural sectional groups.

There are two other specimens without flowers which may
possibly be species of *Phrynium*, one from Sierra Leone in
Don's collection, in leaf only; the other, gathered at Abòh

by Vogel, has a three-valved, three-celled capsule, echinate, like
those of a *Canna*, but with single black arillate seeds in each cell.
1. Costus *afer*, Ker. *Bot. Reg. t.* 683.—C. Arabicus, *Schum. et
Thonn. Beskr. p.* 394?—Sierra Leone, *Vogel, Don ;* Abòh,
Vogel ; both the smooth and the hairy varieties from both
localities.
1. Amomum *grana-paradisi*, Linn.—*Afz. in Ræm. Schult. Syst.
Mant.* 1. *p.* 36 ? Grand Bassa and Abòh, a common plant,
Vogel. (Guinea Pepper).*

These specimens, as far as can be ascertained in the dried
state, where the texture of the flowers is so very delicate,
agree with the plant which Afzelius describes under the
above name as the common *Amonum* of the Guinea Coast,
except that the flowering scape varies from an inch to near
a foot in height, bearing at its summit several large white
flowers, tinged with purple towards the apex, and the pulpous
fruit (perhaps not yet ripe) is not above 1½ in. long.

The other West Tropical African *Scitamineæ* are : *Amomum
grandiflorum*, Sm., *A. Afzelii*, Rosc., *A. latifolium*, Afz., and
Zingiber dubium, Afz., all from Sierra Leone ; and Schumacher
and Thonning mention the *Zingiber officinale*, Rosc., and *Cur-
cuma longa*, Willd., as cultivated.

CVI. Irideæ.

Of this Order there are not as yet, as far as I am aware, any
species published from West Tropical Africa, and I have seen
specimens of but one species, gathered south of the Line by
Curror. It belongs apparently to some unpublished genus
allied to *Gladiolus*, but the flowers are too rotten for examination. There is also in Don's Sierra Leone collection a fragment of another plant, but it has the appearance of a Cape
species, probably from some garden.

CVII. Amaryllideæ.

There are no specimens in the collection, and the following
are the only species recorded from the region. *Hæmanthus*

* See below a note in the Addenda.

multiflorus, Mart. et Nodd., from Sierra Leone ; *H. cruentatus,* Schum. et Thonn., from Guinea, very probably the same species ; *Crinum purpurascens,* Herb., from Fernando Po, *C. Broussoneti,* Herb., and *C. distichum,* Herb., from Sierra Leone ; *C. petiolatum spectabile,* Herb., from St. Thomas ; *Amaryllis nivea* and *A. trigona,* Schum. et Thonn., from Guinea, both evidently *Crina,* and possibly the same as some of Herbert's species, and lastly, *Gethyllis pilosa,* Schum. et Thonn., which from the description must be a *Curculigo,* or some allied plant.

CVIII. Bromeliaceæ.

The Pine-apple is said to be common on the Guinea Coast, wild as well as cultivated.

CIX.? Taccaceæ.

1. Tacca *involucrata,* Schum. et Thonn. *Beskr. p.* 177.—Cape Coast, *Don.*

This species is very near to *T. pinnatifida,* it only differs from the ordinary forms of that species by the admixture of a number of small round or blunt oval segments of the leaves amongst the larger ones, and even these are less pointed than usual. A similar form, but with much larger and more divided leaves was gathered in the Mozambique by Forbes.

CX. Dioscorideæ.

1. Dioscorea (Amphistemon) *latifolia,* Benth. ; glabra, bulbi-fera, foliis alternis sinu lato cordatis longe cuspidatis 7-9-nervibus transverse venulosis punctis oblongis pellucidis, spicis gracillibus subfasciculatis masculis subramosis, floribus solitariis sessilibus bracteatis unibracteolatisque, perianthio sexpartito, antheris filamento æquilongis. — On the Nun River, *Vogel,* (male specimens) ; and probably the same species, Sierra Leone, *Don,* (female specimen.)

Caules teretes, inermes, volubiles. *Folia* longiuscule petiolata, circa 3 poll. longa et sæpius longitudine sua latiora, auriculis

536 FLORA NIGRITIANA.

baseos rotundatis sinu late aperto, cuspide apicis 6-12 lin. longo. *Spicæ masculæ* folia vix superantes, fere a basi floriferæ. *Bractea* perianthio brevior, basi dilatata, acute cuspidata. *Bracteolam* unicam tantum vidi ad unum latus perianthii et hæc ad flores ultimos minima est. *Perianthium* lineam longum, tubo brevissimo, laciniis angustis, exterioribus paullo latioribus, interiores tamen non omnino occultantibus. *Stamina* laciniis breviora, ad basin perianthii inserta. *Ovarii* rudimentum parvum. *Spicæ fæmineæ*, (in specimine Doniano ut videtur ejusdem specie) 8-10 poll. longæ, floribus dissitis solitariis axi fere appressis, perianthii laciniis iis marium similibus. *Ovarium* per anthesin vix lineam longum, mox elongatur, fructum tamen non vidi.

2. Dioscorea (Amphistemon) *præhensilis*, Benth.; glabra, aculeata, ramulis subteretibus, foliis plerisque oppositis ovatis rotundatisve cuspidatis basi sinu lato hastato-cordatis v. summis rotundatis 5-7-9-nervibus, venis reticulatis paucisve transversis, adultis opacis impunctatis, spicis masculis simplicibus fasciculatis folium raro superantibus, floribus sessilibus solitariis, perigonii laciniis lato-ovatis. — Sierra Leone, *Don, Vogel.*

Frutex scandens, ramulis hinc inde compressis, demum striatulis, sæpe glaucis, aculeis recurvis crebris rarisve. *Folia* forma variabilia, sed nunquam profunde cordata et sæpe basi in formam hastatam vergentia, 2-3 poll. v. rarius fere 4 poll. longa, novella punctis paucis oblongis pellucidis mox evanidis notata, rigidule membranacea; venulæ inter costas nunc omnes reticulato-ramosæ, nunc paucæ a costa ad costam transversæ; petiolus longiusculus limbo tamen brevior, ima basi auriculato-dentatus v. nudus. *Spicæ masculæ* tenues, bipollicares, omnes simplices et fasciculatæ, sed interdum (ramulo florifero axillari aphyllo) in paniculas axillares dispositæ. *Bractea* ovata acuta, perianthio brevior, bracteola adhuc minor. *Perianthium* ¾ lin. diametro, laciniis 3 exterioribus orbiculato-concavis æstivatione valde imbricatis, interiores omnino occultantibus, his lato ovatis, æstivatione subvalvatis v. angustissime imbricatis. *Stamina* in centro floris biseriata,

perianthii dimidium æquantia, antheris ovato-oblongis contiguis erectis filamento sublongioribus. *Stirps* fœminea non
visa.

3. Dioscorea *Cayennensis*, Lam.—*Griseb. in Endl. et Mart. Fl.
Bras. Diosc. p.* 33?—Cultivated on the Nun and Quorra
rivers, where the bulbs (tubers?) are the size of a Horse-
chestnut, but compressed, (*Vogel.*)
These specimens, nearly allied to the preceding species, have
however longer male spikes always solitary in the axils, and the
stems scarcely prickly or only in the lower part. They agree
with some old West Indian specimens of a *Dioscorea,* which
appears to me to be the *D. rotundata,* Poit. Dict. Suppl. 3. p.
139, and answer in every respect to Grisebach's more accurate
character of *D. Cayennensis,* except that, as in other allied
species, I find only one bracteola besides the subtending bract
to each flower, or two in the whole, not three.

4. Dioscorea *hirtiflora,* Benth.; caule striato puberulo, foliis
alternis cordato-orbiculatis subreniformibusve acute cuspidatis
membranaceis 5-7-nervibus reticulato-venosis supra glabris
subtus stellato-pubescentibus, spicis masculis laxis subsimplicibus fasciculatis hirtis, floribus solitariis subpedicellatis, perianthio 6-partito, staminibus fertilibus 3 brevibus, sterilibus
3 filiformibus ovarii rudimentum trifidum æquantibus.—On
the Quorra, opposite Stirling, *Vogel.*

Ramuli tenues, ut videtur inermes, pilis brevibus fuscis fasciculatis ramosisve conspersi v. glabrati. *Folia* longiuscule petiolata, 1½-2 poll. longa, et sæpe longitudine sua latiora, cuspide
apicis acutissimo, auriculis baseos rotundatis, sinu in novellis
angusto in adultis lato; venæ parum conspicuæ; punctæ
pellucidæ in junioribus lineares, demum obscuratæ. *Spicæ
masculæ* graciles, 1½-2-pollicares, rhachide floribusque pilis
stellulatis canescentibus. *Flores* demum dissiti, bractea angusta acutissima, bracteola minuta, apertos non vidi, alabastrum ¾ lin. longum. *Perianthii* laciniæ 3 exteriores
ovatæ, extus hirtæ; 3 interiores angustæ, glabræ, consistentia
multo tenuiore. *Stamina* fertilia ad basin perianthii cum
laciniis interioribus alternantia, filamentis brevissimis, an-

theris subglobosis ; sterilia interiora filiformia, fertilibus longiora, infra apicem articulata. *Ovarii* rudimentum divisum
in lacinias tres filiformes staminodia æquantes et iis subsimiles nisi continuas. In parte inferiore spicarum adsunt
interdum flores nonnulli imperfecte hermaphroditi, ovario
infero, ovula nonnulla continente, perianthii laciniis staminibus stylisque imperfectis coronato. *Stirps fœminea* haud visa.
The species is not referrible to any of Grisebach's S. American
sections.

5. Dioscorea *rubiginosa,* Benth.; foliis alternis late cordatis
subcuspidatis 7-9-nervibus supra glabris subtus caule inflorescentiaque ferrugineo-tomentosis, spicis fœmineis fasciculatis.—Sierra Leone, *Don.*

Specimen unicum fœmineum ab omnibus a me cognitis tomento
brevi ferrugineo (e pilis stellatis composito) distinctum. *Folia*
caulina 3-4-pollicaria, floralia dimidio minora. *Spicæ* 3-4-
pollicares. *Flores* dissiti, sessiles, bractea parva lata cuspidata, bracteola minima oblonga. *Ovarium* per anthesin 1½
lin. longum, obtuse trigonum, intus incomplete triloculare,
placentis 3 linearibus utrinque uniovulatis perianthii laciniis
interioribus oppositis. *Stylus* basi breviter simplex erectus,
dein divisus in lobos 3 cum placentis alternantes recurvos
breves canaliculato-dilatatos apice emarginatos laciniis perianthii breviores. *Perianthii* laciniæ extus tomentosæ,
exteriores ovatæ, interiores angustiores.

6. Dioscorea *dæmona,* Roxb. *Fl. Ind.* 3. *p.* 805.—*Wight, Ic. t.*
811.—D. virosa, *Wall. ?*—Aguapim, *Vogel;* a single specimen, with young male inflorescence, apparently the same as
the widely diffused East Indian plant.

7. Dioscorea *vespertilio,* Benth.; glaberrima, foliis trisectis segmentis petiolulatis obovali-oblongis obtusissimis subcoriaceis
nitidis, fructus alis 2 latissime expansis transverse ovato-
oblongis rigide membranaceis, tertia angustissima v. costæ-
formi.—Sierra Leone, *Don;* a single specimen, in leaf, with
a few loose fruits.

Tota glaberrima et lævis, siccitate glauco-nigricans, quodammodo *Stauntoniam* referens. *Petioli* alterni, subtripollicares,

apice divisi in petiolulos tres semipollicares ; infra petiolum
adest sæpe mucro brevis recurva. *Foliorum segmenta* 2-3
poll. longa, supra medium 1-1½ poll. lata, basi et apice
obtusa, margine recurva, sub lente minutissime et creberrime
pellucido-punctata : costa media venæque ab ea divergentes
1-2, alternæ, subtus prominentes ; venæ 2 oppositæ tenuiores
ad basin costæ mediæ convergentes ; venulæ reticulatæ sub-
transversæque. *Inflorescentia* et *flores* desunt. *Capsulæ* axis
1-1¼ poll. longa ; alæ horizontaliter divergentes, 1½ poll.
longæ, parallele venosæ. *Semina* non vidi.

There are besides, in Don's collection from Sierra Leone,
some bunches of capsules of a *Dioscorea*, possibly one of those
above-described, but there are no leaves to identify them ; and
the *D. alata*, Linn. and *D. sativa*, Linn. are both, according to
Thonning, in cultivation.

CXI. LILIACEÆ.

1. Gloriosa *superba*, Linn.; var petalis apice tantum undulatis.
—Grand Bassa, *Vogel.*

The foliage is exactly like that of the Indian *G. superba*, and
of the Mascarene (not W. African) *G. virescens*, Lindl. The
flower is fully as large as in *G. superba*, yellow when young,
red after it is fully out, according to Vogel. Forbes's Madagascar specimen of *G. virescens* is bad, and has two flowers, the
one young, apparently about the size represented in the Botanical Magazine, the other, more developed, fully as large as the
East Indian ones, but not so undulate. Lamark describes a
Senegalese form, with petals broader than any I have seen.
Schumacher and Thonning describe a *G. angulata*, from Guinea,
with petals pubescent at the apex, his other character, the
angularly compressed stem, frequently occurring in all the
varieties. There is, moreover, a South African form, gathered
by Drège, and referred to *G. virescens* by Kunth, which I have
not seen. Whether all these be mere varieties of *G. superba*,
or whether any or all the African ones may belong to a distinct
species, and if so, what are its geographical limits, can only be
determined from better materials than we possess at present.

Of the tribe of *Asphodeleæ* there are but two specimens, both without flowers, in the collections; one from Sierra Leone, *Don*, may be the *Chlorophytum orchidastrum*, Lindl., the other from Grand Bassa, *Vogel*, is wholly indeterminable. The published West Tropical African species are : *Urginea Senegalensis*, Kunth., Senegal; *Chlorophytum inornatum*, Ker., and *C. orchidastrum*, Lindl., Sierra Leone; *Allium Guineense*, Schum. et Thonn., *Ornithogatum ensifolium*, Schum. et Thonn., *Aloe picta*, Thunb., and *Sanseviera Guineensis*, Willd., all from Guinea.

There are two specimens of *Asparagus* in Vogel's collection, both without flowers or fruit, and almost all the leaves loose; the one appears to be the East Indian *A. falcatus*, Linn., or a species closely allied to it; the other is very near to the *A. retrofracta* and *A. Asiatica*.

The *Dracæna fragrans*, Ker., from Sierra Leone and Guinea, *D. ovata*, Ker., from Sierra Leone ; and *Dianella triandra*, Afz., (*Duchekia*, Kostel,), from Sierra Leone, complete the list of known W. African species of this generally extensive Order, which would probably have furnished us with many more from this country, were they easier to collect and to dry.

CXII. MELANTHACEÆ.

Schumacher and Thonning have published a *Helonias Guineensis*, not taken up by subsequent authors. Judging from the description, it is not a *Helonias*, but belongs to some genus, perhaps new, of the tribe *Melanthieæ*.

CXIII. JUNCEÆ.

1. Flagellaria *Indica*, Linn.—*Kunth, Enum. 3. p.* 370.—F. Guineensis, *Schum. et Thonn. Beskr. p.* 181.—Cape Coast, *Vogel, Don;* a common East Indian and Tropical African plant.

CXIV. Commelyneæ.

1. Commelyna *communis*, Linn.—*Kunth, Enum.* 4. *p.* 36.—On the Quorra, at Stirling, *Ansell.*—This single specimen is the only one agreeing with the Eastern form described with unilateral pubescence on the stem.

2. Commelyna *agraria*, Kunth, *Enum.* 4. *p.* 38.—Sierra Leone, and on the Quorra, *Vogel;* St. Thomas, *Don;* agreeing well with Kunth's description as well as with American specimens, and apparently as common in West Tropical Africa as in Tropical America.*

3. Commelyna sp.—A single specimen, without any station given, but a memorandum of Vogel's stating that the flowers are yellowish. The sheaths of the leaves are covered with long hairs; in other respects it agrees with *C. agraria.*

4. Commelyna *Bengalensis*, Linn.—*Kunth, Enum.* 4. *p.* 50.—On the Quorra, at Stirling, *Ansell.*

5. Commelyna *capitata*, Benth.; caule repente minute scabro-puberulo, foliis subsessilibus ovato- v. oblongo-lanceolatis puberulis glabratisve, vaginis ore rufo-ciliatis, spathis turbinato-cucullatis in capitulum terminalem aggregatis margine rufo-ciliatis, pedunculis geminis altero incluso 3-4-floro, altero exserto unifloro.—Cape Palmas, *Vogel.*

Ab affinibus distinguitur pilis ad margines spatharum numerosis et in spathas ipsas sparsis, et spathis 4-5 in capitulum sessilem v. pedunculatum confertis. *Folia* dissita, latitudine varia. *Sepala* lata, tenuiter membranacea, eglandulosa. *Petalum* impar late ovatum. *Capsulæ* valvulæ crassiusculæ nec nitidæ nec striatæ.

6. Commelyna *sulcata*, Willd.—*Kunth, Enum.* 4. *p.* 56.—Sierra Leone, *Don;* Accra and Fernando Po, *Vogel.* Flowers blue or white.

7. Commelyna *nigritana*, Benth.; caulibus basi repentibus minute puberulis glabratisve, foliis sessilibus lineari-lanceolatis, vaginis ore hirtellis, spathis oppositifoliis breviter pedunculatis

---

* Since the above was written I have had before me a number of East India specimens, which lead me to doubt whether the *C. agraria* be really distinct from *C. communis.*

cucullatis acuminatis hirtellis, pedunculis solitariis 3-4-floris (sterili nullo ?), sepalis glanduloso-lineolatis, (petalo impari oblongo ?)—On the Quorra, at Attah, *Vogel.*

Caules semipedales ad pedales. *Folia* 2-3 poll. longa, circiter 2 lin. lata. *Pedunculi* puberuli, 3-4 lin. longi. *Spathæ* reflexæ, semipollicares, extus pilis minutis hirtellæ et prope basin aliis majoribus subpaleaceis hispidæ. *Flores* inclusi, parvi. *Petala* 2, longe unguiculata, calyce subtriplo longiora, angusta; tertium breve sessile.

The specimens are very rotten, and I could only examine one imperfect flower.

8. Commelyna *aspera*, G. Don, *in Herb. Soc. Hort. Lond.*; caulibus repentibus glabris, foliis lanceolatis v. lineari-lanceolatis supra v. utrinque scabro-hirtellis, spathis subsessilibus ad apicem caulis confertis turbinato-cucullatis acuminatis extus pilosis, sepalis glabris, (petalo impari ovato ?)—Accra, *Don;* Confluence of the Niger, *Vogel.*

This may possibly be a mere variety of the American *C. elegans,* Kunth.

There are two other West African species described, *C. umbellata,* Schum. et Thonn., from Guinea, which must be very near our *C. Nigritana,* although the description does not quite agree, and the Arabian *C. Forskalei,* which is indicated also from Senegambia.

1. Cyanotis *lanata,* Benth.; foliis lanceolatis linearibusve inflorescentiis cauleque laxe sericeo-lanatis, spicis lateralibus terminalibusque subgeminis, bracteolis falcato-lanceolatis, sepalis pilosissimis, corolla infundibulari triloba, staminibus vix exsertis.—Savannahs on the Quorra, at Addaenda and Patteh, *Vogel.*

Caules pedales v. longiores, diffusi, subramosi. *Pili* molles, albidi, laxe sericei, ad nodos et vaginas copiosi, in caule foliisque rariores, demum subevanidi. *Folia* 2-3-pollicaria. *Capitula* e spicis geminis formata, pleraque terminalia, adjecto nonnunquam altero axillari. *Folia* floralia complicata, inferius sæpe caulinis conforme, alterum brevius. *Inflorescentia C. cristatæ,* sed bracteæ angustiores valde falcatæ. *Sepala* 2 lin. longa, lanceolata, pilis longis copiosis ciliata.

Corolla glabra (ex Vog. intense cœrulea), tubo calyce paullo longiore, limbi laciniis subæqualibus. *Capsula* calycem subæquans, punctis paucis nigris notata, ad angulos et apice pilis longis ciliata. *Semina* foveolato-rugosa.

2. Cyanotis *longifolia*, Benth.; foliis subradicalibus linearilanceolatis molliter pilosissimis, scapo subnudo foliis longiore, spicis geminis altero breviter altero longius pedunculatis folio florali fultis, bracteis falcatis pilosiusculis.—Congo, south of the Line, *Curror*.

Folia 4-6 poll. longa, basi breviter vaginantia, subcæspitosa. *Scapus* folia longiora breviter superans, parce subunilateraliter pilosus, præter folium unicum pollicare ad ortum pedunculorum aphyllus. *Pedunculus* alter altero duplo triplove longior. *Spicæ C. cristatæ*, nisi minores. *Sepala* linearilanceolata, longe pilosa. *Corolla* breviter infundibularis. *Filamenta* exserta, apice dense barbata.

POLYSPATHA, gen. nov.

Flores irregulares. *Sepala* navicularia, impari duplo latiore, persistentia. *Petala* libera, 2 (v. 3 ?) longissime unguiculata, lamina tenuissima, (impari latiore brevi ?) *Stamina* 3; antheræ æquales oblongæ, loculis connectivo angustissimo junctis (sterilia nulla). *Ovarium* sessile, biloculare, ovulis in loculo solitariis. *Capsula* regulariter bivalvis, valvulis medio septiferis. *Semina* oblonga, lateraliter affixa, ventre longitudinaliter costata, dorso umbilicato-depressa et radiatim multicostata.—*Inflorescentia:* flores plures subsessiles intra spathas (seu bracteas) complicatas secus ramos flexuosos paniculæ sessiles et reflexas.

1. Polyspatha *paniculata*, Benth.—In woods, Fernando Po, *Vogel*.

Caulis basi prostratus, radicans, tum erectus, 1-2 pedes altus, uti tota planta pube minuta scaber. *Vaginæ* pollicares, laxæ, ore breviter ciliato, inferiorum truncato, superiorum obliquo. *Folia* subsemipedalia, ovata, acuminata et acuta, basi in petiolum brevem angustata. *Panicula* terminalis,

544 FLORA NIGRITIANA.

ramis simplicibus approximatis folio parvo fultis v. inferioribus ex axillis foliorum caulinorum superiorum natis, supremo
6-8 poll. longo, cæteris brevioribus; rhachis flexuosa, pubescens. *Spathæ* 5-6-lin. longæ, acutæ, membranaceæ, ciliatæ, costatæ, ad quemquam flexuram sessiles, in rhachin arcte
reflexæ. *Flores* intra spatham sæpius 4-5, bracteis parvis intermixtis, e basi spathæ prominentes. *Sepala* angusta, circa 3 lin.
longa, extus pilosula, carinata et apice fere cucullata. *Petala*
consistentia ita tenui ut formam laminæ in speciminibus siccis
haud rite observare potui. *Stamina* petalis æquilonga.
Ovarium breve, glabrum. *Stylus* calyce duplo saltem longior, apice bis terve involutus, summo apice stigmatosus,
brevissime bifidus et leviter dilatatus. *Capsula* nitidula,
ovata, compressa, obtusissima, 2 lin. longa.

PALISOTA, *Reichb.*, gen. nov.

Flores subirregulares. *Sepala* ovata, obtusa, impari parum majore. *Petala* sessilia, sepalis subsimilia iisque paullo longiora. *Stamina* fertilia 3, petalis opposita, difformia; duo
brevia, filamentis filiformibus anthera oblonga vix longioribus; tertium longius, filamento crasso anthera ovata pluries
longiore; sterilia 2 (v. rarius 3?) brevia, ananthera, pilis
longis articulatis dense barbata. *Ovarium* sessile, 3-loculare,
loculis biseriatim 5-6 ovulatis. *Stylus* apice simplex sub-
penicillato-stigmatosus.
1. Palisota *thyrsiflora*, Benth.—Commelyna ambigua, *Pal. de
Beauv. Fl. Ow. et Ben. 1. p.* 26. *t.* 15.—Cape Palmas, Grand
Bassa and Fernando Po, *Vogel;* Benin.
Caulis basi decumbens, dein adscendens, ramosus, 2-6-pedalis,
glaber v. barbato-villosus. *Folia* conferta, nunc semipedalia,
nunc sæpius pedalia v. longiora, oblonga, acuminata, basi in
petiolum brevem angustata, crassiuscula, supra glabra v. rarius
medium et basin versus barbata v. hirta, subtus pallida, hirtella,
costa media uti petioli vaginæque breves latæ pilis longis dense
barbata. *Thyrsi* terminales, solitarii v. in axillis superioribus pauci, folia subæquantes, floribundi, basi bracteis non-

nullis vaginati. *Bracteœ* ad basin pedunculorum parvæ, membranaceæ, acuminatæ, subcomplicatæ, caducæ. *Pedunculi* (ex Vog. albi) secus rhachin sparsi, horizontales, breves (raro semipollicares) apice circinnatim 2-6-flori. *Flores* vix 2 lin. longi, glabri v. parce hirtelli, sepalis ex Vogel pallide, petalis intensius, violaceis. *Fructus* maturum non vidi, junior subscariosus videtur.

I have no doubt this is Beauvois' plant, notwithstanding the discrepancies in the inflorescence and flowers. The remarkable habit and most prominent points of structure cannot be mistaken. The bearded sterile stamens are probably what he describes as numerous setæ inserted amongst the stamens, and as to inflorescence and colour, there are so many instances in the work in question where, the very bad specimens preserved affording no information as to these points, they were supplied from recollection or conjecture, that we presume it to have been the case with this plant, apparently not uncommon in the region. The genus would perhaps come under *Aneilema*, taken in its widest extent, but the habit is very different from that of any of the somewhat heterogeneous groups of which that genus is now composed.

1. Aneilema *ovato-oblongum*, Pal. Beauv. *Fl. Ow. et Ben.* 2. *p.* 71. *t.* 104. *f.* 1?; scabro-puberulum, foliis brevipetiolatis ovatis v. ovato-lanceolatis, panicula laxa, sepalis glabris, petalo impari cæteris minore sessili, staminibus 3 fertilibus inæqualibus, filamentis 2 longioribus apice barbellatis, 2 (v. 3?) sterilibus, capsulis bilocularibus, loculis dispermis.— Cape Palmas, *Ansell;* on the Nun and in Fernando Po, *Vogel;* Benin.

This is quite near enough to Beauvois' figure to suggest its being the same species, especially as it appears to have a wide range, and to vary in the form of its leaves, and in the presence or absence of long hairs and ciliæ. The petals, according to Vogel, are white; the deep blue colour given to the flowers of all the *Commelyneœ* figured by Beauvois, is evidently a supposition of the artist.

N N

2. Aneilema *Beninense*, Kunth, *Enum. p.* 73.—Commelyna
Beninensis, *Pal. Beauv. Fl. Ow. et Ben.* 2. *p.* 49. *t.* 87 ;
glabrum v. scabro-puberulum, foliis brevipetiolatis ovatis v.
ovato-lanceolatis, panicula densa, sepalis glabris, petalo impari cæteris minore sessili, staminibus 3 fertilibus inæqualibus,
2 sterilibus, filamentis omnibus subimberbibus, capsulis bilocularibus, loculis trispermis.—Cape Palmas, *Ansell ;* Grand
Bassa, *Vogel.*—β. *sessilifolium*, vaginis elongatis, petiolis
brevissimis.—Fernando Po, *Vogel.*

3. Aneilema *lanceolatum*, Benth. ; scabro-puberulum, foliis
breve petiolatis anguste lanceolatis, panicula brevi densa,
sepalis glabris, petalo impari cæteris minore sessili, staminibus fertilibus 3 inæqualibus, filamentis 2 longioribus apice
barbulatis,3 sterilibus, capsulis bilocularibus, loculis dispermis.
—On the Quorra, at Stirling, *Ansell, Vogel.*

Culmi basi decumbentes, radicantes, apice adscendentes, 1-2-
pedales. *Folia* 2-5 poll. longa, 2-7 lin. lata, acutiuscula,
basi vulgo ciliata. *Panicula* pedunculata, ovata, vix pollicaris,
cæterum ei *A. Beninensi* similis.

The above three species, with *A. umbrosum*, Vahl, from
Guinea, (judging from Thonning's description), and *A.
Drègeanum*, Kunth, from South Africa, constitute, as suggested
by Kunth, a distinct group, either generic or sectional, to which
the following character may be assigned.

Flores irregulares, abortu ovarii superiorum cujusve pedunculi
polygami. *Sepala* 3, impari latiore concavo. *Petala* 3, impari sessili, 2 unguiculatis tenuioribus. *Stamina* 3 fertilia,
filamentis 2 lateralibus longioribus apice barbulatis, tertio
glabro ; sterilia 2 v. 3 brevia glabra, anthera parva vacua.
Ovarium 2-loculare, ovulis in quoque loculo 2 v. 3 superpositis. *Stylus* simplex. *Capsula* elliptica, compressa, obtusa, nitida, loculicide bivalvis. *Semina* angulata, rugosa.—
Herbæ Africanæ. *Paniculæ* simplices v. rarius subramosæ,
pedunculatæ. *Bracteæ* parvæ, membranaceæ, concavæ. *Pedunculi* patentes, unilateraliter pluriflori, bracteolis parvis
orbiculatis persistentibus, floribus pedicellatis.

4. Aneilema *simplex*, Kunth. *Enum.* 4. *p.* 71.—Accra, *Don.*— The specimen is very imperfect, but appears to belong to the species described from Guinea by Vahl and Thonning. The *A. Africanum*, Beauv., and *A. æquinoctiale*, Beauv., both from Benin, are not among our collections.

CXV. RESTIACEÆ.

1. Eriocaulon *rivulare*, G. Don, *in Herb. Soc. Hort. Lond.*, acaule, foliis lato-linearibus, vaginis glabris, pedunculis folia superantibus, capitulis depresso-globosis albo-hirtis, bracteis involucrantibus obovato-orbiculatis subhyalinis glabris subciliatisque capitulo multo brevioribus flores stipantibus cuneatis acutis floribusque apice ciliatis, floribus masculis hexandris, sepalis alte connatis, petalis uniglandulosis, fœmineis trigynis sepalis petalisque distinctis, seminibus leviter striatis glabris.—In a rivulet near Freetown, Sierra Leone, *Don.*

Rhizoma simplex, crassum. *Folia* in aqua submersa, in altero specimine 3-4 poll. longa, 2-3 lin. lata, in altero 8-10 poll. longa, 3-4 lin. lata, subtiliter fenestrato-multinervia. *Scapi* in sicco compressi et tenuiter striati, folia paullo superantes. *Capitulum* circa 4 lin. diametro, primo intuitu ei *E. decangularis* simile. *Bracteæ* involucrantes pauciseriatæ, pallidæ, fere hyalinæ, ¾ lin. longæ, multo tenuiores et minus conspicuæ quam in speciebus affinibus Indicis; interiores hyalinæ, uti sepala petalaque apice pilis albis dense ciliatæ, flores æquantes, lineam longæ. *Flores* centrales masculi, in ambitu fœminei pluriseriati, omnes breviter pedicellati. *Sepala* anguste cuneata, in medio stipite inserta, marium fere ad apicem in calycem breviter trilobum connata, fœminum distincta. *Petala* immediate sub genitalibus inserta, marium staminibus breviora, fœminum longiora latioraque eglandulosa. *Filamenta* tria petalis opposita tribus alternis sublongiora; antheræ globosæ, didymæ. *Ovarium* sessile, stylo ad medium trifido. *Semina* ¼ lin. longa.

2. Eriocaulon *radicans*, Benth.; subacaule, foliis longe linearibus gramineis intimo pedunculum vaginante, capitulo hemi-

548 FLORA NIGRITIANA.

sphærico, squamis involucrantibus imbricatis latis obtusis stramineis appresse pubescentibus intimis radiantibus, flores stipantibus hyalinis glabris denticulatis, floribus apice ciliatis hexandris trigynisve, sepalis liberis, petalis glandula lineari marium alte connatis, fœminum apice conniventi-connatis.— Sierra Leone, *Don;* Grand Bassa, *Ansell;* south of the Line, *Curror.*

Rhizoma breve, crassum, simplex v. apice pluriceps. *Folia* semipedalia ad pedalia v. longiora, 2-5 lin. lata, rigidula, opaca, tenuissime multinervia nec fenestrata, glabra v. hirsuta. *Pedunculus* vulgo folia breviter superans, basi usque ad tertiam fere longitudinis folio intimo fere ad apicem convoluto vaginatus. *Capitulum* absque radio 4-5 lin. diametro, squamis exterioribus involucri arcte appressis rigidis, intimis apice radiatim patentibus, lamina interdum subpetaloidea nivea 3 lin. longa, in plerisque tamen speciminibus multo brevior, rarius vix conspicua. *Bracteæ* flores stipantes ovatæ, concavæ, apice sæpius glabræ. *Receptaculum* setis longe virentibus dense hirsutum. *Flores* breviter stipitati, involucrum (radio neglecto) subæquantes. *Sepala* in medio stipite oblonga, hyalina, apice ciliata, fœminea masculis paullo latiora. *Petala* in maribus in corollam turbinato-tubulosam apice ciliatam breviter trifidam connata, intus medio glandula longiuscule lineari nigra munita; in fœmineis angusta, basi libera, superne tantum in tubum brevem connata. *Antheræ* oblongo-sagittatæ, corolla paullo breviores. *Ovarium* triquetrum, stylo ad medium trifido.

The hairs of the leaves and the involucres, and the degree of development of the radiating scales appear to be very variable, possibly there may be more than one species, but the flowers seem to be similar in the few specimens I have seen. The leaves, involucres, bracts, &c., distinguish them all from *E. stellare,* Guill., and some other radiating Brazilian species which resemble it in many points of structure of the flowers.

1. Xyris *laxifolia,* Mart. *Herb. Fl. Bras. p.* 293. *n.* 547.—
Tropical Africa, south of the Line, *Curror.*
I cannot find the slightest difference between these and the

Brazilian specimens described by Martius. They resemble in every respect the common eastern *X. Indica*, except in the oval or oblong opaque spots on the otherwise shining scales of the involucre. Kunth describes a *X. platicaulis*, Poir., from the Mauritius and S. East Africa with the same opaque spot, but that appears to be a much smaller plant.

The only species hitherto published as West Tropical African is the *X. filiformis*, Lam., from Sierra Leone.

CXVI. CYPERACEÆ.

1. Cyperus *polystachyus*, Rottb.—*Kunth. Enum. 2. p.* 13.— Cape Palmas, *Ansell;* Sierra Leone and on the Nun River, *Vogel;* a common species in Africa, Asia and America, found also in South Europe.

2. Cyperus *aurantiacus*, Humb. et Kunth.—*Kunth, Enum. 2. p.* 20.—C. amabilis, *Vahl, Enum. 2. p.* 318?—C. lepidus, *Hochst. in Kotsch. Pl. exs. Herb. Nub. n.* 139.—On the Quorra, at Patteh, *Vogel;* Nubia and Tropical America.

An annual, varying from three to six or seven inches in height. The specimens agree precisely with Schomburgk's n. 221, from Guiana, with Moritzi's n. 1571, from Columbia, and with Kotschy's, from Nubia; as well as with Kunth's character. Vahl's description appears to apply to the same species, although the spikelets in our plant are more numerous, and the squamæ often produced into a very short point. It is a tristylous species, allied to, but distinct from the E. Indian *C. castaneus.*

3. Cyperus *aristatus*, Rottb.—*Kunth, Enum. 2. p.* 23.—In the town of Accra, *Vogel;* Senegal, South Africa and East India; also North America and the Galapagos, if I am right in considering the C. *inflexus*, Muhl., as the same species, and, I believe, in Brazil.

4. Cyperus *elegans*, Linn.—*Kunth, Enum. 2. p.* 28.—Fernando Po, *Vogel;* Tropical America, Africa and Asia.

This appears to be a widely spread species, to which are probably referrible C. *mæstus*, Kunth, and some others of the group of *Diffusi*, as well as one or two of Nees' Brazilian species.

5. Cyperus *Haspan*, Linn.—*Kunth, Enum.* 2. *p.* 34.—Sandy shores of the Nun, *Vogel;* Sierra Leone, *Don;* a common Tropical species in both hemispheres.

6. Cyperus *difformis,* Linn.—*Kunth, Enum.* 2. *p.* 38.—Sierra Leone, *Don;* Asia, Africa and South Europe.

7. Cyperus *coloratus,* Vahl, *Enum.* 2. *p.* 312.—Mariscus coloratus, *Nees.*—*Kunth, Enum.* 2. *p.* 126.—Accra and Cape Coast, *Vogel, Don.*

This species is allied to *C. sphærocephalus,* although the spiculæ, containing from five to eight flowers, are less compressed. The oval heads contain from twenty to thirty of these spiculæ, which are generally of a yellowish hue. The involucral leaves are narrow and slender, sometimes longer than the stem, but usually shorter.

There are a few specimens, from Aguapim, of a variety (or perhaps species ?) with depressed globose heads, and the involucral leaves broader and more numerous.

8. Cyperus *margaritaceus,* Vahl.—*Kunth, Enum.* 2. *p.* 46.— Accra, *Don, Ansell;* on the Quorra, *Vogel.*

This species is certainly closely allied to the East Indian *C. niveus,* but the habit of the plant is more tufted, as described by Thonning, without any, or scarcely any, of the creeping rhizoma of *C. niveus,* and the spikelets are rather larger, although without so many flowers in each.

9. Cyperus *rotundus,* Linn.—*Kunth, Enum.* 2. *p.* 58.—In the town of Sierra Leone, *Vogel;* common in nearly every part of the world.

There is also, from Sierra Leone, a variety with shorter spikelets, more decidedly spicate, and the squamæ not nearly so darkly coloured, which appears to come very near to the *C. lucidulus,* Kunth.

10. Cyperus *sphacelatus,* Rottb.—*Kunth, Enum.* 2. *p.* 63.— Cape Coast, *Don;* Grand Bassa and Sierra Leone, *Vogel, Ansell;* a South American species.

The dark-purple spot on the margin of the squamæ is sometimes wanting, especially in the larger specimens, which are as much as 2 feet high. The spikelets are then of a pale-greenish hue. In the smaller specimens, varying from 6 inches to 1 foot in height, the spot is usually very marked.

11. Cyperus *Papyrus*, Linn.—*Kunth, Enum.* 2. *p.* 64.—Very abundant on the banks of the Lagos, a river near Accra, *Don.* Evidently the same species as the Egyptian *Papyrus.*

12. Cyperus *exaltatus*, Retz.—*Kunth, Enum.* 2. *p.* 70.— St. Thomas, *Don.*

This specimen is precisely similar to the n. 3328 of Wallich's Catalogue from Rajemahl, which has much fewer flowers than the more common East Indian form.

13. Cyperus *ligularis*, Linn.—*Kunth, Enum.* 2. *p.* 79.—Sierra Leone and Grand Bassa, *Vogel;* on the Gambia, *Don;* Tropical America.

14. Cyperus *distans*, Linn.—*Kunth, Enum.* 2. *p.* 93.—Sierra Leone and Fernando Po, *Vogel;* St. Thomas, *Don;* Tropical America, Africa, and Asia.

15. Cyperus (v. Mariscus?) sp.—Fernando Po, *Vogel.* A large species, evidently different from any of the preceding, but the single specimen is too young to determine.

Besides the above, there are seventeen species of *Cyperus* mentioned as inhabitants of West Tropical Africa, viz: *C. intactus,* Vahl, Senegal; *C. patens,* Vahl, Guinea; *C. compressus,* Linn., Senegal, and common in America, Asia, Africa, and South Europe; *C. patulus,* Kit., Sierra Leone and South Europe; *C. articulatus,* Linn., a common Tropical species; *C. bidentatus,* Vahl, Senegal; *C. pustulatus,* Vahl, Guinea; *C. venustus,* Br., Senegal, Africa, Asia, and Australia; *C. radiatus,* Vahl, Senegal, Guinea, Madagascar, and East India; *C. crassipes,* Vahl, Senegal to Benin; *C. pectinatus,* Vahl, *C. scirpoides,* Vahl, *C. polyphyllus,* Vahl, and *C. microstachyos,* Vahl, all four from Guinea; *C. recurvus,* Vahl, Sierra Leone; *C. dilatatus,* Schum. et Thonn., and *C. angustifolius,* Schum. et Thonn., (non Nees.), the two last both from Guinea, and overlooked in Kunth's Enumeratio.

1. Mariscus *umbellatus*, Vahl.—*Kunth, Enum.* 2. *p.* 118.—
Cape Coast, *Don ;* Sierra Leone and on the Quorra, *Vogel ;*
Senegal to Benin, South Africa, and Asia.
Another species, *M. alternifolius,* Vahl, has been described
from Guinea.

1. Kyllingia *aphylla,* Kunth, *Enum.* 2. *p.* 127.—Sierra Leone,
Vogel, Don ; Grand Bassa, *Vogel, Ansell ;* on the Nun,
Vogel.—Senegal to Benin and Tropical America.

2. Kyllingia *monocephala,* Linn.—*Kunth, Enum.* 2. *p.* 129.—
Sierra Leone to Fernando Po, *Vogel, Don,* and others. A
common Tropical species.

3. Kyllingia *polyphylla,* Willd.—*Kunth, Enum.* 2. *p.* 135.—
Bassa Cove, *Ansell, Vogel ;* Cape Coast, *Don.* Very closely
allied to the East Indian *K. gracilis,* and *K. cylindrica,* as
well as to some of the Brazilian species, and possibly a mere
variety of *K. monocephala,* with elongated heads, and more
unequal inner squamæ.

In both the above species, two lateral heads are very frequently developed among the more vigorous of the African specimens, or on the luxuriant stems of tufts, where they are
generally monocephalous. The form of the seeds as well as
their number, the relative proportion of the two inner squamæ
of the spikelets, the ciliæ on the dorsal rib of these squamæ,
and the number of stamens, from one to three, are exceedingly
variable in different spikes of the same plants; and it appears to
me, that many of the species established on similar characters,
are but mere varieties of *K. monocephala.*

I have seen no West African specimens answering the cha-
rater of *K. triceps,* common in other Tropical regions. The
K. bulbosa, Beauv., from Benin, is probably a mere variety of
K. monocephala, which has often a tendency to thicken the
base of the stem. The *K. squamulosa,* Vahl., *K. dipsacoides,*
Schum. et Thonn., and *K. erecta,* Schum. et Thonn., all three
from Guinea, are unknown to me, and may be distinct.

1. Remirea *maritima,* Aubl.—*Kunth, Enum.* 2. *p.* 139.—Sierra
Leone, *Don ;* sea-coast of West Tropical Africa and East
Tropical America.

1. Eleocharis *capitata*, Br.—*Kunth. Enum.* 2. *p.* 150.—Cape Palmas and sandy shores of the Nun, *Vogel;* America, Africa, Asia, and Australia.

These are very vigorous specimens, the spikes sometimes short and ovate, at others cylindrical, and nearly half an inch long.

The *E. atropurpurea,* a mere variety of the above, is found in Senegambia as well as in East India.

The *Scirpus maritimus,* Linn., found over a great part of the globe, grows also in Senegambia.

1. Fuirena *umbellata,* Rottb.—*Kunth, Enum.* 2. *p.* 185.— F. pentagona, *Schum. et Thonn. Beskr. p.* 42?—Sierra Leone and St. Thomas, *Don;* on the Nun River, *Vogel;* Tropical America and Asia.

2. Fuirena *glomerata,* Lam.—*Kunth. Enum.* 2. *p.* 184.— F. Rottboellii, *Nees.*—Accra, *Don;* East India.

There appears to be considerable confusion in the distributed specimens, both of Wallich and of Wight, between this plant and the *K. uncinata,* Kunth, (*K. ciliaris,* Roxb., according to Nees); they are, however, accurately distinguished both by Kunth and by Nees in their published characters. The two species appear equally common in India, but the few African specimens I have seen, belong all to the *F. glomerata.*

1. Isolepis *barbata,* Br.—*Kunth, Enum.* 2. *p.* 208.—Accra, Grand Bassa and on the Quorra, *Vogel;* Senegal to Benin, East India and Australia.

Some of the Grand Bassa specimens are above a foot and a half high, whilst others among the Quorra ones are scarcely two or three inches; but all are alike in the structure of their flowers.

Five other species of *Isolepis* are cited as W. Tropical African, viz. *I. supina,* Br., Senegal, Guinea, and other parts of Africa, S. Europe, East India, and Australia; *I. prælongata,* Nees, Senegal and East India; *I. Micheliana,* Rœm. et Schult., Senegal, North Africa, South Europe, East India; *I. filamentosa,* Rœm. et Schult., Guinea; and *I. Willdenowii,* Kunth, Sierra Leone.

The *Nemum spadiceum,* Desv., from Sierra Leone and the West Indies, is unknown to me.

1. Fimbristyles (Trichelostyles) *muriculata,* Benth.; culmis cæspitosis 5-6-angularibus glabris basi foliatis, foliis lineari-bus planis rigidulis culmo brevioribus, umbella irregulariter decomposita pauciradiata, involucro brevi, spicis ovato-lanceo-latis acutis centralibus sessilibus; squamis carinatis ovatis acutis mucronulatis fuscescentibus trinervibus glabris nitidis, stylo trifido, achenio stramineo-albido undique tuberculoso-muriculato.—Accra, *Don.*

Near *F. quinquangularis* and *F. autumnalis,* but differs from the former in the form of the spikelets and squamæ, and from *F. autumnalis* in the muricate achenia.

2. Fimbristyles *hispidula,* Kunth, *Enum.* 2. *p.* 227.—On the Quorra, *Vogel, Ansell;* Senegal, Guinea, South Africa and Tropical America.—The specimens vary from six inches to three feet in heighth, and are sometimes nearly smooth, though generally the stiff spreading hairs of the leaves and stems are very copious.

3. Fimbristyles *communis,* Kunth, *Enum.* 2. *p.* 234.—Senegal to Benin, and Fernando Po, *Don, Vogel, and others;* Africa, Asia, and probably South America.

4. Fimbristyles *ferruginea,* Vahl, *Enum.* 2. *p.* 291.—F. ferru-ginea, arvensis et Sieberiana, *Kunth, Enum.* 2. *p.* 236, 237. —On the Gambia, *Don;* Grand Bassa and sandy shores of the Nun, *Vogel;* Tropical America, Africa and Asia.

5. Fimbristyles *obtusifolia,* Kunth, *Enum.* 2. *p.* 240.—Grand Bassa and on the Nun River, *Vogel;* Senegal and South Africa.

Schumacher and Thonning have described another species, *F. scabrida,* from Guinea.

1. Abildgaardia *monostachya,* Vahl.—*Kunth, Enum.* 2. *p.* 247. —St. Thomas, *Don;* Tropical America, Africa, East India and Australia.

2. Abildgaardia *pilosa,* Nees.—*Kunth, Enum.* 2. *p.* 248.— Accra, *Don:* and in Vogel's collection without the precise station; Senegal and Guinea.

The *A. barbata,* Beauv., from the banks of the Formosa, and
the *A. lanceolata,* Schum. and Thonn., from Guinea, appear
to be both of them nearly allied to the *A. pilosa.*

1. Lipocarpha *argentea,* Br.—*Kunth, Enum,* 2. *p.* 266.—
Sierra Leone, *Don;* Tropical America, Africa and Australia.

2. Lipocarpha *sphacelata,* Kunth, *Enum.* 2. *p.* 267.—Cape
Palmas, *Ansell;* East India.

The style is generally trifid, but occasionally, in the same
spikes bifid, and the *L. filiformis,* Kunth, from Guinea, is
probably the same species, nor does the Brazilian *L. gracilis,*
Nees, appear to be essentially distinct. In the African specimens, the spikes are usually from 5 to 7 in the head.

1. Hypolytrum *latifolium,* A. Rich.—*Kunth, Enum.* 2. *p.* 271.
—Fernando Po, *Vogel;* Tropical Africa and Asia.

1. Rhynchospora *Wallichiana,* Kunth, *Enum.* 2. *p.* 289.—
Grand Bassa, *Vogel;* Mauritius, East India and China.

These specimens have the heads smaller, and fewer spikelets
than the East Indian ones, but their structure is precisely the
same. The style has two minute lobes at the extremity, which
soon wear down after fecundation.

2. Rhynchospora *aurea,* Vahl.—Kunth, *Enum.* 2. *p.* 293.—
Sierra Leone, *Vogel;* Tropical America, Africa, Asia and
Australia.

1. Scleria *flagellum,* Sw. ?—*Kunth, Enum.* 2. *p.* 339 ?—Sierra
Leone, *Don;* Grand Bassa, *Vogel, Ansell.*—Tropical America and Africa.

The leaves are nearly smooth, and the achenia white, but it
agrees in other respects with Kunth's description.

2. Scleria *reflexa,* Humb. et Kunth.—*Kunth. Enum.* 2. *p.* 340.
—Fernando Po, *Vogel;* Tropical America.

3. Scleria *racemosa,* Poir ?—*Kunth. Enum.* 2. *p.* 344 ?—Grand
Bassa, *Vogel.*—Madagascar.

These specimens answer to Kunth's description. The achenia
are 1¼ lines in diameter, but have not perhaps yet attained
their full size. I have another gathered by Michelin in Senegal, closely resembling this one, but with still larger achenia

much depressed in form, which may be a distinct species.
There is also a *S. verrucosa*, Willd., described from Guinea,
which is not in our collections.

4. Scleria *spicæformis*, Benth., vaginis subalatis foliisque angustis hirsutis, ligula brevi rotundata rigida, panicula breviter spicæformi, spiculis fœmineis unifloris singulis sessilibus
cum masculis pluribus subsessilibus fasciculatis, floribus masculis plerisque triandris, disco subintegro turbinato-paterifor-
mi nudo, achenio ovoideo albo longitudinaliter multistriato.
—Grand Bassa, *Vogel*.

Culmi bipedales, plurifoliati. *Vaginæ* inferiores dense hispidæ,
triquetræ, angulis superne sæpe in alas angustas expansis.
Panicula spicæformis terminalis 1½-pollicaris, densa, axillaris
1-2 brevis, depauperata, sessilis. *Bracteæ* basi hirtæ, rigidæ, pleræque spiculas superantes. *Spiculæ* in quoque fasciculo 3-5, masculæ 4-5 lin. longæ, recurvæ; interior fœminea
brevior. *Squamæ* valde inæquales, siccæ, hispidulæ, acuminatæ, exteriores cujusve spiculæ breviores vacuæ. *Flores*
masculi perfecti 2-3, intimi plures semiabortivi, fœmineus
fere semper in spicula sua solitarius, addito rarius secundo
sterili. *Stylus* ad medium trifidus. *Achenium* fere lapideum,
album, eleganter longitudinaliter multistriatum, superne
obsolete verruculosum.

This species does not precisely come into any of the genera
or sections into which Nees has divided the Brazilian *Scleriæ*,
but if, as appears most convenient, the genus *Scleria* be retained
entire, according to the views of Kunth and most botanists,
this plant cannot be separated from it.

CXVI. GRAMINEÆ.

1. Leersia *disticha*, Benth.; panicula gracili effusa, spiculis
approximatis distichis, paleis margine nudis muticis, carina
brevissime aspero-hirta, staminibus 2 (v. 3 ?)—On the Nun
River, *Vogel*.

Culmi fere *L. oryzoidis*, basi reptantes radicantes ramosi, dein
adscendentes, 1-2-pedales, uti folia glaberrimi. *Folia* latius-

cule lanceolata, tenuia, venulis transversis subfenestrata, ligula
brevi. *Panicula* longe pedunculata, ramis paucis gracillimis. *Spicæ* pedunculo filiformi sustensæ, compressæ, e spiculis 6-12 regulariter distichis nec unilateralibus compositæ.
Glumæ fere *L. asperæ,* sed paullo longiores, tenuiores et
magis compressæ. *Staminum* nonnisi reliquias vidi, quarum
duo tantum in quoque flore observavi. *Semen* jam maturum
ovatum, compressum, fuscum, longitudinaliter striatum.

1. Oryza *sativa,* Linn. (Rice.)—Generally cultivated in West
Tropical Africa. Vogel's specimens are of a bearded variety.

The *Maize, Zea Mays,* Linn., is also much cultivated, but
no specimens were gathered.

1. Paspalum *brevifolium,* Fluegge.—*Kunth, Enum.* 1. *p.* 48.—
P. longiflorum, *Retz ex N. ab E. in Herb. Wight, non
Beauv.*—Sierra Leone, *Herb. Hooker;* Mauritius and East
India.

The leaves are wanting to these specimens, but they are
otherwise perfectly similar to the common East Indian species
distributed by Wight and Arnott under the number 1603, and
by Wallich under that of 8752.

2. Paspalum *conjugatum,* Berg.—*Kunth, Enum.* 1. *p.* 51.—
Sierra Leone to the Quorra, and Fernando Po, *Vogel;* St.
Thomas, *Don;* Tropical Africa and Asia.

3. Paspalum *distichum,* Burm.—*Kunth, Enum.* 1. *p.* 52.—
P. longiflorum. *Beauv.*—Sandy shores of the Nun, *Vogel;*
St. Thomas, *Don;* Tropical Africa and Asia, (n. 8757 of
Wallich's Catalogue.)

There are no specimens of *P. vaginatum,* Sw., said to be
found in Guinea, nor yet of the plant figured by Beauvois
under that name, but which, judging from the figure, is very
different from the common one.

4. Paspalum *scrobiculatum,* Linn.—*Kunth, Enum.* 1. *p.* 53.—
P. Kora, *Beauv. Fl. Ow. et Ben. t.* 85, *f.* 2. (a reference
overlooked by Kunth.)—Sierra Leone to the Quorra, *Vogel;*
St. Thomas, *Don.*—Tropical Africa, Asia and Australia.

5. Paspalum *dissectum,* Linn.—*Kunth, Enum.* 1. *p.* 54. var.

foliis vaginisque villosis.—Sierra Leone, *Don ;* South Africa and East India.

This is the same variety as the one gathered by Krauss, at Port Natal, under the n. 147. Kunth excludes from the species the forms with pubescent leaves, which Willdenow, Trinius and others admit. The species differs little from *P. scrobiculatum,* except in the shorter spikes with a broader rhachis.

The *P. barbatum,* Schum. et Thonn., from Guinea, is said to be likewise very near *P. scrobiculatum.*

1. Olyra *latifolia,* Linn.—O. paniculata, *Sw.—Kunth, Enum.* 1. *p.* 69.—Sierra Leone, *Don ;* Tropical America.

2. Olyra *brevifolia,* Schum. et Thonn., *Beskr. p.* 402.—St. Thomas, *Don ;* Fernando Po, *Vogel ;* Guinea.

This is probably a mere pubescent variety of *O. latifolia,* to which ought to be referred several of the Brazilian supposed species.

1. Leptaspis sp.—Fernando Po, *Vogel.*

This is a single imperfect specimen, insufficient to characterize the species, if it be distinct from *L. urceolata,* Br. It agrees precisely, as far as it goes, with a Ceylon specimen, received from Dr. Gardner under the n. 1045 ; and both appear more slender, with fewer and smaller flowers, than the specimens from Penang, (Wall. Catal. n. 8901), from Malacca, (Griffith), and from the Philippine Islands, (Cuming, n. 1739), all of which agree with Bennett's description and figure of *L. urceolata.*

1. Urochloa *paniculata,* Benth. ; paniculæ ramis gracilibus inferioribus verticillatis, spiculis fasciculatis pedicellatis, glumis glabris superiore margine ciliato floreque masculo mucronatis, hermaphroditi arista flore longiore.—At the confluence of the Niger on Stirling Hill, *Ansell.*

Habitus et folia omnino *U. cimicinæ. Rami* paniculæ numerosiores, tenuiores, floribundi. *Spiculæ* angustiores et minus ciliatæ, sæpius 4 in quoque fasciculo, quarum 1-2 interdum masculæ et uniloræ. *Flos* hermaphroditus, glumam supe-

riorem floremque masculum breviter superans. *Arista* quam
in *U. cimicina* multo longior.

1. Tricholæna *sphacelata,* Benth.; foliis glabris lanceolato-linearibus, vaginarum ore geniculisque barbatis, panicula ramosa
floribunda, gluma superiore valvulaque floris masculi inferiore hyalino-albidis dorso longe piliferis, acumine fusco
glabro antice ciliato sub apice longe setifero.—On the Quorra
at Pandiaki, *Ansell.*

Elatior et omnibus partibus major quam *T. Teneriffæ,* cui cæterum habitu et inflorescentia accedit. *Culmi* et *folia* in
sicco viridia, nec glaucescentia, hæc plana, 3-5 poll. longa,
3-4 lin. lata. *Flores* iis *T. Teneriffæ* duplo majores. *Gluma*
inferior oblonga, pilosa, hyalina, quam spicula duplo triplove brevior ; superior acumine neglecto linea paullo longior,
acumen gluma ipsa vix brevius, apice retusum, nervo dorsali
producto in aristam tenuissimam serrulatam gluma longiorem.
Floris inferioris neutri et submasculi palea inferior glumæ
superiori simillima ; superior brevior, multo angustior, hyalina, margine ciliata; hermaphroditi gluma superior brevior,
paleis angustis hyalinis margine ciliatis.

The *Saccharum Teneriffæ,* Linn., long since transferred to
Panicum by Brown, has nevertheless a distinctive character,
which induced Schrader to establish it as a separate genus
under the name of *Tricholæna.* Kunth did not adopt it, but
retained the plant under *Panicum.* Nees, however, has since
re-established the genus on firmer foundations, adding four South
African species, which with the present one and the *Aira Chi-
nensis,* Retz, (n. 8660 of Wallich's Catalogue) raise the number
of known species to seven. I am unable, therefore, to understand the grounds on which Parlatore, in the first part of the
present work (supra, p. 189) restores the original species to
Saccharum, when it does not at all agree with his own elaborate and accurate character of that genus in his Flora
Palermitana.

1. Isachne *minutula,* Kunth, *Enum.* 1. *p.* 137.—Cape Palmas,
Ansell; Tropical Africa, Asia, and America; and a var. with
minutely pubescent leaves.—Sierra Leone, *Don.*

The Timor plant, which Decaisne identified with Lamarck's *Panicum polygonoides*, is certainly this species, and a true *Isachne*. The flowers are much smaller than in the more common East Indian *I. muricata*, (Nees. in Herb. Wight). It is doubtful which of them is the original *Meneritana*.

1. Panicum *brizoides*, Linn.—*Kunth, Enum. 2. p. 78?*—Cape Coast, *Vogel;* East India.

The lower glume is not ovate, but broad and truncate, as in *P. paspaloides, fluitans,* and *brizæforme,* which are all probably varieties of one species, common in Tropical Asia and Africa, introduced apparently from thence into Tropical America, and both as an aquatic and as a cultivated plant, very likely to be variable.

2. Panicum *falciferum,* Trin.—*Kunth, Enum. 1. p. 80.*—Accra, *Don.*

3. Panicum *Gayanum,* Kunth, *Enum. 1. p. 79.*—On the Quorra at Stirling, *Vogel;* Senegal.

4. Panicum *horizontale,* Mey.—*Kunth, Enum. 1. p. 81.*—Sierra Leone, *Vogel, Don;* Grand Bassa and Fernando Po, *Vogel.*

5. Panicum *distichophyllum,* Trin.—*Kunth, Enum. 1. p. 90. non N. ab E.*—Accra, *Don;* at the confluence of the Niger, *Ansell.*

6. Panicum *Numidianum,* Lam.—*N. ab E. Gram. Afr. Austr. p. 33.*—Abòh, *Vogel.* A single specimen, either belonging to this, or a closely allied species.

7. Panicum *maximum,* Jacq. (Guinea Grass).—Sierra Leone, Aguapim, and on the Quorra, *Vogel;* St. Thomas, *Don.*

This grass, originating from Tropical Africa, is now extensively cultivated in East India and South America. In Vogel's collection there are two varieties, one growing in moist places, and attaining the height of 6 or 8 feet or more, the other about 2 or 3 feet high, with narrower leaves and smaller flowers, was found in dry situations on the Quorra.

8. Panicum *arenarium,* Brot.—*N. ab E. Gram. Afr. Austr. p. 37.*—Cape Coast, *Don;* South Europe, Africa, and East India.

9. Panicum *coloratum*, Linn.—*N. ab E. Gram. Afr. Austr. p.* 38.—On the Gambia, *Don;* South Europe, Africa, and America.

10. Panicum *lætum*, Kunth, *Enum.* 1. *p.* 115.—Sierra Leone, *Don;* Senegal.

11. Panicum *ovalifolium*, Poir.—*N. ab E. Gram. Afr. Austr. p.* 44. *in adnot.*—Fernando Po, *Vogel;* Guinea, Benin, and East India (Wall. Catal. n. 8737).

12. Panicum, a small specimen, which I am unable to identify, and not sufficient to describe.—St. Thomas, *Don.*

13. Panicum *sarmentosum*, Roxb. *Fl. Ind.* 1. *p.* 308; var. foliis glabriusculis.—Grand Bassa, *Vogel.*

This species is allied to *P. frumentaceum,* but the branches of the panicle are much longer, and more developed. It appears to be a generally cultivated plant, for I have specimens from the Mauritius, from the Calcutta Garden (Wall. Catal. n. 8724), and from South America, sent by Tweedie, as " cultivated from African seeds."

14. Panicum *frumentaceum,* Roxb. *Fl. Ind.* 1. *p.* 307.— Abòh, growing in the water, *Vogel;* Africa and East India, often cultivated.

15. Panicum *plicatum,* Willd.—Senegal to Benin, where it attains the height of 6 or 8 feet, or more, *Vogel* and others ; Tropical America, Africa, and Asia.

Under this name I have included a series of specimens from various countries, differing, at first sight, more or less from each other in the degree of hairiness, in stature, as well as, in some measure, in the size of the panicles and flowers, but on examination I have been unable to find any constancy in the characters, by which several species have been attempted to be distinguished. Among the published forms which probably belong to it, may be mentioned, *P. pyramidale,* Lam., *P. arvense,* Kunth?, *P. lineatum,* Schum. et Thonn., and *P. latum,* Schum. et Thonn., from West Tropical Africa ; *P. Nepalense,* Spr., *P. nervosum,* Roxb., and *P. excurrens,* Trin., from East India ; *P. costatum,* Roxb., and *P. plicatum,* Lam., from the Mauritius.

16. Panicum sp., very near to the South African *P. Linden-bergianum.*—Setaria longiseta, *Beauv.*, *Kunth*, *Enum.* 1. *p.* 158.—Fernando Po, *Vogel.*

17. Panicum *glaucum*, Linn.—On the Quorra, *Vogel;* and common almost all over the world.

Taking the genus *Panicum* in its most comprehensive, as well as most definite form, so as to include *Digitaria*, *Setaria* and *Echinochloa*, there are now near 550 supposed species recorded by different writers. It is, however, now known, that a considerable number are spread over all the warmer parts of the globe, and have been described over and over again under different names ; and a careful revision and comparison would probably reduce the number under 300. Until this is done, the accurate determination of any but the best known species is next to impossible, nor has it been ascertained what are the distinctive characters most to be relied on, and how far they vary under cultivation. Nees von Esenbeck has, indeed, sketched out some excellent sections in his work on Cape Grasses, but has not applied them, in any published work, to the *Panica* of any other country. I fear, therefore, that some of the above determinations may not prove quite correct, and there remain 22 published West Tropical African species, which either are not in the collection before me, or I have been unable to recognise. These are : from Senegal, *P. Perrottetii*, Kunth., *subalbidum*, Kunth, *raripilum*, Kunth, and *scabrum*, Lam. ; from Sierra Leone, *P. muscarium*, Trin., *lineatum*, Trin., and *tenellum*, Lam. ; from Benin, *P. setigerum*, Beauv., and *convolutum*, Beauv. ; and from Guinea, *P. regulare*, Nees., *P. pallide-fuscum, sphacelatum, cauda-ratti, subangustum, collare, serrulatum, longifolium, deflexum, sparsum,* and *plantagineum,* Schum. et Thonn. ; and *Digitaria reflexa*, and *D. nuda*, Schum. et Thonn., also from Guinea.

1. Thysanolæna *acarifera*, N. ab E. *Pl. Meyen, p.* 181.—On the Gambia, *Boteler ;* common in East India.

1. Stenotaphrum *Americanum*, Schranck.—*Kunth, Enum.* 1. *p.* 138.—St. Thomas, *Don ;* Guinea, South Africa and Tropical America.

1. Oplismenus *Burmanni*, Beauv.—*Kunth, Enum.* 1. *p.* 139. —Fernando Po, *Vogel;* Sierra Leone, *Don;* Tropical America, Africa, and Asia. (Wall. Catal. n. 8677 et 8678).

2. Oplismenus *Africanus*, Beauv.—*Kunth, Enum.* 1. *p.* 160.— St. Thomas, *Don;* Fernando Po, *Vogel;* South Africa.

1. Gymnothrix *hordeoides*, Kunth, *Enum.* 1. *p.* 160.—Grand Bassa and Fernando Po, *Vogel;* Sierra Leone.

1. Pennisetum *macrostachyum*, Benth.; elatum, foliis scaberrimis, culmo vaginisque glabris pilosisve, spica elongata, rhachi villosissima, spiculis 2-4-nis, involucri setis exterioribus numerosis flores breviter superantibus, interiore unica longa basi plumosa, gluma inferiore minuta, superiore floribus multo breviore, flore neutro unipaleaceo, palea hyalina 5-nervi, stylis basi connatis.—Abòh and Fernando Po, *Vogel.*
Culmus subramosus, basi decumbens et radicans, dein erectus, 6-8-pedalis et altior. *Vaginæ* striatæ, læves v. hinc scabriusculæ. *Folia* 2-3-pedalia, pollicem lata, utrinque pilis brevibus sparsis scaberrima, ad margines serrulato-spinulosa, ad oras vaginarum barbata; geniculi barbati v. lanati. *Spica* subsessilis, ultrapedalis, densiflora, albida v. flavescens; rhachis pilis mollibus albis dense villosa. *Flores* subreflexi. *Seta* interior cujusve involucri 9-10 lin. longa, basi plumosa, cæteræ numerosæ scabridæ fere dimidio breviores. *Glumæ* et paleæ tenues, fere hyalinæ; floris hermaphroditi apice scabriusculæ.

This must be very near to, and possibly a mere variety of, *P. purpureum*, Schum. et Thonn., a Guinea plant, but it has no tendency to assume a purple colour, I have never seen more than a single plumose seta in each involucre, and the inner glume is considerably shorter than the neutral flower.

2. Pennisetum *polystachyum*, Schult. *Mant.* 2. *p.* 456.—P. holcoides et P. barbatum, *Kunth, Enum.* 1. *p.* 163.—Cape Coast and Accra, *Don, Vogel;* Arabia and East India.

These specimens are precisely similar to those of Roxburgh from East India. The leaves are smooth or hairy, the nodi generally without hairs, the spiculæ always solitary in each involucre, the outer glume generally wanting, but sometimes even nearly a line long, the upper one mucronate or very shortly awned, the neutral flower reduced to one palea, the styles free, and in this, as in some other *Penniseta*, I have occasionally observed three of them instead of two.

3. Pennisetum *gracile*, Benth.; glabrum, ramosum, foliis angustis scabriusculis, spica tenui, spiculis solitariis, involucri setis exterioribus paucis interioribusque plumosis pluribus spiculam subsuperantibus, una longissima, flore masculo bipaleaceo, palea exteriore glumaque superiore 5-nervibus.— Sierra Leone, *Don*.

Culmi basi decumbentes, ramis suberectis vix semipedalibus. *Spica* tenuis, 1½-2-pollicaris, purpurascens. *Rhachis* glabra. *Spiculæ* glabræ v. apice scabro-puberulæ. *Gluma* exterior nulla v. minutissima, superior uti palea exterior floris masculi apice mucronata.

The *P. violaceum*, Rich., and *P. Prieuri*, Kunth., both from Senegal, appear to be different from either of the above.

1. Cenchrus *echinatus*, Linn.—*Kunth, Enum.* 1. *p.* 166.—On the Gambia, *Don;* South America, Africa and East India.

Another species, *C. barbatus*, Schum. et Thonn., is described from Guinea.

The *Lappago racemosa*, Willd., common over a great part of the globe, is also found in Senegal. The *Latipes Senegalensis*, Kunth, from Senegal, is unknown to me. Six species of *Aristida* are said to be found in West Tropical Africa, none of which are in the collection. They are: *A. stipæformis*, Lam., *hordeacea*, Kunth, and *festucoides*, Poir., from Senegal; and *A. submucronata, cœrulescens* and *longiflora*, Schum. et Thonn., from Guinea, but probably some of the three latter may be repetitions of the others.

1. Sporobolus *virginicus*, Kunth, *Enum.* 1. *p.* 210, forma parvula, glumis lævissimis v. vix ad carinam scabriusculis.— St. Thomas, *Don*, a common maritime plant in America and Africa.

2. Sporobolus *robustus*, Kunth, *Enum.* 1. *p.* 213.—Grand Bassa, in marshes, where it is gregarious, and attains six or eight feet in height, *Vogel;* Cape Verd Isles, *Ansell;* Ga-

boon Coast, *Middleton;* to 15° south latitude, *Thwaites;*
Abyssinia.

3. Sporobolus *pyramidalis,* Beauv.—*Kunth, Enum.* 1. *p.* 213.
— Accra, *Don;* Cape Coast and Sierra Leone, *Vogel;*
Oware.

4. Sporobolus *commutatus,* Kunth, *Enum.* 1. *p.* 214.—On the
Quorra, at Patteh, *Vogel;* East India.

5. Sporobolus *myrianthus,* Benth.; glaberrimus, culmis gracilibus erectis, foliis convoluto-setaceis, vaginis ore ciliatis,
panicula effusa ramosissima ramulis capillaceis apice unifloris,
spiculis angustis acuminatis, gluma exteriore flore triplo
breviore obtusa, superiore dimidium floris æquante mucronulata, paleis æqualibus, exteriore obsolete mucronulata.—On
the Quorra, at Patteh, *Vogel.*

Habitus *S. minutiflori,* differt statura ut videtur altiore, panicula ampla, 9-pollicari, ramis capillaribus numerosissimis,
foliis tenuibus 2-4-pollicaribus, et præsertim spiculis angustis
subacuminatis fere lineam longis.

6. Sporobolus *minutiflorus,* Link.—*Kunth, Enum.* 1. *p.* 214.—
On the Quorra, *Mac William;* Brazil and East India.

The other recorded West Tropican African *Sporoboli* are:
S. littoralis, Kunth, from Senegal and South America; *Vilfa
helvola,* Trin., Senegal, and Abyssinia; *V. paniculata,* Trin.,
Sierra Leone; *Agrostis extensa,* Schum. et Thonn., and *A.
congener,* Schum. et Thonn., both from Guinea, the latter probably one of the forms of *S. Virginicus.*

The *Agrostis tropica,* Beauv., from Prince's Island (off the
Gaboon Coast) and from the Mauritius, appears to be a true
Agrostis as now limited.

The *Crypsis aculeata,* Ait., belonging to the Mediterranean
region, extends to Senegal. *Triraphis pumilio,* described by
Brown in the Appendix to Oudney and Clapperton's Voyage,
may not, strictly speaking, belong to the Western Tropical
region.

1. Microchloa *obtusiflora,* Benth.; foliis brevibus convoluto-
filiformibus, glumis obtusiusculis, paleis dorso longiuscule
pilosis.—On the Quorra, at Patteh, *Vogel.*

Habitu refert *M. setaceam* (Brasiliensem), et quodammodo

etiam *Oropetium Thomæanum.* *Folia* brevia prioris, sed in sicco magis convoluta, arcuata, semipollicaria v. paullo longiora. *Spica* 1-2-pollicaris, pedunculo e vagina ampliata folio superioris haud exserto, rhachi latiuscula compressa. *Glumæ* quam in *M. setacea* subbreviores, obtusæ v. nervum dorsalem exteriorem in mucronem obsoletum subexserentes. *Paleæ* glumis paullo breviores. Cæterum character accuratissimum genericum *Neesii* in Gram. Afr. Austr. p. 246, et huic speciei optime convenit.

The *Schœnefeldtia gracilis,* Kunth, extends from Senegal to Nubia.

1. Ctenium *elegans,* Kunth, *Enum.* 1. *p.* 275.—Accra, *Don;* Senegal and Nubia.

2. Ctenium *canescens,* Benth.; spicis 2-3-nis, spiculis 4-floris, flore infimo neutro unipaleaceo, cæteris bipaleaceis, quorum altero neutro v. masculo tertio fœmineo v. hermaphrodito, summo masculo, gluma superiore canescenti-hirtello binervi, costis in aristas productis, altera ex apice excurrente, altera a medio dorso divaricato.—Whydah, *Don.*

Folia desunt. *Culmus* apice pubescens. *Spicæ* in altero specimine geminæ, in altero ternæ, primo intuitu iis *C. Americanæ* similes, sed florum structura diversa. *Gluma* inferior (quoad spicam interior) parva, exterior superior flores subæquans, costis 2 (nec 3 ut in *C. Americana*) conspicuis, pilis brevibus rigidis canescentibus. *Flos* fœmineus v. hermaphroditus apice ciliis numerosis albis aristas fere æquantibus onustus. *Palea* exterior florum 3 inferiorum dorso aristata, flos summus masculus, triander, paleis muticis.

1. Dactyloctenium *mucronatum,* Willd.—*N. ab. E. Gram. Afr. Austr. p.* 250.—Cape Coast, *Don, Vogel;* on the Quorra, *Vogel;* America, Africa and Asia.

1. Enteropogon *melicoides, N. ab E. in Herb. Wight.*—Chloris simplex, *Schum. et Thonn. Beskr. p.* 51.—C. distachya, *Kunth, Enum.* 1. *p.* 265 ?—Cape Coast, *Don,* a single specimen with a long solitary spike precisely answering Thonnings description, as well as the East Indian specimens. The aspect of the plant is very unlike that of any *Chloris.*

1. Chloris *breviseta,* Benth., culmo ramoso compresso vaginis-

que glabris, foliis planis subpilosis, spicis 7-11 digitatis, gluma superiore nervo mucronato flosculum hermaphroditum dimidiato-obovatum ad carinam marginisque ciliatum æquante, neutris binis, inferiore truncato setigero superiorem muticum minutum includente, setis flosculorum tenuissimis brevibus.—Cape Coast, *Don, Vogel.*

Near to *C. compressa* DC. in the structure of the flowers, but the ciliæ of the hermaphrodite outer palea are very different, and the arista is but little more than a line in length. In this respect *C. Abyssinica,* Hochst., agrees with it, but the latter species has very different shaped paleæ, and more florets.

The other West Tropical African species described are: *C. Prieuri* and *C. Gayana,* Kunth, from Senegal; *C. pilosa* and *C. Guineensis,* Schum. et Thonn., from Guinea and *C. penicillata,* Pers., from Guinea and East India.

Two *Leptochloæ* are found in the region: *L. mollis,* Kunth, from Senegal, and *L. Arabica,* Kunth, from Senegal, Egypt, Arabia and East India.

1. Eleusine *Indica,* Gærtn.—Sierra Leone to the Quorra and Fernando Po, *Vogel;* South America, Africa and East India to Japan.

Another species, *E. glabra,* Schum. et Thonn., is described from Guinea, as also a species of *Aira; A. bicolor,* Schum. et Thonn.

1. Eragrostis *rubiginosa,* Trin.—*Kunth, Enum.* 1. *p.* 339. (sub Poa).—Poa turgida, *Schum. et Thonn. Beskr. p.* 66 ?—Accra, *Don.*

2. Eragrostis *ciliaris,* Link.—*Kunth, Enum.* 1. *p.* 337. (sub Poa).—Accra, *Don;* on the Quorra at the confluence, *Ansell;* America, South Africa, and East India.

3. Eragrostis *Abyssinica,* Link?—*Kunth, Enum.* 1. *p.* 332 ? (sub Poa).—On the Quorra at Stirling, *Ansell.* If not identical with the Abyssinian plant cultivated in the North East of Africa, it is closely allied to it.

4. Eragrostis *linearis,* Schum. et Thonn. *Beskr. p.* 67. (sub Poa).—On the Gambia, *Boteler;* Guinea.

5. Eragrostis *biformis*, Kunth, *Enum.* 1. *p.* 332 *?* (sub Poa).—
On the Gambia, *Don ;* Cape Coast, *Vogel.*
6. Eragrostis *tremula*, Lam.—*Kunth, Enum.* 1. *p.* 332. (sub
Poa).—On the Gambia and Sierra Leone, *Don ;* Bassa Cove,
Ansell ; Nubia.
7. Eragrostis, near to *E. tremula,* and perhaps a mere variety.
—On the Quorra, *Vogel, Ansell.*

The determination of the five last *Eragrostides* is very
uncertain, the species of this numerous genus are often distinguished by slight, although constant characters, difficult to
describe, and often not alluded to in descriptions, whilst the
number of florets in each spicula, the most obvious character
usually relied on, is sometimes very variable in the same species.
Besides the above, there are six others described from the
same region, viz : *Poa squamata,* Lam., from Sierra Leone ;
P. Cambessediana, Kunth, from Senegal ; *P. fascicularis,* Trin.,
from Congo; and *P. cachectica,* Schum. et Thonn., *P. Hip-
puris,* Schum. et Thonn., and *Eragrostis Hornemanniana,*
N. ab E., from Guinea.
1. Poa *mucronata*, Poir.—*Kunth. Enum.* 1. *p.* 334 *?*—On the
Nun River, *Vogel.*

The specimens are young, but as far as they go they agree
with Beauvois' figure and description. Their appearance is
neither that of *Eragrostis,* nor of *Poa* proper, but rather of
Centotheca, although the reflexed bristles of the paleæ are
wanting.
1. Centotheca *lappacea*, Desv.—*Kunth, Enum.* 1. *p.* 366.—
Cape Palmas and Fernando Po, *Vogel ;* St. Thomas, *Don ;*
Tropical Asia and Australia.

The *Elytrophorus articulatus,* Beauv., a common East Indian plant, extends to Senegal.
1. Festuca *rottböllioides*, Kunth, *Enum.* 1. *p.* 395 ?—St.
Thomas, *Don ;* perhaps introduced from Europe.

No *Bambuseæ* are recorded from West Tropical Africa, nor
were any specimens brought by the Expedition, excepting a few
leaves of what appear to be the *Bambusa vulgaris,* from Accra,

without any label; probably from cultivated plants. There are not either any *Hordeaceæ* known to be indigenous, and it does not appear that any of our common grains are cultivated there.

1. Rottböllia *exaltata*, Linn.—*Kunth, Enum.* 1. *p.* 466.—Fernando Po, and common on the banks of the Quorra, where it reaches the height of from 5 to 10 feet, with the leaf-sheaths covered with stinging hairs, *Vogel;* East India and Australia.

1. Manisuris *granularis*, Sw.—*Kunth, Enum.* 1. *p.* 469.—On the Quorra, *Vogel, Ansell*; common in most of the warmer regions of the globe.

1. Perotis *latifolia*, Ait.—*Kunth, Enum.* 1. *p.* 470.—On the Quorra at Pandiaki, *Ansell*; South Africa and East India to Japan.

These are luxuriant specimens, with larger and more rigid leaves than usual, the raceme full 9 inches long, with rather small crowded flowers. The plant varies in all these respects, and the *P. hordeiformis*, N. ab E., cannot well be distinguished as a species.

1. Saccharum *spontaneum*, Linn.—*Kunth, Enum.* 1. *p.* 475. *et pars.* 2. *p.* 385.—S. punctatum, *Schum. et Thonn. Beskr. p.* 46.—On the Quorra, *Vogel;* Guinea and East India.

This agrees with the descriptions of Roxburgh and of Kunth, as well as with some of our East Indian specimens. It grows, according to Vogel, to the height of 6 or 8 feet, the leaves are very rigid, narrow, and nearly plane, with a very broad mid-rib; the axis of the panicle very villous, as in *S. Ægyptiacum*, but the flowers of the latter are nearly twice as large. In both species the outer glume is usually two-nerved, but not always so, the mid-rib being often more or less visible or prominent.

The true Sugar Cane, *Saccharum officinarum*, Linn., is cultivated in Guinea.

1. Imperata *arundinacea*, Cyr.—*Kunth, Enum.* 1. *p.* 477.— I. Thunbergii, *N. ab E. Gram. Afr. Austr. p.* 89.—Sierra Leone, *Don;* Senegal, South Europe and throughout Africa, East India to North Australia.

The *Erianthus*? *repens*, Beauv., from Guinea, *Elionurus elegans*, Kunth, from Senegal, and *Anthistiria glauca*, Desf. from Guinea and North Africa, are not in the collections before us.

1. Andropogon (Heteropogon) *contortus*, Linn.—Accra, *Don;* on the Quorra, *Ansell;* Senegal to Benin, South Europe, and all over Africa and East India ; also in Mexico and some parts of South America, perhaps introduced from the Old World.

The rhachis in this species is smooth, or more or less pubescent, but usually very slightly so, and it appears very difficult to distinguish the *A. Allionii* as a species. As a genus, *Heteropogon*, and several others dismembered from *Andropogon*, ought probably to be adopted, but the whole of this extensive group has been left in such a state of utter confusion in Kunth's Enumeratio, and the new genera of Nees and others, hitherto applied only to so few of the total number, that I have been unable to make use of them on the present occasion, without entering into that close examination and comparison of the East Indian species, which it is understood that Professor Nees has made, and which the limits of the present work do not now admit of. It is to be hoped that that distinguished agrostologist, may shortly publish the results of his labours in his "Glumaceæ Indiæ Orientalis," so often quoted in his Plantæ Meyenianæ and Flora Africæ Australioris.

2. Andropogon *Donianus*, Benth.; culmo ramoso vaginisque lævibus, spicis longe pedunculatis solitariis muticis, rhachibus sericeo-barbatis, spiculis geminis, hermaphroditæ gluma exteriori bimucronata pectinato-ciliata, neutrius pedicellatæ acuminata integra.—Sierra Leone, *Don.*

Culmus rigidus, teres v. leviter compressus, nitidus. *Vaginæ* compresso-carinatæ, striatæ, inferiores equitantes distichæ. *Folia* rigidula, complicata, glaucescentia, superiora brevia. *Spicæ* a vagina superiore 6-8 pollicibus distantes, bipollicares. *Rhachidis* pili patentes, spiculis breviores. *Gluma* exterior spiculæ hermaphroditæ 2-2½ lin. longa, cartilaginea, tenuiter multinervis, costis 2 validioribus submarginalibus eleganter pectinato-ciliatis et in mucrones breves excurrentibus, marginibus ipsis anguste hyalinis inflexis ; gluma superior angusta, rigidule membranacea, sericeo-pilosa ; floris neutri palea hyalina, vix brevissime mucronata, paleæ hermaphroditi minores muticæ ; spiculæ masculæ gluma exterior multinervis acuta submucronata, interior et paleæ hyalinæ.

3. Andropogon (Shizachyrium ?) *pulchellus*, Don *in Herb. Hort. Soc. Lond.*; culmo adscendente compresso superne paniculato, foliis brevibus, spicis solitariis, rhachi pilis longissimis barbata, spiculis geminis, hermaphroditæ gluma exteriore multinervi bicuspidata brevissime ciliolata, floris hermaphroditæ arista tenui spiculæ æquilonga, pedicello masculo spiculam superante, gluma exteriore multinervi setaceo-acuminata.—On the Gambia, *Don.*

Culmus rigidulus, 1½-pedalis. *Folia* 1-3-pollicaria, glauco-rubentia. *Panicula* fastigiata, stricta, ramulis ultimis vaginis foliorum usque ad spicam involutis. *Spicæ* 1½-2-pollicares, paucifloræ, pilis sericeis patentibus quam spicula longioribus. *Gluma* exterior spiculæ hermaphroditæ rigida, fere 3 lin. longa.

4. Andropogon (Schizachyrium ?) *leptostachyus*, Benth.; culmo erecto stricto ramoso, vaginis inferioribus hirsutis, foliis planis, spicis solitariis tenuibus, rhachi ad articulationes breviter ciliata, spiculis geminis, hermaphroditæ gluma exteriore bicuspidata binervi vix ciliata, paleæ arista tenui vix spiculam triplo superante, masculæ pedicello hinc ciliolato, gluma exteriore sub-3 nervi bicuspidata uniaristata.—On the Quorra, *Vogel.*

Culmi 2-3-pedales. *Folia* glabra v. puberula, haud rigida. *Spicæ* haud numerosæ, pedunculo usque ad spicam ipsam vagina folii involuto, 2-3-pollicares, primo intuitu glabræ, ciliæ tamen breves adsunt circa basin spiculæ sessilis et ad unum latus stipitis masculi. *Spiculæ hermaphroditæ* gluma exterior 2½ lin. longa, venis 2 v. rarius 3 parum conspicuis ; palea exterior floris perfecti profunde bifida, arista semipollicari geniculata tenuissima. *Spiculæ masculæ* pedicellus

spicula brevior, gluma exterior vix 2 lin. longa, in aristam tenuissime capillaceam fere ejusdem longitudinis desinens.

5. Andropogon *tectorum*, Schum. et Thonn. *Beskr. p.* 49.—
Sierra Leone, *Don;* Guinea, where, according to Thonning, it
is the commonest grass, overrunning the fields from the shore
to the hills.

6. Andropogon *Gayanus*, Kunth, *Enum.* 1. *p.* 491.—On the
Nun and Quorra, *Vogel;* on the Gambia, *Boteler;* Nubia.

7. Andropogon *pertusus*, Willd.—Kunth, *Enum.* 1. *p.* 498.—
Accra, *Don;* East and South Africa and East India.

The Egyptian and South African forms have been published
as distinct species, in East India the plant varies exceedingly;
possibly there may really be a group of several different species,
but with the materials before me, I am totally unable to extri-
cate their synonymy.

8. Andropogon *schœnanthus*, Linn.—On the Gambia and Sierra
Leone, *Don, Vogel;* East India.

The group of *Lemon-grasses* is another of those widely spread
gregarious and variable grasses among which the number of
real species, one or more, has not yet been properly ascertained.
The Tropical African form is that to which Nees formerly
applied Roxburgh's name of *Iwarancusa*, and in some more
recent publications, that of *A. Martini*, it is the commonest
East Indian form, and probably the original *A. schœnanthus*,
Linn., but certainly not the one Roxburgh distributed as
A. Iwarancusa.

9. Andropogon *arundinaceus*, Willd.—*Kunth, Enum.* 1. *p.* 506.
—On the Quorra, *Vogel;* Guinea.

This agrees with the character as far as it goes. It is a tall-
growing species, allied to *A. Gryllus*, with two rather long male
spikelets on short pedicels at every articulation with the sessile
hermaphrodite one, which is remarkable for its very long rigid
twisted arista, very pubescent in its lower half.

10. Andropogon *verticillatus*, Schum. et Thonn, *Beskr. p.* 50.
—Accra, *Don, Vogel.*

The branches of the panicle have but few articulations, the stipitate male spiculæ are two together, with the hermaphrodite at the extremity, single at the other articulations.

11. Andropogon (Anatherum) *Nigritanus,* Benth.; erectus, elatus, foliis longis, panicula verticillata ampla, ramis longis simplicibus, spiculis geminis, glumis muricatis acuminatis, hermaphroditæ palea breviter aristata.—On the Nun at Atok, *Vogel.*

Habitus A. *muricati* cui et floribus affinis, hi vero breviter aristati. *Culmi* 6-8-pedales. *Panicula* sesquipedalis, ramis numerosis subverticillatis flexuosis 4-6-pollicaribus. *Rhachis* tenuis ad articulationes breviter ciliata. *Spiculæ* dissitæ, rhachi appressæ, hermaphroditæ 3-lin. longæ, masculæ breviores, pedicello glabro v. brevissime ciliolato. *Glumæ* hermaphroditæ coriaceæ, subæquales, 2-3-nerves, dorso pilis paucis 2-3-seriatis muricato-hispidæ, apice in mucronem brevem desinentes, margine hyalinæ. *Paleæ* hyalinæ, exterior floris hermaphroditi glumis paullo brevior, apice bifida et aristam 2-3-lin. longam emittens.

Though closely allied to the common East Indian *Vitiver,* this is evidently a distinct species; it has not been observed whether the root has the same perfume and properties.

The other West Tropical African species of *Andropogon* published are: *A. brevifolius,* Sw., from Senegal and Tropical America; *A. simplex,* Schum. et Thonn., from Guinea, apparently allied to *A. brevifolius;* and *A. canaliculatus,* Schum. et Thonn., and *A. Guineensis,* Schum. et Thonn., both from Guinea, and apparently allied to *A. Gayanus;* besides four species entered in Steudel's Nomenclator as published by Trinius, but as the work is not quoted, I have been unable to find the descriptions. I much regret this, as probably some of them are the same as those above described from Don's collection. These are *A. eucnemis,* Trin., and *A. fulvibarbis,* Trin., from Guinea, (Accra?) and *A. leptocomus,* Trin., and *A. platypus,* Trin., from Sierra Leone.

1. Sorghum *saccharatum,* Willd.?—Cape Palmas, Cape Coast, and on the Nun, *Vogel.*

This is certainly the species so widely diffused in Africa which goes usually under the name of *S. saccharatum*, but I have much doubt whether it be more than a large variety of *S. Halepense*. It grows to the height of six or eight feet, with very ample spreading panicles; the Cape Coast and Cape Palmas specimens are from cultivated grounds, and have the fertile spiculæ about 3 lines long. The Nun specimens, from the inundated banks of the river, are still more luxuriant, and their large spikelets, about 4 lines long, clothed with red-brown hairs, give them a very rich aspect.

Several varieties of the *S. vulgare*, Linn., are generally cultivated in Guinea, as well as the *S. saccharatum*.

There is a specimen of a grass in Don's herbarium from Sierra Leone which I am unable to refer to any genus known to me, but the whole of the flowers having fallen away, with the exception of the remarkable glumæ, I am unable to describe it.

The Ferns and Cryptogamic plants brought by the Expedition are too few in number, and of too little interest, either to give any idea of the cryptogamic vegetation of the country, or to make any detailed enumeration advisable on the present occasion.

ADDENDA ET CORRIGENDA.

THE difficulty of searching out from a great variety of works, both general and special, the published species belonging to a particular region, more especially when the plants are not arranged in such works according to their natural Orders, may have been the cause of a considerable number having been overlooked. Whilst the preceding pages have been going through the press, I have discovered a few which have been omitted in their proper places, or which have been since published, and I take this opportunity of correcting a few material errors of copying or of the press.

P. 122. 1. 28. For *Mysicarpus*, read *Alysicarpus*.

P. 129. Before 96, *Umbilicus,* insert XXIV* *Crassulaceæ.*

P. 172. Omit n. 227. *Dalechampia Cordofana,* the article having been re-written and inserted as n. 217. *Dalechampia Senegalensis,* p. 174.

P. 206. l. 22. For *Hablitzia,* read *Habzelia.*

P. 240. To *Lophira alata,* add the synonym : *Lophira simplex,* G. Don, Gard. Dict. 1. p. 814.

P. 290 and 291. To the *Connaraceæ* add: *Omphalobium nervosum,* G. Don, and *Cnestis racemosa,* G. Don, both from Sierra Leone.

P. 342. Anisophyllum *laurinum.* I had overlooked the mention of this plant under the name of *Anisophyllea laurina,* Br., in Sabine's paper on the fruits of West Tropical Africa, *Hort. Trans. v. 5. p.* 466. It appears from the few notes there given, that what I have described as stipules are minute leaves *nearly opposite* to the apparently alternate leaves, a point which I think is yet doubtful. It is also stated that Mr. Brown has examined perfect flowers, which, it is much to be regretted, have not been described; but a drawing in the possession of the Horticultural Society shows that the ovary of the perfect flower is inferior, with four cells and one pendulous ovule in each, and is crowned by four distinct styles. The second species, my *A. Zeylanicum,* has just been published from a Manuscript of the late Dr. Gardner, in the October number for the present year of Hooker's Journal of Botany, under the name of *Tetracrypta cinnamomoides,* and referred to *Hamamelidaceæ.* The real affinities of the plant are to my mind, as yet very doubtful.

P. 361. l. 5. For t. 387, read t. 4387.

P. 439. After *Campanulaceæ,* add: *Sphenoclea Zeylanica,* Linn., a common Tropical plant, found also in Guinea. It is another of those anomalous species, which has not as yet been clearly connected with any known Order, and which, to cut the Gordian knot, is considered by some as constituting a natural Order of itself.

P. 461. After *Convolvulaceæ,* add : *Hydrolea Zeylanica,* Linn., another common Tropical species, found in Guinea. It belongs to the Order of *Hydroleaceæ*, to be inserted next to *Convolvulaceæ*.

P. 496. To *Polygonaceæ*, add : C. Koch has just published (Linnæa, v. 22. v. 205) a *Polygonum tropicum*, from Senegal, which is probably the same as *P. exiguum*.

P. 255. At the end of *Dicotyledones*, add : *Thonningia sanguinea*, Vahl, from Guinea, belonging to the anomalous Order, *Balanophoreæ*.

P. 428. Add : *Pontederia natans*, Beauv., to the described Monocotyledons of West Tropical Africa.

P. 534. *Amomum Grana-Paradisi*. Since this article has been in the printer's hands I have seen three papers of Dr. Pereira's published in the Pharmaceutical Journal, vols. 2˙and 6, which show that, after a very careful investigation, he has come to the conclusion that the *Amomum* seeds known in this country as *Guinea Grains*, are the produce of *one* species, including the *A. Grana-Paradisi* of Smith and the *A. Melegueta* of Roscoe. This species is evidently the same which both Afzelius and Vogel state to be common all over the coast. As it is more generally known as the *Malaghetty Pepper* than the *Habzelia Æthiopica*, several observations made by Dr. Hooker on the importance of that drug, must be considered as applying more especially to this plant. There is no doubt, however, that the seeds of both these widely different plants and the fruit of a third, as different from either, the *Cubeba Clusii*, p. 514, have all been known more or less under the name of *Guinea Pepper*.

The total number of species enumerated from the collections made in West Tropical Africa by the officers of the Niger Expedition, and by Don and others, amounts to 974, of which Dicotyledons 803, and Monocotyledons 171. The principal Orders are among Dicotyledons : *Leguminosæ* 113, *Rubiaceæ* 97, *Compositæ* 40, *Acanthaceæ* and *Eupborbiaceæ* 37 each, *Convolvulaceæ* and *Urticeæ* (including *Artocarpeæ*) 27 each, *Malvaceæ* and *Melastomaceæ* 23 each ; none of the others reaching 20. Among Monocotyledons : *Gramineæ* 79, *Cyperaceæ* 39, *Commelyneæ* 16, none of the others reaching 10.

The number of additional published species mentioned is very
uncertain, as in so many cases one species is published under
different names by different authors; they may, however, be
stated approximately as follows:

	In Sierra Leone, Guinea, Congo, and Angola.	Additional in Senegal only.	Total.	Grand total, including those here enumerated or described
Phœnogamous plants . .	443	453	896	1870
Dicotyledons . .	306	415	721	1524
Leguminosæ . .	53	98	151	264
Rubiaceæ . . .	37	25	62	159
Compositæ . . .	11	31	42	82
Acanthaceæ . .	7	17	24	61
Euphorbiaceæ . .	11	7	18	55
Convolvulaceæ .	8	10	18	45
Malvaceæ . . .	5	27	32	55
Urticeæ	7	1	8	35
Monocotyledons .	137	38	175	346
Gramineæ . . .	50	23	73	152
Cyperaceæ . . .	26	9	35	74
Commelyneæ . .	3	1	4	20

It must be recollected that the numbers among *Thalamifloræ,
Calycifloræ, Acanthaceæ, Scrophularineæ, Labiatæ, Gramineæ*
and *Cyperaceæ*, bear a much greater proportion to the real
numbers in the country than in the case of most of the other
Orders.

INDEX OF GENERA

CONTAINED IN THE

SPICILEGIA GORGONEA

AND IN THE

FLORA NIGRITIANA

AND OF

SPECIES

TRANSFERRED FROM GENERA, UNDER WHICH THEY HAD BEEN
PREVIOUSLY PUBLISHED.

The Synonymes are Printed in Italic.

Olyra, 558.
Omphacarpus, 237.
Omphalobium, 290, 575.
Omphalocarpum, 441.
Oncinotis, 451.
Oncoba, 220.
Ophioglossum, 192.
Oplismenus, 563.
Ormocarpon, 302.
Ornithogalum, 490.
Ornitrophe, 248.
Oryza, 557.
Osbeckia, 130, 345; *antennina*, 349;
   grandiflora, 347 ; *rotundifolia*,
   348.
Ostryocarpus, 315.
Otomeria, 405.
Ouvirandra, 528.
Oxalis, 115, 269.
Oxyanthus, 387.
Oxystelma, 454.

P.

Pachylobus, 285.
Pachyrrhizus, 311.
Palisota, 544.
Pandanus, 527.
Panicum, 185, 560.
Papaver, 98.
Parietaria, 518.
Parinarium, 333.
Paritium, 227.
Parkia, 329.
Parkinsonia, 323.
Parmelia, 196.
Paspalum, 187, 557.
Paullinia, 247.
Paulo-Wilhelmia, 479.
Pavetta, 414; *lateriflora*, 357.
Pegolettia, 138, 432.
Peltospermum, 400.
Pennisetum, 183, 563.
Pentaclethra, 329.
Pentadesma, 240.
Pentas, 401.
Pentatropis, 454.
Peperomia, 514.
Pepper (Guinea); see Habzelia, 206;
   Cubeba, 514 ; and Amonum, 534 ;
   also Addenda, 576.
Pergularia, 458.
Periploca, 150.
Peristrophe, 168, 484.
Perotis, 569.
Petrocarya, 335.
Phaca, 123.
Phagnalon, 135.

Phallaria, 408.
Phaseolus, 308.
Phelypæa, 167.
Philadendron, 527.
Philenoptera, 319.
Phrynium, 531.
Phyllanthus, 175, 510.
Physalis, 161, 472.
Physichilus, 477.
Picris, 439.
Piper, 514.
Piptadenia, 330.
Piptolæna, 447.
Pistia, 527.
Plantago, 171.
Platanocarpum, 380.
Platostoma, 488.
Plectranthus, 488.
Plectronia, 410.
Plectrotropis, 311.
Pluchea, 137, 432.
Plumbago, 169, 490.
Plumiera, 450.
Poa, 567, 568.
Podalyria hæmatoxylon, 321.
Poivrea, 337.
Pollichia, 153, 471.
Polycarpæa, 104, 374.
Polychlæna, 233.
Polyechma, 477.
Polygala, 103, 221.
Polygonum, 174, 496, 576.
Polyspatha, 543.
Polystachya, 530.
Pomatium *dubium*, 385 ; *spicatum*,
   394.
Pontederia, 576.
Porana, 469.
Portulaca, 129, 373.
Potamogeton, 181, 528,
Pothomorphe, 514.
Pouchetia, 395.
Pouzolsia, 518.
Premna, 485.
Prevostea, 469.
Prieurea, 344.
Prosopis, 330.
Prostea, 249.
Psidium, 358.
Psorospermum, 241.
Psychotria, 418.
Pteris, 192.
Pterocarpus, 314.
Pulicaria, 432.
Pupalia, 494.
Pycnocoma, 508.
Pythonium, 527.

END.

LONDON:
Printed by Schulze and Co., 13, Poland Street.

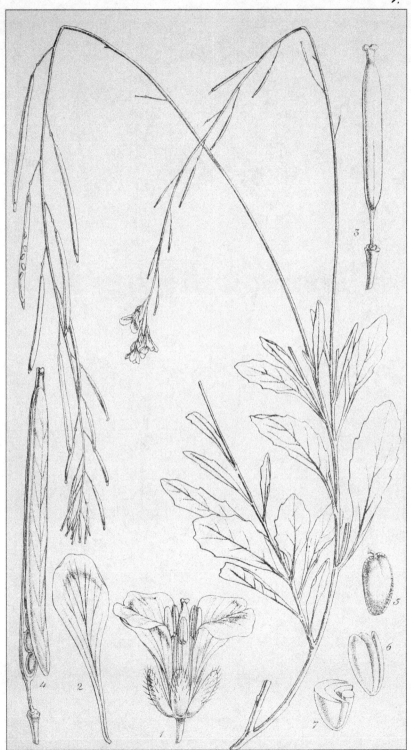

Madeley litho 3 Wellington St Strand.

Sinapidendron gracile, Webb.

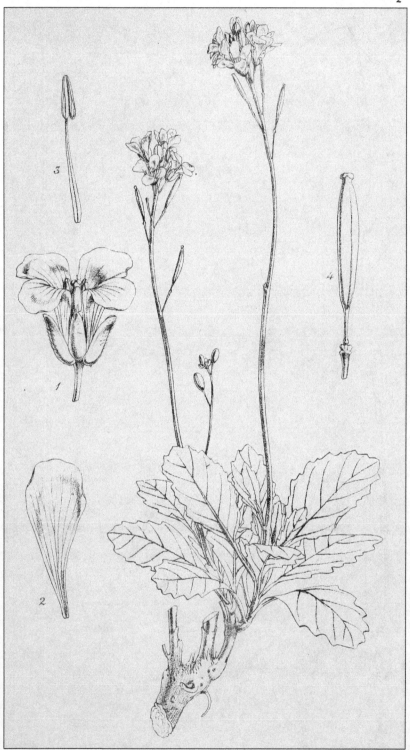

Fitch del.

Madeley litho 3 Wellington St Strand.

Sisymbrium Vogelii Webb.

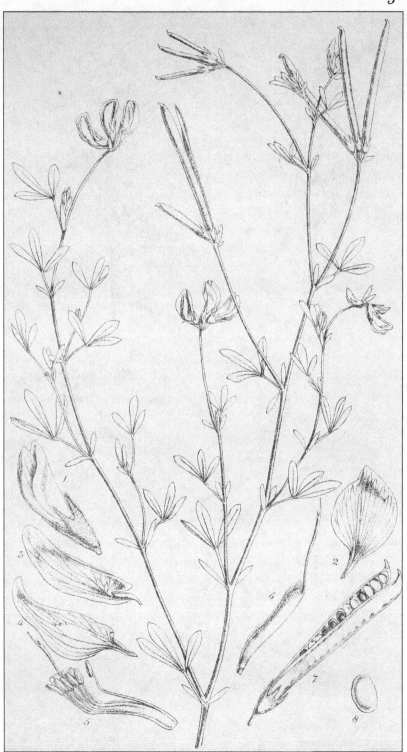

Madeley litho 3 Wellington St Strand.

Lotus Brunneri Webb.

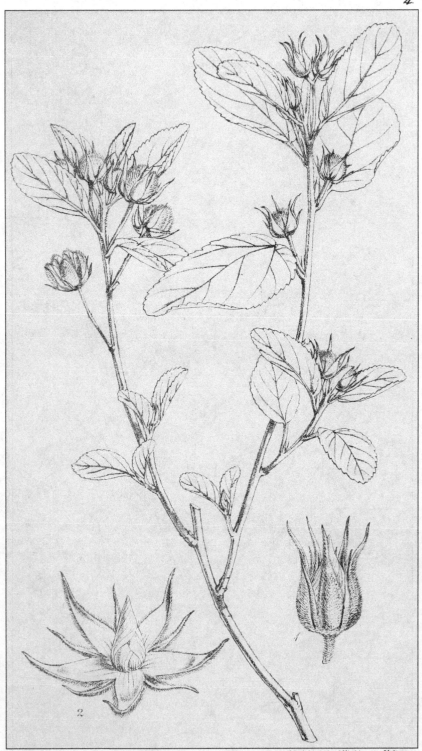

4.

Madeley litho 3 Wellington S.t Strand.

Melhania Leprieurii. Webb.

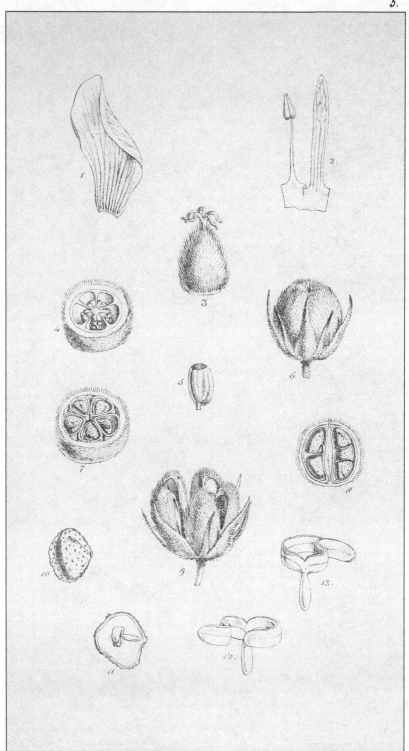

Melhania Leprieurii. Webb.

6

Fitch, del. et lith.

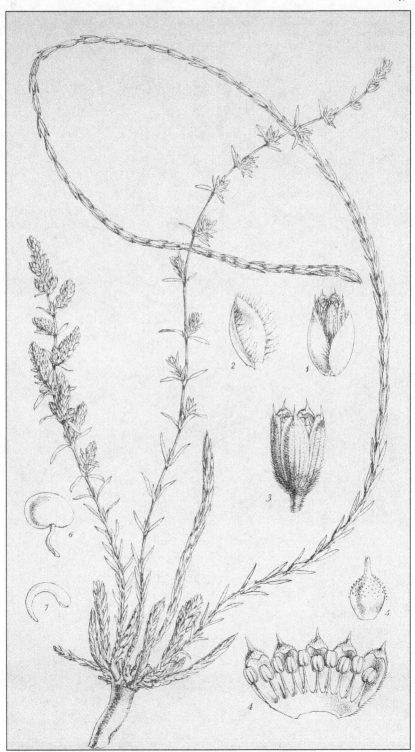

7.

Madeley litho 3 Wellington St Strand.

Paronychia illecebroides. Webb.

Phaca Vogelii, Webb.

9

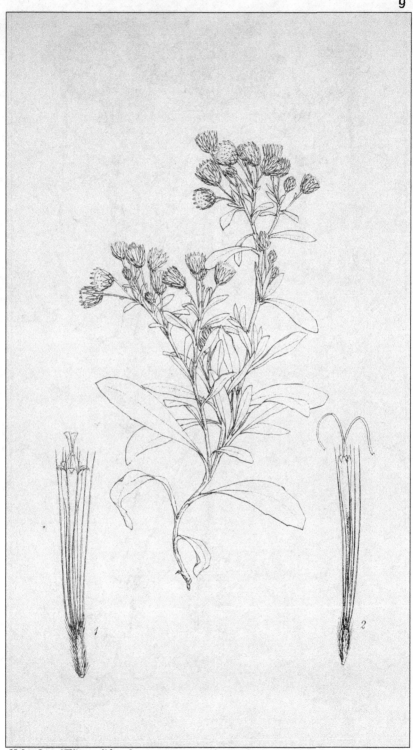

Phagnalon melanoleucum, Webb

Sonchus Daltoni, Webb.

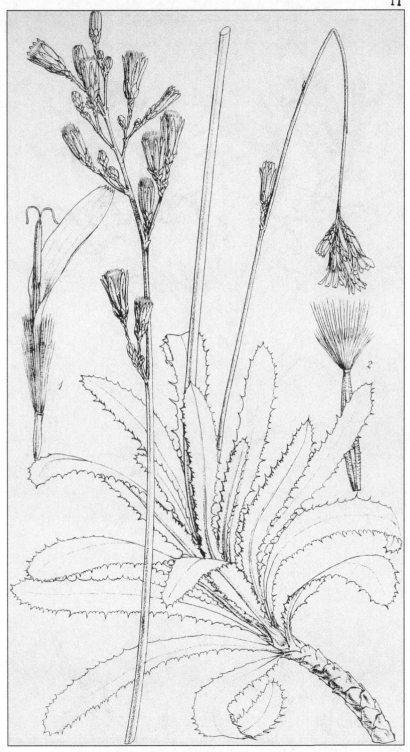

Rhabdotheca picridioides, Webb.

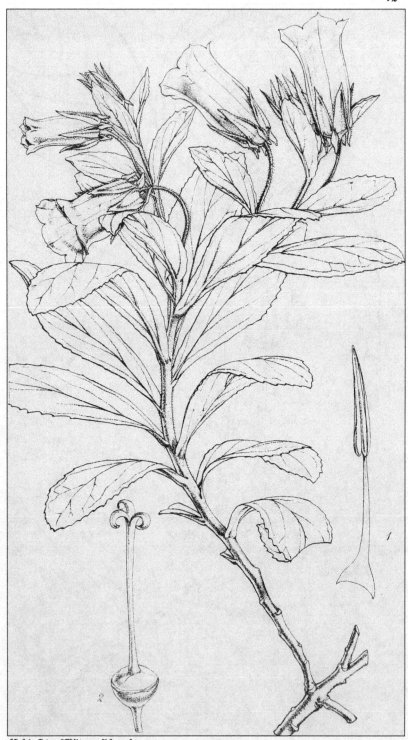

12

Madeley Printer 3 Wellington S.º Strand.

Campanula Jacobæa, C. Sm.

Sapota marginata, Dne.

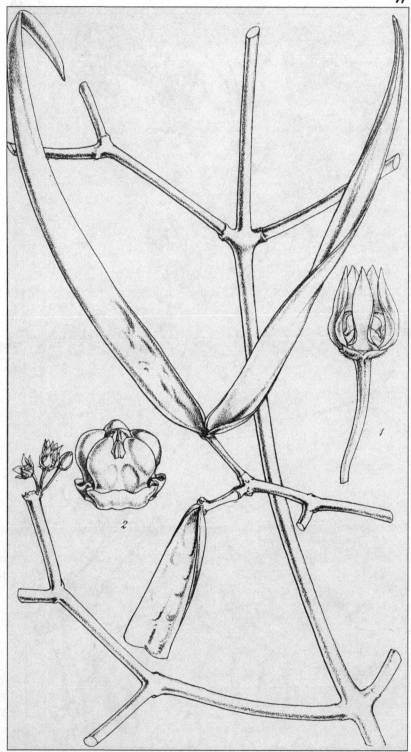

W Fitch del.

Sarcostemma Daltoni, Dne.

15.

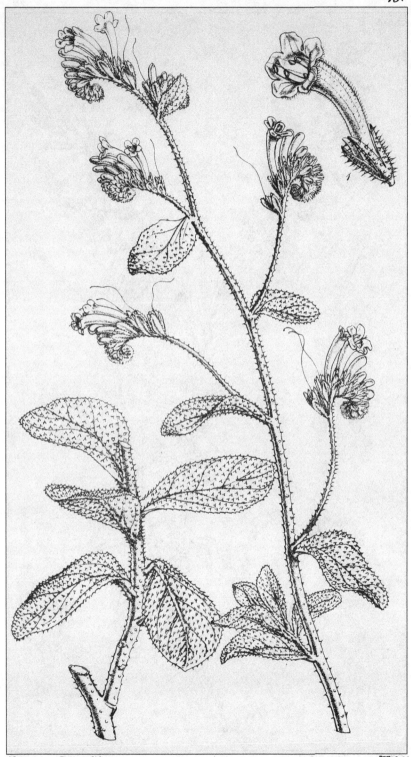

Madeley, Printer 3 Wellington S.t Strand.　　　　　　W.Fitch del.

Echium stenosiphon, Webb.

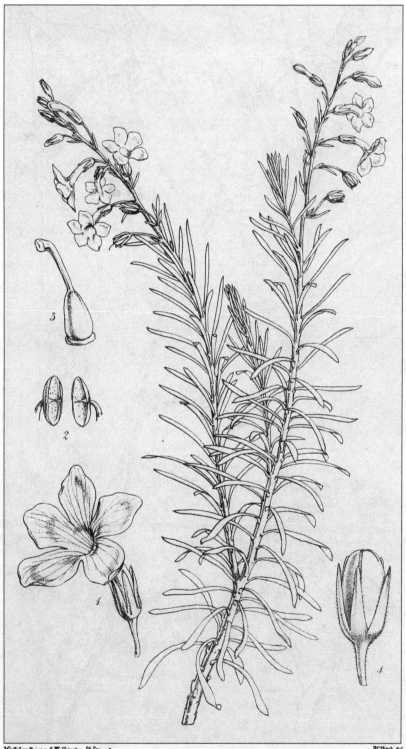

Madeley, Printer 3 Wellington St Strand W Fitch del.

Campylanthus Benthami, Webb.

17.

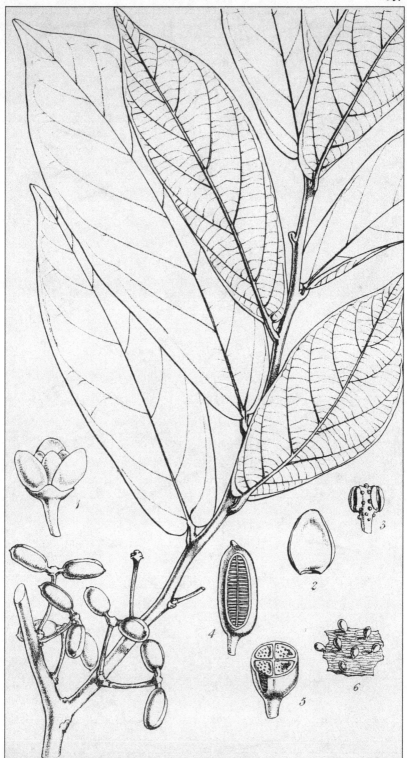

Madeley, Printer, 3 Wellington S.t Strand.

W Fitch del.

Uvaria Vogelii, Hook. fil.

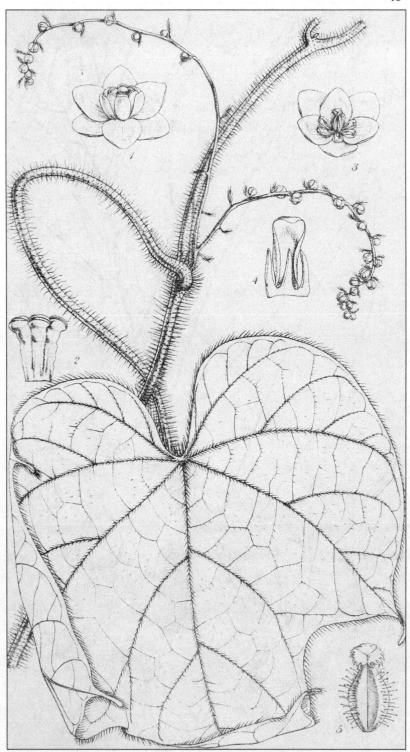

Madeley Printer 3 Wellington S.º Strand.

Cocculus macranthus. Hook. fil.

19 & 20.

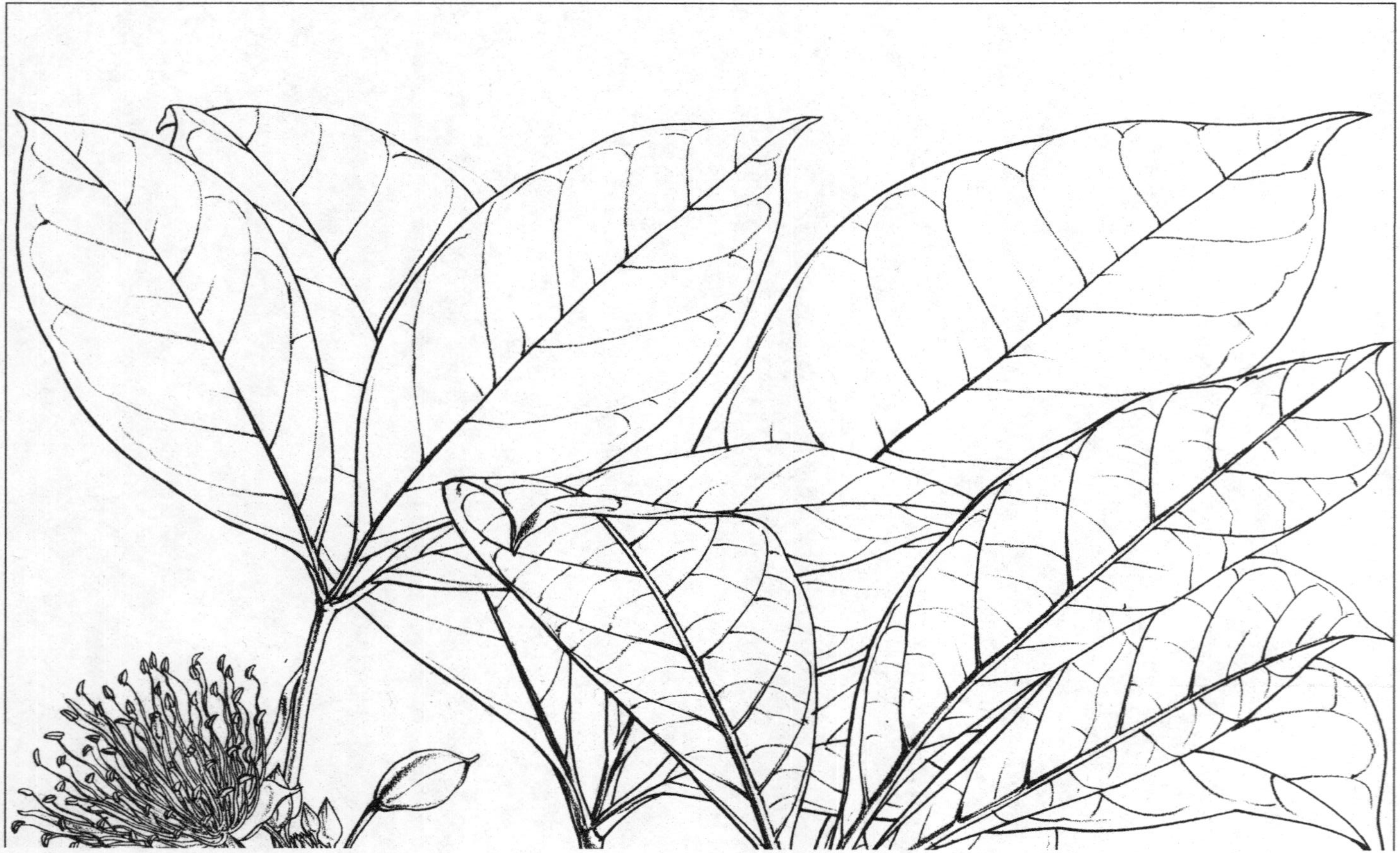

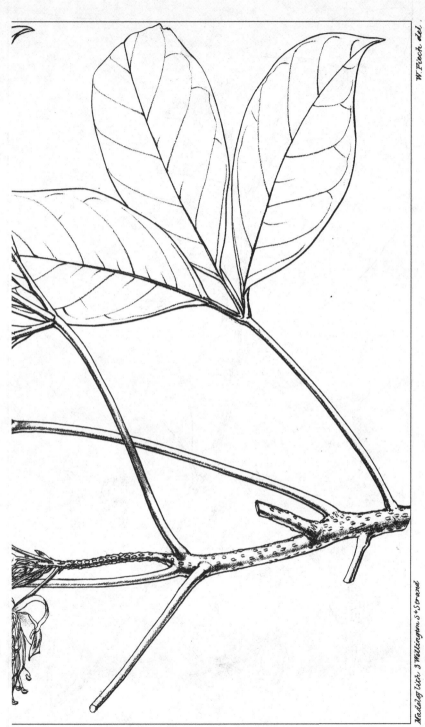

Ritchiea erecta, Hook. fil.

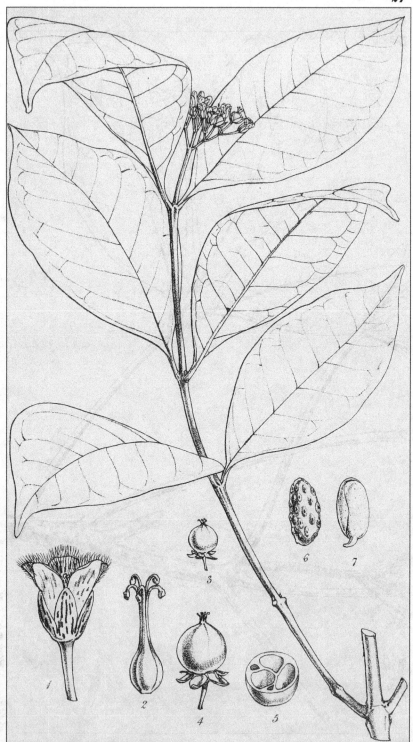

Madeley Printer 3 Wellington St Strand

W Fitch del.

Psorospermum tenuifolium, Hook. fil.

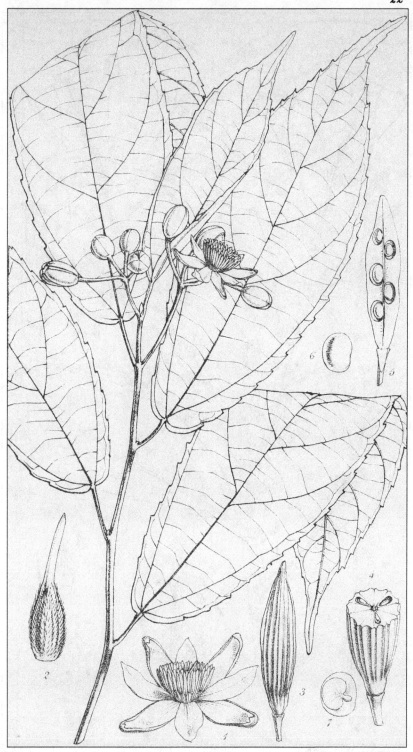

Glyphæa grewioides, Hook. fil.

Ochtocosmus Africanus, Hook. fil.

24.

Madeley Printer 3 Wellington St Strand

W Fitch del.

Acridocarpus corymbosus. Hook fil.

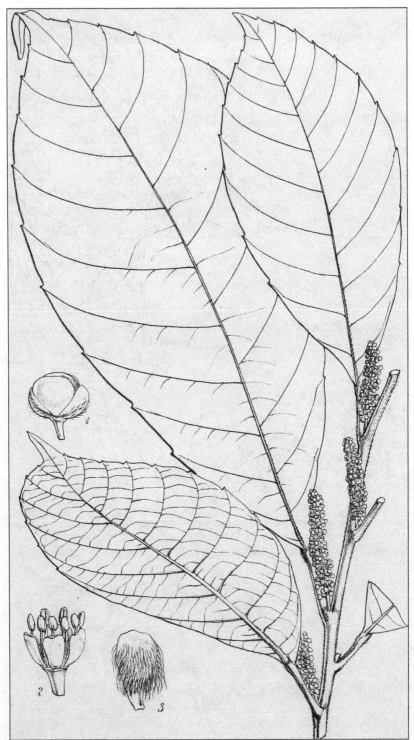

Madeley Printer 3 Wellington St Strand

W Fitch del

Schmidelia monophylla, Hook. fil.

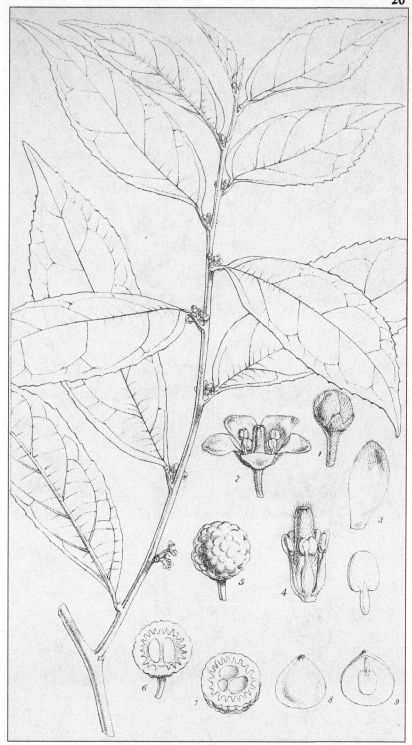

Microdesmis puberula, Hook. fil.

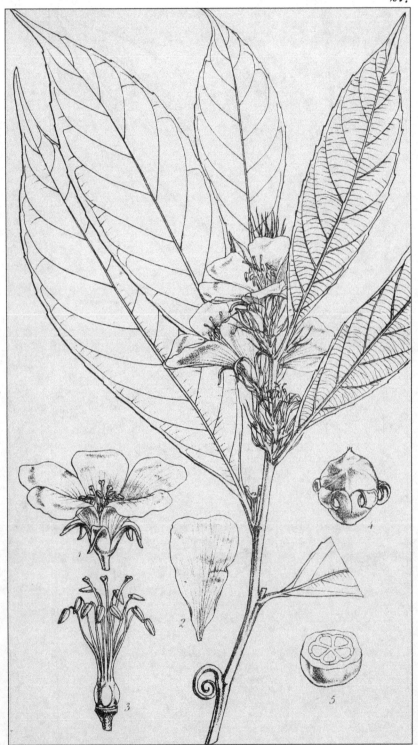

Madeley Printer 3 Wellington S.t Strand

W.Fitch del.

Hugonia Planchoni, Hook fil.

Madeley,Printer 3 Wellington S.t Strand

W.Fitch del.

Apodytes Beninensis. Hook. fil.

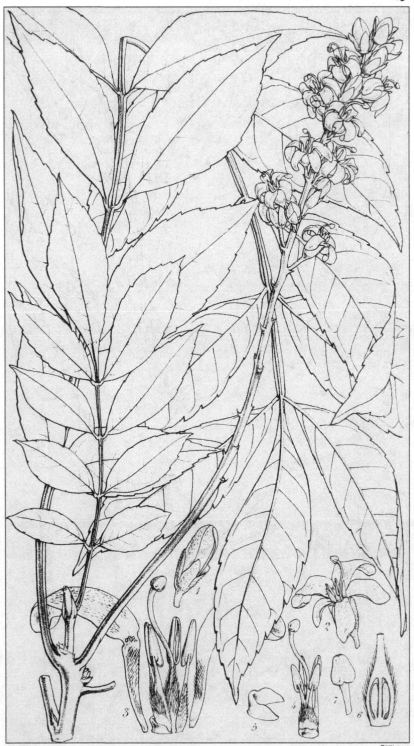

Natalia Paulliniodes Planch.

Madeley Printer 3 Wellington St Strand.

W Fitch del.

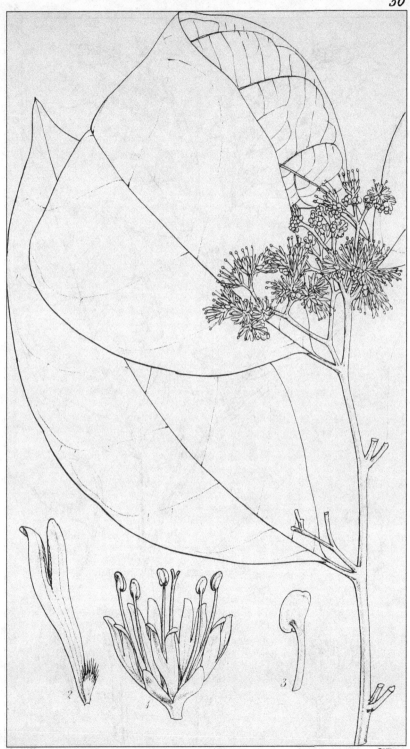

Chailletia floribunda Planch.

Madeley, Printer 3 Wellington St. Strand W. Fitch del.

Leucomphalos capparideus, Benth.

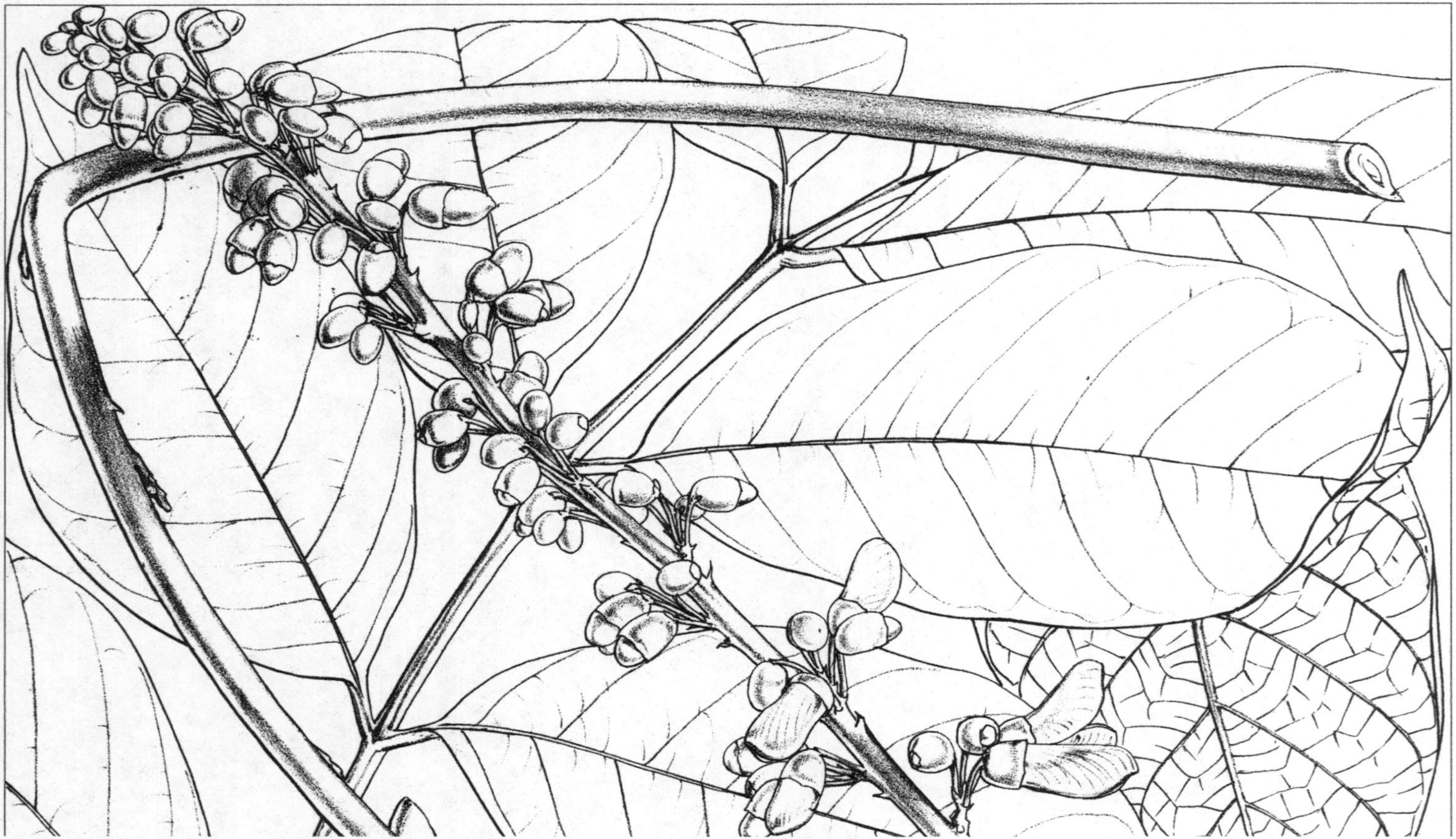

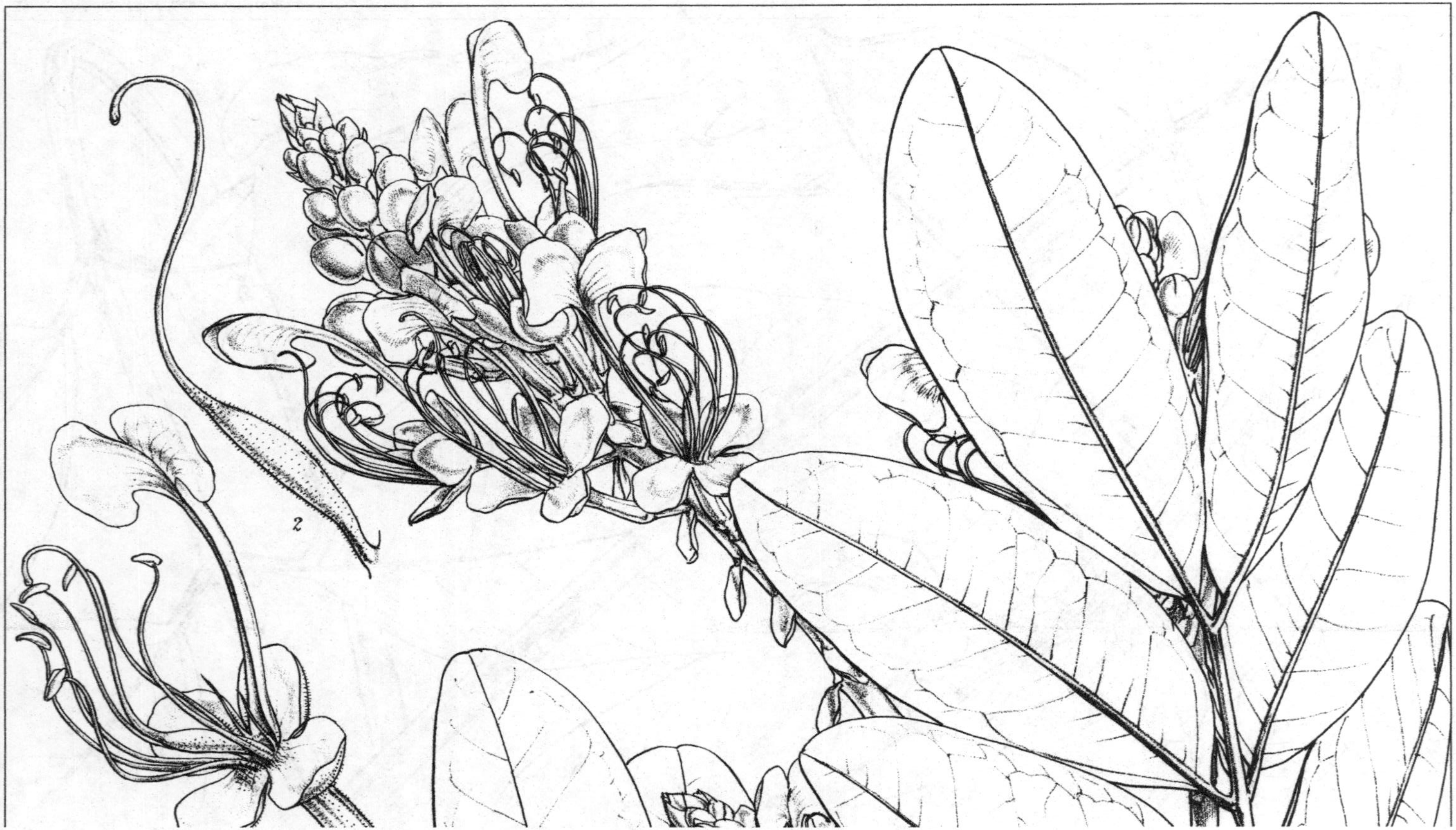
2

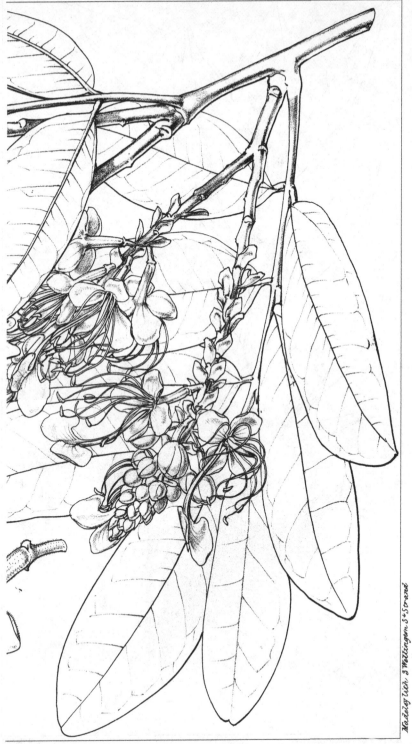

Aratia huntanta: Vaa

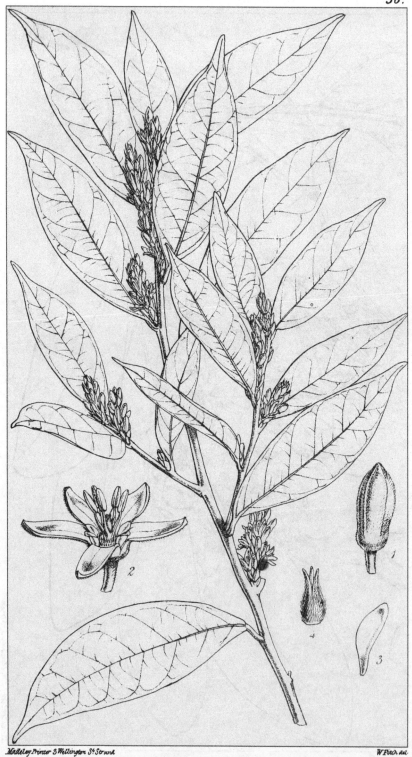

Clistanthus Polystachyus.

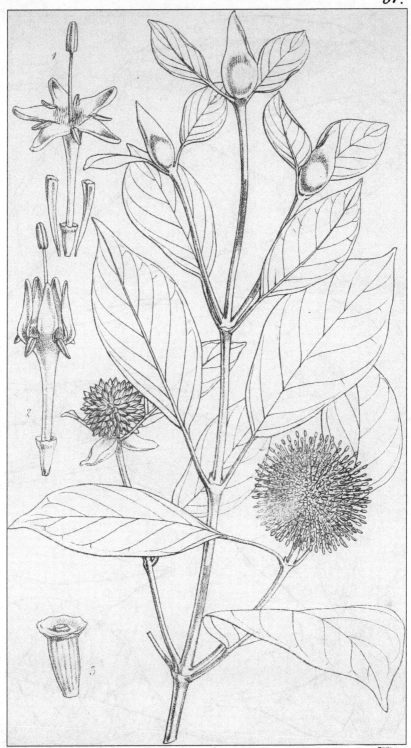

W. Fitch del.

Platanocarpum Africanum, Hook. fil.

38&39.

W. Fitch del.

Symphonia Vandlii Hook. 657

Vincent Brooks Day & Son, lith.

40 & 41.

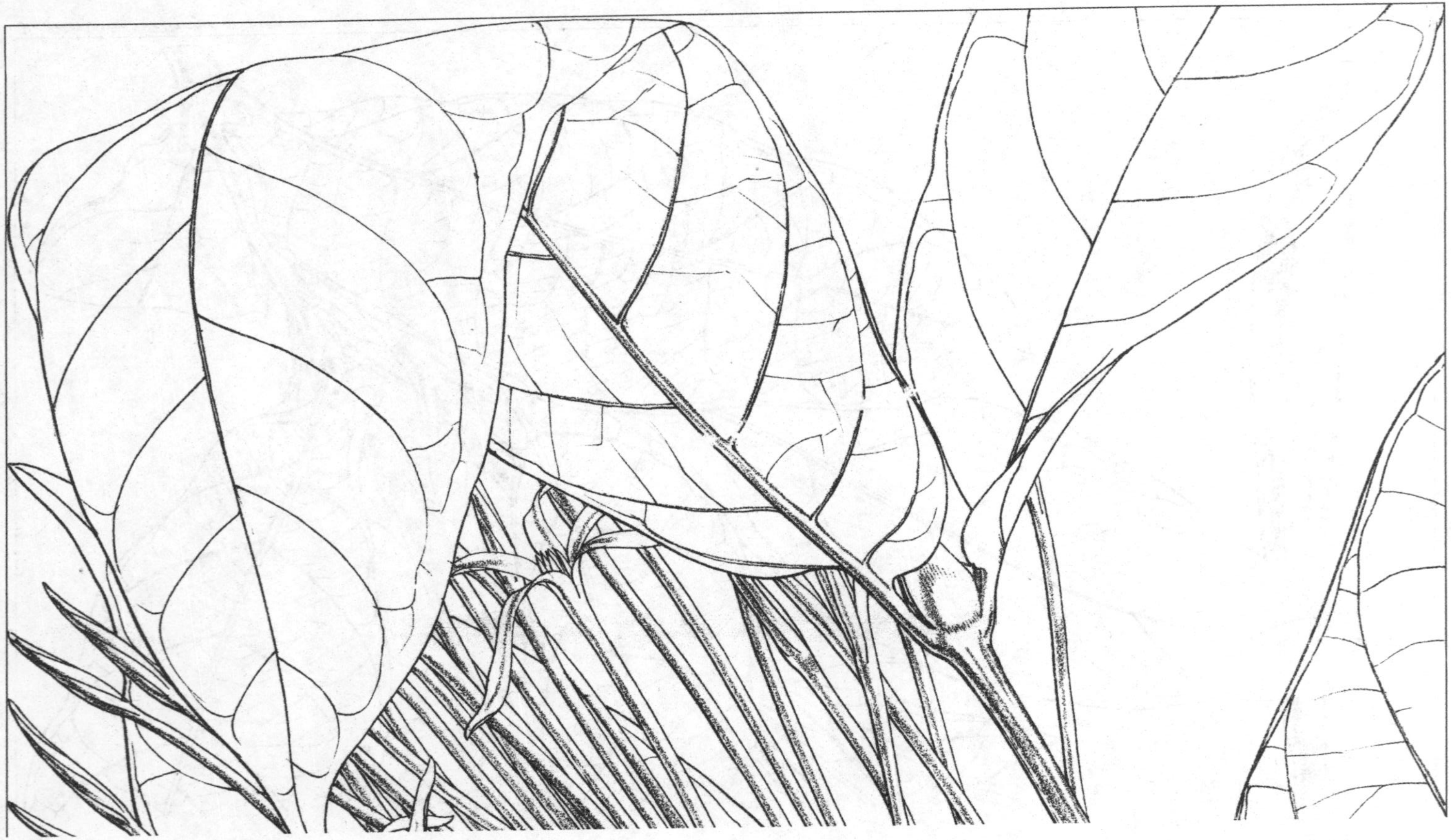

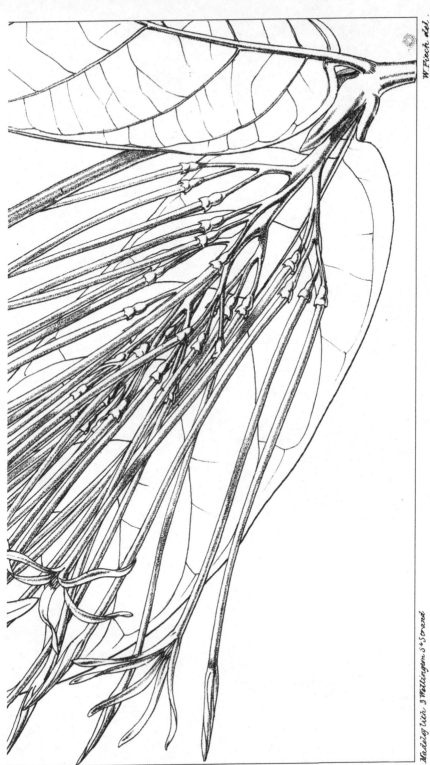

Oxyanthus formosus, Hook. fil.

Madeley lith 3 Wellington St Strand

W.Fitch. del

Uncaria Africana, G.Don.

43 – 4.

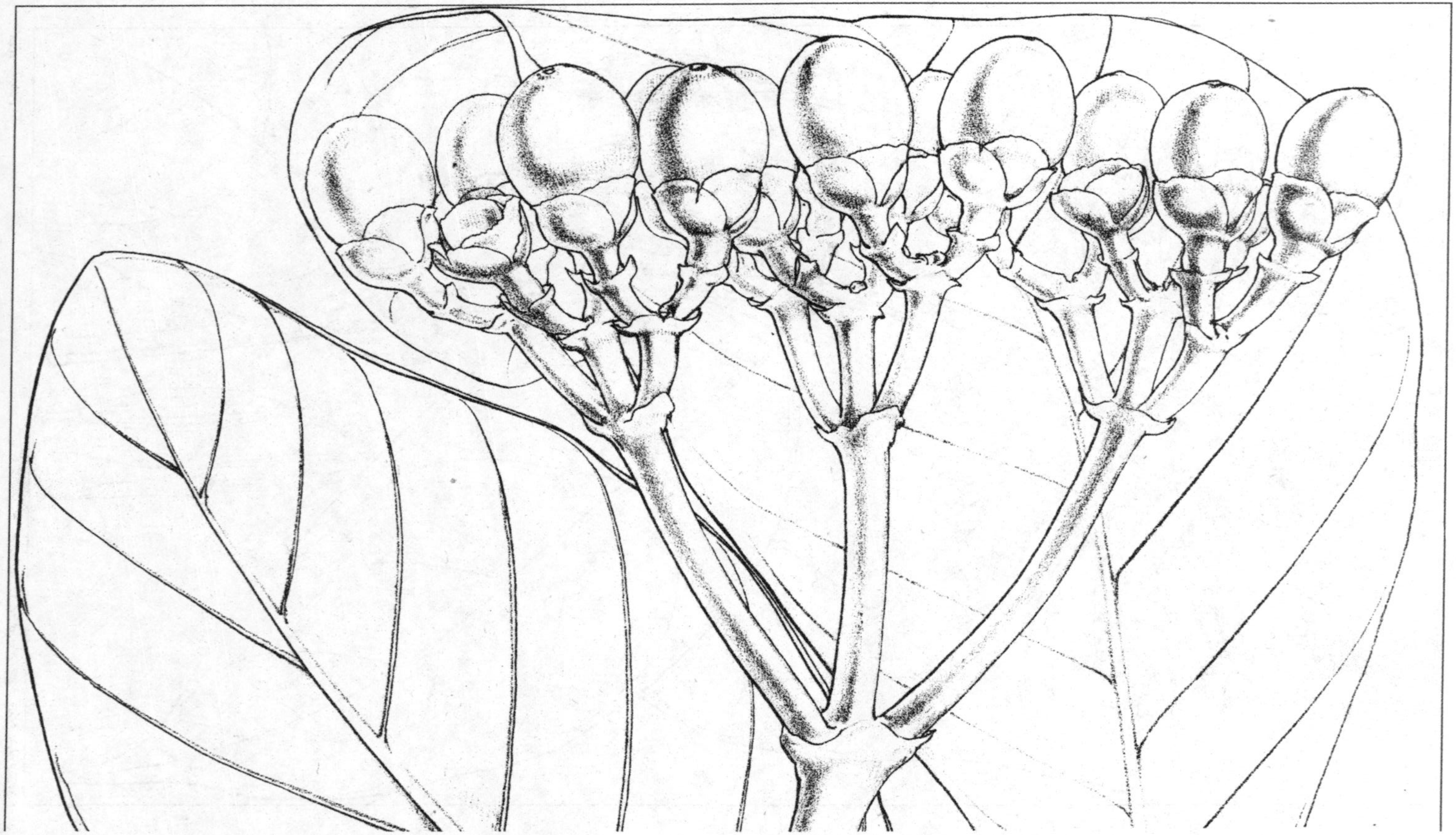

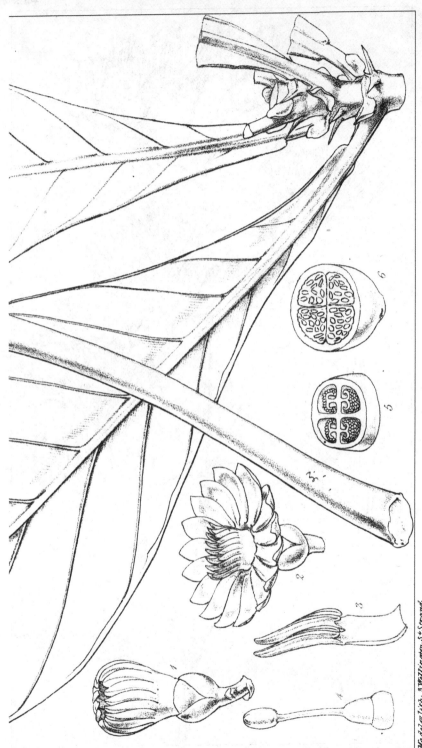

W. Fitch del.

Hanhart lith. 3 Wellington S.t Strand.

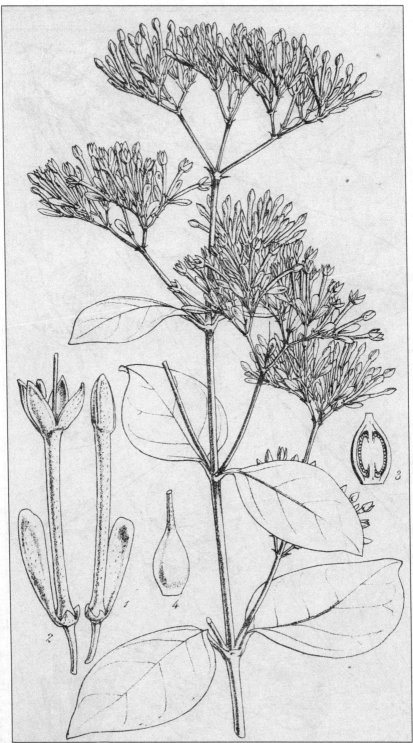

Madeley Printer 3 Wellington St Strand

W Fitch del.

Usteria Guineensis, Willd

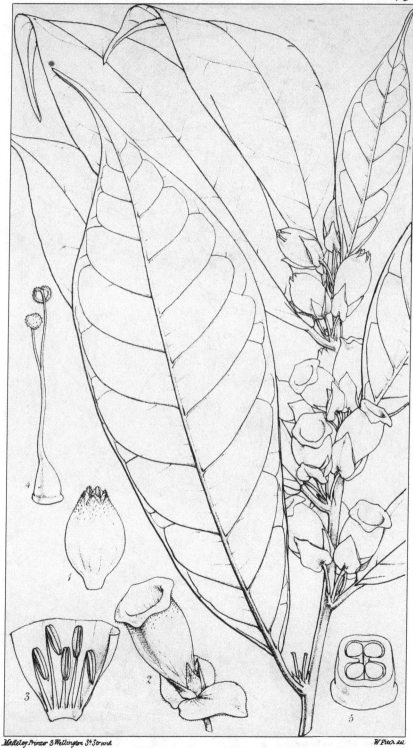

Madeley, Printer 3 Wellington St Strand W Fitch del.

Codonanthus alternifolia Planch.

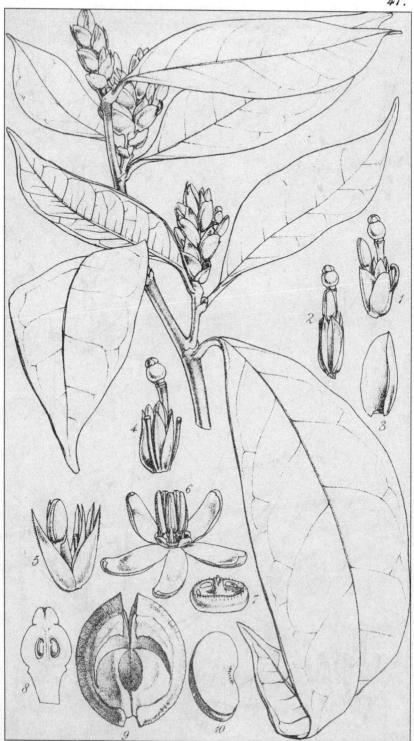

Madeley Printer 3 Wellington S! Strand

W Fitch del.

Amanoa strobilantha, Planch.

W. Fitch del.

Dicranolepis disticha, Planch.

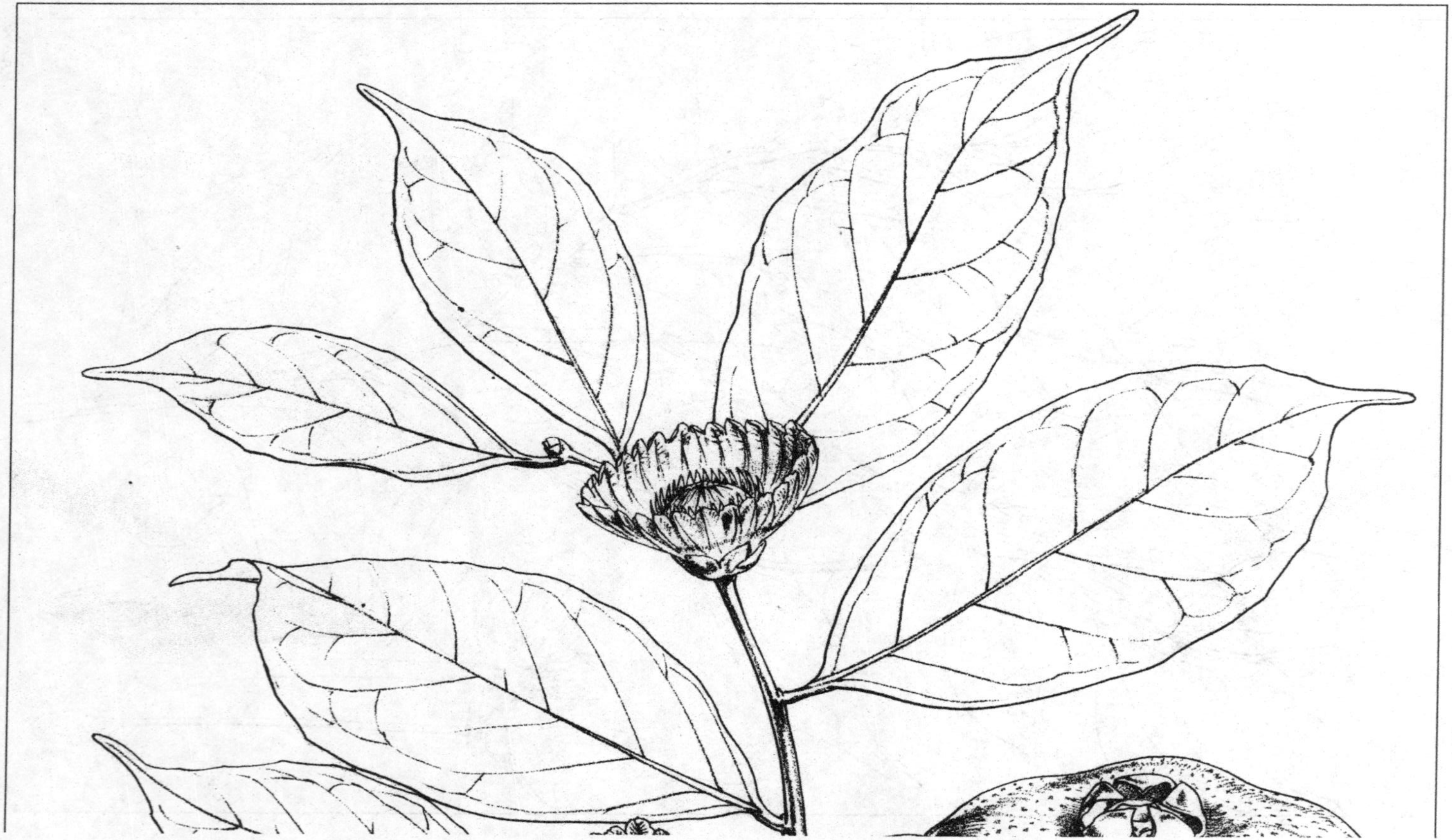

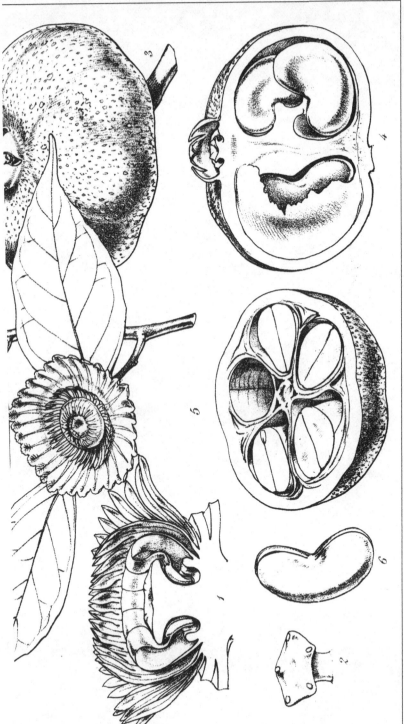